Science Networks · Historical Studies
Volume 2

Birkhäuser Verlag
Basel · Boston · Berlin

Ivor Grattan-Guinness

Convolutions in French Mathematics, 1800–1840

From the Calculus and Mechanics to Mathematical Analysis and Mathematical Physics

1990

Birkhäuser Verlag
Basel · Boston · Berlin

The Series "Science Networks. Historical Studies" is published simultaneously by Birkhäuser Verlag, Basel, and VEB Deutscher Verlag der Wissenschaften, Berlin.

I. Grattan-Guinness
Middlesex Polytechnic
Enfield
Middlesex EN3 4SF
England

CIP-Titelaufnahme der Deutschen Bibliothek
Grattan-Guinness, Ivor:
Convolution in French mathematics 1800–1840 : from the calculus and mechanics to mathematical analysis and mathematical physics / Ivor Grattan-Guinness. – Basel ; Boston ; Berlin : Birkhäuser.
ISBN 3-7643-2240-3

Vol. 1. The Setting. – 1990
(Science networks historical studies ; Vol. 2)
ISBN 3-7643-2237-3
NE: GT

The original edition is published by Birkhäuser Verlag, Basel
Distribution worldwide with the exception of socialist countries:
Birkhäuser Verlag, Basel.
Distribution in socialist countries:
VEB Deutscher Verlag der Wissenschaften, Berlin

Printed in GDR
ISBN 3-7643-2237-3 (Vol. 1)
ISBN 3-7643-2238-1 (Vol. 2)
ISBN 3-7643-2239-X (Vol. 3)
ISBN 3-7643-2240-3 (Set)
ISBN 0-8176-2237-3 (Vol. 1)
ISBN 0-8176-2238-1 (Vol. 2)
ISBN 0-8176-2239-X (Vol. 3)
ISBN 0-8176-2240-3 (Set)

VOLUME ONE

THE SETTINGS

Chapter 0

Table of Contents

As this book is long and quite complicated, the reader is recommended to study this table in more detail than might normally be considered, in order to appraise the general structure and order of the material. The title of every section and sub-section is given here, and the contents of each chapter is repeated at its head.

Chapter 3

The 18th-century heritage: the calculus around 1800

Chapter 6

Progress in mechanics, 1800-1815

Chapter 12

The consolidation of heat theory, 1816-1825

Chapter 13

The entry of Fresnel: physical optics, 1815-1824

Chapter 14

The entry of physicist Ampère:
electricity and magnetism, especially their connections, 1820-1826

Chapter 15

The entry of Navier and the triumph of Cauchy: elasticity theory, 1819-1830

Chapter 16

Work for the workers:
engineering mechanics and its instruction, 1800-1830

Chapter 18

Towards the 1840s: epilogue

Chapter 19

Chorus

Chapter 23

Indexes

Chapter 1

Propaedeutic

1.1 Gobbets

I presume that few who have paid any attention to the history of the Mathematical Analysis, will doubt that it has been developed in a certain order, or that that order has been, to a great extent, necessary—being determined, either by steps of logical deduction, or by the successive introduction of new ideas and conceptions, when the time for their evolution had arrived. And these are causes which operate in perfect harmony. Each new scientific conception gives occasion to new applications of deductive reasoning; but those applications may be only possible through the methods and the processes which belong to an earlier stage. [...]

Now there is this reason for grounding the order of exposition upon the historical sequence of discovery, that by so doing we are most likely to present each new form of truth to the mind, precisely at the stage at which the mind is most fitted to receive it, or even, like that of the discoverer, to go forth to meet it. Of the many forms of false culture, a premature converse with abstractions is perhaps the most likely to prove fatal to the growth of a masculine vigour of intellect.

Boole *Equations*, preface[1])

[...] the sciences are like a beautiful river, of which the course is easy to follow, when it has acquired a certain regularity; but if one wants to go back to the source, one will find it nowhere, because it is everywhere; it is spread so much [as to be] over all the surface of the earth; it is the same if one wants to go back to the origin of the sciences, one will find only obscurity, vague ideas, vicious circles: and one loses oneself in the primitive ideas.

Carnot *1783a*, conclusion

It is interesting thus to follow the intellectual truths of analysis in the phenomena of nature. This correspondence, of which the system of the

---

[1]) The method of citing works by catchwords or dating codes is described in detail in § 1.5.1.

world will offer us numerous examples, makes one of the greatest charms attached to mathematical speculations.

Laplace *Exposition*$_2$ (1799), 171

Geometry, which should only obey Physics, when united with it sometimes commands it. If it happens that the question which we wish to examine is too complicated for all the elements to be able to enter into the analytical comparison which we wish to make, we separate the more inconvenient [elements], we substitute others for them, less troublesome, but also less real, and we are surprised to arrive, notwithstanding a painful labour, only at a result contradicted by nature; as if after having disguised it, cut it short or altered it, a purely mechanical combination could give it back to us.

d'Alembert *1752b*, x-xi

I have often thought that an interesting essay might be written on the influence of race on the selection of mathematical methods.

Rouse Ball *1889a*, 123

Are the humanistic and scientific approaches different? Scientists can calculate the torsion of a skyscraper at the wing-beat of a bird, or 155 motions of the Moon and 500 smaller ones in addition. They move in academic garb and sing logarithms. They say, 'The sky is ours', like priests in charge of heaven. We poor humanists cannot even think clearly, or write a sentence without a blunder, commoners of 'common sense'. We never take a step without stumbling; they move solemnly, ever unerringly, never a step back, and carry bell, book, and candle.

Velikovsky *1951a*, 66

1.2 Micropaedia

1.2.1 Introduction

This book describes two parallel extensions of areas of knowledge in the early 19th century. The 18th-century differential and integral calculus (including some special functions and integrals, and the summation of series) became subsumed under mathematical analysis, a general theory of the calculus, convergence of infinite series, and functions, using limits as the unifying concept. At the same time 18th-century mechanics was supplemented by the mathematisation of heat diffusion, electricity and magnetism and by considerable advances in the mathematisation of optics, to become a spectrum of disciplines which I call 'mathematical physics'. Figure 121.1 indicates the principal parts of mathematical analysis and mathematical physics, and also shows the broadenings of each from the calculus and mechanics respectively. Each half of the figure is a synoptic table, an 18th-century *encyclopédiste* manner of presenting knowledge which was then still widely used.

The terms 'mathematical analysis' and 'mathematical physics' are used in their modern senses to refer to the classical theories. Thus the categories used in Figure 121.1 should need no special explanation; however, an exception applies to the five divisions into which I have sorted out mechanics. 'Molecular' refers to studies of the supposed fine structure of matter, and its modes of (inter-)action; 'corporeal' is my name for the mechanics of "ordinary-sized" objects on this planet; 'engineering' deals with the analyses of artefacts, structures and instruments; 'planetary' is concerned with the mechanics of a planet (or, more strictly, a heavenly body) when its shape and motion(s) are central to the study at hand; and, finally, 'celestial' covers what is also called here 'mathematical astronomy', especially when the heavenly bodies are taken to be point-masses. 'Celestial mechanics' is preferred as a name when it is used alongside one of the other divisions; or with all four together, which are collectively called 'terrestrial mechanics'.

The Figure indicates some of the major topics which fell under each of the divisions. The distinctions between them were not always sharp—for example, elasticity theory was often molecular but sometimes "purely" corporeal—but they furnish a working scenario.

I focus on the mathematical aspects of mathematical physics, especially those related to mathematical analysis; and I concentrate on France, where almost all the major developments were made in the period 1800–1830. Indeed, it is the period of French *domination* of mathematical analysis and mathematical physics. After 1830 these subjects became much more multi-national, and require a sepa-

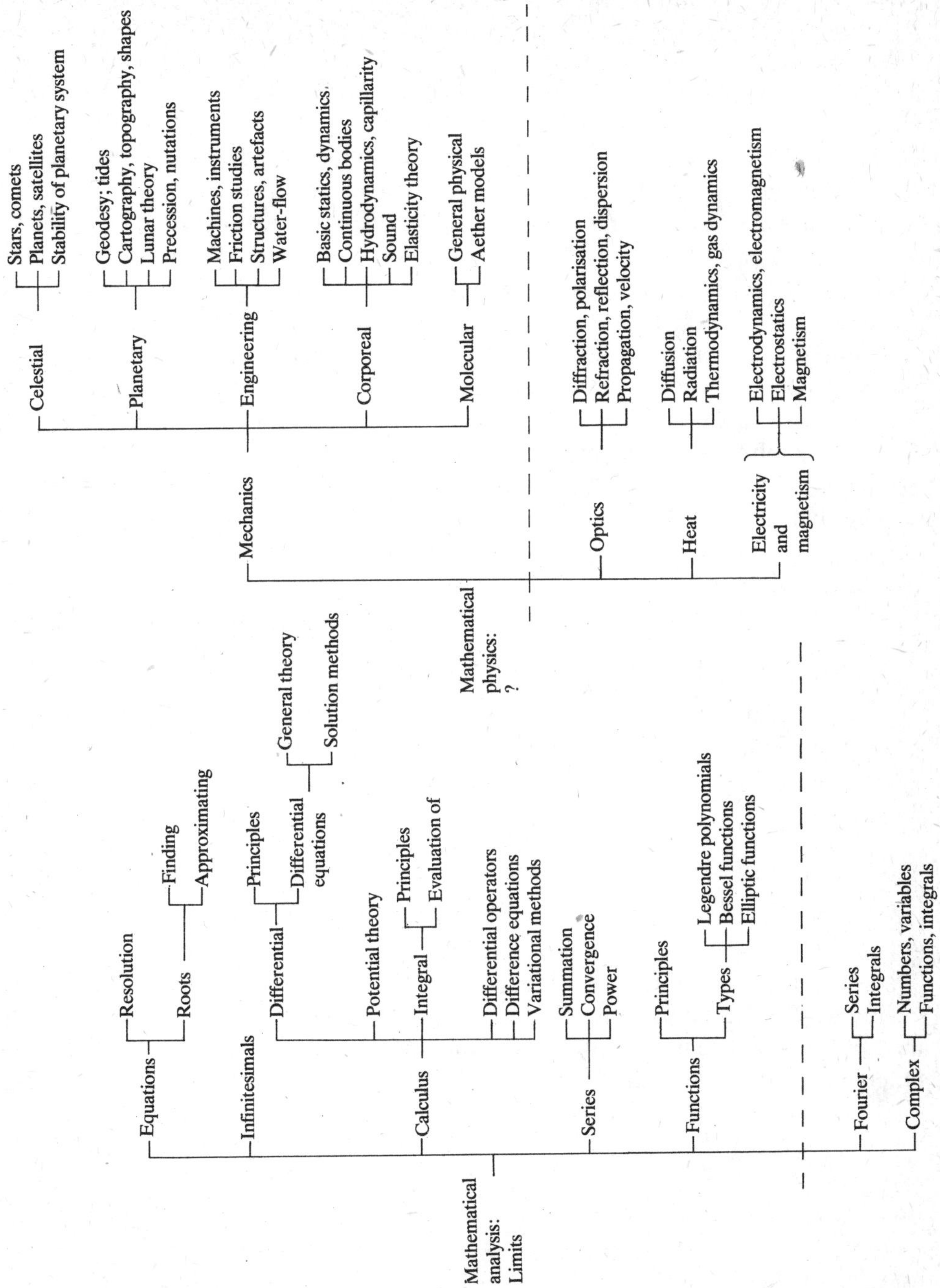

Figure 121.1 Structure of mathematical analysis and mathematical physics at 1830. The part of each diagram above the dotted line roughly indicates the situation as of 1800

rate study: Chapter 18 is a preliminary study of the French parts up to the early 1840s. Most of this book deals with the research; but education plays a significant role, both as an inspiration for some of the new ideas and also as sources of study and teaching for our figures.

Figure 121.1 indicates that a large number of subjects and topics is involved. I have attempted to cover them all, because of the numerous inter-connections between them, especially in, or via, the mathematics: the whole is much larger than the sum of its parts. But it is an exceptionally complicated whole from various points of view, which are examined in turn in this introductory chapter.

The rest of this section contains a factual summary of the principal events and developments. The account serves as an introduction to the details of the story, and also indicates how the subject matter is distributed among the later chapters. The other sections of the chapter deal respectively with aspects of philosophy and of historiography, and with certain textual matters.

I conclude this opening with a few points about terms and organisation. Chapters are divided into numbered sections and sub-sections; I use the symbol '§' to indicate either sections or sub-sections in cross-references. Interludes appear occasionally as supplements to certain (sub-)sections, and carry their number. Works are cited by dating codes ('*1809b*') or catchwords ('*Mechanics*') in the text; the system is described in more detail in § 1.5.1, and the references are given in the bibliography (Chapter 22). Single quotation marks are used to enclose quotations or to mention words and symbols: double quotation marks serve as scare-quotes to warn of an ironic or unusual designation of words or phrases. Finally, 'Revolutionary period' covers any time between 1789 and 1815, whether Revolutionary proper, Republican or Imperial.

1.2.2 Institutions and individuals around 1800

Chapter 2 is a survey of the principal institutions and individuals in France around 1800. The centralised structure of science and education in Revolutionary France meant that virtually everything and everybody was in Paris, the town which dominates our entire story.

After the Revolution science education was rejuvenated by the founding in 1794 of the *Ecole Polytechnique* as an engineering academy (§ 2.3). Monge was very influential in the design of the courses; he, Lagrange and de Prony were among its original teachers, and were assisted by younger men such as Lacroix, Fourier (until 1798) and Hachette. The graduating examiners in these early years were Bossut and Laplace; the latter was succeeded in 1799 by Legendre.

After passing through the *Ecole Polytechnique*, a *polytechnicien* usually went to a specialist school. The most important school for this story is the *Ecole des Ponts et Chaussées*, which started in the mid 18th century and to which the *Ecole Polytechnique* was partly conceived as a successor (§ 2.4.1). From 1798 de Prony was its *Directeur* as well as a *professeur* at the *Ecole Polytechnique* (§ 2.5.3). The *Collège de France*, and the *Université* (with its *Ecole Normale*) are also described, in § 2.4.4.

Research was usually presented to the mathematical and physical *classe* of the *Institut National* (as the *Académie des Sciences* was known during the Revolutionary period) (§ 2.2). Its *Mémoires* was the most prestigious journal for members of the *classe* to publish in. The posts of *secrétaires perpétuels* were established in a reform of 1803; Delambre covered "mathematical" sciences. Among other societies (§ 2.2.4), the *Société Philomatique de Paris* played a particularly useful role. The *Ecole Polytechnique's* own *Journal* was important for papers in mathematics and its applications (§ 2.3.3).

Papers usually were published in journals, some of which were associated with institutions; but quite a few works appeared as pamphlets, a form of publication then used more often than it is today. Textbooks (at various levels) were published in profusion (§ 2.2.4). Commercial publishers put out almost all of this material; their rise in importance during this period is very remarkable.

§ 2.5 provides brief profiles of seven of the principal early figures—Coulomb, Lagrange, Laplace, de Prony, Lacroix, Monge and Legendre—together with a quintet of noteworthy though more minor figures. § 2.6 contains a more rapid survey of several of the others, including the emergence of the new generation of *polytechniciens* in the early 1800s. It ends with a map of the most pertinent area of Paris at that time (Figure 263.1).

1.2.3 The calculus, 1800–1815

Chapter 3 commences the technical part of the story with a survey of the state of affairs around 1800 in the calculus and related topics. Lagrange's view of the calculus was influential, especially from the authoritative presentation in his *Théorie des fonctions analytiques* of 1797 (§ 3.2.2–3). Here functions were treated as algebraic formulae: in particular, the value of "any" function $f(x + h)$ was held to be expressible as a power-series in the increment h on the value x. The 'derived functions' of f at x were then *defined* in terms of the coefficients of h in the expansion, and were claimed to be calculable by purely algebraic means without recource to limits or infinitesimals. The integral was defined as the inverse of the first derived function. In addition to Lagrange's view, Euler's older theory of differential coefficients, using infinitesimals, was still advocated (§ 3.2.4). There was also some interest in the theory of limits (§ 3.2.6).

Differential equations formed the centre of mathematical attention (§ 3.3). Functional or power-series forms of solution were preferred. Much attention was given to the generality of a solution, and to the existence of singular solutions. Closely related studies involved methods of summing series and evaluating integrals (§ 3.4).

Difference equations and interpolation also gained attention, especially from de Prony (§ 3.5). He taught the former in detail at the *Ecole Polytechnique*, and also directed the compilation of a vast set of logarithmic and trigonometric tables during the 1790s.

Among works published around 1800, special mention is made of Lacroix's textbooks. In particular, the first edition of his three-volume *Traité du calcul différentiel et du calcul intégral* (1797–1800) provided an authoritative survey of much of the mathematics described above (§ 3.2.7). One notable feature of his work was his interest in the history of mathematics, and in § 3.2.8 I compare his treatment with those in histories of mathematics published at that time.

Chapter 4 surveys those developments made between 1800 and 1815 which continued the traditions described in Chapter 3. It begins with notice of major figures who emerged during this period (§ 4.2): the *polytechniciens* Binet, Biot, Poinsot and Poisson; and Ampère, who began teaching at the *Ecole Polytechnique* in 1804. A noteworthy new periodical, the *Annales des mathématiques pures et appliquées*, began to appear in 1810, edited by Gergonne (§ 4.2.6).

In the calculus Lagrange's view was prominent. A set of his lectures appeared in several editions (§ 4.3.1); and the second edition of his *Fonctions* was published in 1813, the year of his death (§ 4.8.2). His views on the calculus were extended in various ways (§ 4.3); from foundational points of view by Ampère and Servois, and as a theory of operators by Arbogast, Jacques Français and Brisson. His treatment of the calculus of variations was examined by Ampère, again from foundational points of view (§ 4.5). His approach to mathematics was attacked by the interesting peripheral figure Wroński, who "refuted" all major mathematicians' thought (§ 4.3.8). Both Parseval and Poisson contributed to the general theory of differential and difference equations (§ 4.4).

There was also considerable interest in the theory of equations, especially for its bearing on problems in differential equations and their use in mechanics (§ 4.6). Again Lagrange was the dominant figure, especially with the second (1808) edition of his *Traité de la résolution des équations numériques*. The problems included not only resolution of equations but also the presence of complex roots, and the existence of, and approximation to, real roots. In addition to Lagrange, significant contributions were made by Budan, Cauchy, Corancez and Fourier. A related topic was complex numbers; but here developments were in a spirit foreign to Lagrange, for they saw the introduction of the geometrical representation by Argand and others (§ 4.7.1).

Another fertile area of application of the calculus was to differential geometry and the analysis of curves and surfaces. Here the *polytechnicien* Lancret wrote two notable papers before his early death in 1807 (§ 4.7.4). In addition, analytical geometry became established, to the state of textbook writing (§ 4.7.3).

Interlude 480.1 offers some notes on progress in other countries. In England Newtonian fluxional calculus was still dominant. In Italy a continuing mathematical tradition provided a steady stream of minor but useful contributions.

1.2.4 Mechanics and engineering, 1800–1815

Chapters 5, 6 and 7 offer a service to the history of mechanics similar to that provided for the history of the calculus in the two preceding chapters. Chapter 5 surveys the principal themes in terrestrial and celestial mechanics around 1800.

Lagrange's *Méchanique analitique* (1788) was the most influential treatise in terrestrial mechanics (§ 5.2.1–4). He extended his algebraic approach to the calculus by expressing his mechanical principles in variational form. The relationship between the 'principle of virtual velocities', which he espoused in his treatise without proof, and other basic principles such as composition of forces and the lever, were discussed around 1800 by him, Fourier, Carnot, Monge, Francoeur and de Prony (§ 5.3). As in the calculus, Euler's approach to mechanics (§ 5.2.5) was advocated as well as Lagrange's, and mixtures of the two approaches appeared in the writings of Carnot, Francoeur and de Prony (§ 5.2.6–8).

Laplace's *Mécanique céleste* attempted to give an authoritative presentation of celestial and (parts of) planetary mechanics (§ 5.4). Volumes 1 and 2 were published in 1799, as was the second edition of his introductory *Exposition du système du monde* (§ 5.4.2). In volume 1 he described the motions of the heavenly bodies in general, especially the effects of the perturbing forces (§ 5.4.4) and the stability of the planetary system (§ 5.4.5).

Volume 2 was largely devoted to planetary mathematical physics, which is the concern of § 5.5. The selection of topics here includes the motion of a planet around its centre of gravity, and theory of the tides. Attention is also paid to the theory of attractions, where he and Legendre had been rivals in the 1780s over 'Legendre functions' (as we now call them); the related topic of potential theory, which had also found application in terrestrial mechanics; the decimal system of weights and measures, and the determination of the meridianal arc by Méchain and Delambre; and the current interest in cartography and geodesy, especially in Puissant.

Chapter 6 continues to around 1815 developments in corporeal, planetary and celestial mechanics. Much of the material was covered for the students of his old school by *polytechnicien* Poisson, in the first edition of his *Traité de mécanique* of 1811 (§ 6.2.3). Another graduate, Poinsot, also made notable contributions: he introduced his theory of the couple into statics in 1803 (§ 6.2.1), and two years later he looked afresh (as did Ampère) at the principle of virtual velocities (§ 6.2.2). An important addition to variational mechanics was the theory of the Lagrange and Poisson 'brackets', developed by them between 1808 and 1810 as an offshoot of Poisson's extension of the proof of the stability of the planetary system (§ 6.3).

Laplace published volumes 3 and 4 of his *Mécanique céleste* in 1802 and 1805 (§ 6.4.1), devoting much of them to the perturbations of planets and satellites, and to comets. These topics, and also meteorites, concerned him and Lagrange, and also Biot and Poisson (§ 6.4). In general the French sought "exact", even though complicated, solutions to astronomical problems, rather in contrast to German interest in compact, approximate answers (Interlude 641.1).

In planetary mechanics the shape of the earth was studied in different ways. Arago and Biot carried out much research using the "simple" pendulum, an instrument which Laplace and Poisson showed to be far from simple in the fine details of its motion (§ 6.5). Meanwhile, cartographer Puissant continued his work by directing the preparation of the new map of France (§ 6.4.9). A gradually developing use of statistics is evident, principally from Laplace: in addition, Legendre introduced the least squares criterion in 1806 (§ 6.4.7-9).

Chapter 7 concentrates largely on physics and molecular mechanics, in particular a programme initiated by Laplace around 1805 (§ 7.2.1). Some valuable successes were scored in capillary theory and on the velocity of sound (§ 7.2), and they were continued in a way in a dispute over the analysis of vibrating surfaces during the early 1810s between the Laplacian Poisson and the non-Laplacian Germain (§ 7.3). But the most substantial achievements came in optics, under Laplace's view that light should be treated as a molecular phenomenon. The studies were stimulated by Malus's discovery in 1808 of the polarisability of reflected light (§ 7.4.3) and his work on double refraction (§ 7.4.5). Arago and Biot had priority and "property" disputes in 1806 and 1813 over aspects of optics (§ 7.4.1, 7); the resulting deterioration in their relationship affected the scientific community substantially in later years. Laplace himself contributed to optics by studying atmospheric refraction (§ 7.4.1): this was in fact his first exercise in the new tradition. His programme is often now called 'Laplacian physics'.

Electricity and magnetism also came under scrutiny, especially from Biot (§ 7.5). But the most substantial mathematical study of this period was Poisson's two major papers of 1812 and 1813 on electrostatics (§ 7.6). By this time Laplace's programme was in a healthy state: its later fate is a major concern of our Volume 2.

Chapter 8 complements the preceding chapters in describing a few of the developments in engineering during the period 1800–1815. Engineering was a major preoccupation of French science, especially because of the education of public works and military engineers at the *Ecole Polytechnique*, and their further training and employment in the specialist schools and *Corps* (§ 2.3–4). In this book attention is confined to aspects of engineering where mathematics was used to a significant degree, or where concepts or theories were introduced and developed which have a major bearing on the other aspects of mathematical physics.

One topic subjected to mathematical investigation was earth pressure theory, and its application to embankments and fortifications. Coulomb's 18th-century work was influential, and a sequence of studies by de Prony and others culminated in a distinguished study by Français (§ 8.4.5).

Another major interest concerned hydrodynamics, especially hydraulics and water flow. Here Girard directed a major canal project, including a port for Paris (§ 8.5.5). Partly in this connection, there was discussion of different ways of measuring water-flow, by de Prony and others (§ 8.5.3–4). A related topic was shipbuilding, where in the early 1810s Dupin's early research work included the stability of floating bodies (§ 8.3.4) and the flexure of timber (§ 8.4.6).

An important common factor in many of these studies was Monge's first love, descriptive geometry. He tried to elevate this practical subject to the status of major theoretical discipline, and so adjoined to it various considerations inessential to the engineer or surveyor (§ 8.3.2). It reached its heyday in the 1800s and 1810s; but Monge's ambitions were not realised, for reasons which are discussed in § 8.3.1.

This completes Volume 1 of the book, with its septet of chapters covering the period 1800–1815. Engineering does not receive extended attention again until Chapter 16, although relevant aspects occur *passim* in the intervening seven chapters, to which I now turn.

1.2.5 Heat theory and mathematical analysis, 1805–1815

From now on the fairly settled state of affairs in the calculus and mechanics was disturbed by new innovations. Volume 2, containing the next eleven chapters, is concerned with the changes, which took place up to 1830, and their impact during the next decade.

Chapter 9 contains an account of Fourier's work on heat diffusion and the Fourier series and integral methods of solving differential equations. His main work was done between 1805 and 1812; but I describe it in *this* Volume, and devote to it a separate chapter, partly because he worked in isolation from the Paris community (he was a *Préfet* in Grenoble for most of that period) and partly because after 1815 his ideas made an impact on the emergence of both mathematical analysis and mathematical physics.

Fourier's work is of especial significance for mathematical physics because it was the first large-scale mathematical treatment of a phenomenon which (surely) lay *outside* the realm of rational mechanics. We see his first, unsuccessful work, of perhaps 1802 (§ 9.2.2); then we follow his obtaining the heat diffusion equation, and finding solutions by trigonometric series and other functions (§ 9.2.2-9).

There was a controversy. The diffusion equation was closely studied both as a valid model of the phenomenon and for the manner of its derivation (§ 9.3.1). The Fourier series were worse than new; for they had been discredited in the 18th century by Lagrange and others. Further, Fourier's calculus was Eulerian, in contrast with Lagrange's style (§ 9.2.4).

However, Fourier eventually secured the approval of Laplace, who led him towards his (new) Fourier integral solution in 1811 for heat diffusion in an *infinite* body and helped him obtain and win a prize problem for heat diffusion in 1811 at the *Institut* (§ 9.3.2). But publication of his heretical work was still not possible, and it is only around 1815, when we see him at last permanently in Paris (but in social difficulties, as a former friend of the deposed Napoléon) that he is beginning to publish and publicise his work (§ 9.4). A surprising omission is the Laplace transform, despite the presence of all the factors needed for its rapid development at that time (Interlude 934.1).

Having now arrived at 1815, the first duty of Chapter 10 is to summarise the principal biographical and institutional changes during the Bourbon Restoration (1816-1830) after the fall of Napoléon (§ 10.1.1-2). In addition, the early career is sketched of Cauchy (§ 10.1.3), who dominates this and the next chapter in their tale of the emergence of mathematical analysis.

The technical part of the chapter describes an intensive competition between Poisson and Cauchy, with parts played also by Laplace and Fourier in the area of complex variables and linear partial differential equations. Cauchy took up the naive methods of evaluating definite integrals by putting i into the integrands, and inaugurated a critical phase in 1814 with his first ideas on complex variables (§ 10.2). A year later he and Poisson produced extensive studies of hydrodynamics in which Cauchy extended his complex-variable methods by finding Fourier's integral theorem for himself (§ 10.3). Then the competition ensued over solving linear differential equations by traditional methods with power series and functional forms (Poisson) or new ones involving Fourier analysis and complex variables (Fourier and Cauchy) (§ 10.4). Despite much ingenuity, Poisson came in a definite third.

During the time of the Bourbon Restoration, Cauchy, a fanatical supporter, was in a remarkable prime, producing masses of superb mathematics. Chapter 11 is dominated by his further progress with real- and complex-variable analysis. After discussion of work on the general theory of differential equations by him and Ampère (§ 11.2), his *Cours d'analyse* at the *Ecole Polytechnique* (published in 1821) is described, with its elevation of the theory of limits to serve as the basis of the theory of functions, the convergence of infinite series, and several related topics (§ 11.3). Then more competition ensued with Poisson, this time over the definition and use of real- and complex-variable integrals (§ 11.4). Cauchy's further progress is recorded in § 11.5 with an account of his 1823 *Résumé* of his calculus teaching at the *Ecole Polytechnique*, including a recently discovered second part on existence theorems for differential equations. These works established his real-variable calculus; the complex-variable integral was extended in an 1825 pamphlet on integration along complex paths, together with a few succeeding papers in which he introduced the word 'residue' and tried to systematise the theory as it had progressed so far (§ 11.5.7–9).

§ 11.6 is a miscellaneous section, describing various studies of the mid 1820s on topics described earlier in the chapter. They include the convergence of Fourier series, Ampère's thoughts on the foundations of the calculus, Laplace's study of the convergence of the expansion of the distance function in celestial mechanics, and Poisson's proof of the reality of latent roots for certain differential equations.

1.2.6 The new areas of mathematical physics, 1815–1827

The focus now switches to applied mathematics: in particular, the next three chapters deal with developments in the new areas of mathematical physics, outside mechanics. They deal, in turn, with heat theory, physical optics, and the connections between electricity and magnetism.

Chapter 12 picks up several points from Chapter 9 on Fourier. Firstly, § 12.2 describes some ideas of the 1810s on the physics of heat, especially Dulong and Petit's criticisms of Newton's law of cooling. Laplace worked between 1816 and 1821 on the velocity of sound; and, as usual, Poisson took it up (§ 12.3). Laplace and Fourier looked further into the cooling of the earth; Laplace gave pointed praise to Fourier in 1820 (§ 12.4.4). Fourier's book on heat diffusion appeared in 1822, at around the time that Poisson was publishing long papers on Laplacian heat theory (§ 12.5). The account ends suitably with a disagreement of the mid

1820s between these two on the principles of heat radiation (§ 12.5.4). Note is then taken of the fact that mathematical heat theory was well established in science by the mid 1820s (§ 12.6).

Chapter 13 deals with a battle over physical optics between the Laplacians (mainly Biot) and Fresnel, *polytechnicien* and *ingénieur des ponts et chaussées*. Fresnel began in 1815 by examining diffraction (an area rather neglected by the Laplacians), and eventually won a prize at the *Académie* four years later (§ 13.2.6). The clash with his opponents was explicitly stated, for from the start he employed a wave theory (of *longitudinal* vibrations of the tiny particles of the aether), the interference principle, and Huygens's principles of secondary sources of vibration. On the mathematical side, he also produced at equation (1325.8) his 'Fresnel integrals'.

In the succeeding years Fresnel looked intensively at reflection, refraction and polarisation; and he also modified his whole approach in 1821 by converting to a theory of the *transversal* mode of vibration of particles in the aether (§ 13.3.4). His major suite of achievements was his ellipsoids of elasticity, by means of which he was able to restate known properties of double refraction, and indeed also extend them (§ 13.4.2-4). In his professional capacity he applied his knowledge to the design of lighthouse lamps (Interlude 1361.1).

As was mentioned in § 1.2.4, optics was the Laplacians' strongest area, and they came back with competing ideas or questions. In 1819 Poisson deduced from Fresnel's diffraction theory that there would be a white spot behind a circular diffractor, a prediction which Arago (who had left the Laplacian cause) verified (§ 13.2.7). Two years later there was a (non-)discussion between Poisson and Fresnel on the possible motions of 'fluids' (§ 13.4.1). Biot challenged Fresnel about double refraction at that time (§ 13.4.2), and also proposed that the interference principle could be common ground between the rival theories, if it were taken to be an empirical law (§ 13.2.7). But Fresnel won—on physical principles, mathematical reasoning and experimental precision—and Laplace praised him in the most explicit way at the *Académie* in 1822 (§ 13.4.5).

Chapter 14 completes this trio with an account of Ampère's mathematicisation of electrodynamics and electromagnetism. Oersted's discovery of the action between a magnetic needle and an electrical wire was a great surprise; but when Arago repeated it before the *savants* at the *Académie* in September 1820, soon after Oersted's publication, Ampère quickly found that action also took place between wire and wire (§ 14.2.1), and he devoted the bulk of his research effort for the next six years to the theory of, and experiments on, the new phenomena.

Early on Ampère took several decisions of various kinds to guide his work. All the phenomena were electrical (so that magnetism had to be explained as a particular form of electricity); electricity itself acted through the aether; the action between two elements of wire lay along the joining line, following some law which had to be determined; and the main effort was to be put into studying equilibriate situations, where *no* motion occurred between wires. The mathematical study of the action followed an interesting course of desimplification: a trigonometric form (1423.1) of 1821 for straight elements; a differential form (1426.2) in the following year for curved elements; and a line integral (1441.1) in 1824 for action between element and wire. He and his follower Savary applied these and related expressions to a variety of cases, and from two particular configurations in which equilibrium obtained they were able to deduce an inverse square law between strength of action and distance of separation (§ 14.3.2). Ampère also offered a theory of magnets in which a dipole was construed as a tiny electrical solenoid (§ 14.4.1).

The Laplacian contributions to the new theory was largely confined to experiments on electromagnetic intensity carried out by Biot and his assistant Pouillet (§ 14.2.4, § 14.4.2). One of Biot's proposed laws was mistaken, as Savary pointed out (§ 14.3.2). But in 1824 Poisson produced an extensive analysis of magnetic action (§ 14.4.3), and Ampère had to convert its principal ideas into his own theory based on solenoids. This was achieved in a remarkable transformation of surface integrals (Figure 1446.1), in his book of 1826 which contained his definitive presentation of the whole theory. Poisson gave no attention to electrodynamics, but he studied the process of magnetisation in a sequel paper of 1826 (§ 14.5.2).

By that time all the non-Laplacians had published their major statements on the three new areas of mathematical physics, and the Laplacian tradition was well into decline. Chapter 14 ends with Laplace indicating in 1825 some major questions for physics.

1.2.7 Elasticity theory and energy mechanics, 1800–1830

The next two chapters describe a pair of major developments within mechanics itself, though certain points of contact with the new areas also obtained. Important in both stories was *polytechnicien* Navier, whose biography begins Chapter 15 (§ 15.2.1–3). He began his career by publishing editions of engineering treatises, especially in 1819 the first volume of Bélidor's *Architecture hydraulique* of 1729. Appointed that year to the *Ecole des Ponts et Chaussées*, he submitted to the *Académie* a string of important papers on elastic bodies and viscous fluids: the latter yielded his form of the 'Navier-Stokes equations' (§ 15.2.4–8). In addition, Navier applied his knowledge to the design of a suspension bridge (then a new form) for Paris; but unfortunately it fell down (§ 15.3.2).

In reaction to one of Navier's original papers Cauchy came in 1822 to a new conception of elasticity theory, essentially the stress-strain model. Five years were to pass before he provided the details; but then they appeared in a long string of papers on the isotropic and anisotropic theories, some using molecular models but others assuming "pure" continuity of the substance, and with a group of applications around 1830 to Fresnelian optics. § 15.4–6 are devoted to this story, together with Poisson's competing Laplacian treatment, howls of priority from Navier, and independent contributions from Lamé and Clapeyron.

Engineering dominates Chapter 16, which concerns a new general approach to energy mechanics adopted by some of the engineer *savants*. The point is explained in § 16.1, and then in § 16.2 two new *polytechniciens* who joined Navier in this enterprise are introduced: Coriolis and Poncelet.

Up to the late 1810s the concept of work, while used fairly often in mechanics, was not exploited systematically as a basic notion for the subject. Some figures are discussed in § 16.3; Euler and Daniel Bernoulli, and then Lazare Carnot, Coulomb and Hachette. But with Navier's Bélidor edition of 1819 much progress was made, to be continued by Coriolis and Poncelet with books in 1829. These studies are described in § 16.4, which includes Coriolis being hardly honest about his knowledge of Poncelet's concurrent publications.

§ 16.5 describes some case studies of the 1820s where the concept of work was prominent. These include de Prony's introduction of his brake to measure machine performance, Poncelet's design of an undershot water-wheel, and Fourneyron's turbine. The section includes also Navier's application of these ideas to gas dynamics and to the flight of birds, and discusses the lack of interest in Sadi Carnot's book of 1824 on the 'motive power of fire'.

§ 16.6 rounds of the story with some of the educational and social aspects. An important thread in it is the development of ergonomics, which Dupin highlighted in his lectures at the Paris *Conservatoire des Arts et Métiers* in the 1820s. His public lectures there on social topics involving industry and the working conditions of the working class gave fresh grounds for studying the concept of work. Some other educational examples are described, including the newly founded (1829) *Ecole Centrale des Arts et Manufactures*, and the *Association Polytechnique* for workers' training (1830), where Comte was active. A concluding § 16.7 describes the varied roles for work concepts in mechanics, engineering, education and society.

1.2.8 On the aftermath: towards the 1840s

By 1830 all the major changes had taken place, and were recognised as such; however, this story cannot thereby terminate abruptly. The last two chapters deal with this problem, by accommodating topics left aside for some chapters, noting the impact of all the new theories, and organising a dignified retreat from the scene as the century advances.

Chapter 17 begins by catching up with the changing social situation. New figures included Duhamel, Liouville, Clapeyron and Lamé from the *Ecole Polytechnique* (the last two in Russia for the 1820s), and the *normalien* Cournot; foreign visitors comprised Abel, Dirichlet, Libri, Ostrogradsky and Sturm. Of new journals Crelle's *Journal für die reine und angewandte Mathematik* (1826) is of note: the French sometimes contributed to it.

Mathematical analysis is taken up in § 17.2, with Abel and Dirichlet applying Cauchy's mathematical analysis, and Abel and Jacobi introducing elliptic functions (as the aged Legendre gracefully acknowledged). The theory of equations and inequalities concerns § 17.3, with Fourier dominant; the Sturm sequence (1829) was a variant on a theorem of his, and his advocacy what we recognise as linear programming aroused some interest. Its lack of success was largely due to want of linear algebra, which ironically was nascent in a paper of 1829 by Cauchy (§ 17.3.6).

In corporeal mechanics (§ 17.4) Cauchy wrote on linear moments and momental ellipsoids, while Poisson and others studied the motion of fluid in a cylinder. Heat diffusion is covered in § 17.5, with Fourier's last contribution taken along with the first ones of several of the new men named above. The *lack* of interest in electricity and magnetism is also noted (§ 17.5.4), in contrast to the foreign achievements of Green and Ohm (Interlude 1754.1).

Celestial and planetary mechanics advanced (§ 17.6), under the leadership of Laplace. In his final years he increased the use of statistics, and studied again the rotation of the earth. Poisson joined him here, and also disputed the foundations of potential theory with Ivory (in England). Other figures considered the great inequality of Saturn and Jupiter, and also lunar theory.

Engineering (§ 17.7) was marked by interest in the economic aspects of construction, such as Girard on lock systems for canals and Brisson on the building of bridges: Navier thought in general about financial investment in engineering projects. Outside Paris, Lamé and Clapeyron in Russia thought up and solved the location problem in operational research, studied the stability of arches, and analysed friction in gearing systems: Poncelet at Metz also contributed to the latter two topics.

The final Chapter 18 surveys the years immediately after 1830 in two progressive stages of brevity. § 18.2 covers all topics in the new decade, noting institutional changes after the 1830 Revolution and the strengths (and weaknesses) of research and teaching effort. A notable feature is the way the new figures (and also Coriolis) made up combinations of ideas taken from their competitive predecessors. § 18.3 sees out the 1830s and looks further forward, in still shorter order: the deaths of many of our main figures are recorded, and the story ends with continuing uncertainty over policy at the *Ecole Polytechnique*.

1.2.9 Considerations and information

Thus the main story comes to an end: a period (1800–1815) which largely built upon late-18th-century traditions (while of course adding fundamentally to it) was followed by a similar time (1815–1830) of major changes in pure and applied mathematics and then in engineering, which were joined to the old thought in the ensuing years (1830–1840). Chapter 19, which completes Volume 2, considers this scenario from various points of view: communal, institutional, educational and technical. In § 19.2 the *Académie* and the *Ecole Polytechnique* are given especial attention in a survey of the institutions: also noted are the *cumul* system of multiple appointments, the competitiveness, and the publishing industry. § 19.3 begins by noting that the community divides by its research interests into two groups: those oriented around general applications and pure mathematics, and their companions with a special concern with engineering questions. The careers of Laplace, Biot and Poisson are looked at in a little detail, and a group of algebraic thinkers is detected among the minor figures. Then § 19.4 considers the broadening into mathematical analysis and mathematical physics illustrated in Figure 121.1, and picks up some of the philosophical questions which are going to be outlined in § 1.3-4. Metatheory is introduced in § 19.5, in the form of a self-appraisal of the study; and the story ends in § 19.6 as it began, with a sextet of gobbets.

Volume 3 of the book provides information and data of various kinds. Chapter 20 contains eleven manuscripts or obscurely published passages. Chapter 21 contains three tables. The main one is the third, a chronology showing for each year the principal publications and important dates in the careers of the main figures, and also indicating major events, especially those connected with relevant institutions and journals. The first table re-organises some of this information in the form of career outlines for each major figure, including full names and dates, main appointments held between 1795 and 1840, and where appropriate his years of study at the *Ecole Polytechnique* and his further training. The second table lists the successive holders of principal posts there, and also the "important" students.

Chapter 22 contains the bibliography of all the literature (both primary and secondary) cited in the text: a few particular manuscripts are included. Finally, Chapter 23 contains three comprehensive indexes.

1.3 On the philosophy of mathematics

In this section I point out six problems concerning either the mathematical or the physical aspects of mathematical physics, or the interactions between them. It is their philosophical character that demands their attention in this opening chapter, for they may occur at any point in the ensuing text. Some strategies related to differential equations are outlined in § 5.1.2, at the beginning of the account of mechanics.

1.3.1 How it means: similarity of structure between mathematical and physical theories

A principal concern of this book is the use of mathematics in the study of physical phenomena. The question then arises of the *extent* to which the mathematics can be interpreted in terms of the physical context—in other words, *how* the mathematics means in the physics. An aspect of the problem may be described by the word 'structure-similarity'. A large-scale example is the broadening of mechanics and the calculus into mathematical physics and mathematical analysis respectively, illustrated in Figure 121.1; but mostly we shall be concerned with small-scale cases—a particular piece of mathematics used in a specific type of physical problem (see my *1987d*).

For example, consider Fourier series as used in the study of heat diffusion. The structure of (say) the mathematical expression $\sum_r a_r \cos rx$ is that of a (hopefully convergent) series of periodic functions. The sum itself represents the temperature of a point in the body; but does this imply that heat is to be thought of as composed of an infinite series of basic states? As we shall see in § 9.2.5, Fourier was reluctant to make such a claim; and his view is quite reasonable, since the mathematical structure does not *have* to be interpreted as physical structure also. On the other hand, such structure-similarity may be quite satisfactory in some other physical interpretation; for example, a Fourier-series solution to the wave equation, where the periodic functions refer to harmonics in a study of acoustics (§ 3.3.3).

Questions of this type can be asked about *any* use of mathematics in a physical context; and the answers can be perplexingly varied, as the simple example above shows. These questions become especially interesting in this period, because of the rise of complex variable and differential operator methods to solve differential equations. We shall not always find coherent answers presented by our figures, or even always conscious awareness of the questions; but they occur often, and also relate to some other problems that are raised in this section.

The problem of structure-similarity is, I think, the one intended by those who, both then and now, disparage the use of mathematics in physical theories as abstract and divorced from physical reality. In such a simple form the criticism cannot be taken seriously; for in what way is abstraction suddenly achieved by moving from (say) 'the velocity of the particle is 7' to '$v = 7$'? A mathematical expression may be little or no more abstract than a verbal one. I think that the intended criticism partly concerns structure-*dis*similarity, as just discussed, and partly the problem to be raised next.

1.3.2 Physical guidance of mathematical investigations

> One would have to have completely forgotten the history of science so as not to remember that the desire to know nature has had the most constant and the happiest influence on the development of mathematics.
>
> Poincaré *1898a*, 86

Part of the mystery of mathematics is its ability to inhabit worlds of its own (including worlds where structure-dissimilarity occurs). Hence there is a temptation for the mathematics to "run away" with itself, as it were; that is, having been introduced in connection with a physical problem, it is then pursued to lengths or in forms which in fact are of no physical interest or even sense. *Pace* Poincaré, the influence is not *always* constant or happy.

Thus for example, we shall see in § 12.5.2 Poisson analyse the cooling of an annulus under certain conditions which are to represent the distortion of shape suffered by astrolabes when used in sunny weather; but his formulae involve simplifications of the physical situations which could be in error to the order of the distortion itself, the formulae cannot be numerically evaluated without mathematical approximations which again may be of the order of the effect, and the accuracy of thermometers rendered the problem impractical (for the time, anyway). Again, in § 15.2.4–7 we shall find Navier defining some parameters in his studies of elastic bodies and viscous fluids which arise satisfactorily in the mathematics but which are hard to interpret as physics; for they are unclear in units and dimensions, an aspect of mathematical physics which Fourier had recently revived (§ 9.2.2). Oversimplification of the physical model is another danger. For example, de Prony and other engineers considered from this point of view various methods of measuring the flow of large quantities of water (§ 8.5.3). Sometimes the difficulties were of a conceptual kind; for example, the limited role assigned to physical considerations in Lagrange's algebraic treatment of mechanics aroused some reservations (§ 5.2.8, § 16.3.3).

As these examples show, there are various routes to this notional kind of mathematical physics. While by their nature the results obtained were of little, or at best problematic, use, and thus had limited significance for progress in mathematical physics, they shed valuable light on the ways in which the subject was being conceived and practised, and thus sometimes on the intellectual competition between different mathematical methods.

The position of mathematical physics, and the problem of its guidance by physical considerations, can best be appraised by setting forth the following spectrum of activities in mathematics and physics:

132.1 Mathematics pursued with no apparent interpretations in physics in mind (and none sought), although it may play a role in interpretable mathematics in some way.

132.2 Mathematics pursued with no interpretation in physics involved but with an awareness that such interpretations exist. Indeed, they may have motivated the mathematics in the first place.

132.3 Mathematics pursued with an interpretation in physics initially in mind, but where the guidance of the physical problem has been lost, in ways just described.

132.4 Mathematics pursued with an interpretation in physics in mind and where the guidance of the physical problem provides adequate control on the mathematics produced.

132.5 Theoretical analysis of (certain) physical phenomena, where the mathematical component is small or even non-existent. Such work is often foundational in character, or concerned with physical parameters.

132.6 Experimental work in physics, designed to test the (mathematical and non-mathematical) theory—including the theory of any instruments involved—that has been worked out already.

132.7 Engineering constructs, machines, and instruments: that is, the design of equipment and structures to fit particular situations or types of situations.

This study is principally concerned with activities § 132.2-4, with some attention to the other four types; and this sub-section has emphasized the pitfalls of slipping from § 132.4 into § 132.3.

1.3.3 The place of experientiability in formulating physical theories

> Everyone makes for himself a clear idea of the motion of a *point*, that is to say, of the motion of a corpuscle which one supposes [to be] infinitely small, and which one reduces by thought in some way to a mathematical point [...]
>
> Poinsot *1834a*, opening

Another problem, which relates to the previous ones, is the question of whether physical theories should be formulated in terms of experientiable (or empirical) concepts, or whether non-empirical concepts be admitted. Laplacian physicists tried if possible to formulate theories in terms of inter-molecular forces, not necessarily obeying Newtonian laws of attraction. Among alternative views, the wave theory of light was not formulated in substantially more empirical terms than was the Laplacian theory of molecules (Chapter 13), and Ampère's theory of electrodynamics was no more replete with self-evidence than was the model due to Biot (Chapter 14). On the other hand, Fourier's theory of heat diffusion was notable for its empirical cast (Chapter 9) and the engineering discussed in Chapter 16 was closely associated with positivism.

This problem relates to the double tasks that science sometimes sets itself, and the tensions that can arise between them (see my *1986d*). There is the task of exploring the fine structure of the world, where the fineness seems to elude our direct experience and so scientific theories cannot rely on it; and the task of studying the experience-sized phenomena that happen around us. The clash of views was basically between the empiricists (who wished to restrict theorising to experiential categories) and the trans-empiricists (such as the Laplacians, who were not so confined).

The role of experientiability is of especial importance in this study, since the calculus plays a major role here, usually in a form employing differentials—that is, infinitesimally small quantities and variables. How could their use be reconciled with the physical phenomena to which they refer? The question arose for the empiricists in an obvious, and rather irresoluble, manner; and it also faced the trans-empiricists, especially in the following way. During this period they often spoke of very small objects (usually 'molecules'), and sometimes also of very very small parts, very very very small parts, ... in an hierarchy of tiny things (see Ampère in § 18.2.7, for example); and they faced the difficulty of relating them to (orders of) infinitesimals (dx, and perhaps also d^2x, d^3x, ...) (see my *1988b*). The issue was not confined to the differential calculus, for a companion set of questions arose in the integral calculus: whether the sum of inter-molecular actions could be expressed as an integral (Laplace in § 7.2.1, for example) or as sums (Cauchy and Poisson in their elasticity theories of § 15.4–5). The problem as a whole exemplifies the issue of structure-similarity outlined in § 1.3.1: the quotation from Poinsot at the head of this sub-section states one of the simplest cases.

1.3.4 Science as one or sciences as many?

Laplacian physics also serves as an example for the question raised here—namely, the problem of whether the sciences are to be regarded as unified or whether some are definitely autonomous from others. As I mentioned above, the fall of Laplacian physics meant a fall for the hopes of unification, for a time at least. In particular, Fourier was a non-Laplacian in another respect, for he claimed the autonomy of heat diffusion from mechanics on the grounds of the irreversibility of heat processes (§ 12.4.5). (Note that Fourier's position was another example of the failure of large-scale structure-similarity, in sense of § 1.3.1, for he could claim autonomy for heat diffusion while also aware of the interpretability of his solution methods both inside and outside mechanics: mathematical versatility does not entail physical versatility.) After the decline of Laplacian physics in the 1820s, and indeed beyond the period studied here, the French seem to have been uncertain about, or even uninterested in, the possibility of unification.

In between versatility and autonomy sits analogy; and in various places we see theories developing in one science S_1 in imitation of theories already produced for science S_2. The principal philosophical question is the *extent* to which the analogy is pressed; whether S_1 is just treated *like* S_2, or whether each is in some way reducible to the other or both to something else. Elasticity provides interesting examples; for the elasticity of the aether may be like that of a membrane (§ 15.6.5).

1.3.5 Are physical theories true or hypothetical?

My next problem concerns the question of whether a theory is regarded as resting on *true* foundations, obtained by some means (usually regarded at this time to be induction from basic facts) or whether a theory is a hypothesis, with possibly insecure foundations. In addition, hypotheses concerning the constitution of a particular kind of substance or action in the physical world were distinguished from laws, which were regarded as empirical descriptions of the pertaining phenomena, shorn of such hypotheses. For example, in optics both corpuscular and waval hypotheses on the nature of light could affirm Huygens's law of double refraction of a ray of light in certain crystals (§ 7.4.5, § 13.4.4). Figures such as Fourier in heat diffusion (§ 9.2.2) were disinclined to adopt any hypotheses of this kind at all. The questions posed above now turn into the issue of the 'truth' or otherwise of laws, and whether or not their confirmation by an hypothesis turned them into 'truths'. The status of a theory as law or hypothesis could be open to discussion.

It is clear that a complicated intellectual scenario is at hand here, with various combinations of position pertaining; thus plenty of scope for disagreement was available, and also for rather unclear positions being adopted. Among the differences is the readiness (or not) to speak of hypotheses at all: the engineers are notable for using hypotheses, perhaps because they lived closer to the physical world than did their more theoretical colleagues. Often the issue was argued at the level of theory formation rather than ontology (secondary rather than primary causes, to use the Aristotelian terminology still sometimes employed at that time).

The distinction between these views of theories is not to be identified with the distinction between induction and deduction; for an inductivist will need deduction (whatever that may be) to articulate the consequences of his theory that he has established as "true". This method was clearly stated by Hamilton (*1833a*, 314):

> The science of optics, like every other physical science, has two different directions of progress, which have been called the ascending and the descending scale, the inductive and the deductive method, the way of analysis and of synthesis. In every physical science, we must ascent from facts to laws, by the way of induction and analysis; and we must descend from laws to consequences, by the deductive and synthetic way. We must gather and groupe [*sic*] appearances, until the scientific imagination discerns their hidden law, and unity arises from variety; and then from unity must reduce variety, and force the discovered law to utter its revelations of the future.

One might in fact, adjoin hypotheses, concerning specific features of the phenomena, to the "true" principles which have already been established. Laplace's treatment of the ebb and flow of the sea, for example, seems to read in this spirit (*Exposition*$_2$ (1799), 248–249):

> The sole hypotheses which I permit myself, are that the sea inundates the entire earth, and that it experiences only slight obstacles in its motions: my entire theory besides is rigorous and founded on the principles of the motion of fluids. In drawing thus closer to nature, I have had the satisfaction of seeing that my results drew closer to the observations [...]

Sometimes such hypotheses are held to be established as "true" after (alleged) verification by experiment or experience. This is the only way that I can find to interpret the phrase 'veritable hypothesis' which often occurs in Laplace (§ 6.4.7) and some others: for example, in Laplace's further discussion of the sea, where he stated that 'The researches that I have done on the oscillations of the sea, have given me the means of determining the effect of the reaction of the waters, in the veritable hypotheses of nature' (p. 275).

As the issue is one of physics rather than of mathematics, I shall not discuss it to the extent to which some historians of physics have welcomely gone. But there are aspects of the problem where mathematics plays a role. For example, when Cauchy advocated mathematical analysis based on limits in the 1820s, he contrasted it with the 'inductions of algebra' which he wished to replace (§ 11.3.4). Perhaps we see structure-similarity at work again; hypothetico-deductivism in physics, "logico"-deductivism in mathematics. On the other hand, Cauchy appears to present the foundations of his mathematical analysis as secure in some mathematical sense; so once again the structure-similarity may only apply to a certain extent. Again, as Hamilton's remark above exemplifies, distinctions between analysis and synthesis were sometimes held to bear upon the issue, although the discussion was often muddled (§ 3.2.5).

1.3.6 On mathematical *Denkweisen*

My next point is specifically concerned with *how* the mathematics was carried out, rather than physical problem(s) to which it was being applied. It is a problem which the German word 'Denkweisen' captures best; perhaps 'styles of thinking' is the nearest, though clumsy, English equivalent.

Many *Denkweisen* have appeared in the development of mathematics. During the period under study, there were three major forms, which I call the algebraic, the geometrical and the analytical. A *Denkweise* is not to be identified with its parent subject; for example one might do geometry in an algebraic way (Carnot is a good example: § 4.7.1). There was universal agreement that the mathematics of mathematical physics was centred around differential equations and their solutions, and/or on concomitant theories such as series and integrals; but there were marked differences in the *Denkweisen* employed in the study. An algebraist works only with formulae, and he extols their generality; diagrams are to be avoided, and also dependence on geometric configuration. By contrast, a geometer welcomes such configurations, even if he does not draw diagrams; he may also like theories cast in terms of empirical concepts (which connects us up with § 1.3.3). In further contrast, an analyst seeks autonomy from algebra and geometry in techniques based on, or using, limits.

Thus, for example, to an algebraist an integral is a formula produced by the inverse of the operation of differentiation (§ 3.2.3), to a geometer it is basically an area (§ 9.2.6), and to an analyst it is the limit of a sequence of partition-sums (§ 11.5.3). The difference is borne out also in different methods of proof; for example, of the law of composition of forces (§ 5.3.2, § 17.4.1).

The brief history of these three *Denkweisen* is that around 1800 the algebraic was dominant because of Lagrange's influence, but by the 1830s the analytical was attracting the loyalty of most major newcomers. However, the geometrical, which goes back to the Leibniz-Euler tradition of the calculus, maintained a distinguished status throughout the period (and after), especially because of its unsurpassed flexibility in aiding *the discovery of new results*.

I stress the question of *Denkweisen* for several reasons. Firstly, it seems to indicate a figure's way of working, and not just convenient means of *presenting* his work; for example, the same *Denkweise* will usually appear in his manuscripts as well as in his publications. Secondly, he seems to retain the same *Denkweise* throughout his career (although he uses other styles if he wishes; for example, in § 10.4 we see the geometrical Fourier and the analytical Cauchy both adept at handling differential operators). Thus, thirdly, contemporaries may display different *Denkweisen*, which may play a role in their competitions and controversies. Fourthly, these differences can appear in education also; in particular, in § 5.2 I point out a variety of different *Denkweisen* for mechanics that were taught at the *Ecole Polytechnique* at the same time, which must have bewildered the students. Fifthly, *Denkweisen* bear on other problems such as structure-similarity; for example, if the integral is treated as a sum (either a geometrical infinitesimal sum or an analytical limit of partition-sums), then it may be interpreted physically as, say, a work function (§ 16.4) but not if it is treated as the inverse of a differential. Sixthly, its concern with the method rather than the content of the mathematical physics is borne out, for example, by noting that the theoretician Lagrange and the engineer de Prony (partly) were both algebraists (§ 3.2.1, § 3.5.3) while the engineer Monge and the theoretician Fourier were both geometrical (§ 3.3.5, § 9.2.6)—and moreover, the influences of Lagrange on de Prony and of Monge on Fourier were quite explicit. Finally, though also tentatively, it may even bear on aspects of a figure's mental profile. I note without further comment that Lagrange's strong preference for algebra over geometry seems to have a parallel in his preference for music over the theatre.[1])

To sum up, *Denkweisen* involve both epistemological and methodological aspects of the work of a *savant*, and in instances such as Lagrange, maybe psychological ones also. The first two of these aspects are often recorded in this book, but lack of quality evidence forbids the last one to be pressed.

In view of the variety of *Denkweisen* prevelant during this period, it is amusing to notice that *all* mathematicians were called 'geometers'. As Gergonne pointed out in his *1818a*, the Renaissance had 'mathematicians' as they had monkeys, and 'geometer' was introduced in the early 18th century to refer to a person of altogether higher competence. The choice of the word 'geometer' was excellent for that time; but as the calculus and algebra developed, it became steadily less appropriate. Nevertheless, it clung on (until late into the 19th century, in fact), like an outdated convention which, for some deep anthropological reason, had to be maintained. It is in this spirit of irony that one should read Lagrange, the geometer *par excellence*, announcing in the preface to his *Méchanique analitique* that 'One will not find any Figures in this Work. The methods that I expound here do not require [...] geometrical reasonings' (*Mechanics*$_1$ (1788), vi).

1.3.7 On mathematical notations

The handling of mathematical notations has caused me some serious difficulties. On the one hand, my desire to avoid modern analyses of this old mathematics (§ 1.4.3) makes me reluctant to use modern notations; in particular, I have minimised the use of vector and matrix notations, since the period is notable for its failure to develop these theories. On the other hand, the original notations are sometimes cumbersome, obscure or ambiguous to such an extent that imitation is undesirable.

My normal practice is to use notations that are familiar to the modern reader if no serious historical anachronism is involved. For example, I write '$\sin^2 x$' for $(\sin x)^2$ where the original often reads '$\sin x^2$' (although I follow, unwillingly, *our* inconsistency in also writing '$\Delta^2 x$' for $\Delta\Delta x$ and not for $(\Delta x)^2$). $\sqrt{-1}$ was written this way at the time, but 'i' (which actually was introduced in Euler *1794a*) is used here; I also follow the primary literature in (occasionally) using 'i' for a real-valued increment. I put 'e' for the base of natural logarithms, as was done by

[1]) This latter preference is recorded in Maurice *1819a*, 173-174. de Prony showed a partial algebraic *Denkweise* (§ 3.5.3, § 5.2.6) and also a great interest in music (§ 2.5.3, § 6.6.1).

Euler and some others in the 18th century ('c' was normally used during this period); 'ln' for the natural logarithm function; and 'log' for the logarithm to base 10. I have also accepted the neglected suggestion of Eagle (*1958a*, ix) of writing 'τ' for the more cumbersome '$\pi/2$', a suggestion made on the excellent semiotic grounds that since 'π' has two legs, half its value needs only one!

In order to compress statements of series I use summation notations; they are conspicuously absent from most of the primary literature. Fourier seems to have been the first to use them regularly (§ 9.4), and even then not exclusively. The most common convention was to set down the first few terms of an infinite series (and/or the first few members of a sequence of equations) and then put 'etc.' or, less often, '...'. Certain authors, especially Laplace, often used '...' to bridge the gap in a *finite* series when the final terms were also set down, and reserved 'etc.' for *infinite* series. These notations could be used together; for example, Lacroix wrote Laplace's generating function (352.1) as

$$\text{`}u = y_0 + y_1 t + y_2 t^2 + y_3 t^3 \ldots + y_x t^x + y_{x+1} t^{x+1} + \text{etc.'} \tag{137.1}$$

(Lacroix *Calculus*$_2$, *3* (1819), 322). Occasionally '...' indicated a series continued by an algorithm, while 'etc.' marked continuation by preservation of form; for instance,

$$nx^n + (n(n-1)/1.2)\, x^{n-1} + \ \ldots \quad \text{and} \quad \sin x + \tfrac{1}{2}\sin 2x + \text{etc.} \tag{137.2}$$

(or vice versa: the practice was not consistent). Of course the summation notation collapses these distinctions.

In all contexts, when several primary texts are discussed in a short space, I have made a uniform use of notations even if these texts vary between each other. For example, in one z could be a function of x and y, while in another x was a function of z and u.

The notations of the calculus have caused especial difficulties. Nowadays, especially because of the Weierstrassians in the late 19th century, we are accustomed to call dz/dx a derivative (defined as the limit of the difference quotient) and regard 'dz/dx' *as a single notation*. But in the 18th century, and often during the period studied here (including Cauchy: see § 11.5.2), the notation was read in the Leibniz-Euler tradition as a ratio of differentials; literally '$dz \div dx$' (§ 3.2.4). The talk then of 'differential equations' meant what it said, equations relating differentials dz, dx, d^2x and so on—in contrast to the strange modern practise of using the phrase to describe equations relating a function and its derivatives. (Unfortunately, sometimes they also used the words 'differential' and 'difference' interchangeably, even in contexts where differential equations and difference equations were both being discussed.) I denote derivatives (or their analogues) by the notations z' or $f'(x)$, and use 'dz/dx' rather rarely; n-th order derivatives are written '$z^{(n)}$', '$f^{(n)}(x)$' or '$d^n z/dx^n$'. In mechanics I sometimes employ the standard over-dot notation '$\dot{x}$' for differentiation with respect to time.

In partial differentiation the difficulty over '*dz/dx*' extended to uncertainty whether it was to be read as a total or a partial differential, as the following optimistic quotation from J. Cousin (*1796a*, *1*, 252–253) serves witness:

> To represent the differential of an indeterminate function z of the variables y, x and of constants, we are accustomed to write
>
> $$dz = \frac{dz}{dy}\,dy + \frac{dz}{dx}\,dx, \qquad [(137.3)]$$
>
> & to give to the coefficients $\dfrac{dz}{dy}, \dfrac{dz}{dx}$ the name of partial differences of the first order of the function z. One ordinarily gives this name to each of the terms $\dfrac{dz}{dy}\,dy, \dfrac{dz}{dx}\,dx$; if I do not do it thus, then it is only for more clarity.

In other words, *dz/dy* really means $dz \div dy$, unless it is sitting in a total differential as in Cousin's equation (137.3).

Euler introduced the bracket notation '(*dz/dy*)' to denote the partial differential; but it was not universally adopted. Indeed, a plethora of notations were used in partial differentiation in the 18th and 19th centuries (see Cajori *Notations*, *2* (1929), 196–261), with the modern notation '$\partial z/\partial x$' becoming standard only in the late 19th century. However, it appears sometimes in the late 18th century (in Legendre and Laplace, for example) probably because then and also in the 19th century '∂' was the standard handwritten form of '*d*'!

I have chosen to rely mainly on the notation 'z_x', which is both familiar and compact. When several independent variables $\{x_r\}$ are involved I write 'z_{x_r}', which cannot suggest $(z_x)_r$ because r is used *only* as a suffix variable. If several dependent variables $\{z_r\}$ also occur, then I use brackets to write '$(z_s)_{x_r}$' or '$(z_s)_x$'. Brackets also help to distinguish z_{x^n} from $z_{x^{(n)}}$; the former is $\partial^n z/\partial x^n$, the latter $\partial z/\partial x^{(n)}$. When only functions of *two* variables is involved, Euler's notations '*p*' for z_x, '*q*' for z_y, and so on, are sometimes employed.

I use these various notations only when stating or manipulating differential equations; if their *formation* in the context of a physical interpretation is at issue, then the notation and the model, including the use of differentials, is followed. In addition, original notations are preserved in quotations from original sources, even the ambiguous '*dz/dx*'. The only exception from *verbatim* quotation is that when transcribing or translating from manuscripts I render '∂' as such only if it is *explicitly* so indicated.

I trust that these various conventions will not disturb too much a reader who wants to go from my text to the original. He must note that in some original sources there are confusing switches of referent of certain symbols. The principal symbols and their referents are:

D	differential operator; derivation in Arbogast's sense (§ 4.3.6); coefficient in a polynomial; a point; a number; a distance; as '(D)', an equation number
d	differential operator; a distance
Δ	forward difference operator; differential operator; density; variation of arbitrary constant; an angle
δ	Laplacian operator (our '∇^2'); differential operator; an angle; a distance; variations
f or F	a force; a function
$'$	derivation (in Lagrange's sense of the derived function f' of f); distinguishing marks (as in, say, the collection of masses m, m', m'' ...)

Multiple uses of symbols are particularly noticeable in the writings of Lagrange and those whom he influenced. For example, in his discussion of incompressible fluids, an equation (D) contained the differential equations, where 'd' indicated the partial differential operator (the total differentials were indicated by 'D') and 'Δ' represented the density of the fluid (*Mechanics*$_1$ (1788), 443–446). With care the reader will not be confused by these practises, although he may wonder at the cast of mind that produced them.

Finally, I have introduced two indubitably modern notations, without incurring mathematical anachronism. Firstly, I use '(a, b)' and '$[a, b]$' respectively to indicate an author's employment of (in effect) an open or a closed interval of values. I mention if he *knowingly* worked with something like the difference between the two: in certain contexts the matter is important. Secondly, regarding definitions, '$A := B$' means that A is defined by B, and '$A =: B$' that B is defined by A. During our period no special symbols were used for definitions, and I have used this notation only when it is *quite clear* that a definition was intended in the original text, and not (as is often the case) that some relationship was introduced without the definitional situation being clarified. In addition, I occasionally use cursive capitals (such as '$\mathscr{A}$') to refer to a body, say, or its surface.

1.4 On the historiography of mathematics

1.4.1 The history of the history of mathematics of this period

> Mathematical reasoning is generally admitted to be an essential feature of the development of science during the past four centuries. Yet with the recent emergence of history of science as an independent discipline, there has been a marked tendency to omit any serious critical analysis of such reasoning.
>
> Brush *1970a*, 115

Brush's judgement expresses the principal historical difficulty that I have faced. Unfortunately no major synthesising study of French mathematics of this period has ever been attempted, and some parts of it have hardly been studied at all; further, many of the historical and philosophical questions outlined in the previous section have not been taken up in the historical literature. To understand why our ignorance of these developments is so profound, I shall briefly outline some features of its historical study—the history of its history, in other words.

There were some historical studies carried out during the period 1800-1840, a fact which is duly noted on occasion; but the first concerted efforts occurred in the period 1880-1920, when some very useful work was done. The interest was taken mostly by Germans, especially in connection with the *Encyklopädie der mathematischen Wissenschaften* (1898-1935). This massive project, containing over 12,000 pages when it was finally completed, was initiated by Felix Klein and others in the mid 1890s as a report on the development of all aspects of pure and applied mathematics up to contemporary work (see Lorey *1916a*, 223-236). Although the main interest of the authors was in recent progress, several articles provide useful information on many aspects of mathematical physics in the early 19th century; those by H. Burkhardt and P. Stäckel are especially noteworthy. However, in accordance with the guidelines for articles given to authors (and reprinted in von Dyck *1904a*) information is all that the articles give: historiographical questions are almost entirely ignored. Results are usually written down without discussion of their proofs, for example, and possibly using anachronistic notations; only rarely is attention paid to the motivations of the historical figures, their positions in their communities and relationships with colleagues.

The *Deutsche Mathematiker-Vereinigung*, which had been founded shortly a few years before the inauguration of the *Encyklopädie*, was closely involved in the choice of authors, and several additional valuable articles were published as reports in the Society's *Jahresbericht*. The same strengths and weakness can be found in those reports as in the *Encyklopädie* articles (sometimes the same authors are involved), and also in other secondary literature of that time; for example, in the relevant volumes of the series *Ostwald's Klassiker der exakten Wissenschaften* of editions of primary texts, the fourth (1908) volume of M. Cantor's *Vorlesungen über die Geschichte der Mathematik*, the relevant articles in the series *Abhandlungen zur Geschichte der Mathematik* that he edited, the editorial prefaces to some volumes of the edition of Euler's *Opera omnia*, the histories of particular branches of the subject by historians such as E. Mach and I. Todhunter, and the mathematical remarks appended to textbooks by authors such as H. Lamb, A. E. H. Love and Lord Rayleigh.

Another relevant source that was prepared, or at least started, during the late 19th century were editions of the works of Cauchy, Fourier, Lagrange and Laplace: unfortunately, as editions they are appalling. For example, the notations are frequently modernised; in particular, $\partial y/\partial x$ is introduced to make the old partial differential look like a partial derivative (§ 1.3.7). The original references, usually vague, are at best only partly filled out; and little editorial attention is paid to mistakes in passage work. Omissions are sometimes extraordinary; for example, the so-called *Oeuvres complètes* of Laplace lacks his first three published papers and his first book (see Stigler *1978a*). Again, Fourier's *1816a*, the paper where first he published the diffusion equation, the Fourier series, the Fourier integral, and the notation '$\int_a^b$', was omitted from his edition! (The edition for Fresnel is much better: indeed, it contains so many variant texts that it is not always easy to recall which version one is currently reading.) Perhaps it is just as well for the study of other figures that historians have to go back to the original publications.

Between 1920 and 1950 very little significant work was done on the history of our period. Since 1950 there has been a considerable increase in interest in the history of science in general, and especially during the last fifteen years some relevant studies have appeared. Most have been written by historians of physics, with an excellent grasp of the *physical* (including experimental) aspects of the subject and of the various educational and social factors; but, as Brush noted, the *mathematical* aspects have been passed over in silence, or merely recorded without any attention paid to historical or historiographical questions. In other words, for these writers mathematics is merely a *bank*: one goes to it when necessary for items which are not *themselves* worthy of historical study. I do not understand why historians take this view, since in their work they note the increasing use of mathematics in physics in the early 19th century. I have read several articles in which words like 'mathematicisation' appear even in the titles, and yet no significant discussion of any mathematics occurs in the text. Thus for modern historians of science this book raises the question of what mathematical physics actually is, in addition to issues concerning its history (and that of the pertaining mathematics).

To sum up, the available historical literature is either strong on mathematics but weak on historiography, or strong on physics and historiography but mute on mathematics. It is the purpose of this book to fill some of the huge gaps that exist between these two extremes by both providing details of the mathematical aspects of the subject and exploring some of the historiographical issues that surround them. In the rest of this section I shall outline in some detail the constraints which I have had to impose in order to bring the project down to a practicable size.

1.4.2 A selective, suggestive history

> Like all great travellers, I have seen more than I have remembered, and remembered more than I have seen.
>
> Disraeli

As my summary of the history in § 1.2 has indicated, we are concerned with a massive amount of work, together with various kinds of attendant complications. The primary literature covers many thousands of pages: it includes over 200 relevant papers each from Cauchy and Poisson, for a start. There is enough material for at least 30 doctorates, or for a definitive history of several thousand pages—and indeed, the importance of the developments could warrant studies on this scale. But most of them remain unwritten. Thus, in my opinion, it is not possible for anyone to attempt even an approximation to a definitive history; it certainly is beyond my capacities. What is offered here is a selective, suggestive history, which does no more than open up this vast, neglected area of the history of science. The central parts of the story have been described in some detail, but the rest is exhibited only by a few examples or case studies. Even then the scope is enormous, and I have undoubtedly overlooked many interesting points of detail, interconnection or anticipation.

I have treated only occasionally those branches of mathematics whose points of contact with the ones described here were infrequent, even though they are of considerable importance *as* mathematical topics: probability and statistics, and various aspects of geometry and algebra, are the main branches neglected. Concerning level, very little is said here on the literature for school-teaching.

Looking further afield, I rarely dwell on the basic conceptions of physical theories; little attention is paid to experiments, instruments or observations, or to the epistemological and methodological issues which were then sometimes given a prominent role by physicists. Thus this book makes a limited contribution to the histories of physics, astronomy and engineering.

Again, the treatment of the social aspects of the history is confined to comments on individual biography, the formation of groups in clique-conscious Paris, and the development of scientific and higher educational institutions. Finally, I have been perfunctory in treating foreign, and even provincial French, figures; their work is normally considered only with respect to its bearing on Parisian thought. (This seems to have been the view which the Parisian historical figures themselves often took of outside contributions!) An important exception is Fourier's early work on heat diffusion, which was developed during his Grenoble prefectureship but which heavily influenced Paris later. Foreign visitors to Paris (§ 17.1.2), such as Dirichlet and Ostrogradsky, are also included, and a few interludes do look further afield.

On the other hand, I have *not* confined my attention only to the most "successful" developments. They naturally capture much of the space available; but in order to indicate *why* they gained success, and not merely *that* they did so, it is necessary to describe some of the contemporary competition. Hence, for example, Chapters 3–8 describe several matters for the period 1800–1815 which rapidly fell away afterwards, because of their failure to match "successful" rival approaches. This forgotten work has to be recalled in order to give a balanced picture of the period; and in several cases the neglect has been undeserved.

Although selective with respect to all the historical details, this history does attempt at least to touch on all the main developments, and so must be regarded more as an encyclopaedic than a "logical" history. Hence it belongs more to the spirit of the times that it describes than to that within which it was written.

This point needs much emphasis. It is normal now for a historian of science to take some particular aspect of the period and topics under study, and suborne its purpose strictly to argue for the end intended: in this subject area, for example, one might argue that numerical methods were more important than general theory for the progress of mathematical physics. But such histories must await another generation; my main purpose here is to describe 'what happened'. This is the sense in which the account is *encyclopédiste*, a type of history which is not *à la mode* today. Thus to the reader's question 'where are you going?', which he may understandably ask, the answer is: 'I am enjoying the scenery'. To his reaction that he cannot see the forest for the trees, I would reply that the trees are the principal feature.

Naturally, this reply is too simple: the considerations raised in this section, and the outline of the story in § 1.2, show that certain lines of direction and choice have been taken; they are dealt with in more detail in Chapter 19. Nevertheless, the differences between the intentions of this study and the style of normal modern histories are quite substantial.

In addition to the difficulties peculiar to this history itself, there are the usual problems consequent upon the fragmentary character of the information. In some ways history is like archaeology: the historian picks over the surviving ruins of an edifice in space-time. The fragments that he finds are almost always publications, manuscripts and artefacts. All else has vanished: the more intimate and confidential thoughts and hopes, the cracks and edges of personal contact, the conversations and discussions at meetings, and so on. Now some important aspects of the history may depend essentially on these events, for which no record survives. There is nothing to be done about this problem, except on occasion to point to their probable bearing on the incomplete story that the available sources provide.

In other, more mathematical, words, the historical space is insufficiently defined. I use the word 'space' deliberately, for I regard the historian as working with a space in the modern mathematician's sense: a multi-dimensional region, whose dimensions are determined by the historical and historiographical factors which he brings to his studies. Historical figures are like mass-points; influences between them are like forces of attraction and repulsion; and more general influences resemble fields. A community is thus a collection of mass-points, usually in some sort of equilibrium but vulnerable to substantial disturbance. I find the analogy useful, although I would not take it further and play Cauchy, for example, and try to set up the differential equations to represent the phenomena. Cliometrics has not yet advanced so far.

1.4.3 On the pre- and post-history

I have not attempted any extensive discussion of the 17th- and 18th-century work which our historical figures inherited. This is a serious historiographical problem, for the 18th century contains highly relevant pre-history; but there are no comprehensive or authoritative histories of the work of that period. In fact, the situation with regard to the history of 18th-century mechanics and calculus is rather similar to that applying to the early 19th century: most of the available literature is either very technical but historiographically naive, or else historiographically sensitive but technically anaemic. But I have made only a small effort to fill this gap, and so have avoided facile generalities about the 18th century.

The situation is somewhat alleviated by a fact which is an important theme of

Chapters 3 and 5: that various publications of around 1800 summed up "the state of the art" for many aspects of the 18th-century work and were a major influence on the understanding of the 19th-century figures, even if that understanding can *now* be exposed as incomplete or imbalanced. But I am not at all content with my treatment of the 18th-century achievements, which hopefully will receive one day the attention that they deserve.

With regard to the *post*-history of this study, so very much happened that even a summary account cannot be given. In addition, a philosophical issue is involved. One can interpret some mathematics from the past in two different ways: in terms of its own historical genesis, including its *pre*-history, and the authors' motivations and intentions; or as a formal treatment in terms which were developed later (as part of the post-history, therefore) and are of importance today. (I note *en passant* a variant of the formal treatment, where a piece of mathematics is interpreted in terms of a *still older* approach.) Thus an historical study will point out ways in which the mathematics *lacked* later ideas, while a formal treatment would insert them.

I have no objection to the pursuit of formal analyses: they can provide useful insight for further scientific research. But I do protest at the mathematicians' normal practise of *identifying* such treatments with the history of mathematics. On the contrary, the two kinds of interpretation may be quite different. For example, a formal treatment will ignore the "unsuccessful" developments, whereas, as we have just noted, a historian needs to consider them. Again, a formal treatment will stress the *similarities* between past and (more) recent work, while a history will emphasize the *differences* (as in Cauchy's establishment of mathematical analysis, for example, where some emphasis is laid in Chapter 11 on the limited degree of rigour achieved). Finally, a formal treatment will not be concerned with chronology, whereas in history chronology poses such difficulties that I devote the next sub-section to them.

1.4.4 Chronological or topical order?

As the summary history in § 1.2 shows, I have often used a markedly chronological approach. In one sense, this might be obvious; events follow each other in time. But in technical histories such as this, there is a temptation to take each major topic in turn to describe its development over the period under study, thus laying down a series of parallel progresses, like furrows in a field.

It would have been easier to write this history in such a way; but I have respected chronology for two reasons. Firstly, the treatment by topic may well beg the question of whether a topic existed *as such* throughout the period. A particularly interesting aspect of the historiography of science is the embryology of a topic—its growth to that state from a disparate collection of results (or observations or experiments). Thus its early, pre-topical, states must be appraised in terms of the *other*, motivating, topics pertinent at that time. Secondly, one of the most important features of the mathematical aspects of mathematical physics is its versatility, the ability of a piece of mathematics to be used in a variety of different physical interpretations. Often we shall find a mathematical (pre-)topic inaugurated in one physical context and soon used and extended in others; a process which the agricultural metaphor above would express in terms of ploughing *cross*-furrows, when attention to chronology is essential.

On the other hand, a purely chronological treatment would be quite chaotic—almost a random jump from one theme to another. My compromise policy has been to try to follow chronology over broad areas of mathematical analysis or mathematical physics, and also within a chapter on closely related topics; but in places I have had to go back some years when changing topics between sections of a chapter. In addition, a few larger regressions occur when changing chapters, for reasons which are explained *in situ*. But overall a chronological drive forward is preserved, so that most sections can be read in two ways; both as a description of the events at hand, and as a pointer forward to a later section, maybe in a later chapter. In particular, many of the examples described in a given section were chosen out of many possible candidates because they are used at least once later. Sometimes a forward reference is provided, if it helps clarify the limitations of the description. Inevitably some topics are set aside for quite a large part of the book (for example, optics from § 7.4 to Chapter 13).

Two minor types of infringement of chronology must be noted. Firstly, I have sometimes outlined the history of a minor topic entirely within a section, even if the time-span of the chapter is exceeded; for example, § 4.3.6 quickly takes the history of Arbogast's 'calculus of derivations' past 1840. Secondly, some interludes have chronological independence from their pertaining sections.

The chapters into which the total story has been divided have been chosen in order to give each one a certain integrity, both with regard to the period and topics chosen. In addition, the organisation of Chapters 3–8 needs some special explanation. I have described the developments in the calculus in Chapters 3 and 4 followed by the parallel developments in mechanics in Chapters 5 and 6, rather than use the more strictly chronological order 3-5-4-6, mainly because the reader then needs to hold on to two chapters' worth of material rather than four, and because Chapter 4 contains some mathematical details that are pertinent to Chapter 5. The studies of Laplacian physics and of engineering were then placed at Chapters 7 and 8, because they presumed theories from both calculus and mechanics.

To sum up, *there is no optimal ordering of material in a history as large and complicated as this*. I am fairly satisfied with the order followed, and see some advantages in the reader following it. However, other orders are possible, and many readers will only want bits here and there anyway; so the book may readily be treated as a collection of essays and used accordingly. Guidance is provided by the detailed table of contents, the cross-references in the text (which are very numerous, but which can be ignored by the uninterested), the index and the lengthy chronological summary given in § 21.3. A further aid to chronology is the bibliography, where the items for each author are laid out this way (apart from multi-volumed works).

On a point of detail, for the period 1793–1805 I give the dates as used in the Revolutionary calendar (of 12 30-day months, each divided into three *décades*, 'complementary days' in September completing the year), with the ordinary date provided in brackets. (According to Pastoret *1827a*, Laplace was involved in the abolishment of that calendar.) Conversion rules between the systems, discussed in Francoeur *1822b*, are provided in many encyclopaedias.

Finally, I must warn that in many situations it is hard to determine chronology accurately. Undated manuscripts are not the only case; publications also have problems. For example, journals are available today in bound volumes, carrying a date on the title page; but parts of a journal may have been published over several years. Again, some journals do not indicate the date of reception of the paper, so that the time of its composition may be unclear. (The *procès-verbaux* of the *Académie* (§ 2.2.3) can be a valuable aid here, since they often record the submission of both manuscripts and of printed numbers of journals.) In terms of the archaeological analogy of § 1.4.3, seriation can be difficult to determine, especially in a community like that one where a month may be a long time. I discuss a closely related aspect of this problem in the next sub-section.

1.4.5 Complications of the primary literature

The primary literature is immensely complicated during the period 1800–1840. The major papers would often be planned for publication by the *Académie* (or the *classe* of the *Institut*); but the delays could be considerable. Hence the author, or some other figure(s), would write one or more summary papers, for relatively rapid publication in the many ancillary journals (which are listed in § 22.3). In addition, there may have been a report on the major paper at the *Académie* if the author was not a member, a notice of it in the yearly account of the *secrétaire perpétuel*, a review or two in certain journals, and possibly some translations into other languages. The *Royal Society catalogue of scientific papers* admirably lists much of this literature in its bibliographical and index volumes, although it displays an irritating habit sometimes of inventing titles, and it omits all of the engineering journals.

No study of early-19th-century French science has come to terms with the complications of the primary literature. In particular, histories usually dwell only on the main papers; but the ancillary publications are often historically more important. The reports, in occasionally describing a text of somewhat different content from the publication, suggest that re-writing of the paper occurred before publication; and, like the notices and reviews, they supply reactions of others at the time. The summaries shed interesting light on what the authors held to be the principal results or features of the main paper; and they were often the sources of influence on others because they appeared first. Sometimes a summary will vary a little between its different publications.

Another reason for the importance of these ancillary sources is the extremely competitive atmosphere that they reveal as prevailing during this period. Motivations to a problem can include a desire to surpass results of "colleagues", with resultant polemics—not only involving fundamental disagreements about mathematical method or physical conceptions but also charges of plagiarism and claims of priority. Thus both the complications of the literature and the closeness of the chronology appear as intimately connected difficulties.

The complications sometimes become exasperating. It is often important to distinguish one text from another; but a reference can be as vague as 'the paper he read at the *Académie* some years ago'. (The practice of listing references at the end of a paper was urged only in the late 19th century, and then took decades to become widespread.) The French were often careless over names, referring to foreigners such as the British 'Athood' or 'Yvori', or even to compatriots such as 'Fourrier' and 'Perceval' (Parseval). Vagueness like this can be important in a society where there were either two or three engineers called 'Dubuat' (or, indiscriminately, 'Du Buat'); where two Binet brothers and a certain Dinet held teaching posts at the *Ecole Polytechnique* in the 1810s; where three of the early workers in electromagnetism were Felix Savary, Felix Savart, and his uncle Nicholas Savart; and where papers of the late 1820s signed 'A. C.' may have been written by Augustin Cauchy, Auguste Comte, Alexandre Corréard or Antoine Cournot!

1.4.6 Manuscript sources

I have made much use of various manuscripts in the preparation of this book, not only to aid in the solution of problems of chronology but especially to provide details of institutional developments and of individuals' thought, personal relationships, lecture courses, and so on. Some of the manuscripts have been published since the period which I am treating; but many remain unpublished, and in Chapter 20 I present a selection of transcriptions from some interesting documents.

The principal locations of manuscripts that I have used are in or near Paris: in the *Académie des Sciences*, the *Archives Nationales* (but see below), the *Bibliothèque de l'Institut*, the *Bibliothèque Nationale*, the *Ecole Nationale des Ponts et Chaussées* and the *Ecole Polytechnique*. Those in the *Académie* and the two *écoles* provided information concerning their respective institutions, which are the most important ones for this study. With regard to individuals' personal manuscripts, these locations also yielded between them (much of) the *Nachlässe* of Ampère, de Prony, Fourier, Francoeur, Germain, Lacroix, Lagrange, Liouville and Poinsot, and useful collections for Delambre, Duhamel, Navier, Poncelet and Saint-Venant. The *dossiers personnels* in the archives of the *Académie* furnished information of some sort on most figures. Several other Parisian archives have also been used.

There are also substantial collections of relevant manuscripts in two foreign centres: the *Sammlung Darmstaedter*, in the *Staatsbibliothek* (West Berlin); and the David Eugene Smith Historical Collection, in Columbia University (New York). The Wellcome Historical Library (London) is also worth a mention; and several other non-French archives have been used. All archives are named in § 1.5.2–3: their addresses are given in *World of learning*.

The quality and use of manuscripts have both been limited in various ways. Firstly, the collections vary not only in size and importance but also in cataloguing and organisation (some are in a *very* chaotic state), so that information may have been missed. Secondly, the collections contain many draft versions of papers, proof-sheets of publications, lecture notes, and so on, but relatively few letters (not very surprising, since most figures were in Paris most of the time). Thirdly, I have far from exhausted information in the collections used; between them they *could* occupy an historian for much of his working life, especially one resident in Paris and thus able to consult them frequently. Fourthly, I have not fully worked on manuscript collections which play more minor or peripheral roles in the story: in particular, the mountainous but poorly organised collections in the *Archives Nationales* (founded in the mid 1790s, by the way, and housed in its present site since 1808). They tend to concentrate on aspects of biography, national policy and institutional decision, none of which is a major concern here; and while I found many worthwhile files there, I have no doubt that important materials lie buried there for a later historian to find. Finally, although many documents have survived, much is lost, or at least not found: the lack of major collections for Arago, Biot, Coriolis, Dupin, Fresnel, Lamé, Legendre, Poisson and Sturm is regrettable, while most of Cauchy's and Laplace's *Nachlässe* are known to have been destroyed.

To sum up, as usual I have tried to open up an aspect of the history without attempting to offer any definitive account. The main sources are cited by letters abbreviating the name of the archive (as listed in § 1.5.2), followed by the number of the manuscript collection: for example, '*BI*, ms. 954'. An important exception is the archives of the *Ecole Polytechnique*, where a special type of reference explained in § 2.3.0 and § 2.3.2 is used.

Some individual manuscripts are treated like publications for reference purposes, and are cited in the manner to be described in the next sub-section. If a manuscript is published in Chapter 20, reference is given, in text and bibliography, to the appropriate section.

1.5 On the organisation of the text

1.5.1 Bibliography, references, numerations

Individual items, whether published or manuscript, are cited wherever possible in the form *'1818a'*, where the use of italics and the succeeding *'a'* (even if only one work for the year by the author is used) should safely distinguish publication from year. The year in question is usually the (apparent) *year of publication* (or writing, if a manuscript), apart from a few important publications where it is essential to stress the year of composition. For works published in more than one edition and/or volume over a number of years, and for manuscripts of uncertain date, an English catchword is used, with the edition indicated by a subscript and the volume number rendered in italics, as in the example

Lacroix *Calculus*$_2$, *3* (1819), 315-319.

For works which have been reprinted and/or translated several times, I cite by section or article number, or even by theorem or equation, wherever possible; otherwise page numbers of an accessible version are given. In the last case, when the choice has been between the original page numbers and those in an *Oeuvres*, I have used the latter on grounds of accessibility; but, recalling the comments of § 1.4.1, the choice has been made most reluctantly.

Many items are anonymous, or at least their authors cannot be determined for certain. When the item concerns a particular person or institution—for example, when it is a review of a book—then I give the reference in the manner described above, within the list of works for that person. However, in order to emphasise that it is not a work by him, I borrow Lagrange's use of the prime to cite such "derived" items in the form *'1820b''*. (No works by that person will be cited in the form *'1820b'*.) The heading 'Anonymous' only covers items of a more general character.

The bibliography itself uses various other conventions and abbreviations. Readers wishing to make substantial use of it should consult the prefactory explanations given in § 22.1-4. It is inevitably very large, so that a reader who wishes to use the literature cited in a particular passage will have to look from author to author and take out the required references for himself. This is inconvenient; but no better way seemed available to handle such a mountain of publication than to furnish one collective list.

I number equations by their sub-section number: for example, (433.12) is the 12th equation in § 4.3.3. If (say) § 5.1 has no sub-sections, the equation number takes the form (510.4). Similarly, if reference is made to a brief unnumbered passage set at the head of § 5.1, then it is called '§ 5.1.0'. Tables, figures and interludes also take the sub-section number, as Figure 121.1 already exemplified. Cross-referencing in the book is aided by the presence of sub-section numbers in the running-heads.

As shown already by this chapter, I preserve the names of various French institutions; for example, the various *écoles* (which themselves are collectively so named) and the *classe* or *Académie*, with their publications the *Mémoires* and *Savants étrangers* (§ 2.2.2). I thereby gain the advantage that the English equivalents of these words can refer to schools, academies or institutes in general, or to specific such institutions outside Paris. When referring to the modern institution, I add the adjective 'National' where appropriate; but for the older versions I incorporate 'Impérial' or 'Royal' only when making some pertinent point.

I often call the historical figures *'savants'*, for the word carries the connotation of scientist and scholar to which many of them aspired. I retain the names of posts, such as *'professeur'*, *'examinateur'* and *'secrétaire perpétuel'*; and I also employ *'polytechnicien'*, to refer to a graduate of the *Ecole Polytechnique*.

Almost all translations from the French are mine; they follow the original text more closely than is habitual today. The chief case of another translation being used is Bowditch's of Laplace's *Mécanique céleste* (§ 5.4.3).

To end with the beginnings, most chapters start with a section called 'Plan of the chapter'. However, in a few cases these sections are called 'Prefaces', because they launch out with a sub-section or two of material of relevance also to some later chapter(s) before outlining the local plan.

1.5.2 Footnotes and abbreviations

As far as possible I have dispensed with footnotes. Economic collapse dominates over technological sophistication nowadays, so that footnotes are condemned to appear cheaply as useless end-notes at the back of a book. The contrast with the *primary* literature is striking; although it was printed at a time of much less technological capacity, footnotes were almost always *foot*notes. (In addition, the mathematics *was* usually well set; my compliments extend to the typesetting of mathematics of that period, although not the paper then used, which was often poor, though expensive.) The method of citing literature just described was adopted largely to avoid the use of footnotes; but in addition I have retired from the 'footnote game', where scholar X discusses at length the work of scholar Y in many lines of detailed small type, slightly disagreeing with Y from six out of ten points of view .... I am grateful for those works which were useful, and annoyed at the time wasted in reading others which turned out to be not worth reading. Secondary literature is discussed only where some substantive point is at issue—and it is cited only when it is of effective use (so that many pertinent scribbles have been set aside).

Throughout the footnotes, tables, interludes, the bibliography, and in references given in the text, the following abbreviations are used (with 'National', 'Impérial' or 'Royal' assumed as appropriate):

AN *Archives Nationales*
AS *Académie des Sciences*
BI *Bibliothèque de l'Institut*
BL *Bureau des Longitudes*
BN *Bibliothèque Nationale*
BS *Bibliothèque de la Sorbonne*
CF *Collège de France*
CI first *classe* of the *Institut*
E(N)PC *Ecole (Nationale) des Ponts et Chaussées*
EN *Ecole Normale*
EP *Ecole Polytechnique*
MU *Moniteur universel*
OB *Observatoire*
PF *Faculté des Sciences* in Paris
SP *Société Philomatique*
UN *Université de France*
VT *Chateau de Vincennes, Service Historique de l'Armée de la Terre*

When these institutions are mentioned in the text, however, the names are given in full: such is the old-fashionedness of this historian. In addition, the three main archive collections outside Paris mentioned in § 1.4.6 are cited as follows:

Darmstaedter	*Sammlung Darmstaedter, Staatsbibliothek* (West Berlin)
Smith	David Eugene Smith Historical Collection, Columbia University (New York)
Wellcome	Wellcome Institute Library (London)

1.5.3 Permissions and acknowledgements

For granting me access to their archives, and where appropriate also for allowing the publication of texts (mostly in Chapter 20), I am grateful to all the institutions included in the two lists of abbreviations, and in addition to these: in Paris, the *Ecole Centrale des Arts et Manufactures*, the *Maison d'Auguste Comte*, the *Ecole Nationale des Beaux Arts*; in the British Isles, The British Library, the Royal Society, the Royal Greenwich Observatory, and the Trinity Colleges of Cambridge and Dublin; in the German Democratic Republic, the *Akademie der Wissenschaften* and the *Staatsbibliothek*; in Italy, the *Accademia* in Turin and the *Biblioteche Riccardiana* and *Medicea-Laurenziana* in Florence; in Switzerland, the library of the *Université de Génève*; and finally, in the United States of America, the American Philosophical Society (Philadelphia).

Contributions towards the expenses of the researches in Paris were received from the British Academy Small Grants Scheme and the Royal Society Grant for Research in the History of Science. Much of the early planning was carried out while visiting the Institute for Advanced Study, Princeton, U.S.A.: a major product of the period was the long survey paper *1981a*, of which this book is a fulfilment.

Almost all the required primary and secondary literature was found in the remarkably rich libraries in London. I am especially grateful for the facilities made available at the Royal Society; University College London also earns an honorable mention in this despatch.

Chapter 16 has already appeared in a shortened form as my paper *1984b*. I am grateful to the publishers, Taylor and Francis Ltd., for tolerating the measure of material common to the two texts.

Among those who have read the manuscript, I owe a special debt to J.L. Greenberg, who has put up with the text and revisions over several years, and never failed to give them his time and attention. Comments, or at least pertinent advice, were also received from E.J. Aiton, C. Billoux, U. Bottazzini, M. Bradley, J. Buchwald, R. Cooke, J.J. Cross, C.C. Gillispie, T. Guitard, R. Home, N. Kipnis, J. Langins, J. Lützen (who also translated for me some passages from Norwegian), M. Panza, R. Taton and C. Wilson.

Two further people need a special mention. As my research assistant at Middlesex Polytechnic, N. Guicciardini transcribed some of the documents in Chapter 20, prepared much of the chronological table (§21.3), and carried out preparatory work to enable me to compile the bibliography. Last, and most of all, my wife gave up much time to type out the manuscript, and to help me in Paris in several secretarial ways.

Anybody can scribble a book: many people do. To publish a long and expensive academic tome in today's economic climate takes much greater courage. I am indebted beyond expression to *Birkhäuser Verlag* and the *VEB Deutscher Verlag der Wissenschaften*, and also to E. Hiebert and H. Wussing as editors of their new series, for taking on this monster. In particular, the Swiss show again their admirable international spirit, in agreeing with their German and American editors to publish a book by an English author on French science. There could be no hope of it being published in the author's home country, and also probably not in the country upon which he has written. It is a fortunate circumstance that the book appears around the time of the bicentenary celebrations of the events which set in train the story that it tells.

March 1988

Chapter 2

Institutions and individuals around the 1800s

2.1 Plan of the chapter

The purpose of this chapter is to provide some social, institutional and personal background against which the technical concerns of the six following chapters are set. First, I deal with the principal institutions, in the forms in which they were set up around 1795 and developed up to the late 1800s. The mathematical and physical *classe* of the *Institut* is the main topic of § 2.2, the *Ecole Polytechnique* similarly dominates § 2.3, while § 2.4 deals with the *Ecole des Ponts et Chaussées* and the *Ecole Normale* and then with various other educational institutions. In § 2.5 the emphasis switches to five individuals—Lagrange, Laplace, Lacroix, de Prony and Legendre—whose work was not only of importance around 1800 but also carried influence well into the 19th century; two figures, Coulomb and Monge, whose work bore especially on engineering; and a quintet of minor figures. Finally, in § 2.6 a survey is provided of the various generations active in the 1800s, and also of the new emerging profession of science: Figure 262.1 is a schematic outline of the various institutions described, and Figure 263.1 a selective map of the parts of Paris where they were located and where most of the *savants* normally lived.

Various studies have been made of aspects of late-18th-century French science, and they will be cited where appropriate. I express here gratitude to Gillispie *1980a* for its grasp of the Parisian parts of the whole field. Only a limited amount of detail is provided in this chapter: most statements are susceptible to desimplification.

2.2 Research at the *Institut* and elsewhere

2.2.1 The place of the "mathematical" sciences

> The volumes of the mathematical and physical Classes [of the *Institut*] are chiefly confined to natural history, chemistry, and medicine. There are only two mathematical memoirs [...]; for the mathematical members of the *Institut National* publish their works themselves.
>
> Bugge, 1800 (*1969a*, 96)

This comment by an astute Danish observer of Parisian science, while rather exaggerated, expresses an important aspect of the situation of the calculus and mechanics (including engineering) around 1800. Most of the major publications in these fields were textbooks and treatises; and while they sometimes contained new results, their chief purpose was to summarise and consolidate knowledge which had been developed in earlier decades.

Bugge's use of the terms 'mathematical' and 'physical' needs explanation. "Mathematical" sciences include all the topics with which we are concerned (and also experimental physics!); the "physical" sciences include natural history, chemistry and medicine. This use of the adjectives expressed a tradition of the 17th and especially the 18th century, in which one would almost *expect* mathematics to be absent from the "physical" sciences (see Kuhn *1976a*; and Heilbron *1979a*, ch. 1). Optics was perhaps the only science where mathematics and experimental traditions co-existed with some comfort. The revised scenario applying by the 1820s is described in § 19.3.1.

I must comment also on my use of the word 'research'. It was not as widely employed then as it is today; but it is not too anachronistic, for the essential idea of adding to knowledge was present then. Differences lie rather in attendant philosophies of science, and in the relationships between research and education, rather than in the basic notion of research itself.

2.2.2 The mathematical and physical *classe* of the *Institut*

In August 1793 all the Paris academies were abolished. The *Académie Royale de la Marine*, which had operated at Brest with 75 members (see Mascart *1919a*, 188–211), was never reopened; but the *Académie Royale des Sciences* became the first *classe* of the *Institut National des Sciences et des Arts*, devoted to 'mathematical and physical sciences'. Opened in 1795, it met in the Louvre, but in 1806 it moved to the building of the former *Collège des Quatre-Nations* on the left bank, where it has remained to this day (square 8H of Figure 263.1).

The small quantity of mathematical papers that were to appear in its *Mémoires* contrasts with its organisation, in that, like the old *Académie*, the lists of its ten *sections* began with the mathematical sciences. Six places were assigned to each *section*, and their members were supposed to live in Paris; provincials could expect election only as *associés non résidants*, for whom six places were also assigned to each *section*.[1])

---

[1]) The principal source of data on the *classe* of the *Institut*, and its restored form of 1816 as the *Académie Royale des Sciences* (§ 10.1.1), is *Institut de France 1979a*. On the history of the pre-Revolutionary *AS* and the *CI* up to the 1803 reform (which is described in a moment) see Hahn *1971a*; and for the period of the Terror, see Outram *1983a*. For the general social and institutional situation of science in France at this time, see Crosland *1967a*. The minutes of the meetings of the *CI* and *AS* are discussed in § 2.2.3.

Comments on the membership of the *classe* are restricted to those full members with whom we shall have most concern later. The first *section*, entitled 'Mathematics', comprised Borda (who was succeeded on his death in 1799 by Lacroix), Bossut, Delambre, Lagrange, Laplace and Legendre. The second *section*, 'Mechanical Arts', included Carnot, Monge, de Prony and Bonaparte (who replaced Carnot in 1797 when he was ousted in a political coup: he was re-elected in 1800). The third *section*, 'Astronomy', contained Lalande and Méchain; while the fourth *section*, 'Experimental physics', had Coulomb and Cousin among its members. The other six *sections*—concerned with chemistry, natural history, botany, medicine and related sciences—take us into the "physical" sciences, whose concerns were dominating the pages of the *Mémoires*. Lagrange was the first *Président*, with Laplace as *Vice-président* (and *Président* for part of 1796).

The "mathematical" and "physical" sciences each had a *secrétaire*, elected annually from among the full members. The duties for "mathematical" sciences were taken in two-year stretches between 1795 and 1803 successively by de Prony, the physicist L. Lefèvre-Gineau, Delambre and Lacroix.

In addition to the usage of the words 'mathematical' and 'physical', the membership of *sections* is sometimes surprising in its own right. For example, the ingenious experimenter and instrument designer Borda, and the astronomer Delambre, were both "mathematicians"; presumably their current interest in mathematical tables (§ 3.5.4) was the determining factor. Often these surprises were caused by the happenstance of the next vacancy in a particular *section*; but one might have expected a better organisation of a complete reform.[1])

In 1803 Bonaparte decreed a reorganisation of the *Institut*. The mathematical and physical *classe* retained its premier place; but the other two *classes*, covering humanities and the arts, were reorganised into three, and each *classe* was given more autonomy (see *PV, 2*, 621-626). Within the first *classe* the distinction between "mathematical" and "physical" sciences was formalised by splitting it into two *divisions*. The "mathematical" *division* comprised, in order, *sections* entitled 'Geometry' (the pre-Revolutionary posh (§ 1.3.6) name being restored in place of 'Mathematics'), 'Mechanics' (the old 'Mechanical Arts', also rechristened in more aristocratic style), 'Astronomy', 'Geography and Navigation' (a small *section* of three members, transferred from the old *Classe des Sciences Morales et Politiques*), and 'General Physics' (the old *section* 'Experimental Physics', now

[1]) When a place became vacant, a candidate would sometimes publish a list of his publications, or perhaps a more extended *Notice sur les travaux*, in his support. These leaflets and pamphlets can be a useful source of information (see Crosland *1983a*). Some libraries hold a reasonable selection: for example *AS, BI, BN* and the Science Museum Library, London (a collection once owned by the mineralogist and *CI/AS* member A. Brongniart).

Table 222.1 Notices of the *secrétaires (perpétuels)* for "mathematical" sciences, *an* 7 (1798)–1807.

The list is confined to the *full* publications that I have found, and the summaries in the official newspaper *Moniteur universel*. I have excluded brief notices, often of individual meetings, which appeared in these and other publications.

Period covered[a] *an*	*trimetre(s)*	Author(s)	*Mag. enc.*	*Esp. journ.*[h]	J = *J. phys.* M = *Mém. soc. sav.*[b]	*MU* (summary)
4	[all?]	Lacépède and de Prony[c]	(*an* 4), *3*, 534–543	(*an* 5), *3*, 263–288		
5	[all?]	Lacépède[c]	(*an* 5), *6*, 528–552	(*an* 5), *5*, 235–243		
6	1st	de Prony	(*an* 5), *4*, 97–106	(*an* 6), *5*, 265–269		
	2nd	Lefèvre-Gineau	(*an* 6), *1*, 264–270			
	3rd	Lefèvre-Gineau	(*an* 6), *4*, 97–103	(*an* 6), *9*, 103–109		
	4th	Lefèvre-Gineau		(*an* 7), *3*, 208–213		
7	1st	Lefèvre-Gineau	(*an* 7), *1*, 227–231	(*an* 7), *5*, 197–200		
	2nd	Lefèvre-Gineau	(*an* 7), *2*, 541–546	(*an* 7), *9*, 173–179		
	3rd	Lefèvre-Gineau				
	4th	Lefèvre-Gineau	(*an* 7), *5*, 530–535	(*an* 7), *12*, 192–197		
8	1st	Lefèvre-Gineau	(*an* 7), *5*, 535–542	(1800), *1*, 189–196		
	2nd	Lefèvre-Gineau	(*an* 8), *1*, 84–89			
	3rd	Delambre		(1800), *10*, 161–170	M *1* (*an* 9), 195–203	
	4th	Delambre	(*an* 8), *4*, 236–242		J *53* (1801), 380–389	
9	1st	Delambre	(*an* 9), *1*, 352–363[d]	(1801), *2*, 213–216		(*an* 9), 554
	2nd	Delambre	(*an* 8), *6*, 370–371[d]	(1801), *5*, 151–152		(*an* 9), 906
	3rd	Delambre	(*an* 9), *2*, 80–94	(1801), *8*, 166–176		
	4th	Delambre	(*an* 9), *3*, 379–393	(1801), *11*, 203–210	J *1* (*an* 9), 422–431	

Period covered[a] an	trimetre(s)	Author(s)	*Mag. enc.*	*Esp. journ.*[h]	J = *J. phys.* M = *Mém. soc. sav.*[b]	*MU* (summary)
10	1st	Delambre	(*an* 9), *5*, 118–123	(1802), *1*, 154–158	J *2* (*an* 10), 114–117	(*an* 10), 479, 484
	2nd	Delambre	(*an* 9), *6*, 231–240	(1802), *6*, 189–194	J *2* (*an* 10), 321–327	(*an* 10), 1 289
	3rd	Lacroix	(*an* 10), *1*, 506–516	(1802), *9*, 172–182		(*an* 11), 252–253
	4th	Lacroix	(*an* 10), *3*, 362–371	(1802), *12*, 145–152		(*an* 11), 398
11	1st	Lacroix[c]	(*an* 10), *4*, 510–521	(1803), *1*, 176–189		(*an* 11), 511–512
	2nd	Delambre[e]	(*an* 11), *2*, 77–88			(*an* 11), 1 293–1 294
	3rd	Delambre				
11 / 12	4th– / 3rd	Delambre	(*an* 13), *4*, 177–193	(1805), *8*, 17–195	J *61* (1804), 81–103	(*an* 13), 1 147–1 148, 1 150–1 151, 1 154–1 155
July 1805– June 1806		Delambre[f]	(1806), *4*, 437–449		J *63* (1806), 183–191	(1806), 906–908
1806	2nd	Delambre	(1807), *1*, 452–474[g]	(1807), *2*, 169–187		(1807), 36–38

[a] Sometimes the periods are not too clearly indicated. Note that *an* $k = 179(k+1) - 179(k+2)$.

[b] On this rare journal, see §2.2.4.

[c] This report covers the "physical" sciences also.

[d] *Sic* for chronology!

[e] Delambre was then serving in the new office of *secrétaire perpétuel*.

[f] This report overlaps with Delambre *1806b* (published in the 'Histoire' and so not listed here), which covered the second semester of 1805, and 1806.

[g] This journal was used again during the turmoils of 1814 and 1815, publishing reports there for 1813 (in (1814), *1*, 241–288), 1814 (in (1815), *1*, 104–148, 241–262) and 1815 (in (1816), *1*, 96–152, 254–335: part in *Corr. EP*, *3* (1814–16), 405–408). Later Delambre prepared a collective report *1818a* for the 'Histoire', covering the years 1813–1815.

[h] Many of the reports in this journal are very obscurely titled; thus it is hard to correlate them exactly with other reports. Further, the journal varied somewhat in its use of Revolutionary and orthodox calendars. On it, see §2.2.4.

seemingly conceived more widely). The memberships of each *section* (including the new one) were unaltered; however, there were changes in both personnel and conception of the *associés non résidants*, for the position was renamed 'correspondant' and made available to foreigners as well as to French provincials.

Another important and related change was the appointment to each *division* of a *secrétaire perpétuel*, outside the *sections*, to hold the post until death, in contrast with the changes made in the appointee as *secrétaire*. Delambre was appointed to the "mathematical" position, thus creating a vacancy in the *section* for geometry, to which the young Biot, already an *associé non résidant* since 1800, was elected.

The positions of *secrétaires perpétuels* were both onerous and influential. In addition to transacting with national and foreign bodies in the name of the *classe*, they effectively edited the *Mémoires*, the journal of papers contributed by members of the *classe*. An important duty was writing the yearly notice of the *division* for which the *secrétaire* was responsible. These notices are useful for the historian; they contain reports of work presented by outsiders and summaries of members' papers, and are often among the earliest forms of publication of a piece of research. From 1807 they appeared regularly in the 'Histoire' department of the *Mémoires* (with extracts being published elsewhere), and sometimes contain substantial accounts of the work reported. Between 1795 and 1807 they are of less quality; often briefer in treatment, variable in the period covered (usually three months but sometimes longer periods), and inconsistent in place of publication. However, they give useful information on the "mathematical" sciences around 1800; and as they are rather hard to find, I list the principal publications for 1795-1807 in Table 222.1.

The principal publication was the *Mémoires*, with the 'Histoire'; a bibliography of its papers pertinent to engineering (and including applied mathematics) published between 1666 and 1809 appeared in Lesage *1810a, 1*, 167-252. The *classe* had other publications. In particular, papers submitted by non-members and approved by members were supposed to appear in a journal which I shall refer to as the 'Savants étrangers'; but, as it had no influential backers, it appeared rarely—in 1806, 1811, 1827, and thereafter more regularly.

2.2.3 The meetings of the *classe*, and their *Procès-verbaux*

The duties of the *classe* were basically similar to those of the old *Académie*: to receive and (if sufficient initiative was shown) report on papers submitted by non-members,[1]) and on publications that were sent in; to receive, without refereeing, members' own papers, for intended publication in the *Mémoires*; to elect new or replacement members; to propose prize problems, and examine machines and instruments; to advise on various governmental and educational matters (including the elections to certain chairs); and so on. On occasion there was a bit of science, when a member would perform an experiment of current interest. Until the Revolutionary decimal calendar was terminated at the end of 1805, they met every half-*décade*, or five days, beginning at 5.30 or 6.00 in the evening. The meetings normally lasted two hours, and often were quite busy, although sometimes lack of business brought an early conclusion.

There seems to have been some sagging of activity in 1809; several summer meetings finished early, and in July a commission, including Laplace and Legendre, was set up to discuss means of increasing activity (*PV, 4*, 227–229). Interestingly, it mentioned the large number of visitors inhibiting members from making contributions, and the many other societies in Paris where they read papers and journals where they could publish quicker than with the *Institut*. However, the situation seems to have remained unsatisfactory in various ways; for in December the anatomist J. Tenon, as *Président* at the time, complained that 165 reports on papers submitted since 1802 were still awaited, and that the library remained deficient (p. 293).

The meetings must have had some resemblance to a ritual, appropriate to the costume (illustrated for the Imperial period in Hahn *1971a*, 307) that members had to wear; but membership of the *classe* brought them considerable prestige—and also 1,500 *francs* a year, which was worth roughly a modern student's grant. The *secrétaires perpétuels* were each paid 6,000 *francs* a year.

In 1835 the business of the *Académie* (as it had become) was published as the *Comptes rendus* (§ 18.2.1). The minutes of the meetings for the period 1795–1835 appeared in ten volumes between 1910 and 1922, and was supplemented in 1979 by a comprehensive index volume. These minutes are a very important historical source; so it would be meet to make a few general remarks about them.

Gaston Darboux was the *secrétaire perpétuel* for "mathematical" sciences in 1910, and in his prefatory advertisement *1910a* he explained the difficulties that had been faced while preparing the texts for publication. The written texts 'have been transcribed by employees who were often ignorant', he reported, 'to whom it happened more than once to distort the proper names, and even the sense of phrases'. Much checking was done before publication, but the texts still cause

[1]) In my text, I shall distinguish between members 'presenting' and non-members 'submitting' papers. But I shall not normally discriminate between the main paper and any summary version that may have been read out at the meeting.

some problems in use. Firstly, one of the most valuable sources is the reports on papers submitted by non-members, but it is usually not indicated that in many cases these reports were published at the time, either in the *Mémoires* (or 'Histoire') of the *classe* or (more often) in some other journal; I have tried to give all such references in the bibliography. Secondly, the texts often refer only vaguely to a paper (or instrument), so that it can be difficult both to identify and to trace it through its various appearances in the *classe*'s proceedings. Indeed, many items of business petered out; reports on papers were often never presented, for example. Finally, unfortunately the volumes do not contain the minutes of the public meetings of the *Institut* as a whole, when many of the prize problems were chosen and the winners announced.

However, despite such difficulties the minutes are of inestimable value, often providing the *only* source of information on many matters (apart from the manuscripts of the minutes, which are still preserved in the archives of the *Académie*, and which were used for further checking in the preparation of the recently published index volume). Details of chronology are especially welcome, for the reasons discussed in § 1.4.4. I use the special abbreviation *'PV'* (for *'Procès-verbaux'*) to cite this source. A useful survey of those parts of the meetings relating to mathematics, as recorded in the first four volumes (covering the period 1795-1811), is provided by Boulanger *1914-1917a*.

Although one naturally thinks of the *classe* (or *Académie*) as the premier scientific centre, its role as a forum of research discussions and the propagation of new knowledge needs careful attention. Throughout the period under study we shall often find it slow at publishing papers (although the manuscripts seem to have been available for consultation with the *secrétaires perpétuels*), so that the *Mémoires* 'became more a repository than a living journal' (Hahn *1971a*, 306). (However, in later years and decades it came into its own, when the next generations of researchers (and even historians) read the full statement there of the ideas which they wished to develop.) Around 1800 the status of the *classe* as even a research centre was limited. Thus it would be appropriate to conclude this section with a few comments on other societies and publications relevant to the calculus and mechanics.

2.2.4 On publications and organisations outside the *Institut*

A list of the main journals is given in § 22.3. One of most important ones is the *Journal de l'Ecole Polytechnique*, which I shall discuss in § 2.3.3 when describing the school.

Among various other journals associated with organisations, the most interesting is the *Bulletin* of the *Société Philomathique* (later, 'Philomatique') *de Paris*. Begun as a general scientific society in Paris in 1788 under the splendid name 'Société Gymnastique', its survival in the 1790s was *helped* by the events of the Revolution; for soon after the closing of the *Académie* in 1793 it recruited several former *académiciens*, including Lacroix, Laplace, Monge and de Prony. However, although it published four volumes of *Rapports généraux annuels* between 1798 and 1801, it barely managed to sustain its *Bulletin*, which circulated at first as manuscripts and then as printed folios. Then in 1799 it reprinted some of this early material, together with current works in page-numbered parts, and during the 1800s the *Bulletin* settled down to the publication of monthly issues, with a year's volume containing around 200 pages. Even then publication had to be suspended for the winter of 1809-1810 (see Biot *1810c*). From 1814 a formal commission of six departmental editors administered the *Bulletin*. They either wrote the articles themselves or communicated items sent in by other authors; sometimes not even the commission's minutes make clear who actually wrote a piece. Despite all these difficulties, the *Société* became an important organisation, because of its (normally) rapid publication of summary papers of new work; its *Bulletin* then was one of the major scientific serials.[1])

I note briefly a few other organisations; some are discussed in more detail in later chapters.[2]) In astronomy the *Bureau des Longitudes*, formed in 1795 (§ 5.4.1), published an *Annuaire*, and also the *Connaissance des tem(p)s*, an annual volume of navigational tables which sometimes contained research papers. Engineering was an interest of the *Société d'Encouragement pour l'Industrie Nationale*, which was founded in 1801 and began to publish a *Bulletin* in the following year (§ 16.2.1). In the context of military engineering, the *Ministère de la Guerre* published journals (§ 5.5.1, § 8.2.2).

Four other journals which we need note were publishers' enterprises: the *Journal de physique, de chimie et de l'histoire naturelle*, founded in 1771 under a slightly different title beginning 'Observations ...' (see McLellan *1979a*); the *Magasin encyclopédique*, a general journal which put out six volumes a year from 1795; the *Esprit des journaux*, a Belgian serial of a type similar to the *Magasin*, publishing around 300 small pages at about monthly intervals, including many notices of

[1]) The early history of the *SP* is described in Berthelot *1888a*. Its surviving records are kept at *BS*: the most pertinent to this study are the surviving minute books for the period 1788-1824 (mss. 2081-2086), the minutes of the *Bulletin's* commissions, 1814-1827 (mss. 2094-2095), and *passim* in a rather disappointing collection of letters (*carton* 133). On the confused early history of the *Bulletin*, see the microform edition for the period 1791-1805, including the manuscripts, published by the house of Hachette in 1977.

[2]) The most useful single source of information and further references is Crosland *1967a*. On professionalisation in French science see Crosland *1975a*, and on patronage see Fox *1973a*.

the *classe* and from time to time pertinent articles and book reviews; and finally, the Swiss *Bibliothèque britannique*, founded in 1796, which published articles and translations on British science but also covered material in France and indeed other countries. (A similar *Bibliothèque germanique* was planned around 1804 to cover northern Europe (*PV*, *3*, 173), but it seems not to have been launched.) The official newspaper *Moniteur universel* (also known for a time as the 'Gazette nationale') contains information of interest; not only summaries of the notices of the *secrétaires perpétuels* (Table 222.1) but also shorter articles, reports of papers submitted by non-members to the *classe*, and even some book reviews. *A propos* of the complications of the primary literature, I have relied almost entirely on the *Moniteur universel* for newspaper sources, as it was the most prominent one and reprinted items from other papers; but they should be fully searched in a more detailed study.

In addition to journals, there were of course many books, at three fairly distinct intellectual levels: rather elementary texts, for aspirants to schools such as the *Ecole Polytechnique*; a level which we might compare with the modern undergraduate text, for use at the *Ecole Polytechnique*, the specialist schools and the science *facultés* of the *Université*; and full-scale treatises such as Lagrange's *Méchanique analitique*, which were not intended as educational texts but which might be read by the brightest students or used by the teacher as a pedagogic source. Occasionally these books were also published in a journal, especially the *Journal de l'Ecole Polytechnique*. In this history we are concerned with the upper two levels; a few remarks on the more elementary instruction are given in § 2.4.2.

Almost all of this publication, of both books and journals, was carried out by commercial houses, whose development is a most remarkable feature of French science after the Revolution. The principal firms around 1800 for the calculus and mechanics were Baudouin, Bernard (succeeded by Klostermann), (Firmin) Didot, and Duprat (succeeded over the years by Courcier, *veuve* Courcier, Bachelier, Mallet-Bachelier, and finally the Gauthier-Villars of today). Several other houses were involved, both at 1800 and later in our period.

It seems that the vast scale of the industry depended on the sales of books at the lower of the two levels just described. Apparently their print-runs were often between 1,500 and 3,000 copies for each edition, and sometimes a good deal more. The monographs and treatises normally came out between 500 and 1,000 copies, although this was also exceeded: print-runs of journals were of this same order. Unfortunately, almost no work has been done on the industry, and only tid-bits will be provided in this book.[1])

The sketches given in this section are part of a larger picture, in which in many intellectual areas, both in Paris and in the provinces, there was a considerable increase in the number of societies and forms of publications. de Prony and several others attempted in 1801 to survey these activities throughout the Republic by launching a bi-monthly *Mémoires de sociétés savantes et littéraires de la République Française*. It contained some of the notices of the *secrétaires* of the *classe* (Table 222.1), and is a unique guide of its kind. Unfortunately it folded rather suddenly in 1803, after publishing its 13th number, but it conveyed well the expanding panorama of scientific organisations. In the next two sections I describe further parts of that panorama, when I consider some of the principal institutions in higher education.

2.3 Education at the *Ecole Polytechnique*

The importance of the *Ecole Polytechnique* for the history of science education has long been recognized, and its history continues to be the object of interest and research.[1]) It is, in fact, a most complicated story, with many detailed changes in its aims and their execution; and the available sources, even archival ones, sometimes provide contradictory information. *A propos* of the school's archives, I examined them in detail when they were kept in an extraordinary chaotic system vaguely oriented around calendar years; but a reorganisation was effected in 1983-1984 based on a systematic classification. I have cited all documents in the new form, writing (say) 'VII 2c 1' as an abbreviation for 'Title VII, section 2, paragraph c, article 1' (some titles are less structured than this). I have not had the opportunity to reconsult every document, but trust that the ref-

---

[1]) (to p. 86) Dhombres *1985a* contains much information on mathematics textbooks in France between 1775 to 1820, and indicates the contemporary bibliographies available. *BN* holds a collection of publishers' catalogues, under the callmark Q^{10}. Large files of printers' declarations for the period 1817-1835 are held at *AN*, F^{18} 43-119: they indicate print-runs and sometimes other information. F^{18*} II 1-182 contain the registers of these declarations. Delalain *1899a* lists publishers and printers operating in Paris between 1788 and 1813, while Hatin *1866a* surveys the periodical literature in general. Some information on print-runs, especially for the period 1811-1814, is to be found in the serial *Bibliographie de l'Empire Français.*

[1]) On the general social context of establishing the school, see especially Bradley *1979c*, Crosland *1967a*, ch. 4, Fayet *1960a*, and Williams *1953a* and *1956a*. General histories of the school begin with Fourcy *1828a* (note the anonymous reviews Fourcy *1829a'* and *1829b'*, and also Hachette *1828a*), Pinet *1887a* and *EP 1894-1897a*. For the early years, see also Langins *1979a* and *1987a*, and Paul *1980a*, ch. 2. The little-known report Leverrier *1850a* has some valuable historical information in places, laced with the author's desire to make the school of his day more oriented to practicalities (§ 20.11); part of it was reprinted in *MU*. Information on school life and procedures in the early years of the school is difficult to find; but some is given in de Chambray *1836a* and its reply Bugnot *1837a* (the occasion of their writing is described in § 18.3.3).

erences are all correct.[1]) A few special types of reference are described in § 2.3.2. A list of the principal teachers and students up to 1840 is given in § 21.2.

2.3.1 The school's founding and early organisation

The school opened its doors in December 1794 as a central part of the reconstruction of French higher education. It was originally conceived as a successor to the *Ecole des Ponts et Chaussées* and the *Ecole du Génie* at Mézières, and originally was called the 'Ecole Centrale des Travaux Publics'. The change of title occurred in the following year and reflected some changes of purpose;[2]) but the aim was preserved of providing a basic three-year training in topics pertinent to public-service and military engineering. It was seen as preparatory for more specialist education in other engineering schools (called *écoles d'application*), rather than the *sole* educational institution, as had originally been conceived.

This revised conception seems to have been the brainchild of J. E. Lamblardie, who was the first *Directeur* of the school. (On this post, see below.) The novelties of the conception (not all due to him) included the choice of students solely on merit (as determined by the entrance examination) and provision to them of free tuition and a subsistence allowance. The curriculum is summarised in § 2.3.4.

There were many discussions of the structure and utility of the school, and the balance between theory and practice in its instruction; we shall note some later examples in § 10.1.1, § 11.7.2, § 16.3.5 and § 18.3.3. An interesting early case occurred at the meeting on 10 *vendémiaire an* 5 (1 October 1796) of the Council of Five Hundred (the principal parliamentary house under the 1795 constitution). J. F. Barailon, a doctor who was elected an *associé non résidant* of the *classe* that year, and who was involved in reforming the school system, recalled its founding thus: 'The corps of military engineers and of artillery being insufficient, consid-

---

[1]) On the history of new classification of the archives see Bayle and Billoux *1985a*, and my *1986a*. The most pertinent parts of the archives are these: I 1 (decrees on the organisation of the *EP*); I 2 (its purpose); II (admission, passage and graduation of students); III 3b (teaching scientific subjects); VI 1b2 (*dossiers personnels* on teachers and examiners who were not graduates); VI 2a2 (*dossiers personnels* of students, including those who were later teachers and/or examiners); VIII (publications of the school); X 2c/6-7 (the *Registres de la marche des cours* and *d'instruction*, in which the contents of the lecture courses were recorded from 1805); X 2c/2 (the *Registres matricule*, where the students' marks were recorded); and X 2c/2b (the *Registres* of staff employment). Other important manuscripts and the founding of the school are kept at *ENPC*, ms. 2094; and *AN, passim* in F^{14} 11055 and F^{17} 1139, and elsewhere (see Antoine *1967a*).

[2]) The new name seems to appear first in print in Prieur *1795a*, 29; see also *EP 1796a*, v-vi, and Prieur's letter *1818a* in the *MU*, where he recalled various features of the founding of the school. On these changes, see Langins *1987a* and *1987b*.

ering the number of our enemies, and our armies nearly reduced to nothing by emigration', hence the school was founded, and 'known now under the name of *polytechnique* [...] One would be able to call it *encyclopédique*'. A keen critic, he complained about the number of teachers and administrators, and the variety of subjects taught: whether 'such divergent sciences can all enter the head in the same instant. As for me, I do not believe it' (Barailon *1796a*). He praised the teachers themselves; but, recalling the recent failure of the *Ecole Normale* (§2.4.2), he asked for a commission to reform the school, whose utility he continued to doubt.[1])

Until 1804 the head of the school was called *Directeur*. Various people held this post, beginning with Lamblardie for the first academic year: Monge was in charge in 1797-1798 and 1799-1800. Monge was also the first *Professeur* of descriptive geometry, where he was assisted especially by Hachette. Lagrange was the first *Professeur* of analysis in 1794-1796 and 1799; he was replaced during his absence by de Prony (who was also *Professeur* of mechanics from the start), and after his retirement in 1799 by Lacroix. Fourier assisted Lagrange and de Prony from the start, until he joined Bonaparte's Egyptian campaign in 1798 (§2.6.2); he was succeeded by the textbook writer J. Garnier, who himself was followed in 1802 by the young *polytechnicien* Poisson.

In some years there were two *professeurs* for analysis and two for mechanics, the latter taking over from the former for each year of the course. Often they alternated between first and second years, and sometimes changed subject; and these variations, coupled with the habit of the records to refer to a teacher in the third person, can make it rather hard to find out who taught which courses in the early years.

As regards salaries, at first *professeurs* received 6,000 *francs* per year (about 40% of a salary to provide reasonable living standards) and Fourier and Hachette got 5,000 *francs*. From December 1800 Monge, and also the chemist Berthollet, taught without salary.

The appointments to the school were adorned with special terms. I have used *'professeur'* above, and shall continue to do so; but the term *'instituteur'* was employed throughout the Revolutionary period: *'professeur'* was often used to refer indifferently to this senior post or to the assistant level, which was called *'répétiteur'* when it was introduced in 1798. In addition, there were supplementary appointments (*'suppléant'*, *'adjoint'*, *'per interim'* and so on), and also supervisory posts called *'inspecteurs'*. Quite a number of people held positions at these junior levels.

---

[1]) See especially the discussion of the Five Hundred of 24 *nivôse an* 6 (14.1.1798), reported a few days later in *MU*, (*an* 6), 479-480, 482-483. Barailon, a regular pamphleteer, wrote at greater length on the *EP* in *1798a*: his views are appraised in Langins *1987a*, 95-96.

Examinations were required for admission to the school, for progress of the students through the years, and for graduation into the *écoles d'application*. The posts of examiner, and their holders, are discussed in § 2.3.5.

A library of several thousand volumes was quickly built up by purchase and access of materials confiscated by the Republic (see Bradley *1976a*). In addition to provision of prescribed textbooks, some 18th-century literature was acquired, and it may have been from these copies that the students obtained their knowledge of earlier work. From 1795 until about 1807 the librarian was F. Peyrard, a distinguished scholar who produced French editions of Greek mathematics which were widely praised at the time; then the post went, as an additional duty, to E. Barruel, who was *répétiteur* in physics and chemistry; and then from 1818 to his death in 1842 the office was held by A. Fourcy-Gaudain, then an *inspecteur* at the school. He wrote the first extended history (Fourcy *1828a*) of the school—still a valuable book, despite the large amount of research that has been done since. Its reprinting in 1987 was a most welcome event; J. Dhombres has provided much additional information to supplement the text.

2.3.2 Changes in organisation, 1799–1805

Hachette fairly summed up the conception of the school: 'although the administration of the *Ecole Polytechnique* has experienced, at various times, changes necessitated by circumstances, this establishment has always conserved the mode of teaching and the general plan of studies adopted at the time of its formation' (*1828a*, 263). In other words, the basic conception of the school—a polytechnical establishment, providing basic training and preparation for more specialist study elsewhere—was right for the times, and earned the compliment of imitation in engineering or military schools that were established in other countries (for example, a reform of the academy at West Point in the U.S.A. in 1802). We should not lose sight of this continuity of purpose through the various reorganisations that occurred during the early years and later.

Several changes were made in 1799, during Laplace's six weeks as *Ministre de l'Intérieur* (§ 2.5.2). The length of the course was reduced from three years to two, although individual students were allowed to take longer if necessary (for medical reasons, for example). The relationship between the school and the other specialist schools was finally settled in a form which will be outlined in § 2.3.5.

The *Conseil de Perfectionnement* was instituted in 1800, as the body to direct admissions, teaching and examination, and academic standards in general. Hitherto the school had been largely administered by a *Conseil d'Instruction*, which had to cope with both administrative matters and course development; following a suggestion of Monge (Fourcy *1828a*, 66–67), for four years it had alternated its half-*décade* meetings between these two areas. Now it concentrated on more day-to-day educational affairs, leaving long-term and more general questions to the new council (and administration, discipline, and so on to other councils). Reconstituted each year, the membership of the *Conseil de Perfectionnement* consisted of the *Directeur*, the *examinateurs permanents* (§ 2.3.5) and nominees from the *Conseil d'Instruction*, the *classe* of the *Institut*, and some ministries. The minute-books of these two councils show that they both met about every half-*décade* in their respective early periods, but less often thereafter: the *Conseil d'Instruction* about fortnightly or less, for example (when the orthodox calendar was restored in 1806) and the *Conseil de Perfectionnement* monthly or less, apart from the beginnings and ends of the academic year. These minute-books, kept as bound volumes in the school's archives, contain some valuable information not available elsewhere. I cite them as *'EP, CoP'* or *'EP, CoI'* respectively followed by the volume and page number (or dates of meeting for unfoliated volumes); they have also been used to check all sorts of details concerning the school and its staff and students.

The *Conseil de Perfectionnement* published an annual report, and also the annual programme containing the syllabi and timetable; I cite them by the year of publication, followed by 'rep' and 'prog' respectively. The programmes were made available to the students; and they and the reports were also sent to the *préfets* of *départements* of France and to the major *lycées*, to aid the school's admissions scheme (§ 2.3.4).

This council was very influential, especially in the planning and content of syllabi. At the beginning Monge's views were important:[1]) in many ways its initial profile as a practical engineering school resembled the defunct *Ecole du Génie* at Mézières, where he had taught between 1769 and 1784. But soon, and especially with the inauguration of the *Conseil de Perfectionnement*, a more theoretical character became established, both on the balance between courses (I give some evidence in § 2.3.4) and also in the emphasis on general courses on engineering and machines rather than special courses on particular theories. Laplace was a major influence here: part of his motivation may have come from his personal dislike of Monge.

In December 1804 Citizen Bonaparte put a crown on his head and instantly became Emperor Napoléon—at a ceremony attended by Sergeant Arago from

[1]) The degree of Monge's importance was disputed. Compare, for example, Navier's enthusiasm in *1820a* with Prieur's more cautious view in *1818a*. On students' life at the time, see de Chambray *1836a*, 33–59.

the school's students,[1]) Laplace, and Fourier as *Préfet* of a *département* (§ 9.1.1). Around this time the little man militarised the school, with attendant changes in its balance between military and public works studies. It fell now under the *Ministère de la Guerre*, although the *Ministère de l'Intérieur* continued to be consulted on questions such as the appointment of graduation examiners, since many graduates still went into public works appointments. This was arguably the most *profound* change effected. School costume became school uniform.

A new post of *Directeur des études* was introduced, to supervise the teaching in general; it was second-in-command at the school. The first holder was Gay de Vernon (the *Professeur* of fortification). The position of *Directeur* was re-titled 'Gouverneur'; and instead of the frequent changes of occupant the soldier and statesman G. Lacuée held the post until 1814. During this time Napoléon tried to make the military aspects of the school still more prominent, even decreeing in 1807 that it be replaced by the 'Ecole Napoléonienè'; but for once he did not get his own way (see Bradley *1975a*).

At this time the school changed site. It had been housed from 1794 in a building adjoining the *Palais Bourbon* (square 3J of Figure 263.1), where the legislature met (and still does); but late in 1805 it moved to the *Collèges Navarre* and *Boncourt* (square 11D), where it stayed until transferring south of town in 1976, to its luxurious new site at Palaiseau.

2.3.3 The school's periodicals

The school established a *Journal Polytechnique* in 1795, soon to be renamed as the 'Journal de l'Ecole Polytechnique'. Printed by the Republican/Imperial/Royal print shop, it was distributed by the school in association with some publisher. A principal purpose of the *Journal* was to publish the lectures courses for the use of the students (and others); its early issues were dominated by such material, and also contained decrees concerning the school's organisation. Suspended for financial reasons in 1800, it was soon re-started. A clumsy double-numbering system was used, in which several *cahiers* made up an approximately bi-annual volume. (For reasons explained in § 5.2.7, volume 4 comprised *cahier* 10 (1810) and *cahier* 11 (1802), and *cahiers* 7 and 8 of volume 2 did not appear until 1812.) Poisson, who was in effect joint editor with Hachette from about 1808, stated in *1814b* that the *cahier* system was to be dropped; but both systems were maintained, and sometimes a volume would still contain more than one *cahier*. The

[1]) See *EP*, *CoI*, *4*, meeting of 23 *brumaire an* 13 (14.11.1806). Pinet *1887a*, 52 shows Arago receiving the school's flag from the Emperor at a later ceremony.

chronology of the first twenty *cahiers* (1795-1831) is indicated in Bailhache *1975a*, 202.

During the 1800s the *Journal* appeared less regularly and frequently than desired, and became virtually a research journal (especially in pure and applied mathematics); so in 1804 Hachette launched a new periodical, the *Correspondance sur l'Ecole Impériale Polytechnique*. Appearing every few months (see Hachette's announcement *1806a*), it contained news of the school, its organisation, and *polytechniciens'* careers, and also papers more suitable for the educational level—indeed, some contributions were written by the students or recent graduates. Three volumes had appeared by 1816, when Hachette retired (or rather, was retired, as we see in § 8.2.1).

Both the *Journal* and the *Correspondance* seem to have been quite widely circulated. Documents in the school's archives reveal, for example, the cost of printing 2,400 copies of *cahier* 13 of the *Journal* in 1804, and quotes for 1,000 and 1,500 copies of *cahier* 14 in 1808. When the *Correspondance* was launched, at least 1,000 copies seems to have been proposed (*EP*, VIII 1). Its distribution to students over the years led to various reprintings: I have seen an 1836 reprint, and there may have been later ones. Nevertheless, both periodicals are now rather hard to find—an instructive example of the way in which even widely circulated material tends to disappear over the decades.

2.3.4 Student admission and instruction

The school was intended as an elitist institution: the best students were to be recruited from the whole of the country (or Empire, as and when it became so extended). As we are concerned with some of the tiny minority of the *polytechniciens* who later became *savants*, I shall not comment on the student community as a whole,[1] but speak of the manner of their admission and instruction.

The initial enrolment in 1794-1795 was nearly 400. It was reduced to 360 by an early decree, and until the Restoration the annual enrolment was usually between 100 and 160. The reduction from three to two years' instruction in 1799 (§ 2.3.2) also reduced the total student numbers substantially. Many staff appointed for the large first year were released.

The body of admitted students in a given year constituted a *promotion*: the list was issued 'in order of merit' according to marks obtained, as a first flavour of

[1]) Marielle *1855a* is a fairly reliable source of data on *polytechniciens* to 1853; for a study of the period 1794-1914 see Shinn *1980a*. Pinet *1902a* contains information on the more prominent *polytechniciens*. Paul *1980a*, ch. 2, contains some useful comparisons of the syllabi over the period 1795-1805. Further references are given in later footnotes.

the examination system to come. In order that entry could be made irrespective of means, no fees were charged: indeed, a modest annual salary of 1,200 *francs* was paid. In the summer of 1799 the survey of enrolment found that 160 students were 'without fortune', 75 'presumed in affluence' and 39 'presumed rich', and that most of the families were from artisan and 'artist' classes (*EP, Col, 2*, fol. 38). Around the same time, in a sequence of robustly phrased 'observations' addressed to the Council of Five Hundred, the *Conseil d'Instruction* rejected accusations of being an 'exclusive privilege': on the contrary, 'this establishment will present the true Palladium of the rights of the poor against the abuse of power and the prerogatives of the rich', and a student can *'have proved to have an irreproachable conduct, a morality and assured principles, practiced eye and hands, and finally his head furnished with the fruits of his habitual communications with the great masters, the great examples and the great conceptions'* (*EP 1799a*, 9, 8: see also Fourcy *1828a*, 175-178).

At first admission was by a jury composed of men such as Lagrange, Laplace and Legendre, working only from notes made available to them by examining commissions—a system that was criticised by Fourier (see Challe *1858a*, 127). The situation seems to have been improved by the appointment of three admissions examiners. They would take one third of the Empire each, and spend several weeks in the summer and early autumn visiting several principal towns in their area, to receive recommendations and test applicants. Paris was dealt with separately: it seems to have been regarded as a promotion to be admissions examiner there. In the 1800s an extra examinership was instituted, and France was divided into quarters. These examiners were often young when appointed; the post was a useful source of income (at first, 1,500 *francs*) for an aspiring academic. For example, Biot and Poinsot, both members of the initial enrolment at the school, took on this work within a few years of graduation.

The students were divided into *brigades* of around 18 members, each one led by an older or a prominent student. Between 1799 and 1806 a year's entry was known as the 'first division', from which the students passed to the 'second division'. However, in 1806 someone with a talent for confusion transposed 'first' and 'second'.[1]) This absurd change has endured to this day, despite both the resulting differences between divisions and years (the word 'année' was indeed often used to bring clarity) and also the long-standing custom sometimes to present a two-year syllabus split into divisions by *chronological* order, so that its first/second division was studied by the second/first *division* of students.

---

[1]) The change slipped in surreptitiously, for the academic year 1806-1807; see *EP, Col, 4*, meeting of 31. 10. 1806; and *EP 1806* rep, 7.

The mathematical and engineering syllabi covered the calculus, mechanics, descriptive geometry (including analysis applied to geometry), architecture and fortifications, and 'stereotomy' (perspectives and projections of geometrical figures). The 1804 reform brought in, then or later, courses on machines (§ 16.3.5) and engineering drawing, and in 1811 a course in basic astronomy and geodesy was launched. The other syllabi covered general physics and chemistry. There was also instruction in personal hygiene, and some humanities subjects (grammar, literature and, later, foreign languages—usually English and German).

In 1805 Lacuée introduced the *Registres de la marche des cours* (from 1809, *Registres d'instruction*), in which a brief note of the contents of each lecture was made at the time, often by the *sous-inspecteur* present in the lecture hall to maintain discipline: a sheet of such notes was initiated for each course. They provide valuable information on the progress of a course, although they do not record fine technical details of content (*EP*, X2c/6-7).

Students had access to textbooks, and lecture notes printed in the *Journal*. In addition, manuscript versions of some courses were available, and they often show in detail both the level and the frequency of the instruction. Fourier's lecture notes are a specially valuable source. In 1796 he taught courses in differential and integral calculus, roots of equations, complex numbers, Euclidean geometry, stereotomy, dynamics, statics and hydrostatics (see my and Ravetz's *1972a*, 6-10). The surviving dated notes, which are certainly incomplete, show that he was giving at least one lecture every teaching day; and for one course he wrote out *two* sets of notes for use by the students.

During those early years the curriculum was heavily weighted in favour of Monge's descriptive geometry and related subjects; for example, in 1801 the first-year courses included 153 hours for it, with 60 for the calculus and 40 for mechanics (*EP 1801* prog, 20-21). However, during the 1800s the balance was changed in both years so that time alloted to descriptive geometry fell below that for the calculus and mechanics (§ 8.4.1). Laplace's influence on the *Conseil de Perfectionnement* was a major source of change, which exemplifies the tussle between theory and practice mentioned in § 2.3.1.

The structure of a subject would be presented to the students (in the annual teaching programmes, for example) as a synoptic table, the *encyclopédiste* style of the time for depicting the structure of knowledge (§ 1.2.1). Figure 234.1 shows a typical example, from the 1795 'Programme': both layout and terminology have been followed.[1])

[1]) Among student lecture notes for the early years of the *EP*, see especially the sets of Clermont-Tonnerre for 1800-1801 (*Wellcome*, ms. 1666-1668) and of Atthalin for 1802-1803 (*EP*, III 3b).

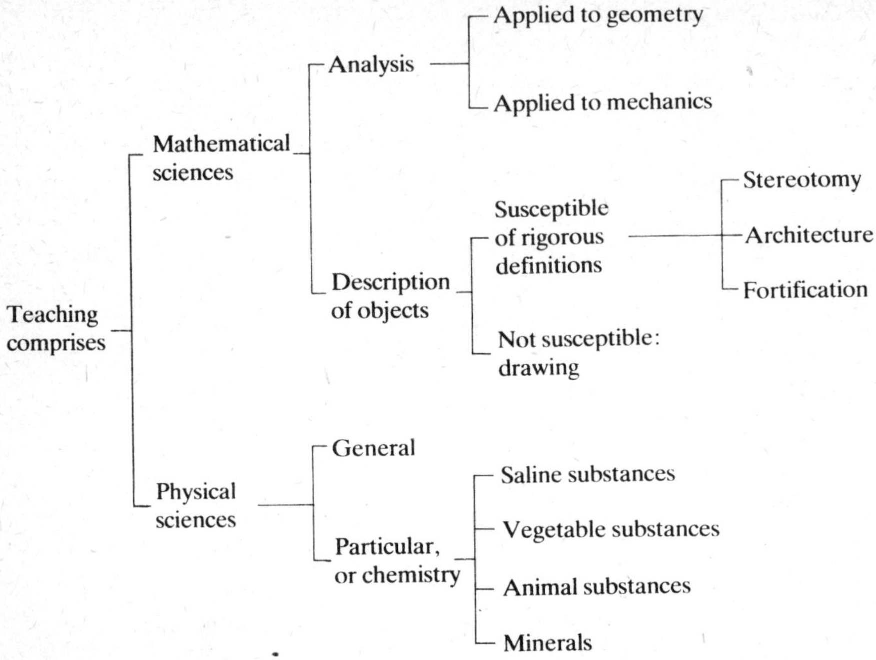

Figure 234.1 A typical synoptic table, indicating the instruction at the *Ecole Polytechnique* in 1795

The courses were supposed to follow exactly the order of topics as presented in the annual programme. However, in a curious contrast, teachers seem to have been given some choice in the treatment of these topics. Presumably this is one factor which explains the different versions of calculus and mechanics that were taught there, as we find in the next four chapters.

The timetable for the students was heavy. Classes were held every day except Sunday, which had arrangements of other kinds. For many years study began at 6.00 a.m. with two hours' work before breakfast. The academic year usually ran from October or November to June or July, with campaigns arranged for the summer.

Lectures were delivered for the usual period of one hour; but from 1804 the period of teaching was extended to $1\frac{1}{2}$ hours, with $\frac{1}{2}$ hour used for questions and revision of the previous lecture (*EP 1804* prog, 9). At times there were difficulties with students; for example, Lacroix complained of inattention in 1802 and 1803, and Hachette experienced class disorder in 1809 (*EP*, IV 4). On each of these occasions, which happened long ago, the lecturers demanded—and received—apologies from the students.

2.3.5 Student examination and graduation to specialist schools

The graduation examiners were originally two 'permanent' ones, for mathematics; from 1798 two temporary examiners, on two-year renewable appointments, were added, to cover respectively 'physics' (including chemistry), and descriptive geometry with graphic arts. (The adjectives 'permanent' and 'temporary' soon became commonly used, although they were unofficial.) The initial *examinateurs permanents* were Laplace (for *Artillerie, Vaissaux* and *Géographes*: see Langins *1987b*) and Bossut (for *Génie, Ponts et Chaussées* and *Mines*); they were replaced respectively by Legendre in 1799 and Lacroix in 1808. Their task was to decide, usually in August and September, on the passage of students from the first to the second year (and until 1799, into the third year), and the final order of graduation and selection of specialist school and service. The examinations were oral, together with practical work where appropriate: in addition, oral or written 'interrogations' took place with staff rather frequently. (The examiners were also present on occasion, until in December 1827 the students' request to exclude them was accepted (*EP*, III 3a).) During the 1800s the position of *examinateur permanent* was raised in importance, in that his annual salary was raised from 3,000 to 6,000 *francs*, which was 1,000 *francs* more than a *professeur's* current stipend although his efforts cannot have been greater than the *professeur's* task of teaching a year's course and supervising the assistants.

When students were admitted to the school, they had already made a commitment to enter one of the public or military services upon graduation, although changes in their choice could be made during their years there and/or a list of preferences could be submitted. Upon graduation the student usually entered a specialist *école d'application*, which was chosen (by his superiors) from his list and noting the national requirements for recruitment, competition to the *école* in question, and his place in the final ranking by order of merit. On occasions the particular needs of the different *écoles* would be considered (see, for example, Lacroix's report of 11 *nivôse an* 10 (2 January 1802) in *EP, Col, 2*, fols. 109–110v). Some courses at the school contained special lectures for students destined for a particular service, where this was felt to be necessary. After passing out from his *école d'application*, the graduate usually started his career in a public works *corps* or in the military forces—although, being a *polytechnicien*, he was more likely to be an engineer administrator than someone who got his hands dirty.

Carnot *1796a* issued a decree on the proposed scheme of *écoles*, but it was rather slow to develop, and was partly modified in the interim. The main *écoles* seem to have been as follows. In Paris there were set up (or reformed) schools for

Mines, Ponts et Chaussées (§ 2.4.1) and *Génie Maritime* (where Laplace was an *examinateur* until it moved to Brest in 1801). In the provinces were to be found a school of *Artillerie* at Fontainebleau, set up in 1802 but moved seven years later to Versailles, near Paris (Poisson was to be an *examinateur* here); another such school was continued at La Fère, and one for *Artillerie et Génie* at Metz (§ 2.4.4). In addition, several minor establishments were organised in various parts of the country: for *Artillerie*, in towns with a substantial army corps; for *Hydrographie*, on vessels in the harbour of towns with important facilities for the navy and/or commerce; for *Cavalerie* at Saumur, in the Loire valley; various centres for the *Administration des Poudres et Salpètres*; and so on. Most of the principal *écoles* took *polytechniciens*; several had their recruitments so restricted. None of them has been well studied by historians, many not at all: even the existence of the system is little known.[1])

When the founding year graduated in 1798, one of their number, A.-M. Héron de Villefosse, inaugurated a reunion banquet of *polytechniciens* which took place from time to time. A school song was written, which began:

Mes bons amis, célébrant les plaisirs
De l'Ecole Polytechnique;
Mais écartons de nos joyeux loisirs
Son attirail scientifique.
Le Z, l'X et l'Y
M'inspirent grand respect.[2])

Maybe this song is the origin of the students' practise of referring to the school as 'l'X'; it came to carry the double connotation of the saturation in mathematics there and the unknown future to follow.

[1]) The building in Paris known then (and now) as the *Ecole Militaire*, near the *Champ de Mars*, seems to have been used largely as a barracks during our period. In addition, a military *Ecole de Mars* existed in tents by the *Bois de Boulogne* from July to October 1794.

The best available survey of the *écoles d'application* is given in Artz *1966a*, ch. 3, with extensive further references provided: on the *Ecole de Mars*, for example, see pp. 167–168.

[2]) I am indebted to W. A. Smeaton for a copy of this song and other similar material, which he possesses, in printed form, in his copy of Fourcy *1828a*. Héron was elected an *académicien libre* of the *AS* in 1816.

2.4 Other schools, mostly in Paris

In this section I look first at the *Ecole des Ponts et Chaussées*, the specialist school which is most pertinent to this study. Then I recall the *Ecole Normale*, in its launchings of both 1795 and 1810, the latter being associated with the founding of the *Université Impériale de France*. Finally, there is a short review of some other schools.

2.4.1 The *Ecole des Ponts et Chaussées*

This school was founded in 1747, and is among the historically most distinguished engineering schools in Europe. The lack of any detailed study of its history is particularly painful.[1])

The school appears to have remained open during the Revolution, under a decree of 1791, and maintained some of its older procedures; in particular, the best of the older students continued to teach the younger ones. The course lasted for three years, with twenty students in each year. Among changes made in 1791, recruitment was by competition, with work to be sent to the school. However, its standards must have fallen, for some of its students failed to qualify in 1794 for entry into the new *Ecole Centrale des Travaux Publics*.[2])

The subjects taught were algebra, geometry, application of algebra to geometry, stereotomy, architecture and draughtsmanship—all topics which were to go in to the *Ecole Polytechnique* syllabi (§2.3.4). Graduates normally entered public works engineering—for example, as engineers of particular kinds for a *départe-*

---

[1]) Serbos *1964a* briefly covers the 18th century; Dartein *1906a* proceeds in more factual detail into the early 19th century. Bradley *1984a*, chs. 2-3 has the most information, centred on de Prony as *Directeur* up to 1839; some of these data for the years around 1800 are given in her *1985a*. Konvitz *1987a* has information on the teaching of cartography (pp. 135-138).

The school's archives has a mass of relevant manuscripts: see especially the large files in *ENPC*, mss. 2629 *bis* and 2432. There are also many research manuscripts by various engineers. The school catalogue *1886a* of manuscripts is very useful, Boinet *1914a* rather less so. Important materials for our period are held at *AN*, especially F^{14} 2148 and 11055-11057.

[2]) A conflation of names, remarkable even by the standards of that time, needs to be indicated here. For at the time of the creation of the *Ecole Centrale des Travaux Publics*, discussions were under way to convert the *EPC* into an *Ecole (Nationale) des Travaux Publics*, possibly incorporating the military school of Mézières, then about to be moved to Metz (§2.4.4). See Antoine *1967a*, correcting Guillaume *1904a*.

ment in France[1])—but as the Revolution developed, they were drawn more and more towards military service.

From its founding in 1747 until his death in 1794, the school's *Directeur* was J. R. Perdonnet. In his last years he was aided by Lamblardie, who succeeded him, and then became the first *Directeur* of the *Ecole Polytechnique* (§2.3.1) before moving back in 1795. This range of duties seems to have led to Lamblardie's early death in 1797, in his 50th year (see de Prony *1798b*). The 80-year-old engineer A. de Chézy, who had been the school's first teacher of mechanics and hydraulics, was brought out of retirement to assume the directorship, but he died during the following year. The next appointee is an important figure for our story; de Prony, who had been a brilliant student at the school in the late 1770s, and had then pursued a successful career which had included assistantships to Perronet and de Chézy (§2.5.3). Like Lamblardie, he was active also in the *Ecole Polytechnique*; but he shouldered the double burden with such strength that he remained in his offices in both schools until his death in 1839.

One of de Prony's early tasks was to reform the syllabi of the school, given the existence of the *Ecole Polytechnique*. This reform was closely related to the 1799 re-organisation of that school when, as was noted in §2.3.4, its relationship with the *écoles d'application* was sorted out (more or less). de Prony introduced a three-year course in which chemistry, mechanics (machines and hydraulics), stereotomy and architecture remained the principal subjects (see *ENPC 1799a*); but the syllabi were designed as continuation of the corresponding courses at the *Ecole Polytechnique*. Student enrolments varied between 20 and 35: they were recruited only from the *Ecole Polytechnique* for many years.

Our interest in this school is basically due to the fact that of the *écoles d'application* this one taught subjects most closely related to our story. Hence we shall find several of the leading *polytechniciens* passing through it before commencing their research careers. However, the instruction itself seems to have made much less impact than that of the *Ecole Polytechnique* until Navier took a chair in 1819 (§15.2.3). Before his arrival, the three or four *professeurs* were all rather minor figures, as was the *Inspecteur* (who was concerned with discipline and administration). Monge had taught descriptive geometry there, but only in 1795 (*ENPC*, ms. 150).

---

[1]) The *Corps* and *Administration PC* have been studied by historians. Debauve *1893a* concentrates on the social and also the legal aspects of their public works activities in the 18th and 19th centuries. Petot *1958a* describes the administration up to 1815, with considerable emphasis on their work during the Revolutionary period: he also provides some information about the school. Many files are held at *AN*, F[14]: see especially 2146–2149 for the *Corps*, and 2152–2341 for *dossiers personnels* of its members of this period (and later). Brunot and Coquand *1982a* furnish notes on many of these engineers, and a survey of these dossiers (pp. 757–892); they also transcribe the texts of organisations of 1755, 1791 and 1804 (pp. 646–684). At *ENPC*, see ms. 2636.

Connections with the *Ecole Polytechnique*, its daughter school, remained strong. For example, in 1796 it had been sited in a building close to the *Hôtel des Invalides* (square 3H of Figure 263.1); but in 1807 it moved to the building attached to the *Palais Bourbon* which the *Ecole Polytechnique* had left two years previously (§ 2.3.2). At that time it tried to imitate its daughter in another way, for in 1806 and 1807 the publisher Bernard announced that in the next January the first volume of the school's *Annales* would appear, as the first in a four-volume annual programme.[1]) However, as we shall see in § 18.2.1, this journal did not start until 1831.

2.4.2 The *Ecole Normale*, 1795

At the time of the Revolution there were around a score of universities in France; and when they were suppressed a variety of alternative proposals for school and university equivalent education, apart from the engineering institutions just described, was proposed. The particular venture of concern here, the *Ecole Normale*, was a failure; and yet it sowed the seed of an educational idea which was of some importance to our story later. It was established in January 1795 in Paris in the *Muséum d'Histoire Naturelle* (square 12C of Figure 263.1) as a kind of "crash-course" institution for training teachers. The teaching was usually set at the lowest of the three levels that were indicated in § 2.2.4; in particular, the 'elementary lectures' on mathematics by Lagrange (*1795a*) and Laplace (*1795a*: see Dhombres *1980a*) were of this kind. The level overlapped somewhat with that at the *Ecole Polytechnique* (see de Prony *1796c*); in fact, both Lagrange's and Laplace's lectures were reprinted in 1812 as a double *cahier* of that school's *Journal*. Among other teachers at the *Ecole Normale*, Monge taught descriptive geometry, assisted by Hachette and Lacroix; and Haüy lectured on physics (§ 7.1.1).

The purpose of the *Ecole Normale* was to produce teachers who could immediately help to found similar schools in their own *départements*. Thus Monge concluded his lectures with a remark about the 'advantages that the arts and industry will extract from the establishment of a school of descriptive geometry in each of the principal towns of France' (quoted from the reprinting in *Geometry*, art. 131). The impracticability of the conception of this school forms a strong contrast to

---

[1]) I have seen this information in Bernard's catalogues bound at the back of the *BN*'s copy of Fischer/Biot *Physics*$_1$ (1806) and one of the *EP*'s copies of Francoeur *Mechanics*$_4$ (1807); doubtless they can be found elsewhere. Maybe it was intended as some sort of successor to the *Recueil polytechnique* cited in § 8.2.2. On the sites of the *EPC*, see Dartein *1906a*, 94–105.

the basic soundness of that of the *Ecole Polytechnique*. Fourier was a pupil there, and he described the teaching: 'numerous rounds of applause made themselves heard [...] when the *professeur* takes his place', he reported to a friend, 'the lectures are almost always interrupted and terminated by clapping' (Challe *1858a*, 115).

It is not surprising to record that the *Ecole Normale* collapsed within four months of its opening. The 1,200 students were not able to comprehend the teaching that was being offered to them (some of which was, in fact, at too high a level); the exceptionally cold winter was no help, either. The main residue that it left were ten volumes of *Séances des écoles normales*, which contained the lectures of Lagrange, Laplace, Haüy, Monge and others (together with questions from the students and the answers); some essays and lectures were added after 1795 by other authors.

Much of the planning of school-level and related education was in the hands of the *Comité d'Instruction Publique* on its creation in 1793, where it broadly followed suggestions of Condorcet. The scheme of teacher training centred on the *Ecole Normale* and its planned provincial imitations was intended to provide teachers for the system of *écoles centrales*, which was launched in the mid 1790s to rejuvenate secondary education. The failure of the *Ecole Normale* was probably one cause of the abolishment of the *écoles centrales* in 1803, for a substantial component of science had been included in the syllabi. These *écoles* were replaced by *lycées*, in which the emphasis on humanities subjects was noticeably greater. I am not discussing primary or secondary education in this study, but I must mention that the *écoles centrales* and the *lycées* provided employment for new graduates from the *Ecole Polytechnique* and/or the specialist schools with academic aspirations (§ 4.2).

In addition, there was a massive programme of textbook writing at the lowest of the three levels indicated in § 2.2.4. The subject matter covered basic arithmetic, algebra and geometry, and in some cases the calculus and parts of mechanics. New editions and/or printing of some books appeared regularly in print-runs of thousands, especially those designated for instruction programmes. Especially favoured were the books of Lacroix, and reprints of the *Cours de mathématiques* of Bézout, which first appeared in the 1760s and came out in various forms and with varying titles until the mid 19th century (Bézout *Course*).[1)]

As *Ministre de l'Intérieur*, Lucien Bonaparte circulated a letter *1800a* to all *professeurs* of mathematics at *écoles centrales* concerning their teaching. He recom-

---

[1)] Lacroix himself wrote an interesting survey of educational philosophy and practise *(Education)*: it first appeared in 1805. On the publication of textbooks in mathematics from Bézout through Lacroix to Cauchy, see Dhombres *1984a*, drawing on bibliographical sources of the time. Liard *Education*, *1* (1888) is a detailed, if somewhat polemical, account of higher education from 1789 to 1795.

mended to them Lagrange's and Laplace's lectures at the *Ecole Normale* as model material. He also reported that the *Ecole Polytechnique* staff had complained about the gap between the attainments and the requirements of new recruits to the school. He asked that algebra be taught as an 'abridged writing' for processes of general reasoning, and that geometry teaching be directed more towards the requirements of descriptive geometry rather than the conic sections, which 'in the past has been overloaded with minutely detailed propositions, for the most part of pure curiosity'. The letter is an interesting document, for it reveals familiarity with mathematics and its teaching. I wonder who wrote it; perhaps his predecessor as a *Ministre*, Laplace?

2.4.3 The *Université Impériale*, 1808; and the *Ecole Normale* again

In March 1808, after two years of advocacy and planning led by the chemist A. F. Fourcroy, a significant innovation was made when Napoléon issued a decree founding the *Université Impériale de France*. This body was run by a *Grand-Maître*, assisted by a *Conseil Impérial de l'Instruction Publique* as an administrative body. Delambre was appointed treasurer of the *Université* (at a tidy salary of 15,000 *francs* per year) and so belonged to this *Conseil* (as did Legendre also).

Modelled in part on the University of Turin, which Napoléon encountered during his Italian campaign of a decade earlier (§2.5.1), the *Université* was responsible for all school education in the Empire. It covered the primary and the secondary levels (where the *lycées* were generally of a higher standard than the *collèges communaux*); in addition, it provided what we recognize as undergraduate-level teaching, in the *facultés*. The Empire was divided up into around 30 regions, in each of which school and (perhaps) *faculté* education was run by an *académie* headed by a *recteur*. Five types of *faculté* were created: medicine, law, theology, letters, and the new discipline of science. *Professeurs* were appointed to the *facultés*: in addition, around 30 *inspecteurs* (including Ampère and Poinsot) were appointed to examine the teaching and advise on staff and textbooks. These posts were in two grades, for a specific *académie*, or across several of them: the post of *recteur* lay in rank between them. Black costumes were required, for both staff and students.

The *Académie de Paris* ran all five *facultés*, of course, and in April 1811 those for theology, letters and science were set up in the *Collège de Plessis*. This building was across the road from the *Sorbonne*, one of the old universities, which had lain partly derelict since 1793 and was partly used as lodgings and by artists.

These buildings were quite close to the *Ecole Polytechnique* (squares 9E-10E of Figure 263.1), and the *Collège* was contiguous with the *Lycée Impériale*, the current name for the old *Collège Louis-le-Grand*. Today a *collège* of that name stands on the site of both buildings, in an edifice constructed in the mid 19th century.

In the *Faculté des Sciences, professeurs* were chosen from those already teaching in major institutions, together with two picked from the *lycées*, so we see some of the usual faces: Lacroix (calculus, also the founder dean) assisted by Francoeur ('higher algebra'), Poisson (mechanics), Biot (astronomy), and Haüy (mineralogy). J. L. Gay-Lussac was given the physics chair, and Hachette took over part of this post as an *adjoint* for descriptive geometry. Separate courses in *sciences mathématiques* and *sciences physiques* were run, with the physics syllabus common to both.

These men preserved their chairs for many years, although from 1810 to the mid 1820s Biot often handed astronomy over to his assistant C. Dinet (a fellow founder-student at the *Ecole Polytechnique* in 1794) and made his own contributions to, or version of, the physics course. Each course required between $1\frac{1}{2}$ and 4 hours per week, in 60- or 90-minute periods, for which the *professeur* received between 4,000 and 5,000 *francs* annually. The degrees of *bachelier, licence* and *docteur* were awarded; the latter was attained by writing two short theses and passing the vivas conducted by a group of *professeurs*, who themselves were awarded the degree *ex officio* upon appointment.

The *Université* was intended to train teachers, administrators, and functionaries of all kinds for the country; for the specific task of producing teachers for quality *lycées* and *facultés* Napoléon resurrected the *Ecole Normale*, as a special institution in Paris to which high-quality students were to be admitted and given accelerated education. Although part of the *Université*, it lay outside all the *académies*. It opened in 1810, in the *Collège de Plessis*; but in 1813 it moved to a disused seminary south of the *Panthéon* (square 10C of Figure 263.1). The enrolment had been planned at 300 students annually: it began with 50, and the number fell still lower (sometimes even down to 15) for the next 30 years.

The *normalien* students took *faculté* courses in their chosen subjects; but they also had special courses of their own, which were sometimes taught by the same *professeurs* (for bonusses on their salaries). In addition, the brightest students taught the others, a system which led later to the post of *maître des conférences*: indeed, the creation of the post is often attributed to the *Ecole*, although we saw in § 2.4.1 that the *Ecole des Ponts et Chaussées* had used a similar system in the 18th century. The course of study at the *Ecole* lasted two years, and was followed in one of its own two *Facultés: Lettres* and *Sciences*.

It all sounds very ambitious, but in fact much was of modest import. The *Uni-*

versité was centred on school-level teaching, and much of the work done in the *facultés des lettres* and *sciences* (including the *Ecole*) corresponded to that of a teacher training college today. Further, in Paris the *Faculté des Sciences* was inferior to the other four there in various ways (the numbers of students enrolled, for example). The doctorate, too, was a modest affair as compared with our expectation of such a degree (some theses will be described). Thus even *normaliens* were of little importance for our study during the period 1800–1830 (though we note in § 18.2.1 the beginning of the rise).[1]) Napoléon had planned for a *Palais de l'Université* to be built by the Seine, just north of the *Champ de Mars* (with a *Palais des Archives* next door); but neither building was ever started.

2.4.4 Some further schools

As far as I can tell, the other *écoles d'application* mentioned in § 2.3.4 did not make sufficient impact on our story to merit lengthy description here. Probably the most noteworthy was the *Ecole de l'Artillerie et du Génie* at Metz, comprising the *Ecole du Génie* of Mézières which was moved in the mid 1790s to Metz (about 250 kilometres east of Paris) to reduce the danger of attack from the British, together with the *Ecole d'Artillerie* at Chalons sur Marne (between Paris and

[1]) The best available study of the early years of the *UN* is Aulard *1911a*: he covers developments up to the fall of Napoléon in 1814, using to good effect various archival sources (on which see below). For briefer surveys of longer periods, see, for example, Liard *Education, 2* (1894), 65–124; Ponteil *1966a*, bk. 2; Fox and Weisz *1980a*, pt. 1; and Palmer *1985a*, 306–315. Shinn *1979a*, 271–291 describes the difficulties faced by the various *facultés des sciences* during the 19th century; Weiss *1982a*, ch. 2 technical training to mid century.

On *PF*, there is a light survey of its mathematics in Hermite *1889a*. See the files on *professeurs* at *BS*, ms. 1643; numerous files in *AN* (where the archives are scattered), especially AJ[16] 5120–5828 (including minutes of *professeurs'* meetings in 5120) and F[17] 20001–21894 (including *professeurs'* careers and pensions, degree certificates, and so on).

On *EN* (in both versions) see Dupuy *1884a* and *1895a*, both *passim*; the volume *EN 1884a*, which contains lists of *normaliens* and other data (as well as Dupuy's first article); Zwerling *1976a*, ch. 2; and many files in *AN*, especially 61AJ 1–432 (its own archives), F[17] 1562 (1795 version) and 4149–4265 (1810 version, personnel, examinations and so on), and AJ[16] 48–49 (1810 version, student admissions).

As usual, *MU* has useful information on several of these developments. See, for example, Fourcroy's proposals in (1806), 644–646, 660–661; the initial decree in (1808), 309–312; appointments to the *PF* in (1809), 723–724, and to many of the *académies* on pp. 1296–1298; the installation of the three *facultés* at the *Collège de Plessis* in (1811), 419–420; and the transfer of the *UN* administration in 1814 from the *Palais Bourbon* to a building on the left bank (on the *Rue de Lille* of today, at square 6I of Figure 263.1) in (1814), 728. On the *lycées*, see, among other things, the table of appointments in (1809), 1406–1408 (but misnumbered 1386–1388) and Napoléon's decree in (1811), 1233–1236.

Various aspects of the history of French education are studied in volumes currently under publication by the *Institut National de la Recherche Pédagogique* (Paris). Some bibliographical works are included.

Metz), which was absorbed in 1802. The Metz school comes into its own for us in § 16.4–5 in connection with Poncelet's teaching there in the 1820s; but it stands out also as a pointed critic of the *Ecole Polytechnique*. For example, in 1810 it sent in a memorandum, stating that it found too much analysis, physics and chemistry in the heads of its *polytechnicien* students, and too little geometry, design skills and practical studies; and for once the Paris school paid a little attention and made a few changes (§ 4.3.2).[1] All these schools had *professeurs* and *examinateurs* of their own, which sometimes meant more jobs for our principal boys.

The *Collège Royal*, renamed *Collège de France* in 1795, was an institution founded in the 16th century to provide relatively high-level education for its own sake: it was housed then, as now, in a building next to the *Collège de Plessis*. Unusually, only elementary mathematics was taught, by mathematics *Professeur* Mauduit; but more advanced topics, especially the calculus, came from Cousin, *Professeur* of 'general and mathematical physics' (as his chair was called from 1791). In 1800 Cousin died, to be succeeded the following year by Biot, who held the chair until his death 61 years later; Lalande taught astronomy until his death in 1807, when Delambre took over; and 'general and experimental physics' belonged to Lefèvre-Gineau until the mid 1820s (§ 13.6.1).[2]

It is much harder to find information on the mathematics teaching at all these other institutions than at the *Ecole Polytechnique*—little has been done, and many archives are in deplorable states—but apparently the *écoles d'application*, and eventually the *Collège de France*, went on to higher levels than the *Ecole Polytechnique*, while the *Ecole Normale* and the *Université* were generally more elementary in the teaching. Further, in all cases, while mathematics occupied a respectable place, by no means did it enjoy the dominant position won at the *Ecole Polytechnique*—a feature worth pondering.

---

[1]) An interesting file of documents, including reflections by *EP Examinateur permanent* Bossut showing sympathy with Metz, is held at *EP*, IIIa; for quotation and discussion, see Fourcy *1828a*, 293–316. The Metz school was a regular and outspoken critic of the *EP* over several decades. Its archives are held at *VT*, X^0 50–100: I have only made a small use of them.

[2]) Almost all of the *CF* files are in *AN*, F^{17}: see, for example, 13551 on Council minutes, 13555 for the dossiers on chairs, and 3855 on the accounts. The *Collège*'s own archive contains in series C XII *dossiers personnels* on the professors, of little interest. On this period, see Sédillot *1870a*, esp. pp. 119–124.

2.5 Major and minor figures

Hitherto many persons have been named, in connection with their appointments and interests. In this section I provide some notes on five figures who held a prominent position in this story around 1800, and for much longer: in turn, Lagrange, Laplace, de Prony, Lacroix and Legendre. Then Coulomb and Monge are introduced, as men with a special concern for engineering. Finally, five more minor figures make a collective bow. Some career details, already mentioned in connection with institutions, are repeated in this section.

2.5.1 Lagrange: scientific dictator

Lagrange was born Luigi de la Grange Tournier in Turin in 1736. In terms of birth, then, he was Italian; but relative to the community in which he lived, or at least that part of his career with which I am concerned, he is a—indeed, *the*—French mathematician. He was submitting papers to the Paris *Académie* already in the 1760s, but spent twenty years at the Berlin Academy before coming to Paris in 1787, where he remained until his death in 1813. During this period he published his principal treatises, on which our interest in his work will largely focus. He is the dominant figure for the next four chapters.[1])

Lagrange's treatises on the calculus were written in response to the educational needs of the Revolution. They form a good example of the three levels of educational material which I mentioned in § 2.3.4 as characteristic of the period; for his lectures at the *Ecole Normale* were at the elementary level, his *Theórie des fonctions analytiques*, delivered at the *Ecole Polytechnique*, stood at the intermediate level (although at the top end of it), and his *Leçons sur la calcul des fonctions*, delivered in 1799, was his most advanced book. In mechanics he published one work: the treatise *Méchanique analitique*, in editions of 1788 and 1811-1815 (two volumes, the latter posthumously edited).

In all these works, as also in his papers, Lagrange usually followed certain physical and mathematical principles upon which he had decided in his youth, and

[1]) Numerous articles on his life have been written, many derivative from obituaries such as Delambre *1816a*, Cossali *1813a* and Maurice *1819a*; among those of note are Loria *1913a*, and Sarton *1944a* and *1950a*. An excellent bibliography of Lagrange's publications is contained in Taton *1974a*. The *Nachlass* is in *BI*, mss. 901-916 (§ 4.8.1); manuscripts on the preparation of the *Oeuvres* are held at ms. 2737 and also at *AN*, F^{17} 3247. A box of interesting materials is kept at *EP*, VI 1a2. L. Pepe (Ferrara) is working on early materials found in Italy. The main collections of correspondence published so far are in his *Works*, *13-14*, and Euler *Works*, (4A) *5-6*. Other letters, manuscripts and secondary literature on specific matters will be cited *in situ*, both for Lagrange and for other figures.

to which he held to the end of his life. It is in that sense that I see him as a scientific dictator: one who dictates the form of the science involved. An important theme in this history is the reception of his views by the new generation of *polytechniciens*.

Lagrange does not seem to have been a successful teacher. Fourier, while a student at the *Ecole Normale* in 1795, noted that 'his voice is very feeble, at least in that he does not become heated; he has a very marked Italian accent and pronounces the *s* like *z* [...] The students, of whom the majority are incapable of appreciating him, give him little welcome, but the *professeurs* make amends for it' (Challe *1858a*, 116). Bugge heard him at the *Ecole Polytechnique* in 1799 and felt that 'whatever this great man says, deserves the highest degree of consideration; but he is too abstract for youth' (Bugge *1969a*, 39). He taught at the school only for a few years (§ 2.3.1), and one can imagine that he did so as a duty rather than with enthusiasm. de Prony recalled in *1835a* that the negotiations with the greatest mathematician in the world to accept a professorship at the school were carried out in a very diplomatic way, allowing him to teach when and what he wanted, in order to balance the interests of the Revolutionary government against the guarantee of *no* teaching responsibilities that the old regime had given him in 1787.[1]) His friend the Swiss-born mathematician F. Maurice recalled soon after his death that he preferred to work alone and disliked giving directions to students. Maurice also recorded that Lagrange planned the morrow's work each day, with the hardest parts reserved for the morning.[2])

Having secured Lagrange's participation in its educational innovations, the new regime used him to further the public image of science. At least one consequence was farcical. In 1798 Talleyrand ordered that a visit be paid to Lagrange's 90-year-old father Tournier in the Piedmont (which was then being added to the French Republic). The *Moniteur universel* reported the ridiculous event that followed, where the deputation and the old man toasted the Republic and the famous son, who had not seen his father for 32 years but could do so now if he

[1]) de Prony's article *1835a* is a good example of the surprising places where information can be found; for it is an article for a biographical dictionary on the Italian mathematician V. Brunacci, to which he added lengthy footnotes about Lagrange and the *EP*. As he died in 1839, I imagine that he felt near to the end of his life, and saw this article as a last opportunity to record such reminiscences. At all events, I am very grateful to him; I quote an extract in § 3.2.4.

[2]) I draw here on a strange and unknown 'letter' published in the *MU* as a supplement to the first published version of Delambre's obituary *(1814c)*. I cite it as Maurice *1814a*: he claimed authorship in his article *1819a* on Lagrange, and I do not doubt him. However, the only signature attached to it, coming at the end of a brief covering note, is 'L.B.M.D.G.'. I take this to be by Lagrange's friend the chemist L. B. Guyton de Morveau, partly on the slight anagrammisation of his initials and partly because he added a few notes to Delambre's full obituary of Lagrange (Delambre *1816a*, lxxviii-lxxx: unfortunately omitted from the reprint of the obituary in Lagrange's *Oeuvres*). I discuss the 'letter' in my *1985a*, and quote from it in § 3.2.5 and § 19.4.4. See also Delambre's letter of 11. 11. 1814 to an editor, kept in *Wellcome*.

wanted to ...[1]) Poor introverted Lagrange, having to read this rubbish about himself in the official newspaper. I have found no record that he took up the invitation reported there.

2.5.2 Laplace: scientific politician

> I have seen much of Mr. de la Place, he is aimiable and a great geometer, but he is tremendously punctilious and hasty, he hardly hears except himself.
>
> J. Trembley to C. Bonnet, 8 November 1786
> (E. Trembley *1948a*, 110)

Had Laplace's father been alive in 1798, Laplace would have been delighted at the prospect of a public visit by a delegation, especially the toast to himself. While we naturally, and correctly, bracket Lagrange and Laplace together in terms of their intellectual pursuits, they were utterly different in conduct.

Pierre Simon Laplace was born at Beaumont-en-Auge in Normany in 1749 into a bourgeois family. Upon coming to Paris in his 20th year, he quickly began the trio of interconnecting scientific pursuits which were to mark out his career: differential equations and their solution, theoretical astronomy, and probability.[2]) His membership of the *Académie* began in 1773, and he taught at the *Ecole Militaire*. By the mid 1790s he was writing and re-writing his astronomical and related mathematical researches into the *Mécanique céleste*, which we begin to examine in §5.4.

The contrast between Lagrange and Laplace is that between dictator and politician. We shall see that, whereas Lagrange stuck to the same conceptions throughout his career, Laplace was more ready to change his mind. There are so-

---

[1]) See the correspondence between Talleyrand and Eymar in *MU*, (*an* 7), 389, 464.

Lagrange had a reputation for meanness; but it seems rather that he had to support his brother in Turin. Some interesting letters to his brother survive in the Lagrange box at *EP*, VI 1a2.

[2]) The interconnections between these topics is the principal theme of Gillispie *1978a*, which is by far the most important study of Laplace hitherto published. It contains not only a substantial intellectual biography but also an extensive bibliography of primary and secondary literature. By contrast, the editions of his works are very unsatisfactory: the seven-volume 1840s *Oeuvres* (see Arago *1842a*) contains only reprints of his major books (I have not used it in this study), while the fourteen-volume edition published between 1878 and 1912, called *Oeuvres complètes*, lacks his first three papers and his first book (see Stigler *1978a*, and also the contemporary manuscripts in Henry *1886a*). It also displays the errors of scholarship described in §1.4.1. Interesting materials on the preparation of these editions are held at *AN*, F^{17} 3247; and in *AS*, *dossier personnel* for Laplace.

It is known that most of Laplace's *Nachlass* was lost during a loot in 1871 and a fire in 1925. However, significant manuscripts appear in Gillispie *1980b*. Hahn *1982a* catalogues his correspondence.

cial analogues, too; a tendency to swing with the political pendulum (to use a scientific analogy suitable to the context). This impressionable style of conduct displeased some of his colleagues: Lacroix, for one, smelt the odours of opportunism (see Taton *1953b*). But it gained him not only great power in the *Institut* and *de facto* leadership of the *Bureau des Longitudes* and the Paris *Observatoire* but also, when Bonaparte siezed power in 1799, the post of *Ministre de l'Intérieur*—at which he lasted only six weeks, although he put through a reform of the *Ecole Polytechnique* (§ 2.3.2). He was concerned with the organisation of the school soon after its founding, and came to exercise his influence chiefly through the *Conseil de Perfectionnement* (to which he belonged as a *classe* nominee) on its founding in 1800. He never taught at the school, although he examined from 1794 to 1799.

From the mid 1800s Laplace and his neighbour the chemist C. Berthollet ran the *Société d'Arcueil*. It was a group of young scientists, mostly *polytechniciens*, whose research was often carried out in the Arcueil mansions of the two leaders (see Crosland *1967a*). Here, however, it was Berthollet's chemists who dominated the group: Laplace could only count Biot and Poisson as important workers in his own major areas of research. This group is discussed further in § 7.1.1.

While Lagrange and Laplace took different attitudes to careerism, they seem both to have been cold and reserved in personality. Maybe they applied this feature to each other: according to Cournot (*1913a*, 83-84),

> One day, after [Laplace] had invited Lagrange to dinner, Lagrange asked: 'Will it be necessary to wear the costume of a senator?' in a mocking tone, of which everyone sensed the malice, except the amphitryon senator.

2.5.3 de Prony: scientific engineer

In contrast to the fame of Lagrange and Laplace, de Prony is today almost forgotten; yet in his day he was as well known as either. Until the rise of the new generation of important *polytechnicien* engineer scientists in the 1810s, he was the leading engineer and engineering educator, not only in his research but especially by his prominent role in the *Ecole Polytechnique* and the *Ecole des Ponts et Chaussées*.[1]) He was even more active than Monge in these areas.

---

[1]) The most useful writings on de Prony are the biography Bradley *1984a*, the obituary Dupin *1840a*, the biographical article Parisot *1846a* and the centenary articles in Walckenaer *1940a*. His library and *Nachlass* are held at *ENPC*, and are catalogued *passim* in the school's *1886a*: they include his own statement of career to 1824 at ms. 1786 (compare his printed 'Note' *1804g*). See also the bound volume of family manuscripts *BN*, ms. n.a. fr. 15778, and the collection of professional letters at *AN*, F^{14} 2304^2. There is no edition of his works.

The Riche family, living at Chamelet in the Beaujolais area of southern France, owned some land nearby at a place called 'Prony', and Gaspard Clair François Marie Riche, born in 1755, added its name to his own. I shall call him 'de Prony', for he seems to have preferred this form; some of his publications carry 'Prony' or 'Riche de Prony'. Expected to follow the family career in law, he was determined instead to be an engineer, and he studied at the *Ecole des Ponts et Chaussées* from 1776 to 1780. He was made a member of the *Corps*, but also assisted at the school at the request of Perronet and de Chézy, becoming *Inspecteur titulaire* there in 1790. His most significant research of this period was in the theory of the thrust of arches, the restoration of the port of Dunkirk, and hydraulic machines. A lover of music (he published a few compositions and arrangements, and wrote on musical temperament), in 1782 he married a pianist from the well placed Fréminville family; her brother was also in the *Corps*. He and his wife inherited a large property at Asnières, which he always retained; and they held cultural and musical *soirées* in Paris until her death in 1822. In 1791 he was nominated chief engineer to the Pyrénées-Orientales *département*; but rather than be out of town, he contrived to be nominated *Directeur* of the cadastral survey based in Paris, for which he organised a huge set of tables (§ 3.5.4).

de Prony's rather high social position caused him some difficulties during the Revolution, but he seems to have gained some protection from Carnot. The reform of higher education greatly improved his fortunes in ways which we have already seen. He appears to have been an effective teacher and a charming person. Bugge attended his lectures at the *Ecole Polytechnique* on hydraulics in 1799, and reported: 'This extraordinary man has the most impressive and captivating delivery, which can possibly be conceived' (Bugge *1969a*, 40). Maria Edgeworth, writing home from Paris in 1802, found that 'He has a great deal of knowledge, and a good nature which makes him ready and willing to communicate it. He is extremely fond of music and plays upon many instruments. His wife has the happiness to possess the same talent' (Edgeworth *1979a*, 57). In addition to his responsibilities in his two schools, which continued throughout his life, he was also very active in the *Corps des Ponts et Chaussées* and in the *classe*. He was in charge of a small *Ecole des Ingénieurs Géographes* from 1795 until its closure in 1803, and was appointed an extra member of the *Bureau des Longitudes* in 1801. At that time he helped to develop the *Société Philomatique* (of which his younger brother, the naturalist Claude Riche (1762–1797), was active in the foundation). He was a prolific writer throughout his career; several of his publications were pamphlets, which are now rare.

de Prony was unusual among our major figures in developing contacts and interest outside France. From his early days he was in touch with British engineers, although he does not seem to have visited that country. But he went several times to Italy—one of his major writings concerned the draining of the Pontine marshes between Naples and Rome (§ 8.5.4)—and his library contained many Italian engineering treatises.

Although he played such a prominent role in public works and education, de Prony does not appear to have brought any political or social ideology to his activities. For example, in 1797 he and Monge nominated Bonaparte to the *classe*; but in the next year he declined Bonaparte's invitation to join the scientific party on the expedition to Egypt,[1]) thereby losing for ever Bonaparte's favour despite all his contributions to engineering and education and his wife's friendship with Empress Josephine. 'I am good only at that', he said of his scientific work. On his death in 1839 his *Nachlass* and library was left to his niece, the widow of the diplomat and mathematician L. Corancez; later she gave them to their true home, the *Ecole des Ponts et Chaussées*.

2.5.4 Lacroix: scientific educator

Lacroix was a unique figure in French science during our period. Although his contributions to research were almost non-existent, he maintained a high reputation by the quality and range of his textbook and treatises, and his historical and philosophical writings on mathematics and mathematical education. The principal textbook writer of his age for pure mathematics, some of his works were reprinted until the late 19th century.[2])

Born in 1765 into a family of modest circumstances in Paris, Sylvestre François Lacroix soon found his passion for mathematics, and was supported by Monge. But he secured a deeper personal and intellectual influence in Condorcet (on whom he wrote the notice *1813a*). He taught with, and for, him at a free school in Paris in the late 1780s, when they also prepared together an edition of Euler's *Lettres à une princesse d'allemagne*; and he was elected as Condorcet's *correspondant* to the *Académie* in 1789. His work shows the same *encyclopédiste* tendency of compiling comprehensive accounts of branches of human knowledge. He also fell under the philosophical influence of Condillac, especially with regard to language; for example, in his writings one sometimes sees the same concern with signs, and thus the distinction made between a symbol and its referent. (The normal situation in late 18th-century algebra was to conflate this dis-

---

[1]) In § 2.6.2 I note some of the *savants* who did go to Egypt. For de Prony's version of his refusal, see Edgeworth *1979a*, 127; and on his relations with Bonaparte, see Dupin *1840a*, 340-344.

[2]) No major study has been written of Lacroix: the most useful studies are Taton *1953a, 1953b* and *1959a*. His manuscripts—a very interesting collection, sorted out by L. Lalanne—are held at *BI*, mss. 2396-2403. A file on his *UN* career is kept at *AN*, F^{14} 20143.

tinction.) By contrast with these positive influences, his relationship with Laplace was somewhat strained; as we noted in §2.5.2, he disliked opportunism, of which his own career, which lasted until his death in 1843, shows no sign.

Lacroix was not involved in the setting up of the *Ecole Polytechnique*: he taught at that time at the *Ecole d'Artillerie* in Besançon. But he obtained a position at the *Ecole Centrale des Quatre Nations* in Paris when the *écoles centrales* were set up in the mid 1790s, assisted Monge at the *Ecole Normale* at that time, and was also appointed to the *Commission d'Instruction Publique* (much influenced by Condorcet, as was noted in §2.4.2). In 1799 he both succeeded Lagrange as *Professeur* at the *Ecole Polytechnique* and the deceased Borda in the *classe* and from 1804 taught at the *Lycée Bonaparte* in Paris; in §2.4.3 we saw his involvement with the *Faculté des Sciences* at its founding.

Of the three levels of educational material mentioned in §2.2.4, most of Lacroix's books were written at the intermediate level; indeed, several were sold as an integrated eight-volume *Cours complet* (see his *1811c*), covering arithmetic, algebra, trigonometry, geometry, and 'algebra applied to geometry' (§4.7.3). However, he also wrote a comprehensive three-volume *Traité* on the calculus, of which the first edition (1797–1800) is our concern in Chapter 3 (especially §3.2.7) and the second edition (1810–1819) is noted in Chapter 4 (especially §4.8.1).

For a man whose competence as a textbook writer was so great, Lacroix's ability as a research mathematician is astonishingly limited. I think that here and there in his books he contributed some touches of his own, but otherwise his research work seems to be limited to two short papers published in the *Bulletin* of the *Société Philomatique* (*1798a* on singular solutions to partial differential equations, and *1803a* on the resistance of fluids): there are also a few research manuscripts. In June 1796 he presented to the *classe* a paper that he had offered to the old *Académie* in 1787, in which he tried to correct the details of the sun's apparent orbit. He also claimed priority for some results presented in 1775, over work published by Delambre in the mid 1780s.[1])

In a report with Lagrange to the *classe* Delambre examined the matter in some detail and finally decided of the later paper that 'if this paper has become more or less useless, [...] it is nonetheless true that at the time when the work which it completes had been started, it was much more interesting' (Delambre *1796a*, 49). He ended his report with a polite hope that Lacroix would return to astronomy, 'to which he could render great service'; but they surely both guessed that Lacroix's research days were probably over.

---

[1]) See Lacroix's letter of 6 *prairial* an 4 (25. 5. 1796) to Laplace, as *Président* of the *CI* (*BI*, ms. 2397). On his work, see Wilson *1980a*, 281–283.

2.5.5 Legendre: mathematician

The men discussed so far each played a notable role in some aspects of the development of French science, and pursued careers within it—not only in the financial sense but also in the exercise of influence on others in the newly emerging profession. Legendre is an example of a man for whom career*ism* seems to have been less important, and devotion to mathematical study the main activity in life. This is not to deny him his prominence in the profession; but the difference is worth noting, for it manifests itself also in the later generations.[1])

Legendre also differs from his colleagues described above by the profile of his mathematical interests. Although he did not command the eminence of Lagrange or Laplace, the range of his interests was, if anything, slightly wider than theirs, including substantial work in number theory and Euclidean geometry (which were not then widely studied and are not of concern here). We shall note some of his considerable contributions to differential equations (§ 3.3.5–7), Legendre functions and the shape of spheroids (§ 5.5.2), elliptic integrals (§ 3.4.4, § 10.2.2), interpolation (§ 6.4.5), numerical mathematics (§ 3.5.4) and weights and measures (§ 5.5.4). We shall also see him reporting to the *classe* or the *Académie* on submitted manuscripts from time to time: he seems to have been quite active there. Among his educational activities, he was an *examinateur permanent* at the *Ecole Polytechnique* from 1799 to 1816, succeeding Laplace and being followed by de Prony.

Details of Legendre's life are scarce, as he apparently wished: according to Poisson's *éloge*, 'Our colleague has often expressed the desire that, in speaking of him, it would only be the matter of his works, which are, in fact, his entire life' (*1833a*, 61). Even his place of birth in 1752 is disputed, between Paris and Toulouse; but there seems little doubt that he was in Paris from an early age and rarely left it before dying there in 1833. His family was fairly wealthy, and thus gave him some measure of financial independence, although he seems to have lost much of his fortune later. In his early years, however, he did not need to pursue a salaried career; for example, before the Revolution he taught for only a few years at the *Ecole Militaire*. He had studied at the *Collège Mazarin*, where he took his thesis at the early age of 18 years; however, in a tradition of the time, it seems only to have been a list of results to be proved and literature to be read rather than an original contribution (see Legendre *1857d*). His research career was

---

[1]) The most useful obituaries and *éloges* of Legendre are Maurice *1833a* and Parisot *1842a*: Elie de Beaumont *1864a* is not very good. Little significant study has been achieved since, although items relating to parts of his output will be cited when appropriate. There is neither an edition of his works nor, apparently, any *Nachlass*.

launched in the 1780s with a prize paper *1782a* on projectiles for the Berlin Academy which drew him to Lagrange's attention, and was followed by several distinguished contributions to analysis, of which we note some in Chapter 3. He was elected *adjoint* to the Paris *Académie* in 1783, promoted to *associé* two years later, and was placed in the *section* for 'Mathematics' when the *classe* of the *Institut* was formed in 1795. He moved to the *section* for 'Geometry' in the 1803 reform, and served as *Président* of the *classe* in 1805-1806. Among his (few) other posts, he succeeded Lagrange at the *Bureau des Longitudes* in 1813. He was the last of his generation, and he seems to have felt it: in his late seventies he told Crelle in a letter of 1831 that Lagrange had lived 77 years and 77 days and Laplace 78 years less 18 days (*Darmstaedter*). His own tally was to be 80 years, 114 days.

2.5.6. Two distinguished engineer *savants*: Coulomb and Monge

Among the late-18th-century *savants* involved in engineering, Coulomb and Monge exercised great influence, Coulomb largely for his research work, Monge particularly at the educational level. Their pre-Revolutionary careers show various similarities, although they do not seem to have been intimate friends.

Charles Augustin Coulomb was born in 1736 at Angoulême in western France, not far from Bordeaux. His father's administrative work for the King brought the family soon afterwards to Paris, where he studied briefly at the *Collège Royal* and then at the *Ecole du Génie* at Mézières in 1760-1761. For the next 20 years he passed his career as an army engineer in Martinique and provincial France, and also carried out many of his researches in engineering mechanics which will make him a major figure for Chapter 8. His work was acknowledged by the *Académie* in 1774 by election as *correspondant* for Bossut, and seven years later he gained full recognition, and a need for Paris residence, by election as an *associé*. He continued his research in mechanics, and also began a series of experiments in electricity and magnetism which are described in § 7.5-6. He was relatively unaffected by the Revolution; in particular, his only involvement with the educational reforms was as an *inspecteur général d'instruction publique*, a post which he held from 1802 until his death in 1806.[1])

Gaspard Monge was born in 1746, ten years after Coulomb, at Beaune near Di-

---

[1]) Gillmor *1971a* is a good biography of Coulomb (Truesdell's review *1974a* is harsh): see also the information in Heyman *1972a*. Gillmor lists most of the extant manuscripts on pp. 300-303; there are also a few items in *ENPC*.

jon in Eastern France. He also studied at the *Ecole du Génie* at Mézières, from 1765 to 1766, and taught there until 1784, succeeding Bossut in 1769. Elected in 1772 as a *correspondant* at the *Académie* for Bossut (two years before Coulomb's election), he became an *associé* at the 1785 reorganisation. He then took a prominent place in the *classe* upon its formation in 1795, and in educational reforms, as we saw in § 2.3.

Monge's career differs from Coulomb's basically in that he was employed as a teacher, whereas Coulomb was a professional engineer. His intellectual interests were rather more mathematical; he was concerned with differential equations and the geometry of surfaces (§ 3.3), and we shall see in § 8.3 his commitment to descriptive geometry. But the two men had some common interests in engineering, especially forts and fortifications, and also in shipping matters, in that during the 1780s Coulomb wrote important reports on canals and the water supply while Monge was heavily involved in naval education.

When Coulomb died in 1806 Monge was at the height of his power and recognition in Napoleonic France; but his health started to fail soon afterwards, and he lived long enough to die in disgrace in Bourbon Paris in 1818, stripped of his place in the reorganised *Académie* of 1816. But his name and work was undimmed in the eyes and minds of many *savant* engineers.[1])

2.5.7 Five further figures

We deal first with four of the minor figures of this generation mentioned at the head of this section. After briefly outlining their biographies, their contributions to our story will be described and contrasted. The fifth figure comes separately at the end.

The oldest of the group is Charles Bossut, born in 1730 at Tartaras in the Loire in western France. Brought up by his uncle after the early loss of his father, he studied at a Jesuit college in Lyons and was an *abbé* until 1792. He taught or examined at the *Ecole du Génie* at Mézières until it moved to Metz in the mid 1790s, when he became an *examinateur permanent* at the *Ecole Polytechnique* until 1809, five years before his death.

Jean-Charles Borda was born into the nobility in 1733 at Dax in south-western France. After studying at Mézières (with Bossut, among others) he made his career in the navy, becoming both a captain and an inspector of vessel construc-

[1]) Of the various studies of Monge, the most relevant is Taton *1951a*, which includes a list of manuscripts on pp. 384-393. Among his followers, Dupin published a biography *1819b*.

tion. He also worked in engineering and physics, and on weights and measures. He died in 1799, when our story officially starts; but his works left their marks upon it.

Réné-Just Haüy, born in 1743 to a poor family near Beauvais, north of Paris, won a scholarship to the *Collège de Navarre* in Paris. Like Bossut he became an *abbé*, but retained this position throughout his life, which ended in 1822. He taught at various Paris institutions, including the *Ecole des Mines* and the *Faculté des Sciences*.

Jean-Baptiste-Joseph Delambre was also born into a poor background, in 1749 at Amiens, somewhat north of Haüy's Beauvais. Most of his scientific knowledge was self-taught, but after he attracted the attention of Lalande (whom Bossut hated, incidentally) he began a successful career in Paris. As was noted in § 2.2.2, he became a founder *secrétaire perpétuel* of the *classe* in 1803, a post which he held until his death in 1822. He taught at the *Collège de France* from 1807, and soon afterwards he joined the *Conseil de l'Instruction Publique*, serving as treasurer until he retired in 1815.

Bossut and Borda worked in hydrodynamics and hydraulics, and also in other areas. Bossut became well-known as a textbook writer; his stature was not that of Lacroix, but his scope was broader, including most aspects of mechanics and the history of mathematics as well as the calculus. He also contributed to the Diderot/d'Alembert *Encyclopédie* and its successor *Encyclopédie méthodique*. Borda concerns us chiefly for his instruments, especially his repeating circle for geodetic observations (§ 5.5.4), his design of the simple pendulum (§ 6.5.2) and his 'mouthpiece' for smoothing the flow of water from orifices (§ 8.5.3). Haüy worked chiefly in mineralogy and crystallography; but his general interests in physics, including an excellent textbook, bring him into our tale on occasion. Delambre was largely involved with theoretical and observational astronomy; and he also wrote extensively on the history of astronomy in his final years, some of these works appearing posthumously. He accumulated an extensive library, which was sold after his death.

While these men followed rather different scientific pursuits, they showed several common features. Borda, Haüy and Delambre were all active in the reform of weights and measures in the 1790s (§ 5.5.4), and in that and other connections Borda and Delambre worked on the construction of tables (§ 3.5.4). Their careers also reveal similarities. Each member held at least *associé* status in the old *Académie* by its dissolution in 1793 and received full membership of the *Institut* on its formation in 1795, when Borda and Delambre were also elected to the *Bureau des Longitudes* (§ 5.4.1). Even their lives had likenesses. None of them was born in Paris, but each died there; only Delambre married, and then in his 56th year.[1])

Somewhat separate from these figures, by reason of his important political career, is Lazare Nicolas Marguerite Carnot. Born into a lawyer's family in 1753 at Nolay near Dijon in eastern France, he studied at Mézières and pursued a career as an engineer. However, he met Robespierre in 1787 and entered politics a few years later, and thereafter was largely preoccupied with public life and administration.

Somehow Carnot managed to produce a respectable range of scientific work in the calculus (§ 3.2.6), mechanics (§ 5.2.6) and engineering (§ 16.3.3), this last influencing his son Sadi's ideas on 'the motive power of fire' (§ 16.5.7). His approach is distinguished by its mixture of algebra and geometry (§ 4.7.1), and its concern with foundations of theories and their potential generality. Gillispie *1971a* is the principal study of his scientific contributions.

Carnot was elected to the *classe* in 1796, but lost his place the following year after a political *coup d'état* in the government. He went into exile in Switzerland until 1800, when Bonaparte siezed power. He was re-elected that year, but was removed in 1816 after the Restoration. He went into exile again, to Magdeburg in Germany, where he died on 2 August 1823. The *Moniteur universel* reported the news ten days later, in $3\frac{1}{2}$ lines.

2.6 New men and a new profession

This chapter began with institutions and then passed to individuals. It ends here mid-way, by briefly looking at groups of more minor figures around 1800: more attention will be given to them in later chapters when their work first takes a prominent place in our story.

---

[1]) Only Borda has been studied in detail: the obituary Lacroix *1800a* prepared for the *SP*, and the biography Mascart *1919a* (which also contain brief summaries, not always accurate, on many others' lives). For the others, see Doublet *1914a* for Bossut, A. Lacroix *1944a* for Haüy, and Cohen *1971a* for Delambre. Among the *éloges*, see Delambre *1815a* on Bossut. Delambre left most manuscripts, of which most were found only recently and are kept in *OB*, mss. 295-321 (see Chapront-Touzé *1986a*): others are kept at *Wellcome*; *BI*, mss. 2033, 2041 and 2042 (among Bertrand's papers); *BN*, ms. fr. 14758; and *Smith*. The *dossiers personnels* at *AS* may also be consulted for all these figures.

In the background is Condorcet, who died in 1794 in his 52nd year. While he worked in the calculus and mechanics, his achievements were limited (Lagrange held them in contempt, according to Maurice *1814a*, 227); his greatest contributions seem to lie in probability theory, which is not a major concern in this book. The main *Nachlass* is held at *BI*, mss. 848-885; a small pile mainly concerned with celestial mechanics has recently been found, and is housed in *OB*, mss. 219-245.

2.6.1 Three generations

The majority of the figures described in § 2.5 were at least in their fifties at 1800. Some other figures were near the ends of their lives: Borda died in 1799 and Coulomb in 1806.

Lacroix, born in 1765, belonged to the next generation. Its other members include Fourier and Hachette, both of whom were appointed to teaching posts at the founding of the *Ecole Polytechnique* (§ 2.3.1). Fourier is chiefly of interest for his work in heat diffusion (Chapter 9), though in § 4.6.3 we note some earlier work in equations and in § 5.3.1 a contribution to mechanics. Hachette was always principally concerned with the engineering aspects of science. He was the most important follower of Monge, and collaborated with him on teaching at the *Ecole Polytechnique* and the writing of textbooks (§ 8.2.1).

Members of the third generation were born between 1775 and 1790, and most of them studied at the *Ecole Polytechnique* during its early years. The first, large, recruitment of 1794–1795 (§ 2.3.1) was very distinguished, for among our major later figures we find Biot, Francoeur, Malus and Poinsot; and interesting, minor, contributors such as Brisson and Lancret. The immediately following years were graced by the presence of J. F. Français (*promotion* of 1797), Poisson (1798), Plana (1800), Dupin (1801), Navier (1802), Arago and Mathieu (1803), Binet and Fresnel (1804), Cauchy (1805), Petit and Poncelet (1807), Belanger and Coriolis (1808), and Olivier and Pontécoulant (1811).[1]) We pick up this list again in § 10.1.1.

By the late 1800s some of the earlier youngsters, especially Binet, Poinsot and Poisson, were on the staff of the school. Another staff member was Ampère, recruited in 1804 as a *répétiteur* in analysis after a provincial career, mostly in his home town of Lyon (§ 4.2.5). Although a member of this generation, he was not a student at the school. While his most important contributions to our story are to electrodynamics (Chapter 14), his earlier work is of note; in § 4.3.2-3, § 11.2.1 and § 11.6.2 we examine his contributions to the calculus, and in § 6.2.2 a paper on mechanics.

---

[1]) It is worth noting also J. L. Gay-Lussac (1778–1850), for he features occasionally in this book as a major chemist and physicist. He entered the *EP* in 1797 (with Français) and went on to de Prony's *EPC* (§ 3.5.3). After leaving there in 1802, he returned to teach chemistry at the *EP* for many years. He was also the founder *Professeur* of physics in the *PF* in 1809, and was elected to the *CI* in 1806. His early career was developed in close connection with the *Société d'Arcueil*. For a biography, see Crosland *1978a*.

2.6.2 Career paths in the new profession of science

It is clear that the social structure of science was changing markedly in Paris after the Revolution (see Dhombres *1986b*). The various *écoles*, the *Université*, the *écoles centrales* and *lycées*, the *classe* of the *Institut*, the *Société Philomatique*, the *Bureau des Longitudes*, the growth of commercial publishers; these were sources of jobs, incomes, and participation in scientific organisations for the diffusion of scientific knowledge, and even research. Figure 262.1 presents the main institutions, including three to be introduced in Chapter 16 and one in § 5.5.1.

In some respects, the apparent modernity of Revolutionary Paris is deceptive. For example, the financial backing of the state was at times inadequate or erratically provided ('plus ça change ...'). As a result, private patronage was still important (perhaps less in our areas of science than in others). In addition, career *savants* had to participate in the *cumul* system of holding several salaried posts simultaneously. This led to much time being devoted to teaching and related duties, and travelling from one place to another to (change costume and) fulfil the

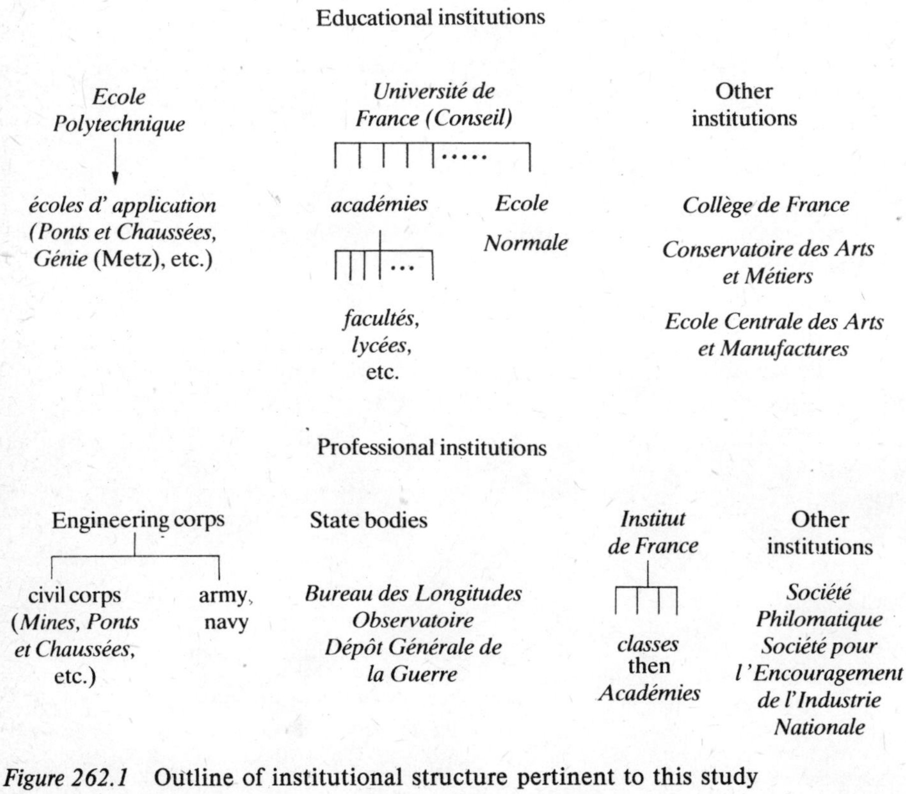

Figure 262.1 Outline of institutional structure pertinent to this study

posts, never mind obtaining them in the first place. It also gave some monopoly of posts to those who had established their reputations, thus further underlining the competitive atmosphere and the (over-)dominating position of certain figures in Paris science (and of Paris in French science overall). The system had its critics, and was discussed from time to time; but it persisted throughout our period of study and affected the careers of almost all of our figures. The consequences will be explored in § 19.2.4.

The situation also meant that people were *individuals*, with several appointments and connections. Thus, for example, when a society or organisation published a list of its members, it gave their private addresses. The contrast with today is very marked, when individuality is flattened out and people are boring imitations of each other.

Another difference of the period from the modern situation is caused by the important place then assigned to military concerns. We have seen them involved in the conception of the *Ecole Polytechnique* and some *écoles d'application*; another example, where science and the military were co-present, was the French expedition to Egypt between 1798 and 1801.[1]) In addition to the military aspects of the conquest and administration of Egypt, there was an attempt to establish French culture there in the form of an *Institut d'Egypte* modelled on the Paris *Institut*. There were also two scientific expeditions, which led not only to research papers presented to the *Institut d'Egypte* (and later to the Paris *Institut*) but also to the birth of modern Egyptology. Of the figures discussed in this study, Monge was closely involved in the choice of scientists, and was in Egypt for the first six months of the expedition. Among those chosen were Girard, Lancret, Malus, Méchain, and especially Fourier, who was *secrétaire* of the *Institut d'Egypte* and leader of one of the expeditions.

Despite these differences from the modern situation, the similarities are very important, and also mark substantial changes from the pre-Revolutionary period. Most of the members of the second and third generations were among the earliest examples of scientists who pursued professionalised careers for the whole of their active lives. Indeed, in the three interweaving cases of Arago, Biot and Poisson we shall find three career*ist* professionals, whose activities were guided by considerations of career as much as of science, and whose friendships and estrangements were strongly to influence not only their own lives but also the structure of the remarkable scientific community to which they belonged (§ 19.3.3).

---

[1]) On the expedition see, for example, Charles-Roux *1837a*. Interesting files on the publication of its scientific and archaeological findings, the vast *Description de l'Egypte*, are held at *AN*, F[17] 1099-1108[2]; see also Gillispie *1987a*.

2.6.3 Paris as a city, 1800–1840

After all this institutional and biographical description it is appropriate to end with some remarks on Paris as a city. I have gathered together here information covering the whole period under study; so that some information is provided on bodies and people who have not yet been discussed in detail, although they were mentioned in § 1.2.

After the Revolution Paris was developed much more as an administrative centre than as an industrial city (although printing and publishing were substantial industries there); presumably for this reason no *école d'application* for industrial and commercial engineering was established there until the late 1820s (§ 16.6.2). Such institutions could have been set up earlier in the provinces; but no such initiative was taken.

During our period Paris was a walled city divided into 12 *arrondissements*; each of these split in turn into four *quartiers*. In 1800 its permanent population was around 550,000; by 1820 it had passed 700,000, and was approaching 900,000 during the mid 1830s. The surrounding area of the Seine *département* added on between 100,000 and 200,000. Most of the city lay to the right bank of the Seine, bounded by a wall flanking the second ring of boulevards of the modern city (from the *Trocadéro* through the *Place de Gaulle* round to the *Place de la Nation*, and then to the *Boulevard de Bercy*). Much of the growth in population was created by filling the area between this ring of boulevards and an inner ring. The left bank area was largely confined to the parts bounded today by the *Boulevard Vincent Auriol* through *Place Denfert-Rochereau* and the *Gare de Montparnasse* to the *Boulevard de Gamelle*.

The differences between the old and the modern Paris are quite substantial, especially due to the reconstructions effected during the second half of the 19th century. However, several of the *quartiers* of principal concern here have been relatively unaffected, with many of the buildings still intact (and in some cases even containing the same institutions) and roads still in place (sometimes even with the same names). On the right bank the *Marais* region was significant, but the most important area lay on the left bank, known then as the *Université* region, bounded by the (then and now) *Jardin des Plantes* to the east, *l'Observatoire* to the south and *Hôtel des Invalides* to the west.

Figure 263.1 is a selective map of the central parts of Paris incorporating these regions, traced from a map which was published in 1825. The selected roads, which are indicated by single lines, are located reasonably accurately, as are the bridges, the *Canal Saint Martin*, the squares and gardens, the Seine and the islands lying within it. (There were then *three* of these islands: the *Isle Saint Louis* of that time was joined by a high wooden bridge to the third one, a store for firewood called *Isle l'Ouvier*, but later on the space between the two islands was filled in.) The names given to various buildings and squares are those of 1825: some had varied before then, and a few have been changed since. The *Ecole Militaire* was then a barracks, not a school; however, it also housed an observatory.

A Cartesian reference system is imposed on the map. If more than one square is cited, then the indication takes the form, say, '3B–3D'; if a *rectangle* of squares is required, '(2B, 5D)' locates opposite corners. This system is used in the lists of Table 263.1.

The first list indicates the principal buildings (marked by a pair of upper case letters), together with the periods of occupancy by our institutions. In addition, periods of lengthy residence by our major figures are shown. Some of them lived "over the shop" of one of their main appointments. For example, de Prony usually resided at the *Ecole des Ponts et Chaussées*, while Arago passed his whole career at the *Observatoire*.

The second list marks out seven preferred areas of private residence by our figures, and also three particularly popular roads, which are indicated by a pair of lower case letters (usually placed at their ends). This list is far from complete, and may not be entirely accurate: addresses are hard to find when a figure did not hold an official appointment, and in any case some of them moved around a great deal, which gives me a space problem for the information. Legendre is the winner, with a long string of addresses usually around squares 5H and 8H; Poisson was also fairly mobile, mostly in the area 8D–8F. Poinsot mostly lived in the north of the part indicated on the map, and so did Girard for many years.[1]

Although Paris grew quite quickly, it was still largely surrounded by forests. These could be dangerous to travel through, so that the city was something of a safe haven (although some areas were bad, and prostitution a major enterprise). Given these and the other circumstances described in this chapter, it is not surprising that Paris was rather *petit bourgeois*.

---

[1]) The principal source of addresses is the annual official *Almanach* (*Royal/Impérial* ... as usual): the private-enterprise Bottin *Almanach du Commerce* is also of use, as it includes a fairly full list of Paris residents. Roads parallel to the Seine were usually numbered from east to west, and other roads away from the river: even numbers lay on the right hand side, odd numbers to the left.

The map chosen for Figure 263.1 was published as a *Nouveau plan routier* of Paris by the house of Lerond. (Many of the scientific publishers mentioned in § 2.2.4 placed their offices in the *Université* area (8E, 9G), by the way.) A variety of maps for the period is available; although slightly late for our purpose, especially recommended is the suite prepared in 1834 by the *polytechnicien* A.-M. Perrot: a page is devoted to each *quartier*, and a modern edition is available which contains much valuable information provided by the editors (Perrot *1960a*). For earlier maps available in the literature, see, for example, Crosland *1967a*, 284 (map of 1796); and a plate in *Corr. EP*, *3* (1804–08).

For statistical information on Paris in general, the chief source for the Restoration period is provided in four volumes of *Recherches statistiques sur la ville de Paris* ... (1821–1829) which were edited by Fourier (§ 17.6.2, footnote). Dupin *1827a* provided a brief survey of the city of that time.

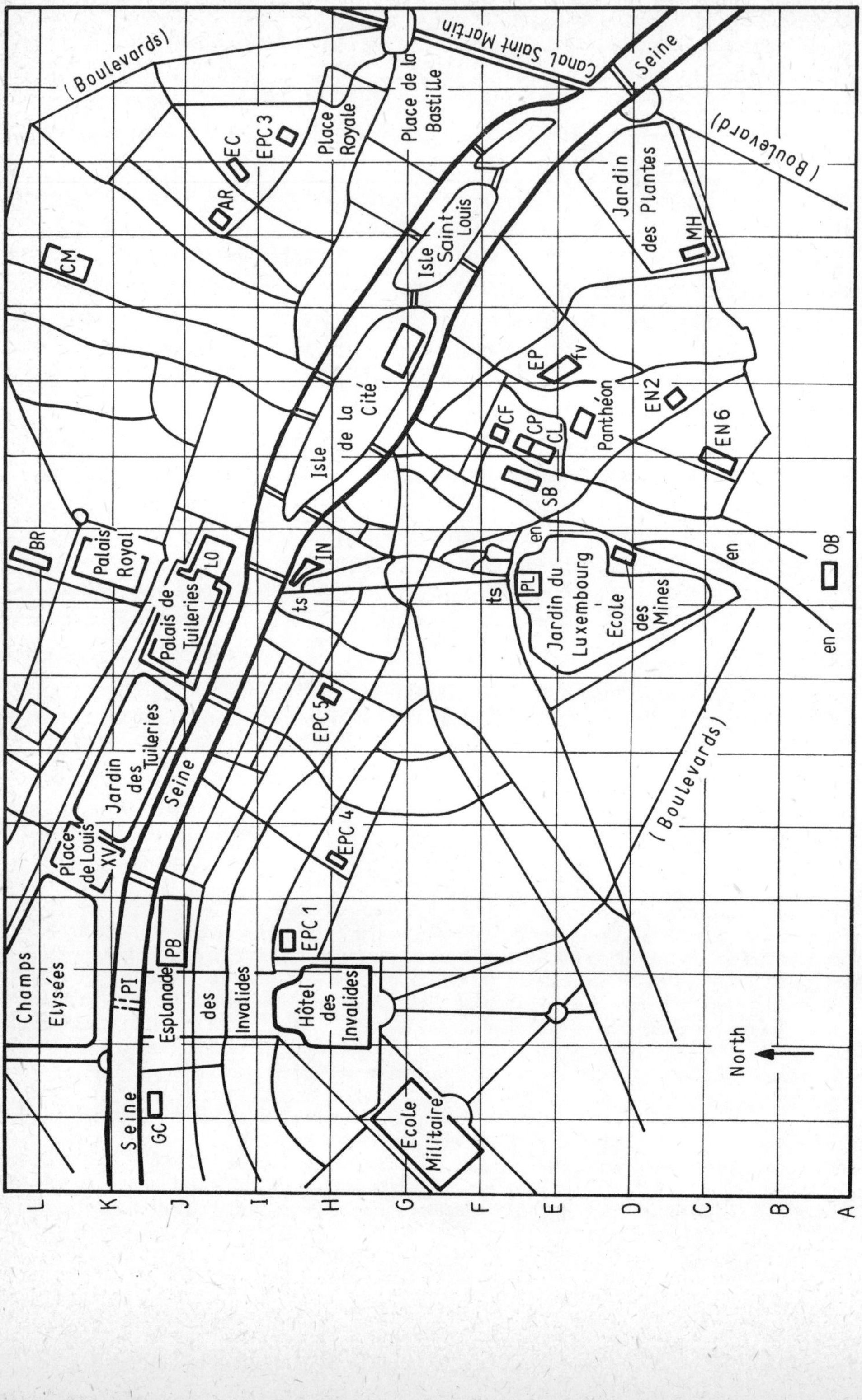

Figure 263.1 Selective map of Paris

Table 263.1 Principal buildings and their institutions; and favoured areas of residence.

* institution is still (partly) resident there; *or* the street is still so named. To save space, only the last two digits of years from the 18th or 19th centuries are given.

Code	Name: map reference	Relevant further data
AR	*Archives Royales* (08-*): 13I	*AN*
BR	*Bibliothèque Royale**: 8K	*BN*
CF	*CF* (1530-*): 10E	Biot (10-62), Savart (32-41)
CL	*Collège Louis-le-Grand**: 10E	*EN*4 (26-28)
CM	*Conservatoire des Arts et Métiers* (99-*): 12K	
CP	*Collège de Plessis*; 10E	*EN*1 (10-14), 5 (28-47) *PF* (11-21)
EC	*Ecole Centrale des Arts et Manufactures* (29-84): 13I	
EN	1 (11-13: in CP), 2 (13-21: 10C), 3 (21-22: in SB), 4 (26-28: in CL), 5 (28-47: in CP), 6 (47-*: 9B)	
EP	*EP* (05-1976): 11D	Binet (16-30), Coriolis (38-43)
EPC	1 (98-07: 4H), 2 (07-14: in PB), 3 (14-29: 14H), 4 (30-45: 4G), 5 (45-*: 6H)	de Prony (98-39)
GC	Gros-Caillou: 1J	Steam engine
IN	*Institut de France*: 8H	*CI* and *AS* (06-*), *BI*
LO	*Louvre*: 8I	*CI* (95-06), Bossut (99-0?)
MH	*Muséum d'Histoire Naturelle**: 12C	*EN* (95)
OB	*Observatoire**: 8A	*BL* (95-*), Arago (06-53)
PB	*Palais Bourbon** complex: 3J	*EP* (94-05), *EPC*, de Prony (07-14)
PI	*Pont des Invalides*: 2J	Navier's scheme (23-26)
PL	*Palais du Luxembourg**: 8E	Cauchy (09-18), Laplace (08-13)
SB	*Sorbonne**: 9E	*PF* (21-*), *EN* (21-22), Poisson (32-40), *BS* (21-*)

Area or street	Residents
Marais (13H, 14I)	Carnot (02-16), Delambre (99-15), Germain (98-23), Girard (16-26)
Between PB and IN (3J, 6H)	Dupin (32-73), Legendre (99-04, 10-13, 18-26, 29-32), Laplace (19-27), Monge (01-16), Poncelet (35-42)
Around *Saint Germain** (6G-7G)	Delambre (18-22), Legendre (16, 27-29)
*Rues Tournon** and *Seine** (ts: 8E-8H)	Lacroix (14-34), Laplace (05-07), Legendre (15, 17), Navier (25-29), Poisson (12-15)
Behind IN (mostly 7G-8G)	Cauchy (18-30, 38-57), Dupin (21-25), Fourier (18-29), Girard (34-36), Laplace (99-05), Navier (16-25, 32-36), Puissant (29-36)
North of PL (Mostly 7F-9F)	Coriolis (32-38), Delambre (16-18), Lacroix (03-38, 40-43), Laplace (14-17), Legendre (06), Poisson (10-12, 22-32)
*Rue d'Enfer** (en: 7A, 9E)	Biot (02-10), Dupin (19-27), Fourier (29-30), Legendre (08), Malus (09-12), Poisson (15-22), Savary (33-41)
Rue Fossés Saint-Victor (11D at fv)	Ampère (18-36), Fresnel (22-27)
West of PL (4D-6E)	Francoeur (25-32), Lacroix (13-17), Laplace (15-17)

Chapter 3

The 18th-century heritage: the calculus around 1800

3.1 Plan of the chapter

I survey the various principal approaches to the calculus in § 3.2; the advocacy of different systems caused a rather confused situation. § 3.3 is devoted to partial differential equations; some methods of solution, and the theory of general and particular solutions. § 3.4 covers some principal series, special functions and integrals, and also the conception of the convergence of an infinite series. § 3.5 takes up the discrete analogue of these studies, with a look at the interest in differences and difference equations.

Throughout the chapter I confine the account to the chief ideas and methods as of 1800: specific cases and neighbouring fields of study are left until later chapters. For example, Legendre functions are considered only when discussing the theory of attractions in § 5.5.1, the calculus of variations is postponed until § 4.5, and the theory of equations to § 4.6.

3.2 The foundations of the calculus in the late 18th century

A rather wide variety of approaches to the calculus had developed by the late 18th century, and both the historical record and the needs of later chapters require that the whole scenario be covered, drawing largely on the main texts published around 1800. Priority is given to Lagrange, for he is the dominant figure of the whole chapter.

3.2.1 Lagrange's courses at the *Ecole Polytechnique*

> The ordinary operations of algebra suffice to resolve problems in the theory of curves.
>
> Lagrange *Functions*$_1$ (1797), 117

When young, Lagrange *1762a* believed that Berkeley's theory of compensation of errors would solve the mysteries of the calculus. Berkeley's view gained some popularity during the 18th century (see Cajori *1919a*), although it cannot possibly work (see my *1970a*). Lagrange abandoned it, and in his *1774a* he presented to the Berlin Academy 'a new type of calculus relative to the differentiation and integration of variable quantities', in which his new approach was outlined. This was the position which he always held later, and taught at the *Ecole Polytechnique*.

de Prony clearly outlined the educational context of Lagrange's lectures in an introductory talk to his own teaching, published in the *Journal de l'Ecole Polytechnique*. 'Mathematical analysis divides itself in two parts', he explained (though not intending the limit-oriented sense of the term 'mathematical analysis' which we have acquired from Cauchy's later lectures at the school (§ 11.3)), 'one which is called *determinate analysis*, considers the relationships which hold between *unknown* but invariable quantities' and led to the theory of equations and also 'to the doctrine of series, which combines intimately with the first part in many respects, [but] however has such relationships with the second, which is called indeterminate analysis'. There one 'considers the relationships between the quantities given by the state of the question, and other quantities susceptible to an infinity of values subject to a common law' (de Prony *1799a*, 215).

de Prony's emphasis on functional relationships between constants or variables captures a principal feature of Lagrange's approach. 'To speak correctly, algebra is in general only the theory of functions', Lagrange told the students in his own introductory essay in the *Journal* (*1799a*, 235); it is a theory which splits into two branches. The first is 'pure algebra' and produces 'primitive functions' from ordinary algebraic operations; the second, the 'theory of analytic functions', considers 'derived functions'. These functions had been obtained in the past by using limits or infinitesimals, but for Lagrange such methods lacked rigour. For example, MacLaurin and d'Alembert used limits, 'but in rigour one can observe that the sub-tangent is not the limit of the sub-secants because nothing stops the sub-secant from increasing even when it has become the sub-tangent'. Consideration of curves 'gave birth to the method of the infinitely small', but one had to transform into other systems (pp. 232–233).

Lagrange published two volumes related to his calculus teaching.[1] The *Théorie des fonctions analytiques*, which we consider in this section, was published in 1797 in three slightly different printings: the print-run was large, at 3,800 copies (*AN*, F^{17} 1084). German and Portugese translations quickly appeared. The *Leçons sur le calcul des fonctions* came out in various editions in the 1800s, as we note in

[1]) Relevant manuscripts include his own notes on the integral calculus (*BI*, ms. 904, fols. 180–195) and some by him and others (*ENPC*, ms. 1323; see also ms. 674). Among printed lecture notes at *EP*, Garnier *1801b* showed influence of Lagrange's approach; de Prony's work is discussed in § 3.2.8. For commentary, see Dickstein *1899a*, Yushkevich *1959a*, Ovaerts *1976a*, Pepe *1986a* and Fraser *1987a*.

§ 4.2.1. The lectures delivered to the *Ecole Normale* were more elementary, although we pick up one result in § 3.5.2. During this period Lagrange was in his sixties.

3.2.2 Lagrange's *Fonctions*: the derived function

Lagrange's *Théorie des fonctions analytiques* was written in haste: the preface did not even get printed (see Pepe *1986b*)! As he recalled in the preface to the second edition, 'this work having been composed as if in a single throw, just as it was printed [...]' (*Functions*$_2$ (1813), 13). In the first edition he hinted at the educational imperatives behind its preparation: 'Finding myself engaged by particular circumstances to develop the general principles of analysis [...]' (*Functions*$_1$ (1797), 5); this was his teaching of 1795 and 1796.

The sub-title of the treatise stated that it contained 'the principles of the differential calculus, freed from all consideration of the infinitely small or vanishing quantities, of limits or fluxions, and reduced to the algebraic analysis of finite quantities'. In its first edition, the first 115 of its 276 pages were devoted to the 'Exposition of the Theory, with its principal uses in Analysis'. A function was announced at the outset as 'any expression of calculus in which these quantities enter in any manner', where this vague word 'any' ('*quelconque*') appeared for neither the first nor the last time in foundational essays on the calculus. Of various forms of function, he laid prime emphasis on Taylor's series, arguing by induction for its generality as follows.

Given the (real-variable) finite increment i on x, then $f(x+i)$ was given by

$$f(x+i) = f(x) + i\,P\,, \tag{322.1}$$

where $P \neq \infty$ when $i = 0$. P was a new function of x and i, just as $f(x+i)$ was; hence (322.1) suggested the relationship

$$P = p + i\,Q\,, \tag{322.2}$$

where p was the value of P, and $Q \neq \infty$, when $i = 0$. Hence Lagrange obtained by induction

$$f(x+i) = f(x) + ip + i^2q + \ldots \tag{322.3}$$

(*Functions*$_1$ (1797), 7-9). P, Q ... were equivalent to orders of infinitesimals, though he did not say so. At times he spoke of 'i becom[ing] zero'—an odd phrase to find in a doctrine avoiding limits.

Lagrange was aware that he must show that there be no fractional powers of i in the expansion of $f(x+i)$, and he offered this proof. Expressions of the form

$ui^{m/n}$, where m and n were integers, were multi-valued. Hence if (322.2) were to contain such a term, 'the developed function $f(x+i)$ would have more different values than the non-developed function, which is absurd. [...] This demonstration is general and rigorous as long as x and i remain indeterminate' (pp.7–8). However, he did not consider *ir*rational powers of i (surely not to do so may beg a question), nor did he worry about the *interpretation* of multi-valued functions and their expansion(s). However, here and some pages further on he noted that (322.2) may not be legitimate when for some value of x a radical may vanish in $f(x)$ and/or certain of its functions $p, q, \ldots$ (pp. 32–33). He also admitted that if a function f and/or $p, q, \ldots$ took infinite values at x, then the expansion of $f(x+i)$ had to include terms with negative powers of i (pp. 38–39).

Lagrange determined the relationship between f and the functions $p, q, \ldots$ in (322.3) as follows. Assuming that x received the double increment $(i+o)$ (both indeterminate), (322.3) yielded

$$f(x+i+o) = f(x+(i+o)) \tag{322.4}$$
$$= f(x) + pi + qi^2 + ri^3 + \ldots + po + 2qio + 3ri^2o + \ldots \tag{322.5}$$

by multiplying out the powers of $(i+o)$, and also

$$f(x+i+o) = f((x+i)+o) \tag{322.6}$$
$$= f(x) + pi + qi^2 + \ldots + f'(x)o + p'io + q'i^2o + \ldots, \tag{322.7}$$

where $f', p', q', \ldots$ were defined (without explaining the primes) from

$$f(x+o) \;= f(x) + f'(x)o + \ldots \tag{322.8}$$

and

$$p(x+o) \;= p + p'o + \ldots. \tag{322.9}$$

(He did not specify the coefficients of higher powers of o.) Then, by equating the coefficients of i, io, i^2o, ... in (322.5,7), he obtained

$$p = f'(x), \qquad 2q = p', \qquad 3r = q', \ldots. \tag{322.10}$$

Now from (322.8) it followed that 'in the same way that $f'(x)$ is the first derived function of $f(x)$, it is clear that p' is the first derived function of p', q' of q and so on. Hence, as '$f'(x)$' was the notation for the first derived function (Lagrange himself actually wrote '$f'(x)$'), '$f''(x)$' was a suitable way of notating the first derived function of $f'(x)$, and so on. Hence (322.3) became, from (322.10),

$$f(x+i) = f(x) + f'(x)i + f''(x)i^2/2! + \ldots \tag{322.11}$$

(pp. 13–14), giving the power-series expansion of the 'primitive function' f in terms of its 'derived functions' $f', f'', \ldots$.

Convergence was apparently assured, since the method 'gives the means of stopping the development of the series at whatever term one would like, and judging the value of the remainder of the series' (p. 42). But he did not produce the differential form of the remainder term which is often named after him; instead, he estimated limits within which it should lie:

$$M(b^{m+1} - a^{m+1})/(m+1) > F(b) - F(a) > N(b^{m+1} - a^{m+1})/(m+1) \qquad (322.12)$$

with m a positive integer, where M and N were respectively the maximal and minimal values of $(F'(z)z^{-m})$ over $[a, b]$ of z (pp. 46-47).

Much of the rest of this part of *Fonctions* was taken up with examples and various important extensions. In particular, Lagrange developed the Taylorian expansion for a function of two variables:

$$f(x+i, y+o) = \sum_{m,n=0}^{\infty} f_n^m(x,y) i^m o^n / m! n!, \qquad (322.13)$$

where 'f_n^m' denoted m derivations with respect to x and n derivations with respect to y (pp. 91-94). The order of derivation was normally held to be immaterial. One important application was to the theory of extreme values for functions of several variables. First-order differentiation of course gave the extrema; distinguishing their types depended on showing the positive or negative definiteness of the higher-order terms of (322.13). Lagrange repeated the first-order analysis on these terms by taking $f_x, f_y, f_{xx}, \ldots$ successively as independent variables and finding necessary conditions in an iterative form which for a minimum with two independent variables may be written:

$$f_{xx} > 0, \quad g := f_{yy} - (f_{xy}^2/4f_{xx}) > 0, \quad h := g_{yy} - (g_{xy}^2/4g_{xx}) > 0, \ldots \qquad (322.14)$$

(pp. 192-196: recall from § 1.3.7 that ':=' in this book denotes 'equal by definition'). He also considered extrema for f under the constraint $\varphi = 0$ by analysing $(f + \lambda\varphi)$ (the 'Lagrange multiplier': pp. 196-198).

3.2.3 Lagrange's *Fonctions*: integration and other aspects

Lagrange treated the integral calculus in much less detail. He conceived of the integral as the inverse of the derived function rather than defining it by some other means. Its calculation 'forms another part of the analysis of functions that one calls *inverse analysis*': it used the same rules 'but taken inversely', which therefore 'is not always applied either with the same facility or with the same success' (p. 90). However, in addition to this inversion theory there had been some development of the theory of integration during the 18th century. Euler *1774a* was perhaps the first systematic study, where he considered double and repeated integrals and their interpretation as volumes, or as expressions for curvilinear areas. In a contemporary paper he also gave the long-hand of what we called a third-order Jacobian; this form was also presented by Euler, Lagrange (in papers on the attraction of spheroids) and Legendre (on multiple integrals).[1]) Legendre's is worth noting as an example of the modes of reasoning then used. To transform the triple integral $\int P dx dy dz$ into variables (p, q, r) given by relationships to (x, y, z) such as

$$dx = A dp + B dq + C dr, \tag{323.1}$$

he took y and z constant, thus reducing the other two relationships to

$$0 (= dy) = A' dp + B' dq + C' dr, \tag{323.2}$$

$$0 (= dz) = A'' dp + B'' dq + C'' dr. \tag{323.3}$$

These gave him the 'Jacobian' determinant to convert from (x, y, z) to (p, y, z) but with a denominator $(B'C'' - B''C')$. He then set $dp = 0$ in (323.2-3) to get into (p, q, z), and then $dz = 0$ in (323.3) to end up with (p, q, r) via the Jacobian, as required. The procedure was algebraic, with no conditions imposed upon any quantity.

Among the general developments of Lagrange's theory in *Fonctions*, there is an interesting passage in which he tried to estimate 'the error that one may commit in restricting oneself to some of the first terms of the series' by proving that

$$\text{if } f'(z) > 0 \quad \text{for} \quad a \leq z \leq b, \text{ then } f(b) > f(a). \tag{323.4}$$

(322.11) suggested to Lagrange that in

$$f(z+i) = f(z) + f'(z)i + f''(z)i^2/2! + \ldots \tag{323.5}$$

he could take the increment i sufficiently small for $(f'(z)i)$ to be majorant in the expansion; so that, since $f'(z)$ was positive, so was $(f(z+i) - f(z))$. He carried out this process over the intervals $[a, a+i]$, $[a+i, a+2i]$,... $[a+ni, a+(n+1)i]$, where i was small enough for

$$f(a+(r+1)i) - f(a+ri) > 0, \quad 0 \leq r \leq n. \tag{323.6}$$

Hence by setting $a + (n+1)i = b$ and summing (323.6) from 0 to n of r, he obtained $f(b) > f(a)$, as required. Lagrange then proceeded to estimate the errors (322.12) (pp.45-50).

(323.4) shows well the scope and limits of Lagrange's work. The proof-method

---

[1]) Euler *1770b*, arts. 45, 123; Lagrange *1762c*, art. 44, and *1775a*, arts. 3-4; Legendre *1791a*, 456-461. Among later writers, the form appears in de Prony *1800b*, 324.

excited the interest of Ampère, for one, as we shall see in § 4.3.3. But its assumptions about the majorance of $f'(z)i$ in (323.5) surely entailed other assumptions about the properties of "any" function f, and his estimates of the error did not lead either to a general remainder term of the Taylor expansion or to a study of its convergence. (However, on pp. 20–21 he did refer to the non-convergence of a series for $\log_a y$ when seeking the expansion in x of $y = a^x$.) Again, although he obtained by algebraic means expansions of this and some other simple functions, he provided no *general* algebraic methods for finding the expansion of "any" function—a point which excited the attention of Arbogast (§ 4.3.6).

In various respects, therefore, Lagrange's views on the calculus lacked conviction. Although his emphasis on power-series expansions was very influential and his term 'derived function' and notation '$f'x$' (or '$f'(x)$') were widely adopted, his philosophy of the calculus *en bloc* was not so readily accepted. Dickstein, in his excellent study *1899a* of the reception of Lagrange's approach, shows that there were various reservations, on both technical and conceptual grounds. Of other views then espoused, the most significant was Euler's, to which we now turn.

3.2.4 Euler's approach to the calculus; a contrast with Lagrange

> [Lagrange's foundations of the calculus] is assuredly a very interesting part of what one might call purely *philosophical* study; but when it is a case of making transcendental analysis an instrument of exploration for questions presented by astronomy, marine, geodesy, and the different branches of the science of the engineer, the consideration of the infinitely small leads to the aim in a manner [which is] more felicitous, more prompt, and more immediately adapted to the nature of the questions, and this is why the *Leibnizian* method has, in general, prevailed in French schools.
>
> de Prony *1835a*, 11

A good place to begin the comparison is with Euler's 'differential coefficient', and its relationship to Lagrange's 'derived function'. One of the main conceptual difficulties in the calculus that Euler inherited from Leibniz and the Bernoullis was the indeterminacy of higher-order differentials, caused by their dependence on the arbitrary first-order partitions of the variables of the problem at hand (see Bos *1974a*). Euler removed this difficulty by introducing his definition of the differential coefficient. If x and y were variables, then the relationship between their differentials could be written

$$dy = pdx. \tag{324.1}$$

Were dx taken to be constant, then $ddx = 0$; hence the higher-order differential ddy was obtained from (324.1):

$$ddy = dpdx + pddx = dpdx. \tag{324.2}$$

In (324.2) p was a variable, just like y; so the relationship between dp and dx was expressed as

$$dp = qdx. \tag{324.3}$$

Hence (324.2) became

$$ddy = q(dx)^2. \tag{324.4}$$

ddy was now determinate; for q was known (differentiability was assumed by Euler), and so was the constant dx. p, q, ... were the successive differential coefficients of y with respect to x (Euler *1755b*, arts. 264–270).

Euler's assignment of a constant differential to x was equivalent to making x an (or the) independent variable for a problem, with (the) other variable(s) dependent on it; thus he gave importance to the concept of the function. He also gave the differential coefficient a status equal to that of differentials themselves. These were aspects of positive influence on Lagrange, and thus of similarity to him (see Yushkevich *1959a*). However, Lagrange removed infinitesimals and replaced Euler's differential coefficients with his own derived functions (322.11).

Euler sometimes called his differentials 'zeros', and the calculus as a theory of 'reckoning with zeros'; but, as Lagrange pointed out in *Functions*$_1$ (1797), 3, this was a poor conception, since if one were reckoning with zeros, then *any* value could serve as the answer. Although historians of the calculus often mention this conception of Euler, it is better to see him as maintaining the Leibnizian conception of differentials as infinitesimal increments.

The distinction between increments and zeros involves the important distinction of *dimension*, between an infinitesimally short line and a point (or an infinitesimally thin rectangle and a line, and so on). It also relates to a difference of *Denkweisen* (§ 1.3.6); for the Leibniz/Euler approach is geometrically oriented while 'reckoning with zeros' is an algebraic conception—unsuitable for Euler, therefore, though consistent with Lagrange's approach to the calculus. Thus, while Euler of course made many contributions to algebra, at both the educational and research levels, his calculus is best understood as drawing on geometrical roots in the sense just described.

The distinction between algebraic and geometrical conceptions also related to practical utility. de Prony expressed the point well in the quotation at the head of this sub-section: Lagrange's theory may make nice philosophy, but it can be hard to use in the field.

3.2.5 'Analysis' and 'synthesis', and their mathematical connections

The basic distinction between analysis and synthesis in proof-methods in mathematics is that in analysis one starts out from the desired result and regresses until apparently impeccable principles are found, while synthesis begins with those principles and derives the result. However, especially during the 18th century mathematicians became accustomed to associate analysis with algebra and synthesis with geometry, although these connections were not clear, least of all in the development and use of the calculus. Lagrange indicated a difficulty at the beginning of a paper *1775a* on the attraction of spheroids:

> Some advantages that algebraic analysis may have over the geometrical methods of the Ancients, that one popularly, although quite improperly, calls *synthesis*, there are Problems where the latter appears preferable, as much for the luminous clarity which accompanies it as by the elegance and facility of the solutions that it gives. It is the same for those [problems] for which algebraic analysis appears in some way insufficient, and where it seems that the synthetic method is the only one capable of reaching [them].

He objected to the association of synthesis with (presumably) Euclidean forms of geometrical reasoning, for they were not relevant either in proof-method or content to problems such as the attraction of spheroids. Thus 'it will sometimes happen, that synthesis, by representing things much simpler than [does] analysis, will lead to a conclusion in a much less complex manner', as Lacroix put it in a lecture on analysis and synthesis to the *Société Philomatique* in 1797 (*1799b*, 199).[1]) However, doubtless with Lagrange's *1775a* among other works in mind,

---

[1]) Lacroix published this essay in the first four editions of his textbook *Geometry*, but then moved it into his *Education* ($_1$(1805) and later editions, pt. 2, sect. 2). He described it as a lecture delivered before the *SP* in 1797, but it was actually read in June 1796 (*BS*, ms. 2082, fols. 199–200). It was never published by the *SP*: I quote from an English translation by J. Toplis, put out in the *Philosophical magazine* in 1804.

he added that 'in the hands of Lagrange, Laplace and Legendre, analysis re-assumes, in these cases, those advantages which it possesses in all others'.

But the association of analysis with algebra was also not sharp; 'for it is not correct to suppose that algebra constitutes exclusively analysis', Lacroix pointed out, since 'it serves likewise to facilitate synthetical demonstrations' (*1799b*, 193). Indeed, when one considers Lagrange's *Théorie des fonctions analytiques* we see that in fact a noticeably "synthetic" approach is being used. The same can be said for his *Méchanique analitique* (§ 5.2.2-3), and also for the development of 'analytic geometry' by him, Lacroix and of Monge, where even that name was sometimes avoided because of the inadequacy of the word 'analytic'.

We shall consider the more general connections between geometry and algebra, including analytic geometry, in § 4.7: we shall even find Carnot proposing another distinction between analysis and synthesis in this context (§ 4.7.1). But let us consider here another factor in the contemporary interest in the matter, which is exemplified in the concern of Lacroix. As was mentioned in § 2.5.4, he was a disciple of Condorcet, who was associated with an influential group of philosophers of that time who came to call themselves '*idéologues*'. They laid great emphasis upon the place of signs in knowledge: their word '*idéologie*' was intended to refer both to ideas in the mind and their mode of representation. We recognise this philosophy as a type of semiotics. Condillac was an especially influential figure: his posthumous *Logique (1780a)* was distributed to all students at the *Ecole Normale* in 1795 (see Albury *1980a*), and his *La langue des calculs (1798a)* described in detail his views on algebra (including the legitimacy as mathematical objects of negative numbers: see Schubring *1986a*). The link from 'logique' to algebra was strong, for algebra was seen as a prime example of knowledge expressed by clear signs; and it was linked in turn to 'analysis', a favoured epistemological concern of the *idéologues*.

With philosophy and mathematics both involved, and from various directions, it is not surprising that 'analysis' and 'synthesis' were having such a hard semantic time. Another distinction was emerging: that synthesis was associable in mathematics with a systematic presentation of knowledge already gained while analysis was a method of finding new knowledge. In an essay 'on analysis and synthesis', Gergonne put the point well:

> In the exposition of truths already discovered, one holds the chain of reasoning in one's hands, and it is only a question of showing to those that one teaches, and to let them run through the various links which compose it, which can be done in a large number of diverse ways [...] But, in the research of new truths, one has not the same liberty. In fact, one knows only one of the two extremities of the chain; and, according

> as if it is the first or the last link which is known, it is analysis or synthesis which one must employ.[1])

He added that it was probably better to give new names to new ideas and preserve 'denominations consecrated by long usage' rather than change the names of traditional techniques, even if the names are no longer appropriate (*1817a*, 369); but in this case the use of 'analysis' and 'synthesis' remained undeservedly widespread. For example, in § 1.3.5 I quoted a passage of 1833 from Hamilton where he over-simply associated "analysis" with scientific induction and "synthesis" with deduction.

Gergonne himself gave one reason why these terms were maintained, despite the disadvantages: 'an author who wants to draw the regards and the attention of the public to his opus, hardly neglects to write at its head: "Analytical Treatise"' (*1817a*, 369). Undoubtedly he had Lagrange's treatises among those in mind; but he may not have known this surprising comment of Lagrange, as reported by Maurice (*1814a*, 228):

> I regarded as quite useless the reading of large treatises of pure analysis: too large a number of methods pass at once before the eyes. It is in the works of application that one must study them; one judges their utility there and one appraises the manner of making use of them.

3.2.6 Limits and infinitesimals in late-18th-century calculus

> [The infinitesimals] neither have nor can have theory; in practise it is a dangerous instrument in the hands of beginners [...] *anticipating*, for my part, *the judgment of posterity*, I would dare predict that this method will be accused one day, and rightly, of having retarded the progress of the mathematical sciences.
>
> Servois *1814b*, 148

Having concentrated hitherto on Euler and Lagrange, in this and the next subsection I consider the position of more minor figures. Hitherto I have tended to stress the difficulties in Lagrange's views on the calculus; so it is only fair to

---

[1]) Gergonne *1817a*, 353–354: Dahan *1986a* discusses this paper. Gergonne first dealt with the distinction between "analysis" and "synthesis" in a little-known paper *1810a* presented to the Academy at Gard.

mention that he had supporters too. One of the most enthusiastic was Servois, who, as we shall see in § 4.3.7, came to some rather interesting ideas in algebra in the course of trying to systematise and extend Lagrange's theory, and so save the calculus from the abominable infinitesimals.

The two figures dealt with here seem to have been influenced by the uncertainty that Lagrange apparently felt about his own position; for they were partly stimulated by a prize problem proposed by the Berlin Academy in 1784, during his time there. A 'clear and precise theory of what is called the infinite in mathematics' was requested, including an account of 'how so many correct theorems have been deduced from a contradictory supposition' such as the infinitesimal.[1]) Although the examiners were not fully satisfied by any entry, the prize was awarded to the Swiss mathematician S. Lhuilier. He advocated limits, following their earlier advocacy by d'Alembert. Mathematical educators of today, who present the theory of limits as if there was something obvious about it, would learn much from tracing the laborious procedures of this thoughtful essay. For example, Lhuilier proved the product theorem for limits

$$\lim_{n\to\infty} p_n q_n = \left(\lim_{n\to\infty} p_n\right)\left(\lim_{n\to\infty} q_n\right) \tag{326.1}$$

and its analogue for quotients, but in both cases his proofs were valid only if the limiting value was approached from above or from below (*1786a*, 18-24, 204-206). Thus he was accepting, but trying to cope with, the criticism of limits which I quoted from Lagrange in § 3.2.1; that if one approaches a value from below, then there is nothing to stop one going through it and out the other side. Only in a revised version of his essay did Lhuilier seem to want to break away from one-sided passage to the limit (*1795a*, 17-21). With regard to the foundations of the calculus, he defined dy/dx as the limiting value (if it existed) of the difference quotient; and, in order to avoid infinitesimals, he urged that the notation be read as a single symbol rather than as the ratio of dy to dx (*1786a*, 139-142). Nevertheless, he still called dy/dx the 'differential ratio' (pp. 32, 44)! He also defined the 'integral ratio' as the ratio of two variables exhibiting a given differential ratio.

Given presentations such as these, it is not surprising that limits did not enjoy much popularity in the late 18th century; for example, Lhuilier's most durable contribution was the notation 'lim' (informally introduced in *1786a*, 24). However, a more substantial influence on the calculus during the early 19th century has its origins at this time, due to the Berlin competition: Carnot, with an essay

---

[1]) The prize problem was announced in *Nouv. mém. Acad. Berlin*, (1784/86), 12-13: the judges' report is in (1786/88), 8.

1785a advocating infinitesimals. Carnot never published this manuscript (it appeared only in Gillispie *1971a*); but a revised version *Calculus*$_1$, published in 1797 just before he lost power and his place in the *classe* (§ 2.2.2), attracted some attention, especially when he published a second edition in 1813.

A notable feature of the successive versions of Carnot's essay is the decline in status of limits in his thought (see Yushkevich *1971a*). In the first version he included several pages on limits (*1785a*, 239–245); in *Calculus*$_1$ (1797), art. 17, he just gave a definition of a limit; in his discussion in *Calculus*$_2$ (1813), arts. 129–135, even the definition was sketchy. It seems that he wanted to maintain the traditional interpretation of infinitesimals as infinitely small increments; he gave prominence to the notion of variable when defining them ($_2$(1813), arts. 14–16, a detail which will arise again in § 11.3.5). But basically, like Lagrange (whose textbook *Fonctions* also had editions in 1797 and 1813) he was algebraic rather than geometrical: he tended to present differentials as objects obeying certain laws rather than as geometrical increments along a line. But he had all the basic components of the Eulerian calculus clearly outlined: the higher-order differentials such as ddx, the difference between ddx and $(dx)^2$, and so on.

So the situation in foundations in the late 18th century was rather confused; various competing standpoints were available, each with its own supporters. I now consider some appraisals of the state of affairs by a few textbook writers and historians writing around 1800.

3.2.7 Appraisals around 1800: Lacroix's large *Traité*

> I see with much pleasure that you are working on a large work on the integral Calculus. [...] The reconciliation of the Methods which you are planning to make, serves to clarify them mutually, and what they have in common contains very often their true metaphysics: this is why that metaphysics is almost always the last thing that one discovers. The spirit arrives at the results as if by instinct; it is only on reflecting upon the route that it and [that of] others have followed that it succeeds in generalising the Methods and in discovering its metaphysics.
>
> Laplace, 1792[1])

---

[1]) Letter from Laplace to Lacroix, 28. 1. 1792 (manuscript in *BI*, ms. 2396); published by Lacroix in *Calculus*$_1$, *1* (1797), xxiv, and by Taton in *1953a*, 591, and *1953b*, 355.

We saw in § 3.2.2 that Lagrange wrote the first edition of his *Fonctions* in haste: he commented himself at one point in the text, 'As our design is not to give a complete treatise [...]' (p. 91). Thus he could conclude his book with relief, by referring to the first volume of 'the new treatise of the differential and integral calculus of Lacroix, where this calculus is presented from all points of view, and developed with all the detail that one could desire'.

Lacroix submitted the plan of his *Traité* to the *classe* in August 1796 (*PV*, *1*, 86), when he was 32 years old. On the last day of 1796, soon after Delambre's dismissal of his astronomy papers (§ 2.5.4), Legendre reported on the manuscript of the first volume, judging it to be a work which 'will make itself conspicuous by the choice of methods, their generality and the rigour of the demonstrations', and he recommended 'that the *Institut national* must welcome and encourage by its approbation the zeal and the talents of c[itizen] Lacroix [...] which were already known to the advantage of the former *Académie des sciences*' (Legendre *1796a*). Soon the folios were being sent from the printers to the staff at the *Ecole Polytechnique*, who praised the book in their lectures (see Lacroix *1797d*). It appeared, like Lagrange's *Fonctions*, in 1797: the publisher was Duprat. A second volume, on integration and differential equations, appeared in 1798, followed two years later by a third on series and differences: they will be noted below where appropriate. Around that time was published a German translation of the first two volumes, prepared by J. P. Gruson (on whom see § 4.3.4).

In his lengthy preface on the history of the calculus, Lacroix recalled that he had begun to collect materials in 1787 and had circulated some early sketches to colleagues for their reactions: Laplace's comments included the encouraging passage quoted at the head of this sub-section. Lacroix reported that printing had started in November 1795 before being 'suspended for several months due to particular reasons' (*Calculus*$_1$, *1* (1797), xxiv); thus he had been able to make only a few remarks on Lagrange's newly published *Fonctions.* He also mentioned on pp. xxi–xxii that he knew of Carnot's views only from a manuscript (presumably either *1785a* or a version of *Calculus*$_1$ (1797)). Otherwise his book was based in the usual way on published sources.

Lacroix kept his text generally free from references (and entirely clear of exercises for the reader), but he adorned his contents pages with extensive bibliographies (pp. xxx–xxxii), showing further his historical knowledge. (In this respect he may have been following an 18th-century tradition which I noted in § 2.5.5 in connection with Legendre's thesis of 1770.) After an introduction on basic notions on functions and series, he devoted two chapters to the principles of the differential calculus, one on the roots of equations, and two on curved lines and surfaces. Here I am only concerned with his account of the foundations of the calculus.

Lacroix began with a general definition of a function: '*Any quantity of which the value depends upon one or several other quantities, is called* function *of these latter, whether one knows or one does not know by which operations it is necessary to pass in order to go back from the latter to the first*' (p.1). This form of definition is often attributed to Dirichlet, for it appears in his *1829a* (§ 17.2.1); but we see it here in Lacroix. Indeed, it is already back in Lagrange's phrase 'in any manner' (§ 3.2.1) and even further back to Euler and John Bernoulli (see Yushkevich *1976a*, 57-62). Where these figures differ is in the types of function that they actually work with; in this respect there was an expansion of conception over the decades. Lacroix, for example, often stayed in and around power series in his introduction, including offering a definition of a limit with the context of the sum of a convergent series in mind (pp. 5-7: he handled conditions for convergence only for simple cases). Again, he just assumed that "any" function could be given a power-series expansion, and even thought it impossible that they could take negative powers of the increment (p. 87).

Obviously in Lagrangian territory, Lacroix defined the differential as the first term in that expansion. Thus if

$$f(x+k) = f(x) + f'(x)k + \ \ldots, \tag{327.1}$$

then the differential $df(x)$ was given by

$$df(x) = f'(x)k, \tag{327.2}$$

where f' was the first 'derived function' of f. He wrote 'dx' for 'k' 'for uniformity of signs', so that (327.2) became

$$df(x) = f'(x)dx \tag{327.3}$$

(pp. 95-97). Higher-order differentials could also be obtained from (327.1); for example,

$$f''(x) = d^2f(x)/dx^2 \quad \text{(that is, } ddf(x) \div (dx)^2\text{)}. \tag{327.4}$$

Hence in Lacroix we see Lagrange and Euler together in that both derived functions and differentials were present, although with differentials introduced and handled in a different way from Euler's reckoning with zeros. We must see this style of working as part of the influence of *encyclopédisme* on Lacroix (§ 2.5.4): there were various methods available, and they should all be presented (see his letter transcribed in § 20.2).

Lacroix extended this treatment to multivariate calculus, but I shall not dwell on these details. There are interesting remarks here and there; for example, on p. 133 he associated the equality of all partial differentials of like kind (for example, $z_{xy} = z_{yx}$) with the commutativity of the differential operators. On pp. 189–190 he showed how 'the method of limits' could be accomodated within the power-series expansion (when the increments on x and $f(x)$ went together to zero, (327.1) showed that $f'(x)$ was left over), although Newton's theory of fluxions was not outlined 'because it draws on the consideration of motion[,] which is foreign to analysis and geometry' (p. 194).

3.2.8 Some other appraisals around 1800

The various treatments of the calculus, with power-series, infinitesimals and limits jostling for attention, is apparent in other publications of the period. For example, when Cousin published a two-volume treatise in 1796, he surveyed all traditions in his brief historical introduction, and in his text actually laid considerable emphasis on limits: 'We can define the differential calculus, *the method of finding the limits of ratios between the differences of variable quantities*; and the integral calculus, which is the inverse of the differential calculus, *the method of going back from the limits of ratios between the differences to the same ratio of the quantities*' (J. Cousin *1796a*, *1*, 151). By contrast, for him Leibniz's infinitesimals were very imperfect: however, his treatment of them was itself unclear, for 'we will call differentials of the first order the terms *dy, dx* of the limit *dy/dx*' *(ibid.)*, which (as was noted in § 1.3.7) leaves ambiguous the interpretation of the notation '*dy/dx*'. His position was similar to Lhuilier's, although he did not cite Lhuilier. By contrast with both these men, Bossut published a treatise on the calculus in 1798 in which he followed Euler's views, including regarding infinitesimals as zeros (Bossut *1798a*, *1*, 94–97), despite the fact that he was a disciple of d'Alembert (see Doublet *1914a*), who had advocated limits!

There was a notable amount of writing on the history of mathematics in the late 18th century (see Günther *1908a*), and those works on or around the calculus also emphasised the presence of different traditions. For example, a French translation of Euler's *Introductio in analysin infinitorum* was published as Euler *1796–1797a*, with notes by the translator J. B. Labey amplifying Euler's conceptions of analysis. A review of the translation stressed its value for the young (Euler *1797a'*), thus indicating a continuing role in education for Euler's approach.

An important event in the history of the history of mathematics was the appearance of the first two volumes of the second edition of Montucla's *Histoire des mathématiques* in 1799 (Montucla *History*). In a review published in the *Magasin encyclopédique*, Lalande *1799a* stated that he had been pressing Montucla since 1792 to bring out these volumes, that printing had started in 1794, and that half of the third volume was on proof. Montucla was still alive at the time of Lalande's writing; but he had been loath to complete the project for personal reasons (see D. E. Smith *1921a* and *1922a*). In fact, he died at the end of 1799, and the remaining two volumes were completed by Lalande himself, with assistance from others. The third volume, dealing with the calculus, was published in 1802: Montucla had in fact completed the short passages on the foundations of the calculus, and Lacroix helped Lalande with the rest.[1]) It is interesting to note how similar in form is this history to Lacroix's treatise, with few references in either; in fact Lacroix's work has a rather fuller bibliography (and far fewer printing errors). While Montucla's comments on foundations amounted to little more than stating methods as and when appropriate, his book is a useful source for the technical developments and applications of the calculus.

Bossut published a general history of mathematics as *1802a.* He surveyed in a non-technical vein both the traditions and the applications of the calculus, although he omitted Lagrange's approach because of a policy of excluding living figures. It is amusing to note that when he did include recent work in his history *1809a* of mathematics to 1808, his book 'aroused against him the self-esteem of living mathematicians, who reproached the author strongly for not having praised them enough' (Bossut *1827a'*)! Historians of recent developments know the feeling: there can be no acceptable account of what has lately happened.

3.2.9 On the interest in foundational questions in the late 18th century

I have dealt at some length on foundational questions of the calculus in this section because, as we shall see throughout this study, there is often an interesting clash between methods used at the stage of creation and those put forward in a later systematisation. Laplace expressed the general point in the remark to Lacroix at the head of § 3.2.7, and de Prony emphasised the particular problems with Lagrange's views in the statement which began this section.

The competition of views on foundations suggests doubts concerning the received historical view: that 18th-century mathematicians were not much bothered by foundational problems. While it is certainly true that they did not write extensively on them, and that Lagrange's interest was far greater than average, I am not convinced that one can draw this conclusion. There are various reasons why one might not write on a problem, and uninterest is only one of them. Other reasons include unawareness of the problem; awareness of it, but an inability to

[1]) On the completion of volume 3, see pp. 245-260, and p. 336 for Lalande's explanatory remarks. Evidence of Lacroix's help appears on, for example, p. 342. On Montucla's life see Sarton *1936a*; and also the review of volumes 3 and 4 in Montucla *1802a'*.

say anything significant about it (the "too-difficult" category) or to put it in a workable form (the "too-vague" category); or awareness of it but a belief that it is basically *solved.* I suspect that this latter option applied to at least some mathematicians of the late 18th century. They could follow Lagrange's views; or if not, use the older theory of Euler, and live with the difficulties of infinitesimals. Had there appeared a major problem in foundations—for example, had anyone noticed Cauchy's counter-example of $\exp(-1/x^2)$ to Lagrange's belief in the Taylor-expandability of "any" function—then I am sure that interest in foundations would have greatly increased. (The counter-example *could* have been found: as will emerge in § 11.4.3, Cauchy did not need any novel techniques to produce it.) Indeed, we shall see in § 5.3.1 there was a considerable degree of attention to a parallel problem in mechanics, where Lagrange's approach was based on the principle of virtual velocities, of which he gave no proof.

3.3 On differential equations

I concentrate on partial differential equations in this chapter because most 18th-century mechanics and 19th-century mathematical physics used them; ordinary differential equations tended to occur more often as by-products than in their own right. However, there are plenty of exceptions to this rule, and both here and later ordinary differential equations will make their due appearance.

3.3.1 The place of linear equations

> The word *linear* is improper; it is relative to Geometry, and in applying it to equations, one has in view a straight line in the equation of which the ordinate (or equally the abscissa) are found only to the first degree: thus one would not know how to regard as linear equations such as $dy + Pydx = Qdx$, which belong most often to transcendental curves.
>
> Lacroix *Calculus*$_1$, *2* (1799), 225

Lacroix's comment both reflects some issues which we have considered in the last section and also raises one which will be of major concern later. In objecting to the use of the word 'linear' to refer algebraically to linear combinations of terms instead of to its traditional geometrical context of straight lines, he exemplified the tension between algebraic and geometrical aspects of the calculus which we noted often in the last section. He obviously held his criticism strongly, for he repeated it word for word in his second edition (*Calculus*$_2$, *2* (1814), 254) and elsewhere; but by then his objection to this use of 'linear' was becoming forlorn, for Fourier's work was beginning to be known, and an age of linearisation of differential equations was born.

Much of the writing in the calculus in the late 18th century was concerned with methods of solving differential equations, general and singular solutions, and related techniques such as summing series and evaluating integrals. It is a massive literature; so I must confine myself here to a few of the most important or representative methods and results, although others will be noted in later chapters when appropriate. Our historical knowledge of these developments is in general very inadequate; the best sources are still the treatises, textbooks and histories mentioned in the last section.

3.3.2 The vibrating string problem: d'Alembert, Euler and the taxonomy of functions

> [...] but the glory of the invention of the integral calculus of partial differentials belongs to Leonhard Euler; this great geometer had integrated in 1734 the first-order equation of the form
>
> $$\left(\frac{dz}{dy}\right) = zf(y) + \left(\frac{dz}{dx}\right)\varphi(x,y),$$
>
> f and φ being signs of functions; y represents a parameter, constant for a determinate curve, and variable from one curve to the other, x and z being rectangular coordinates.
>
> de Prony *Mechanics, 2* (1815), 473

Forms of partial differentiation can be seen in the early history of the calculus (see Cajori *1928a*); but, as de Prony recalled, a major step came in Euler's paper *1740b*. Euler's problem concerned a family of curves, so that, as de Prony also noted, only one variable was of the orthodox kind, ranging along an axis; the other was a parametric variable. In this and related papers Euler took a given function for z_x and sought by 'modular equations' to find a definite form for z_y. Although he was not solving a partial differential equation, his procedures were equivalent to certain solution methods (see Engelsman *1982a*). The formation of partial differential equations to express physical problems, where the independent variables refer to spatial and/or temporal factors and the boundary conditions are also stated, came in the 1740s; but from then on they dominated mechanics, and continued their prominence during the rise of mathematical physics.

One of the earliest physical problems so studied was the motion of a vibrating string; and it became the testing-ground, as it were, for many of the issues concerning the use of partial differential equations to solve physical problems during the 18th century. Thus, although it has been the subject of attention of historians of science both around 1800 and later,[1]) I need to recall some of the principal features of the debate over solution methods.

A uniform elastic string is stretched between two fixed points A and B distance l apart and is set into small horizontal transverse vibration. We take A as origin, AB as x-axis, and y as the axis perpendicular to AB. The 'wave equation' of motion

$$y_{xx} = y_{tt} \tag{332.1}$$

(where I have omitted various physical constants), found by various means in d'Alembert *1749a* and Euler *1749a* and *1750a* (see § 5.2.5), was not disputed; but the form and interpretation of the solution caused a lengthy controversy for the rest of the century which was never really resolved.

Both men obtained the functional solution

$$y = F(x+t) + G(x-t) \tag{332.2}$$

and converted it via the boundary conditions

$$y = 0 \quad \text{when} \quad x = 0 \quad \text{and} \quad x = l \quad \text{for all } t \tag{332.3}$$

into

$$y = F(t+x) - F(t-x), \tag{332.4}$$

where F, of period $2l$, was to be determined from the initial conditions of the problem. (Euler was rather clearer about how the determination was to be executed.) d'Alembert insisted that F and G were "continuous" (to use their own term), whereas Euler wished to admit "discontinuous" functions into the solution (332.2).

The terminology is not the same as ours, and in fact is not easily given precise definitions. A "continuous" function was specified by a single algebraic expres-

[1]) For literature around 1800 and later, see especially Cousin *1796a*, esp. *2*, 199–201; Lacroix *Calculus*$_1$, *2* (1799), 619–621; and Montucla *History*, *3* (1802), 659–667. Later studies include Gibson *1892a*, 137–146; Burkhardt *1908a*, 1–50; Jourdain *1913a*, 669–675; Truesdell *1960a*, 237–300; Ravetz *1961a*; my *1970c*, ch. 1; and Demidoff *1976a*.

sion over its range of definition: hence, given the class of expressions then normally handled, a continuous function was differentiable in our terms. However, there were counter-examples, such as $\sqrt{x^2}$ (as Cauchy pointed out in his *1844a*), so that the nature of the continuity of a function is in fact not clear. A "discontinuous" function was made up of "continuous" algebraic expressions defined over neighbouring intervals and such that the end-point values are equal; for example, x^2 over [0, 1] followed by $(5x - 4)$ over [1, 4]. Such functions were not "continuous" across the joins—indeed, they might well not be differentiable there—but they would be continuous in the sense of being joined up. We shall return to this last notion in a few paragraphs' time.

Euler advocated "discontinuous" functions on the grounds that the initial conditions of the problem would involve the release of the string from a possibly "discontinuous" configuration; for example, it might be held at one point before release with a corner, which (apparently) was not expressible by one formula. However, d'Alembert urged the restriction of F and G to "continuous" functions on the grounds that, to use modern terms, the left- and right-hand derivatives of y at a corner-point x were not equal; hence y_{xx} was not definable, and so could not appear in the wave equation (332.1). In a later, little-known, essay *1780a* he discussed the problems of left- and right-hand derivatives of "discontinuous" functions (see Yushkevich *1975a*).

In contexts such as these we see the types of function Euler actually intended to use, in contrast to the talk of "any" function that we noted in § 3.2.7. The problem of the taxonomy of functions became very complicated by the end of the century, with the various traditions in the calculus leading to their own—often unclear—terminologies (see Yushkevich *1976a*). The vibrating string problem is only one manifestation of the difficulties, and the failure during the century for a definitive position to be reached is typical of the consequences.

The St. Petersburg Academy showed its awareness of these difficulties by proposing in 1787 a prize problem on the matter, phrased in a way which reflected the multiplicity of traditions:

> If the arbitrary functions at which one arrives by integrating [differential] equations in three or several variables represent any curves or surfaces, either algebraic or transcendental, either mechanical, discontinuous, or produced by a voluntary movement of the hand; or if these functions comprise only continuous curves represented by an algebraic or transcendental equation? [1])

[1]) *Nova acta Acad. Sci. Petrop.*, *5* (1787/89), pt. 1, 4–5.

Arbogast won the prize with an essay *1791a* in which he used essentially geometrical considerations both to simplify the taxonomy of functions and also to extend their range beyond "continuity". He converted Euler's sense of "discontinuous" functions into '*discontinuous curves*' which, 'traced by the free movement of the hand, are not submitted to any law for any part of their course; provided that all parts of the curves join together without interruption [...] By *discontinuous functions* we mean functions which represent this sort of curve'. However, if 'the different parts of the curve do not join together', then one had '*discontiguous curves*, because all their parts [...] are not contiguous together, and we give the name of *discontiguous functions* to functions which are supposed to correspond to curves of this nature' (*1791a*, 9-10).

Arbogast's terminology was not too happy, with such similarly sounding words as 'continuous' and 'contiguous' being used together; but the distinction itself was interesting, not only because of its efforts to break clear of the restrictions of continuity—for Arbogast 'in general, all arbitrary functions [...] are not subject either to the law of continuity or to that of contiguity' (*1791a*, 96)—but also to convert to some extent from the algebraic to the geometric approach. (However, Arbogast was basically an algebraic thinker (§ 4.3.6).) In § 9.2.6 we shall see this geometrical treatment of functions taken further by Fourier.

3.3.3 The vibrating string problem: Daniel Bernoulli, Lagrange and trigonometric series

Mention of Fourier brings up the other aspect of the discussion of the vibrating string problem that warrants emphasis here: the proposal of a trigonometric series solution to (332.1)

$$y = \sum_{r=1}^{\infty} a_r \sin(\pi r x/l) \sin(\pi r t/l) . \tag{333.1}$$

The chief advocate of this solution was Daniel Bernoulli, initially in his *1755a*. His argument was physical, being based on the alledged ability of 'all sonorous bodies [to] contain potentially an infinity of sounds and an infinity of corresponding ways of making their regular vibrations'. He was rather careless in his mathematical statement of his position; he made no effort to calculate the coefficients $\{a_r\}$, and did not even write down the t-terms in (333.1). But he asserted that the solution was general, in contrast to a claim by Euler in *1749a*, art. 30 that it was valid only for "continuous" functions.

Euler brought further criticisms in reply to Bernoulli. Perhaps (333.1), 'because of the infinity of undetermined coefficients, is so general that it includes all possible curves', he admitted; however, 'all the curves contained in that equation, although one increases the number of terms to infinity, have certain characteristics which distinguish them from all other curves' (*1755a*, art. 9). For example, they were periodic; and the sine series put forward in (333.1) was also odd. Hence they could not be general.

The manner of interpreting the periodicity of trigonometric series was not understood until Fourier, so I shall postpone further discussion until § 9.2.6. But it is important to note the preference of functional solutions of the type (332.2) over trigonometric series (333.1) until Fourier's work, not only for the wave equation but also for differential equations in general, even though certain trigonometric series—in Euler *1798a*, the general formula for the cosine series—were found from time to time (for references, see my *1970c*, 19).

This attitude to solutions is evident also in Lagrange, who wrote a long early paper *1759a* on the problem. He accepted the functional solution, and also Euler's rather than d'Alembert's interpretation of it; but he obtained it by a different method, which indicates that already he was seeking alternative approaches to Euler's. He used a discrete-masses model for the problem, considering $(n-1)$ masses joined by a weightless cord and satisfying the difference-differential equations

$$d^2y_k/dt^2 = \Delta^2 y_{k-1}, \quad 1 \le k \le n, \quad \text{with} \quad y_0 = y_{n+1} = 0t. \tag{333.2}$$

He solved these equations and then took n in his solution to infinity to obtain the functional solution required. Moreover, the alleged reliance only on algebraic processes yielded the generality to "discontinuous" functions that Euler desired. At one point in his analysis he came close to a full Fourier series solution; but in fact his formula is not very similar to a Fourier series,[1]) and he made no pause to explore these possibilities while en route to the functional solution. As we shall see in § 9.3.4, he did not change his view about such series when Fourier revived them in the 1800s.

3.3.4 Preferred forms of solution

I reported above a strong preference for functional over trigonometric solutions. One of the reasons was that, while the coefficients $\{a_r\}$ in (333.1) were undetermined, the functional solution (332.2) involved functions (F and G there) which

---

[1]) On this matter, see my *1970c*, 15-17. We shall note in § 1.4.1 that Poisson used this work to claim priority for Lagrange over Fourier for the invention of Fourier series, as part of his deliberate attempt to denigrate Fourier's achievements.

could be determined from the functions describing the initial conditions. Since the physical problem was expressed mathematically by the differential equation and its boundary and initial conditions, preferred forms of solution should *explicitly* exhibit the relevant functions, or at least allow access to them.

The solution of differential equations, and the general theory of their solution, was a major interest for 18th-century mathematicians. I confine myself to a selection of methods and topics.[1])

One of the drawbacks to obtaining functional solutions was that they were not always easy to find. Sometimes the form of solution was assumed, and differential equations sought which took it *as* the solution (an example is discussed in Interlude 934.1). More often they were obtained by reducing the differential equation to a total differential or else to a form which could be integrated easily: integrating factors were sometimes used, in the hope of aiding the reduction. However, sometimes none of these possibilities seemed viable; and the most popular alternative was to assume a solution in the form of a power series in one of the independent variables, with functions of the other variable(s) as coefficients, and use the processes of analysis to determine these coefficients. In these contexts came a variety of results in the manipulation of series and difference equations, the occasional functional equation, and a role for solutions to ordinary differential equations. If possible, the resulting power-series was reduced to a finite, or a closed, form. For example, it might be convertible to a Taylor-series expansion, in which case it could be replaced by the value $f(x+h)$ of the corresponding f for the appropriate x and h; for in the 18th century all Taylor series were thought to converge.

In addition to particular forms of solution, there were also various methods of transforming partial differential equations. For example, in his *1774b* Lagrange used the condition for integrability offered by Euler

$$pq_z - qp_z + q_x - p_y = 0\,, \tag{334.1}$$

where 'p' and 'q' were Euler's notations for 'z_x' and 'z_y' (*Integral*, *3* (1770), art. 128), to convert a first-order equation into a linear equation for p in terms of x, y and z via a total differential. Then Lagrange *1781a* and *1787a* discussed

---

[1]) For sources on 18th-century methods in differential equations, Euler's textbook *1755b* on the differential calculus and the multi-volume *Integral* (1768–1770) on the integral calculus may be added to the various texts written around 1800 and discussed in the last section. The information in the appropriate articles in the *Encyklopädie der mathematischen Wissenschaften* is rather sparse; perhaps the most useful articles are Vessiot *1900a* and Weber *1900a*. See also Wallner *1908a*. Some of the many treatises on differential equations have occasional references to primary literature and even short historical accounts. Among recent historical studies of solution methods see Truesdell *1954a, 1955a* and *1960a*, all *passim*; Kline *1972a*, chs. 21–22; and Demidoff *1974a, 1982b* and *1982c*.

(among other things) how the solutions to the first-order equation

$$P + Qq = R, \tag{334.2}$$

where P, Q and R were functions of x and y, could be expressed as a function of two independent solutions u and v of the auxiliary equations

$$dx/P = dy/Q = dz/R, \tag{334.3}$$

which might be solvable by known means. Lagrange's two methods were united in a manuscript of 1784 by P. Charpit into a general method of solving first-order equations. Charpit's paper was never published, but Lagrange saw it in 1793 and improved the rigour of some of Charpit's arguments to produce what is now often called 'the Lagrange-Charpit method'. He presented it in, for example, the *Leçons* (*1806a*, lecture 20).[1])

Part of this theory states that a third solution $\varphi = 0$ of (334.2) is a function of u and v, and the proof rests on showing that the Jacobian of u, v and φ with respect to x, y and z is zero. Slips in Lagrange's proof were repeated until Chrystal *1892a* pointed them out: in particular, that the Jacobian had to be *identically* zero and not in virtue of the (supposed) third solution. As he remarked, mathematicians are not always scrupulous in their logic!

3.3.5 Some methods of solution

A large number of methods of solution of differential equations had become known by 1800; only a tiny portion can be mentioned here. One method is sometimes known today as 'cascades'. It occurred in Euler *Integral, 3* (1770), pt. 1, sect. 2, ch. 5, but became known through Laplace *1777a.* Defining the variable

$$u := z_x + kz \quad \text{with} \quad k \quad \text{constant,} \tag{335.1}$$

Laplace applied it to the general second-order partial differential equation

$$Az_{xx} + Bz_{xy} + Cz_{yy} + Ez_x + Fz_y + Gz + H = 0, \tag{335.2}$$

where the coefficients were functions only of x and y, and reduced it to the

[1]) The manuscript is often said to have been lost, but Arbogast's transcript survives in the Ashburnham Collection, *Biblioteca Medicea-Laurenziana* (Florence) (presumably having been removed from Paris by Libri at some time), and Lacroix's in *AS, dossier personnel* for Charpit (see my and Engelsman's *1982a*). Lacroix reported Charpit's method (especially in *Calculus*$_1$, *2* (1799), 496-497, 513-516, and *Calculus*$_2$, *2* (1814), 548-567). His transcript was rediscovered in 1928 by H. Villat, and its contents described in Saltykoff *1930a* and *1937a*.

form

$$au_{xy} + bu_x + cu_y + eu + f = 0, \tag{335.3}$$

which could be solved by infinite series or by further substitutions which are described in § 4.4.3. Legendre offered an extension in *1789a*, sect. 4, where in effect two new independent variables were introduced by the relations

$$\begin{array}{l} u_x + \lambda_1 u_y = 0 \quad \text{and} \quad v_x + \lambda_2 v_y = 0, \\ \text{with } \lambda_1 \text{ and } \lambda_2 \text{ roots of } A\lambda^2 + 2B\lambda + C = 0. \end{array} \tag{335.4}$$

Equations of the form (335.2) could then be solved without having to be reduced to (335.3).

In this paper Legendre also considered the equation of minimal surfaces

$$(1 + z_y^2)z_{xx} - 2z_x z_y z_{xy} + (1 + z_x^2)z_{yy} = 0, \tag{335.5}$$

which Lagrange had found in another form in *1762b*, appendix. Legendre criticised a solution given in Monge *1787a* on the grounds that it assumed 'some metaphysical principles which geometers do not yet admit' (Legendre *1789a*, 309). Although he did not explain himself, he may have been uneasy about Monge's admittedly unclear ideas about the generation of surfaces. (According to Poisson *1832b*, such objections were formulated by Laplace, and discussed by him with Monge.) Legendre offered an analytical version of Monge's approach, in which x and y were treated as functions of z_x and z_y. A particular type of contact transformation (as they are now called), this procedure goes back to Lagrange *1776a*, art. 53 (with seeds in Euler *Integral, 3* (1770)); it became known at that time as 'Legendre's transformation'. We shall see Poisson use it in § 7.2.6.

Legendre may also have been unhappy with Monge's rather uncritical interpretation of types of solution of differential equations in terms of surfaces and envelopes. For example, in *1786a* Monge took the second-order equation

$$Rz_{xx} + Sz_{xy} + Tz_{yy} = 0, \tag{335.6}$$

of which (335.5) was a special case, and converted it to

$$M + Nz_{yy} = 0, \tag{335.7}$$

where M and N were differential forms. He claimed from various geometrical considerations that the solution to (335.6) took the form

$$U = \varphi(V), \tag{335.8}$$

where φ was an "arbitrary" function, and U and V were solutions of $M = 0$ and $N = 0$ respectively. The reasoning is not completely clear, and it is noticeable that in his folioed lectures on 'analysis applied to geometry' at the *Ecole Polytech-*

nique, Monge used the differential of (335.6):

$$Rdz_{xx} + Sdz_{xy} + Tdz_{yy} = 0, \tag{335.9}$$

and avoided the use of (335.7) (*Analysis*$_2$ (1801), fol. 12). Charpit had treated such matters in an algebraic style in a manner which seems to be free from objections (see Saltykoff *1937a*).

Mention of general equations (335.2) and of the forms of their solutions naturally suggests to us the distinction between equations of the elliptic, parabolic and hyperbolic kinds, according as

$$(4AC - B^2) >, \ = \text{ or } < 0. \tag{335.10}$$

There was some awareness of the distinction in the 18th century. For example, Laplace's transformation of (335.2) to (335.3) in *1777a* was accompanied by consideration of the exceptions to the latter form, and Legendre discussed the matter in more detail in terms of the nature of the roots of (335.4)$_3$ (*1789a*, art. 4). But there seems not to have been much interest in the distinction itself, which did not become a prominent feature of the theory of differential equations until the late 19th century (the terms are due to du Bois Reymond *1889a*, 265).

Having started this selection with one method by Laplace, I end it with another. A powerful algorithm was furnished by his method of successive approximations. He introduced it in an 'Addition' to his *1775a* on particular solutions to differential equations and developed various forms (in, for example, *1776a*, arts. 1–2). It showed its power best in the solution of systems of ordinary differential equations, where few methods were known beyond sums of simple-state solutions for certain linear equations (§ 5.4.5). I follow his presentation in *Mécanique céleste* (*1* (1799), 475–481), but using sufficial variables instead of primes and letter-sets.

Laplace took n ordinary differential equations of order m (he put 'i') between the n functions $\{y_s\}$ of t, each equation expressed in the form

$$y_r^{(m)} + P_r(t, \{y_s, y'_s, \ldots y_s^{(m-1)}\}) = 0, \quad 1 \le r \le n. \tag{335.11}$$

He assumed that the solutions were known in closed form

$$H_r(t, \{y_s\}) = 0. \tag{335.12}$$

He then took the nm equations

$$H_r^{(j)} = 0, \quad 0 \le j \le (m-1), \quad 1 \le r \le n, \tag{335.13}$$

and assumed that they could be re-arranged so that the nm constants of integration of (335.11) were explicitly stated:

$$c_{r,j} = V_{r,j}(t, \{y_s, y'_s, \ldots y_s^{(m-1)}\}). \tag{335.14}$$

Now the differentials of these mn equations,

$$0 = dV_{r,j}, \tag{335.15}$$

being of order m, could be expressed as linear combinations of the equations (335.11):

$$dV_{r,j} = \sum_{q=1}^{n} [F_{r,j,q}(t, \{y_s, y'_s, \ldots y_s^{(m-1)}\}) \, (y_r^{(m)} + P_r)] \, dt. \tag{335.16}$$

Thus the factors $\{F_{r,j,q}\}$ rendering $(335.16)_2$ were given in terms of the known $\{V_{r,j}\}$ by

$$F_{r,j,q} := \text{coefficient of } y_q^{(m)} \text{ in } V_{r,j}. \tag{335.17}$$

Laplace now generated a solution to the next set of differential equations after (335.11), where I replace r by q as the sufficial variable:

$$y_q^{(m)} + P_q + \alpha Q_q(t, \{y_s, y'_s, \ldots y_s^{(m-1)}\}) = 0, \text{ with } \alpha \text{ constant, } 1 \leq q \leq n. \tag{335.18}$$

He multiplied (335.18) for each q by $(F_{r,j,q}dt)$ and used (335.16) to obtain

$$dV_{r,j} + \alpha \sum_{q=1}^{n} (F_{r,j,q} Q_q) \, dt = 0, \tag{335.19}$$

which, when integrated, yielded the solution

$$c_{r,j} - \alpha \int \sum_{q=1}^{n} (F_{r,j,q} Q_q) \, dt = V_{r,j}. \tag{335.20}$$

(335.20) was of the same form as (335.14), and so the process of converting (335.14) to (335.13) to solve (335.11) could be imitated for (335.20) to produce a system of solutions to (335.18).

The solutions would be in closed form if

$$T_{r,j} := \sum_{q=1}^{n} (F_{r,j,q} Q_q) \tag{335.21}$$

furnished exact differentials for (335.19), a situation which would apply if the differential equations (335.20) were linear. However, if exactitude did not obtain but α was small enough for α^2 to be neglected, Laplace used the known solutions (335.12) to (335.11) (where $\alpha = 0$) and the method of solution to generate a sequence of successively approximate solutions to (335.20). The solutions (335.12) of (335.11) supplied the initial values of $\{y_s^{(i)}\}$ and (335.17) furnished the values of $\{F_{r,j,q}\}$ to insert into (335.21), to provide, from (335.20), the second system of solutions

$$c_{r,j} - \alpha \int T_{r,j} dt = V_{r,j}. \tag{335.22}$$

This procedure was iterable; the values of $\{y_s^{(i)}\}$ supplied by (335.22) could be introduced into (335.21) to provide a new sequence of functions $\{{}_1T_{r,j}\}$ to yield the third system of solutions of the form (335.22), and so on. He did not investigate any convergence processes involved in the iteration.

Laplace's method is also iterable in another way; namely, to the system of differential equations

$$y_r^{(m)} + P_r + \sum_j [\alpha_j Q_{j,r}(t, \{y_s, y'_s, \dots y_s^{(m-1)}\})] = 0, \quad 1 \le r \le n, \tag{335.23}$$

which may be solved by considering the terms $\{\alpha_j Q_{j,r}\}$ in succession. He did not mention this possibility; perhaps he felt that the complications already encountered, and further difficulties arising in functions of αt when t was very large (pp. 488–503), were already sufficiently formidable.

3.3.6 Some studies of complete, general and singular solutions

Some of the methods just discussed, especially the Lagrange-Charpit solutions of § 3.3.4 and Monge's treatment of (335.5) and other equations, raised issues concerning the relationship between general and singular solutions. This topic was studied in detail in the late 18th century, and some results were taught at the *Ecole Polytechnique.* Here I shall present a small selection of the more substantial studies and interesting suggestions.

For first-degree ordinary differential equations, the generality of a solution could be appraised by ensuring that the number of constants of integration equalled the order of the equation. Thus methods of solution which proceeded by reducing an equation to equations of lower orders were especially welcome, since not only were the lower-order equations usually easier to integrate but also generality could be secured by recruiting the required constants as the tier of orders was ascended.[1]) Otherwise one would have to obtain these constants "in one go", as it were. Lagrange produced such a method in his *1766a* for linear ordi-

[1]) I suspect that this feature was part of the considerable interest in the 18th century in solving equations of the Riccati type; for Riccati's analysis *1722a* was an early example of reducing a second-order equation to the first order by replacing x by dy/dx as the independent variable. On the history of general and singular solutions, see Rothenberg *1910a*; Demidoff *1973a, 1975a* and *1976a*; and Engelsman *1980a*. Houtain *1854a* contains useful information.

nary differential equations via the construction of an adjoint equation. Interestingly, in generalising this method Laplace reverted to a method of reduction of order when in his early paper *1771a* he first used the cascades method (335.1) to reduce a linear ordinary differential equation of order n to two equations of order $(n-1)$, although the new equations were not necessarily linear.

So the study of general solutions to ordinary differential equations led to difficulties even for basic types of equation. Singular solutions provided even harder a challenge. Indeed, Euler entitled a paper *1759a* on the subject 'Exposition of some paradoxes of the integral calculus', for he gave examples of equations such as

$$(y - xp)(y + (2a - x)p) = c^2 \text{ (constant)}, \quad \text{where} \quad p := dy/dx. \tag{336.1}$$

This equation could not be integrated by known methods, but differentiation with respect to x quickly yielded

$$x = a - ap/\sqrt{(c^2 + a^2p^2)}, \quad y = c/\sqrt{(c^2 + a^2p^2)} \tag{336.2}$$

(where p now served as the parameter), which was a solution of (336.1) but not one contained in its complete solution. In this paper he gave a rule for finding such solutions.

With regard to partial differential equations, Euler's preference for functional solutions had a correlate in his view of the generality of a solution; for he thought that the equations should be reducible to the exact form (when a functional solution would immediately follow, including the functions of integration necessary for a complete solution). Indeed, he declared that if it were not possible to find an integrating factor which converted the equation to an exact form, then 'the proposed differential equation would be absurd and nothing at all is shown' (*Integral, 3* (1770), opening). Monge claimed that the absurdity lay not in the equation but in the assumption 'that their integrals may be expressed by one single equation' (*1787b*, 535: the context is ordinary differential equations, but his remarks seem clearly to apply to partial differential equations). We shall see this problem again in § 11.2.2 when it became 'Pfaff's problem', and an early interest of Cauchy.

For first-order total differential equations in x, y and $z(x,y)$,

$$Pdx + Qdy + Kdz = 0, \tag{336.3}$$

Euler favoured as a necessary condition for integrability the relation

$$P(Q_z - R_y) + Q(R_x - P_y) + R(P_y - Q_x) = 0. \tag{336.4}$$

He liked to convert (336.3) into the form

$$d(z + f) = gdh, \tag{336.5}$$

where f, g and h were functions of x, y, p and q (that is, z_x and z_y), and to proceed from there. Now we noticed at (334.1) he also offered as an alternative condition

$$pq_z - qp_z + q_x - p_y = 0, \tag{336.6}$$

which Lagrange used in *1774b* for the purpose of linearising a differential equation. Lagrange also showed there, by a method of varying constants, that the 'complete solution' to the differential equation would be given by eliminating the constant of integration between the equations

$$N(x, y, z, a) = \varphi(a) \quad \text{and} \quad N_a = \varphi'(a), \tag{336.7}$$

where N was a family of solutions generated by a and φ was arbitrary. The reasoning rested on the condition (336.6) rather than the difference equation itself, and led Lagrange to observe that the search for the general solution solely from the form

$$N(x, y, z, a) = 0 \tag{336.8}$$

could be blocked by the vanishing of a in the course of integration. Thus 'as there are two quantities to eliminate, it would be necessary to have three equations; therefore it would need two more of them beyond the proposed one, and these two can only come from two different differentiations, the one in varying x and the other in varying y' (*1774b*, art. 11).

Lagrange now came to a new conception of the 'complete solution'. In his next paper, *1776a*, he used his new view to create the first systematic theory of singular solutions of differential equations. If an ordinary differential equation took the complete solution

$$V(x, y, a) = 0, \tag{336.9}$$

then conversely it could be recovered by eliminating a from the solution V and its partial derivative V_a. If now a was regarded as a variable, then (336.9) led to

$$V_x dx + V_y dy + V_a da = 0, \tag{336.10}$$

with the additional term $V_a da$. The previous analysis, with a constant, would apply if, from (336.10),

$$V_a da = 0, \tag{336.11}$$

which in turn led to two possibilities. One, $da = 0$, yielded the old case of a being constant. The other, $V_a = 0$, led to the two equations

$$V = 0 \quad \text{and} \quad V_a = 0; \tag{336.12}$$

but since a was now variable, the elimination of a would yield not the complete solution but a 'particular integral' (*1776a*, esp. sect. 4). (Neither here nor in later writings did he seem to consider the possibility that such solutions can arise from infinite values of V_x and/or V_y.) 'Particular integrals' could also be found for higher-order ordinary differential equations (sects. 27–38). Euler's paradox was solved: (336.2) arose from equations such as (336.12).

First order partial differential equations were treated similarly (sects. 39–70). If

$$V(x, y, z, a, b) = 0 \tag{336.13}$$

were a complete solution, then the equation was obtainable by differentiating it partially with respect to x and y and eliminating a and b from

$$V(x, y, z, a, b) = 0, \quad z_x = P(x, y, z, a, b) \quad \text{and} \quad z_y = Q(x, y, z, a, b). \tag{336.14}$$

a and b were now allowed to be variables, so that differentiating (336.13) led to

$$dz = z_x dx + z_y dy + z_a da + z_b db, \tag{336.15}$$

which could be solved by a version of the previous method if

$$z_a da + z_b db = 0. \tag{336.16}$$

Then the differential equation could be recovered by eliminating a and b from (336.14) and a solution

$$a = A(x, y, z) \quad \text{and} \quad b = B(x, y, z) \tag{336.17}$$

of (336.16).

How was the elimination to be executed? If b were taken as a function φ of a, then for Lagrange the elimination led to the 'general solution' of the partial differential equation, with $\varphi(a)$ as the function of integration. If a and b were not functionally related, then the elimination took place in effect between the equations

$$V = 0, \quad V_a = 0 \quad \text{and} \quad V_b = 0, \tag{336.18}$$

and, as in (336.12), the solution would be singular.

3.3.7 Terminology and geometrical interpretations

We must now consider the technical terms. In *1774b* Lagrange followed Euler in calling 'complete' the solution (336.7), with one function of integration. But with his paper *1776a* Lagrange called this form of solution 'general': 'complete' solutions were now given by (336.13), obtained from the new procedure and containing two constants of integration (see sects. 48–56). Elimination of a and b from a complete solution yielded a 'particular integral'; this was to be distinguished from a 'particular solution', which was a special case of the complete solution. Lagrange did not give a name to the solution produced by eliminating the one constant from the general solution, for he did not seem to consider the process.

This terminology is very unhelpful. Both 'general' and 'complete' have a connotation of comprehensiveness, and neither term indicates the relationship to the other. Similarly, the two kinds of 'particular' solution are distinguished only by being called 'integral' and 'solution', nouns which fail to express the differences between them. The situation was rendered still more confusing by the fact that Laplace used 'particular solution' and 'particular integral' in the converse senses from Lagrange (see Lacroix *1798a*; or *Calculus*$_2$, *2* (1814), 373). Fortunately, Lagrange came later to give the name 'singular primitive', or 'singular equation', to singular solutions (see, for example, *Functions*$_1$ (1797), 69). I summarise his and others' terms of the late 18th century in Table 337.1 (on these, see Houtain *1854a*, 973–978).

Table 337.1 Some terms of the late 18th century for solutions to partial differential equations.

	Euler	Lagrange	Laplace, Lacroix	Legendre, Trembley
$V(x, y, z, a, \varphi(a))$	Complete integral/solution	Complete *then* General integral/solution	Complete solution	Complete integral
Elimination of a leads to	Particular integral	Particular integral	Particular solution	Particular integral
$V(x, y, z, a, b)$		Complete primitive/ solution/integral	Complete integral	Complete integral
Elimination of a and b leads to	—	Particular integral *then* Singular primitive	Particular solution	Particular solution
Special case is	—	Particular solution *then* Particular integral	Particular integral	Incomplete integral

In his *1776a* Lagrange pointed out the geometrical interpretation of these solutions. The general solution was the envelope of the surface given by the relationship between a and b (sect. 49); the singular solution was the envelope of all the solutions to the complete solution (sects. 21–22, 43). However, in tune with his algebraic style, he did not discuss these interpretations in detail. Predictably, the man for that job was Monge, especially in his teaching at the *Ecole Polytechnique* of 'analysis applied to geometry' (*Analysis*$_2$ (1801), fols. 6–8). He did not often indulge in the terminology of Table 337.1, but preferred to talk in terms of 'envelopes' and 'characteristics'.

Monge illustrated his position in terms of the example

$$z^2(z_x^2 + z_y^2 + 1) = a^2 . \qquad (337.1)$$

The complete solution was the family of spheres

$$U(x, y, z, \alpha, \beta) := (x - \alpha)^2 + (y - \beta)^2 + z^2 = a^2 \qquad (337.2)$$

of radius a and centre $(\alpha, \beta, 0)$. Eliminating α and β from

$$U = 0, \quad U_\alpha = 0 \quad \text{and} \quad U_\beta = 0 \qquad (337.3)$$

yielded as singular solution $z = \pm a$; two planes parallel to the xy-plane and distant a above and below it. If $\beta = \varphi(\alpha)$, then the general solution of (337.1) was

$$V(x, y, z, \alpha, \varphi(\alpha)) := (x - \alpha)^2 + (y - \varphi(\alpha))^2 + z^2 = a^2 , \qquad (337.4)$$

where the centres of the spheres lay on the curve $y = \varphi(x)$ in the xy-plane. Eliminating α from

$$V = 0 \quad \text{and} \quad V_\alpha = 0 \qquad (337.5)$$

yielded a solution which enveloped the spheres given by (337.4). Further, the two equations of (337.5) represented two surfaces, which for each value of a intersected in a curve. 'The line of contact, of which the equations are [(337.5)], also have a property independent of the curve which characterise the envelope', Monge pointed out in his lectures, 'and we shall call it the *characteristic* of the envelope' (*Analysis*$_2$ (1801), fol. 7, pp. 2–3). In later lectures he showed how to construct different characteristics of the same envelope and their bearing on the general theory of surfaces.

Monge and the *Ecole Polytechnique* make a suitable place to conclude this selection of late-18th-century views on the theory of differential equations. We shall see in § 4.4 that some of his colleagues and students at the school were to take these studies further.

3.4 On series, functions and integrals

The interest in power-series solutions to differential equations was closely allied in the 18th century with the study of series, and their summation. The vast amount of work done during the century on series was also linked with other branches of mathematics (for example, with number theory); I shall confine myself here to some principal results from late in the century, and to the understanding of the time of non-orthodox summation methods and convergence.[1])

3.4.1 Some contributions by Lagrange

First we take a result from Euler's *1770a*, in which he proved that

$$df(x) = \sum_{r=1}^{\infty} (-1)^{r+1} (\Delta^r f(x))/r \qquad (341.1)$$

by (in effect) differentiating

$$f(x+y) = f(x) + \sum_{r=1}^{\infty} yC_r \Delta^r f(x), \quad \text{where} \quad \Delta f(x) := f(x+1) - f(x), \qquad (341.2)$$

with respect to y and then putting $y = 0$. (Strangely, he forgot to exclude periodic functions for which $\Delta f(x) = 0x$.) (341.1) could be cast into an operator form by abstracting Δ^r from f, a procedure which would be rather foreign to Euler's approach; but it was entirely appropriate for Lagrange, who produced a companion result in his *1774a*. Taking the Taylor series

$$f(x+h) = \sum_{r=0}^{\infty} h^r f^{(r)}(x)/r!, \qquad (341.3)$$

he wrote the derived functions in operator form to obtain

$$\Delta f(x) = (d^r f(x)/dx^r)/r! = \exp(h\, df(x)/dx) - 1 \qquad (341.4)$$

(art. 11) where $(f'(x))^r$ and $f^{(r)}(x)$ were assumed to be interchanged. He also gave

[1]) Again, the books published around 1800 still offer among the best guides to 18th-century developments; Lacroix *Calculus*$_1$, *3* (1800) is particularly useful. Among historical studies see Pringsheim *1900a*; Netto *1908a*, 257–302; Wallner *1908a*, 1047–1066; Burkhardt *1910a*; Faber *1935a*; Hofmann *1959a*; Kline *1972a*, ch. 20; and Goldstine *1977a*, chs. 3–4. M. Panza (Paris) is studying these developments.

these extensions, inductively produced:

$$\Delta^n f(x) = (\exp(h\,df(x)/dx) - 1)^n, \tag{341.5}$$

$$\sum{}^n f(x) = (\exp(h\,df(x)/dx) - 1)^{-n} \tag{341.6}$$

(art. 10), and multi-variate analogues (art. 9). He also obtained the multi-variate companion to Euler's (341.1) by re-arranging and taking logarithms (art. 12).

Laplace was an early student of these formulae, especially in his *1780a*, where he provided various general interpolation formulae (§ 3.5.2). He also studied Taylor's and other power series extensively, in connection with his theory of 'generating functions' (352.1). Instead of Lagrange's inductions and analogies, he usually sought for methods out of the calculus itself. For example, in *1776b*, art. 12 he reproved (341.5–6) by defining the increment function $s(\alpha, x)$ of a function $u(x)$ as

$$s(x, \alpha) := v(x, \alpha) - u(x), \quad \text{where} \quad v(x, \alpha) := u(x + \alpha), \tag{341.7}$$

carrying out partial differentiations to yield

$$v_x = s_\alpha; \quad \therefore \quad v = u + \int v_x d\alpha, \tag{341.8}$$

and then successively α-differentiating and -integrating to produce Taylor's series, and also the desired equations from substituting the special case e^x of $u(x)$.

Lagrange's result strengthened the connection between difference and differential equations from a close analogy of structure to a formal relationship. Difference and difference-differential equations—often then called 'mixed' equations, especially by the French (§ 4.4.2)—could be converted to differential form if desired, for example. These sorts of equation were studied occasionally in the 18th century.

Another result by Lagrange which bore both on the calculus and on series now bears his name: the 'Lagrange series', which he first proved in *1770a*, art. 15–16, as a special case of obtaining power-series expansions by generalising upon Newton's relations between the zeros and coefficients of a polynomial. Once again Laplace rederived the series (in fact, in a more general form) by obtaining it as the function $u(x)$, where x itself was a function satisfying the partial differential equation

$$x_\alpha(\alpha, t) = z(x)x_t(\alpha, t), \quad \text{with } z(x) \text{ a known function}, \tag{341.9}$$

similar to $(341.8)_1$ (Laplace *1780a*, art. 8).

Lagrange adopted this approach in his *Fonctions*, in the following form. Assuming the relationship

$$z = x + yf(z), \tag{341.10}$$

then eliminating f yielded the first-order partial differential equation

$$z_x(z-x) - yz_y = 0. \tag{341.11}$$

If the solution for z took the form

$$z = x + \sum_{r=1}^{\infty} a_r(x)\, y^r, \tag{341.12}$$

then direct inspection with (341.10) gave $a_1(x)$ and further passage work produced the following formulae for all the coefficients:

$$a_1(x) = f(x), \quad a_r(x) = (a_1^r)^{(r-1)}/r!, \qquad r \geq 2, \tag{341.13}$$

where 'r' denotes the r-th power and '$(r-1)$' the $(r-1)$-th derived function. Hence (341.12) became after substitution the Lagrange series

$$z = x + \sum_{r=1}^{\infty} y^r (f(x)^r)^{(r-1)}/r!. \tag{341.14}$$

More generally, he obtained by similar means

$$\varphi(z) = \varphi(x) + \sum_{r=1}^{\infty} y^r [\varphi'(x)(f(x)^r)]^{(r-1)}/r!, \tag{341.15}$$

'a very remarkable formula, and of great utility in analysis, above all in the recurrence of series' (*Functions*$_1$ (1797), 100–104).

3.4.2 Views on convergence of infinite series

Infinite series such as (341.14–15) and Taylor series naturally raise the question of convergence. Some modern appraisals of the cavalier style of 18th-century mathematicians in handling infinite series conveys the impression that these poor men set their brains aside when confronted by them. However, their views seem to be more sensible than that (although certainly inadequate for the problem at hand). They are fairly exemplified by Lagrange's remark in § 3.2.2 that we may develop the Taylor series to 'whatever term one would like, and judging the value of the remainder of the series'. Lagrange was interested in the *error term* left by truncation of the series, which presumably was to be 'judged' by comparing the value of $f(x+h)$ with the value of the truncated series. Convergence was taken for granted, so that no formula for the remainder term needed to be calculated.

Lagrange's view should be judged in the context of the fact that the Taylor series is a power series, where good behaviour is more likely to occur than with many other kinds of series. More care was taken at that time with series in general, although even then a full treatment of convergence was usually lacking. For example, in a paper on the 'Euler constant' Euler handled the harmonic series

$$\sum_{r=0}^{\infty} c/(a+rb) \tag{342.1}$$

and obtained upper and lower bounds on the value of $(s_{nm} - s_n)$ (see Euler *1740a*, esp. art. 3). But it would be anachronistic to follow Eneström *1905a* in seeing this analysis as an attempted necessary and sufficient test for convergence using the expression $|s_{nm} - s_n|$, quite apart from the fact that its lack of sufficiency is demonstrated by the counter-example

$$\sum_{r=2}^{\infty} 1/(r \log r) \tag{342.2}$$

(see Pringsheim *1905a*). The same view applies to Euler's apparent anticipation of the 'integral test' for convergence in *1741a*. In fact he was comparing $\sum_{r=m}^{\infty} u_r$ with $\int_m^{\infty} u(x)dx$ for a positive function $u(x)$. The succeeding paper *1741b* shows his further concern with estimators rather than with conditions for convergence. Eneström again fell into post-hoccery in *1894a* when regarding a similar estimation process in MacLaurin *1742a*, arts. 350-354 as an anticipation of a general treatment of convergence (although MacLaurin showed a good understanding of convergence itself).

Euler and MacLaurin are remembered today together in connection with summation formulae. The main one is usually written in the form

$$\begin{aligned}\sum_{r=0}^{n} f(r) = \int_0^n f(x)dx + \sum_{s=1}^{\infty} [B_{2s}(f^{(2s-1)}(n) - f^{(2s-1)}(0))/(2s)!] \\ - \tfrac{1}{2}(f(n) - f(0)),\end{aligned} \tag{342.3}$$

where $\{B_s\}$ are the Bernoulli numbers, given by the expansion

$$x/(e^x - 1) = 1 - \frac{1}{2}x + \sum_{s=1}^{\infty} (-1)^{s+1} B_s x^{2s}/(2s)!.^{1)} \tag{342.4}$$

Neither man examined the convergence of the series: Poisson was to take it up in the mid 1820s (§ 11.6.5).

---

[1]) See Euler *1738b*, art. 2 for his first version of (342.2), and compare MacLaurin *1742a*, arts. 828-835. Mills *1985a* describes the origins of both men's versions.

This work provides a good opportunity to raise the general question of the 18th-century understanding of convergence. Usually mathematicians then knew as well as anyone that

$$u_1 + u_2 + u_3 + \ldots \tag{342.5}$$

meant the first term plus the second, plus the third, and so on, and they realised that the series would not always tend to a particular value. Their difficulties really started with the *interpretation* of their many clever transformations and manipulations of series. If, for example, they could show that

$$\sum_{r=1}^{\infty} (-1)^{r+1} = \frac{1}{2}, \tag{342.6}$$

then in some sense $\frac{1}{2}$ was "the" sum of the series. The sense in question is exemplified by the following statement by Euler, in the summary of his *1760a*:

> Notable enough, however, are the controversies over the series $1 - 1 + 1 - 1 + 1 -$ etc. whose sum was given by Leibniz as $\frac{1}{2}$, although others disagree. [...] Understanding of this question is to be sought in the word 'sum'; this idea, if thus conceived—namely, the sum of a series is said to be that quantity to which it is brought closer as more terms of the series are taken—has relevance only for convergent series, and we should in general give up this idea of sum for divergent series. On the other hand, as series in analysis arise from the expansion of fractions or irrational quantities or even of transcendentals, it will in turn be permissible in calculation to substitute in place of such a series that quantity out of whose development it is produced.

In other words, it was clearly understood that divergent series do not have a sum; but if a calculation produced a value, it had to be "the" sum of that series, not the sum relative to the method of calculation. The sentence that I omitted from the quotation contained the phrase 'No one has yet assigned another value to the sum' of the series in (342.6). This statement shows that Euler regarded the fact as relevant, but it is not; even if no other value has been assigned, $\frac{1}{2}$ is still not "the" sum.

To express the situation in more modern terminology, Euler and his contemporaries were obtaining (usually) correct results in summability and formal power series without having general theories within which the results could be sited. (These theories did not become established until the late 19th century: see Tucciarone *1973a*.) Thus, for example, they used regular and irregular summation methods (that is, methods which yield or do not yield the same sum for a convergent series as does orthodox term-by-term summation) without concern for the distinction—and the facts that many of the series that they handled *were* convergent, and the methods were regular, reduced the chances of detecting the difficulties.

Euler's paper from which I quoted shows well the scope and limits of 18th-century understanding (see Barbeau and Leah *1976a*). It contained a careful distinction of "divergent" series into divergence to infinity (whether or not the terms tend to zero), finite oscillation, and infinite oscillation (*1760a*, art. 1). It also had the 'Euler transformation'

$$\sum_{r=1}^{\infty} u_r = \sum_{s=0}^{\infty} (-1)^s \Delta^s u_1 / 2^{s+1} \quad (\text{if } \Delta u_1 \neq 0) \tag{342.7}$$

(arts. 15-16; compare *1755b*, pt. 2, ch. 1, art. 9), which was used here chiefly to find "the" sum of Wallis's hypergeometric series:

$$1 - 1 + 2 - 6 + \ldots = \sum_{r=0}^{\infty} (-1)^r r! = 0.596\,347\ldots, \tag{342.8}$$

a value which, for the early decimal places, he also obtained by converting the power series involved into continued fractions and then putting $x = 1$ (see Petrova *1969a* and Barbeau *1979a*).

Some mathematicians were chary of becoming embroiled in infinite series and thus possibly with results such as (342.8). An interesting case is J. F. Pfaff's work on the hypergeometric function, which I write in the modern notation

$$F(a, b; c; x) = \sum_{r=0}^{\infty} ((a)_r (b)_r / (c)_r r!)\, x^r, \tag{342.9}$$

where $(a)_r$ is the shifted factorial

$$(a)_0 := 1, \quad (a)_r := \prod_{j=0}^{r-1} (a + j), \qquad r \geq 1. \tag{342.10}$$

Euler had studied the function in the course of developing series solutions to second-order linear ordinary differential equations (*Integral*, *2* (1769), sect. 1, chs. 8-11 *passim*). Pfaff proved certain results, but usually in cases where the function was expressed as a finite series (see *1797a* and *1797b*), so that he failed to find various further properties.

3.4.3 Two responses around 1800 to series

The perception of infinite series is well illustrated by the reception given by the *classe* to work submitted by two minor figures. In a paper of September 1797 (*PV*, *1*, 278) J. F. Callet, who was known for his mathematical tables, argued that the sum of $\sum_{r=1}^{\infty} (-1)^{r+1}$ was $\frac{2}{3}$ rather than $\frac{1}{2}$, since both the series and the sum $\frac{2}{3}$ could be obtained by putting $x = 1$ in

$$(1 + x)/(1 + x + x^2) = 1 - x^2 + x^3 - x^5 + x^6 - \ldots. \tag{343.1}$$

In his report to the *classe*, Lagrange *1797a* argued not against the legitimacy of (343.1) at $x = 1$ but in terms of reading its series as

$$1 + 0x - x^2 + x^3 + 0x^4 - \ldots, \tag{343.2}$$

with successive partial sums 1, 1, 0, ... at $x = 1$; $\frac{2}{3}$ was then 'the mean sum' of (343.1). He generalised the result to a quotient of polynomials whose expansion would yield as 'sum' the quotient of their degrees; but he also admitted that 'the principle upon which it depends is too precarious to be able to be of use in establishing it solidly'. Perhaps he had in mind Daniel Bernoulli's comments in *1772a* on the sum being $\frac{1}{3}$, $\frac{1}{2}$ or $\frac{2}{3}$ according to the presence and position of zeros.

In 1798 and 1799 H. Burmann, one of the German mathematicians of the time who worked on combinatorics, submitted two papers on series. In the one which survives, 'Formulae of development, recurrence and integration' (*1797a*), he sought a formula for the expansion of a function $f(x)$ in a power series of another function $g(x)$. By successively differentiating the expansion to find the coefficients he obtained the series

$$f(x) = \sum_{r=0}^{\infty} \left[\left(\frac{d^r f(x)}{dg(x)^r}\right)\bigg|_{g(x)=0}\right] \frac{(g(x))^r}{r!}, \tag{343.3}$$

and derived various special cases and related formulae. Remarkably, he wrote '$\sum^s$' for s-fold summation, s a positive integer.

In his report to the *classe*, Legendre *1799c* confined himself to this paper, which he found to be the more interesting of the two. He stated the main results, and acknowledged the novelty of handling expansions in $g(x)$ rather than x alone; but he noted that the work was not essentially different from Lagrange's determination of the Lagrange series (341.15) in *1770a* or Laplace's handling of the series in *1780a*. The emphasis of his report was laid on the results themselves; the rigour of the method seems not to have concerned him.

Burmann's result aroused some interest. For example, Legendre's co-reporter Lagrange made some notes on it (see *BI*, ms. 906, fols. 77–80); and the German mathematician J. Soldner restated it, apparently in ignorance of Legendre's published report, in the course of studying generalised formulae for integration by parts (*1809a*, 3, 17–20). Further, J. Herschel attributed to Burmann the notation 'f^{-n}' for the n-th inverse function of f, and the notation

$$\frac{a_1}{b_1+}\frac{a_2}{b_2+}\cdots \tag{343.4}$$

for continued fractions.[1])

These papers of Burmann and Callet exemplify another aspect of French science around 1800. Both reporters recommended the papers for publication in the *Savants étrangers*, and the *classe* decided that the reports be published also. The reports appeared in the 'Histoire' department of volumes of the *Mémoires*; but neither of the original papers reached the print-shop. I hope that their authors were content with the praise.

3.4.4 Some functions and integrals

Functions and integrals were another essential component of techniques for the calculus. I have noted several examples earlier in this section; they occur in fact in all aspects of 18th-century mathematics, including branches such as number theory which are not of concern here. In this sub-section I shall confine myself to some brief further indications of their use.[2])

An appropriate start would be the beta and gamma functions, for they are functions defined as integrals; indeed, they were often then known as 'Eulerian integrals'. They aroused his early attention, in connection with defining a function of a continuous variable x which would take the value $n!$ when $x = n$. In *1738a* he studied both functions, beginning in fact with the beta function

$$\beta(m, n) = \int_0^1 x^m(1-x)^n dx \tag{344.1}$$

---

1) J. Herschel *1820a*, 5, 148; he claimed for himself special cases such as 'f^{-1}' and '$\tan^{-1}$'. He also criticized Burmann's inductive style. He gave no references to Burmann, and I have not been able to find them. Among later workers, Burmann's result attracted Sylvester *1854a*.

2) Lacroix *Calculus*$_1$, *2* (1799) is the most substantial source of the time on these topics, although even his treatment is relatively unsystematic—an indicator of their state of development. Among the secondary literature, there are useful surveys in Wangerin *1904a*, Vivanti *1908a*, and Kline *1972a*, ch. 19.

and eventually producing the desired result in terms of the gamma function (as we now call it, after Legendre):

$$n! = \int_0^\infty (-\ln x)^n \, dx \, (=: \Gamma(n+1)) . \tag{344.2}$$

In various later studies he found the relationship between the gamma and beta functions

$$\beta(m, n) = \Gamma(m)\Gamma(n)/\Gamma(m+n) \tag{344.3}$$

and other properties of both functions (see Davis *1959a*). The now standard proof of (344.3), using a transformation of the double integral, is due to Jacobi *1834a.*

Among functions defined by integrals, elliptic functions were the most widely studied class. The motivations included not only geometrical questions but also, for example, the analysis of the motion of a particle under inverse-square attraction from two sources, the motion of a pendulum, and the attraction of spheroids. In calling them 'elliptic functions' I have been deliberately careless with terminology, for while the literature of the period usually refers to 'elliptic functions', they are elliptic *integrals* to our understanding. The great breakthrough of Abel and Jacobi to invert the relationship between independent and dependent variable (§ 17.2.2-3) is conspicuous by its absence from the 18th century, perhaps nowhere more clearly than in Euler's and Lagrange's incomplete use of analogy from circular functions (Euler *1768a* and Lagrange *1771b*). They both knew that the differential equation

$$dx/\sqrt{1-x^2} = dy/\sqrt{1-y^2} \tag{344.4}$$

took as solution

$$\sin^{-1} x = \pm \sin^{-1} y + \sin^{-1} K, \quad K \text{ constant}, \tag{344.5}$$

which could be converted by trigonometric relations to the purely algebraic form

$$x\sqrt{1-y^2} + y\sqrt{1-x^2} = K; \tag{344.6}$$

and they also saw that (344.6) could be obtained directly by partially integrating both sides of (344.4). Both these forms of solution could be expressed in forms equivalent to the addition theorem for the sine function. Hence, if one were to tackle the integration of

$$dx/\sqrt{F(x)} = dy/\sqrt{F(y)} , \tag{344.7}$$

where F was a quartic, and show that it had a solution of the form

$$a + b(x+y) + c(x^2+y^2) + exy + fxy(x+y) + gx^2y^2 = 0\,, \tag{344.8}$$

in rough analogy with (344.6), then perhaps one may suspect that there were functions involved in the solution of (344.7) which could have properties of addition like those of the sine function. Yet though Euler, Lagrange and others found many properties of elliptic and related integrals, this opportunity was missed, until a Norwegian and a German began to show the quality of mathematics outside France in the late 1820s.

At the end of the century the most prominent enthusiast for elliptic integrals was Legendre, who was to live long enough to see his work eclipsed. His main contribution before 1800 was a substantial systematic treatment of the topic, including a classification of types of integral and reduction theorems and integrals to these types. He presented his work to the *Académie* in April 1792, but its publication was held up by the closure in 1793. Firmin Didot put it out as a short book *1793b*, which probably made little impact at the time: it is now very hard to find.

Legendre's name is perhaps best remembered today for the 'Legendre functions': an account of other developments is postponed until § 5.5.2, where I survey 18th-century work on the attraction of spheroids, which were then their chief motivation. I end this survey with a brief report on the state of knowledge of the other principal transcendental function of the time: the so-called 'Bessel function'.

Bessel's name belongs to our period of study, the early 19th century. We shall note his contributions to the theory while studying Fourier's earlier systematic treatment of $J_0(x)$ in § 9.2.8. The implications of these later developments for the 18th century are correct: there was then not so much a theory as a collection of special properties obtained in different contexts, with connections not necessarily noticed. A substantial study is contained in Euler's study *1766d* of the vibrating circular membrane of radius a (see Bôcher *1893a*). The partial differential equation in polar coordinates took the form

$$z_{tt} = z_{rr} + z_r/r + z_{\varphi\varphi}/r^2\,, \tag{344.9}$$

and when solved by separation of variables (not then widely used a technique for partial equations) led to 'Bessel's equation' for r:

$$u'' + u'/r + (\alpha^2 - (\beta^2/r^2))\,u = 0, \qquad \alpha^2 \text{ and } \beta^2 \text{ constants.} \tag{344.10}$$

Euler naturally assumed a power-series solution for this little-known equation, and obtained an expansion which I write as

$$(J_\beta(\alpha r) :=)\ u(r) = \sum_{k=0}^{\infty} \frac{(-1)^k}{k!\,(\beta+1)_k} \left(\frac{\alpha r}{2}\right)^{\beta+2k} \tag{344.11}$$

(art. 17). He noted that (344.10) reduced to a Riccati equation (art. 16), but did not relate (344.11) to the hypergeometric function (342.9). The condition that the rim be always stationary led him to

$$J_\beta(\alpha a) = 0: \tag{344.12}$$

he had to assume that (344.12) had an infinity of simple real roots $\{p_s\}$ in α (art. 17). He found (essentially) $J_{-\beta}(r)$ as the second solution of (344.10) (arts. 18–19: in this problem the second-kind solution $Y_\beta(\alpha r)$ was not applicable). He then wrote down solutions of (344.9) for different integral values of α in terms of superposition of normal modes, as given by the roots $\{p_s\}$ of (344.12):

$$A \sin(Kt + B) \sin(\beta\varphi + C_\beta)\, J_\beta(p_s r), \quad \text{with } A, B, \{C_\beta\}, K \text{ constants} \tag{344.13}$$

(arts. 20–21).

This work shows Euler in bold conjecture, not always supported by mathematical arguments. The same style appears also in a later paper on the swinging cord, where he not only found the Bessel function again but also asserted that "any" function could be expanded in a series of such functions (*1784a*, art. 14); again he could not justify the claim (which is great contrast to his opposition to Fourier series expansions (§ 3.3.3), a matter noted in § 9.2.6). However, he argued for an infinity of simple real zeros for the function by consideration of infinite products (arts. 23a–31: compare Daniel Bernoulli's earlier and similar claim in *1738b*, art. 17, drawing on his recent study *1738a* of recurrence relations). In these ways do we see that the theory of Bessel functions was in a fragmentary state in the late 18th century—so that Lagrange and Laplace missed them in contexts of mathematical astronomy (§ 5.4.4). The situation was to change with Fourier's analysis of (344.12) for $\beta = 0$ (§ 9.2.8) and Poisson's treatment of (344.9) (§ 12.5.2).

3.5 On differences, interpolation and tables

The connection between the difference and the differential calculus had always been close in the Leibnizian tradition of the calculus, for the differential dx is an infinitesimal case of a forward difference Δx (§ 3.2.4). The link became more formal with Lagrange's formula (341.3) connecting Δx with the differential operator via the exponential function. Difference methods were incorporated in the general context of the calculus, especially by Euler; for example, in some of the methods of summing infinite series and evaluating integrals, topics noted in § 3.4. However, there were some problems peculiar to difference mathematics, and it became enough of a Subject to be written up as such around 1800, especially in a volume of Lacroix's *Traité* and in de Prony's lectures at the *Ecole Polytechnique*. Hence there is the need for this special short section.[1])

3.5.1 Lagrange on difference equations

Much of the theory and terminology of differential equations was imported into difference equations, especially by Lagrange and Laplace, who again were the principal figures of the late 18th century. One of Laplace's earliest papers, *1771a*, was largely concerned with difference equations, and it helped to motivate a major essay *1777a* by Lagrange, in which he tried to extend known results and to simplify existing methods.

Lagrange devoted the first 'article' of his paper to linear equations, beginning with constant coefficients (sect. 1). For ease of reading I shall use subscript notations, and write the equation in the form

$$\sum_{r=0}^{n} k_r y(x+r) = 0. \tag{351.1}$$

The form of solution

$$y(x) = a\alpha^x \tag{351.2}$$

yielded the indicial equation

$$\sum_{r=0}^{n} k_r \alpha^r = 0, \tag{351.3}$$

whose roots $\{\beta_r\}$ led to the solution

$$y(x) = \sum_{r=1}^{n} b_r \beta_r^x, \quad \{b_r\} \text{ constants}, \tag{351.4}$$

of (351.1). Lagrange regarded (351.4) as the 'complete integral' of (351.1), using his term from differential equations (§ 3.3.7), because of the presence of the n

[1]) Lacroix's *Calculus*$_1$, *3* (1800) and de Prony's lectures *1795a* and *1796a* are indeed still the best sources for 18th-century difference mathematics. Bauschinger *1901a* and Seliwanoff *1901a* are very disappointing studies, the relevant parts of Wallner *1908a* and Goldstine *1977a* only slightly less so.

constants $\{b_r\}$; these were calculated from the given values $y(0), y(1), \dots y(n-1)$ of $y(x)$. He also considered the effect on (351.4) of equal roots in (351.3) (sect. 2), in an analysis (using infinitesimals!) which yielded results that he had to modify, in *1798b*, following criticism in Malfatti *1786a*.

Lagrange then turned in sect. 3 to the linear equation with variable coefficients which I write in the form

$$\sum_{r=0}^{n} k_r(x)\, y(x+r) = X(x). \tag{351.5}$$

He began with the reduced equation of (351.5) (where $X(x)$ was replaced by 0) and appealed to its linearity to claim that if $\{p_r(x)\}$ were different 'particular values' of $y(x)$ satisfying it, then its complete integral took the form

$$y(x) = \sum_{r=1}^{n} c_r p_r(x). \tag{351.6}$$

Thus the complete integral of (351.5) itself might be

$$y(x) = \sum_{r=1}^{n} c_r(x)\, p_r(x), \tag{351.7}$$

where the $\{c_r(x)\}$ had to be determined. He began this task by writing, from (351.7),

$$y(x+1) = \sum_{r=1}^{n} c_r(x)\, p_r(x+1) + \sum_{r=1}^{n} \Delta c_r(x)\, p_r(x+1), \tag{351.8}$$

and then assuming, as an 'equation of condition', that the second term on the right hand side was zero. He repeated the procedure for $y(x+2)$ and on up to $y(x+n-1)$, thus producing $(n-1)$ such equations for the n unknowns $\{\Delta c_r(x)\}$

$$\sum_{r=1}^{n} \Delta c_r(x)\, p_r(x+s) = 0, \qquad 1 \le s \le (n-1), \tag{351.9}$$

and also the $(n-1)$ equations

$$y(x+s) = \sum_{r=1}^{n} c_r(x)\, p_r(x+s), \qquad 1 \le s \le (n-1). \tag{351.10}$$

His assumption (351.9) was probably made in imitation of his method of eliminating residuals in his study *1776b* of taking the mean of a set of observations (Interlude 934.1).

One more equation was needed to solve for the $\{\Delta c_r(x)\}$, and it came from taking the form of (351.8) for $(x + n)$:

$$y(x+n) = \sum_{r=1}^{n} c_r(x)\, p_r(x+n) + \sum_{r=1}^{n} \Delta c_r(x)\, p_r(x+n) \tag{351.11}$$

and substituting it and the equations (351.10) into the original difference equation (351.5). After allowing for the fact that each $p_r(x)$ was a solution of the reduced equation of (351.5), these substitutions reduced the desired extra equation to the form

$$b_n(x) \sum_{r=1}^{n} \Delta c_r(x)\, p_r(x+n) = X(x). \tag{351.12}$$

Equations (351.9, 12) could now be solved for the unknowns $\{c_r(x)\}$. Although Lagrange gave no details as to the method, he presented the solutions as

$$\Delta c_r(x) = Q_r(x), \qquad 1 \leq r \leq n, \tag{351.13}$$

n first-order difference equations which had as their own solutions

$$c_r(x) = K_r + \sum_s Q_s(x), \qquad 1 \leq r \leq n, \tag{351.14}$$

where the $\{K_r\}$ were constants of integration and the s-sums were indeterminate. Hence at last, (351.14, 7) gave the complete integral' of (351.5) (sect. 4):

$$y(x) = \sum_{r=1}^{n} \left[\left(K_r + \sum_s Q_s(x)\right) p_r(x)\right]. \tag{351.15}$$

de Prony praised this solution very highly. 'This is how we attain knowledge of the means of integrating any linear equation between two variables z and x', he told the students at the *Ecole Polytechnique*, 'in all the cases where the thing is in general possible, according to the current improvement of analysis' (*1796a*, 491). Perhaps our distance of time allows a little scepticism to mingle with the admiration. Lagrange offered no guidance for finding the particular integrals $\{p_r(x)\}$ of the reduced equation of (351.5) or the solutions (351.13) for the $\{\Delta c_r(x)\}$ (other parts of current knowledge of difference equations, not described here, might have helped). The equations of condition (351.9, 12) were assumed automatically to be consistent—a bold claim considering that, in the usual spirit of Lagrange's calculus (§ 3.2.2), he was working with "any" functions. Even the solution (351.14) for $c_r(x)$ may cause difficulties: $\sum_s Q_s(x)$ could let us in for anything. The rest of Lagrange's *1777a*—the bearing on differential equations, recurrent series, partial difference equations, and so on—continued in the same style. Difference equations of higher degrees were in still more serious difficulties, as Biot and Poisson were to note (§ 4.4.1) as the new century was born.

3.5.2 Laplace and Lagrange on interpolation

It was mentioned in § 2.5.2 that Laplace's intellectual career was an interlinked mixture of calculus techniques, theoretical astronomy and probability. For him interpolation was a combination of the first two concerns, for it related difference equations, the fitting of curves, and astronomical observations; while generating functions

$$u(x, t) := \sum_{x=-\infty}^{\infty} y_x t^x \tag{352.1}$$

related to all three interests. Several principal procedures were outlined in the paper *1780a*. As was noted in § 3.4.1, he proved Lagrange's basic relation (341.4)

$$\Delta u(t) := u(t+\alpha) - u(t) = \exp(\alpha \partial u/\partial t) - 1 \tag{352.2}$$

by a method which did not rely on Lagrange's operator algebra (art. 3). He then generalised (352.2) by defining another difference

$$^1\Delta u(t) := u(t+\alpha_1) - u(t) \tag{352.3}$$

and showing by straightforward means via (352.2) that

$$^1\Delta^n u(t) = ((1+\Delta u(t))^{\alpha_1/\alpha} - 1)^n, \qquad n > 1 \tag{352.4}$$

(art. 3). As he had already reproved Lagrange's companion result (341.6) for sums (art. 4), he also put forward the corresponding generalisation

$$^1\Sigma^n u(t) := \left(\sum_{r=0}^{\infty} u(t + r\alpha_1)\right)^n \tag{352.5}$$

$$= ((1+\Delta u(t))^{\alpha_1/\alpha} - 1)^{-n}, \qquad n > 0.^{1)} \tag{352.6}$$

The equations (352.4, 6) 'contain the theory of interpolation taken with all the generality of which it is susceptible', Laplace claimed with characteristic modesty at the end of art. 6; but within a few years he examined them again in the pa-

[1]) I follow Laplace's notations here, including his use of '∂', which is unusual for the time (§ 1.3.7). Goldstine's account of this paper is very hard to follow, especially when he mis-defines $^1\Delta\ u(t)$ (*1977a*, 201).

per *1782a*. Here he not only introduced the generating function (352.1) and re-proved them with its aid, but he also developed some still more general results, of which a typical example was

$$ {}^{1}\Delta^{n} h^{x} y_{x} = h^{x}(h^{i}(1 + \Delta y_{x})^{i} - 1)^{n}, \qquad n > 0, \tag{352.7} $$

$$ \text{where } \Delta y(x) := y(x+1) - y(x) \quad \text{and} \quad {}^{1}\Delta y(x) := y(x+i) - y(x). \tag{352.8} $$

Lagrange also wrote on interpolation at various times. In *1798d* he analysed a sequence of values $\{T_r\}$ and their differences in his typical algebraic style:

$$ D_0 := T_0, \qquad D_m := \sum_{r=0}^{m} (-1)^r \, mC_r \, T_{m-r}, \qquad m \geq 1. \tag{352.9} $$

If $m < 0$, then the differences became sums. His attachment to operator mathematics is seen in art. 3, for example, when he applied an algebra to the subscripts of the $\{D_r\}$ by claiming that

$$ D_0 + D_1 = D_{0+1} = D_1, \tag{352.10} $$

although (352.9) shows that ordinary multiplication yields

$$ D_0 \times D_1 = T_0 \, D_1. \tag{352.11} $$

He then showed that

$$ \begin{aligned} T_n &= (T_1)^n = (D_0(1 + D_1))^n = (D_0 \exp \ln (1 + D_1))^n \\ &= D^n{}_0 \exp (n \ln (1 + D_1)). \end{aligned} \tag{352.12} $$

Power-series expansions of the term in (352.12), and equations similar to (352.10), led him to various relationships between the $\{T_r\}$ and the $\{D_r\}$ (arts. 3–8). He obtained, as a corollary, the basic formula (352.2) relating differences to differentials by taking the latter to be infinitesimal differences (art. 9). He may have felt that the new proof was more satisfactory than the original symbolic one outlined in § 3.4.1; but since here he regarded $m := \Delta x / dx$ as 'infinitely large' and hence claimed that $m(m-1) = m^2$, the older proof retains much of its appeal.

This algebraic cast of theorems and proofs has to be borne in mind when considering late-18th-century interpolation; for the best known result of the period is atypical in its simplicity of statement and proof. This is 'Lagrange's interpolation formula', as we now call it, which he proved at the end of his lectures *1795a* given at the *Ecole Normale*:

$$ \text{if } u(x) = u_r \quad \text{when} \quad x = r, \qquad 1 \leq r \leq (n+1), \tag{352.13} $$

then $u(x)$ may take the polynomial form

$$u(x) = \sum_{r=1}^{n+1} \left\{ u_r \prod_{s \neq r} [(x - x_s)/(x_r - x_s)] \right\}. \tag{352.14}$$

An especially pleasing feature of (352.14) was that, for a given value of x, $u(x)$ could be calculated directly by logarithms—which leads us naturally to de Prony.

3.5.3 de Prony and the matching of mathematics with physics, 1796

de Prony's interest in differences was exhibited early in his career. He submitted a paper to the *Académie* in 1790 (his 35th year) on the use of interpolation to determine the law of thermal expansion of gases. It was examined by Legendre, who pointed out the limited extent of its originality but gave it the usual kiss of death of a promise of publication in the *Savants étrangers*. So it vanished from sight, as usual; but de Prony published two versions of the method in his treatise on engineering mechanics in 1796 (§ 8.1.1).[1] He also put the second version virtually unaltered in the *Journal de l'Ecole Polytechnique* as the paper *1796b*, in which he outlined for pedagogic purposes an 'experimental and mathematical essay on the laws of expandability of elastic fluids and on those of the expansive force of water vapour and alcohol vapour, at different temperatures'.

de Prony's aim was to bring physics and mathematics together, which meant, among other things, developing those parts of mathematics which most closely related to physical situations. Now in experiments such as the one described in the title of the paper, the physics yielded collections of data, recorded at distinct occasions of time; thus work on the physical laws governing the phenomena required difference mathematics, not the continuous mathematics of the differential calculus. For de Prony this work contributed to the 'explication of effects', to be distinguished from the 'measure of effects'. 'Experiment can only furnish the first prime data on the measure of physical effects', he told the students: calculation was needed thereafter, and 'the method that one employs in this case, is known under the name of *interpolation*' (*1796b*, 25).

de Prony recalled his efforts of 1790, including his own experiments (p. 26); but in this paper he relied for physical theory and experimental results on work carried out in the late 1780s by Guyton de Morveau and Prieur. He came to 'the

[1]) See de Prony *Hydraulics*, *2* (1796), 15-26, 152-196 (respective versions). (Part of ?) his 1790 manuscript, and a transcript of Legendre's report, are kept at *ENPC*, ms. 1730.

veritable form of the function' representing the volume z of a given gas in terms of its temperature x by noting the exponential form of laws in this area of physics. Generalising on this form, 'I was led to an equation of the form'

$$z(x) = \sum_{r=0}^{n} \mu_r \varrho_r^x, \tag{353.1}$$

where the $\{\mu_r\}$ and $\{\varrho_r\}$ were to be determined from the data (p. 27). He recognised (353.1) as the solution to an nth-order linear difference equation

$$\sum_{r=0}^{n} A_r z(x+r) = 0 \tag{353.2}$$

and devoted the mathematical part of his paper by recalling from his analysis lectures (*1796a*, 261–273) Lagrange's method (§ 3.5.1) of solving (353.2) for the $\{A_r\}$ from given values of $z(x)$, forming the indicial equation of (353.2) and finding its roots $\{\varrho_r\}$, and then determining the $\{\mu_r\}$ from Lagrange's interpolation formula (352.13). Among other details, he also pointed out the need to add μ_{n+1} to the right hand side of (353.1) if an even number of data were used, and the adjustments to make if the base points $\{x_r\}$ of the data were not equidistant (*1796b*, 28–35). The first version, on which Legendre had reported, used orthodox elimination methods on the 'simple case' of (353.1).

The rest of the paper was devoted to the application of the results to data for various gases determined by Guyton, Prieur and others. Numbers are everywhere: in equations, text and especially tables, and usually with decimal places, sometimes more than a dozen of them. One admires the industry of the man; the sums must have taken him hours to compute. One respects also his desire to make the mathematics fit the physics. Yet questions also came to mind. Would the 'veritable form' (353.1) make sense as *physical* theory (§ 1.3.1)? May the mathematical calculations not involve errors greater than the experimental ones? Is the linear system to solve for the $\{A_r\}$ stable, for example, and can the indicial equation for the $\{\varrho_r\}$ be solved with sufficient accuracy? de Prony often took six or seven data-points, and so was handling (the equivalent of) third-order minors and solving cubics. It is a wry irony that in 1802 the physical experiments of Guyton and Prieur were criticised, and thus the data from which he had worked was called in question. Further, the critic was the young Gay-Lussac (in his *1802a*, in which he proposed his law (726.6) on the thermal expansion of gases), at the dawn of his career: he may well have heard de Prony lecture on this topic during his days at the *Ecole Polytechnique* (one cannot be certain, as the lectures were optional), and in 1802 he was studying at de Prony's *Ecole des Ponts et Chaussées* ...

3.5.4 de Prony and the great tables

> These tables will not serve in the usual cases, but only in extraordinary cases.
>
> Delambre to Blagden (Lefort *1858a*, [146] or *1858b*, 999)

de Prony's tabulation of results and calculation of numbers to several decimal places is highly characteristic of his mathematical interests, especially during the 1790s. An important stimulus came in 1791 when, as was recorded in § 2.5.3, he got himself appointed *Directeur* of the cadastral survey in connection with a revision of land taxes. In the following year he established a *Bureau du Cadastre* to prepare a vast new set of logarithmic and trigonometric tables to aid in the calculations.

The project seems to have made very slow progress for several years; but when it was eventually completed in 1801, de Prony prepared a 'Notice' *1801b*, which was published in a pamphlet and again in part in the *Mémoires* for 1804, in which he described how the project had been designed. The first *section* of the *Bureau* comprised five or six mathematicians, led by himself and Legendre and including also Carnot and Prieur, to prepare the mathematical formulae and the initial data. The second *section* contained several calculators, led by Garnier, who deduced the values and the differences of various orders from the equations provided by the first *section*, and prepared a page of tables for calculation. This work was done by the third *section*, made up of between 150 and 200 assistants (de Prony *1832a*, 67–68), who filled out the tables by making the required additions and subtractions. (According to Dupin *1824b*, 23, many of the calculators had been hairdressers, thrown out of work by the reduction of hair-styles 'as the geometers say, *to its most simple expression*'!) This *section* passed its work back to the second *section*, who checked the results by means of formulae determined by the first *section*. The work was doubled in all *sections* by the use of pairs of different equations for each table. Thus two copies were produced, each of which could be used as a check on the other. When the project was completed, some of the calculators were transferred to the *Bureau des Longitudes* to work on astronomical tables (Bigourdan *Bureau*, pt. 1 (1928), A25–A28), while much of the equipment went to the *Dépôt Général de la Guerre* (Berthaut *1902a, 1*, 244).

de Prony was inspired to devise his procedure by reading Adam Smith's ideas on the division of labour in his *An inquiry into the nature and causes of the wealth of nations* (see de Prony *1820c*, 7). According to Maria Edgeworth, he realised that 'I will make my calculations as one makes pins' (Edgeworth *1979a*, 151). Lalande reported that *Bureau* had been producing 600 results a day in 1794 (*1803a*, 743-744).

One of de Prony's principal supporters was Delambre. He described the project in one of his notices of the activities of the *classe* (*1801b*, 195-198) and also in a separate report *1801a* (which appeared with de Prony *1801b*). He was concerned himself with interpolation and the construction of tables, initially for their use in astronomy and then in connection with the determination of the meridian (§ 5.5.4: see his *1793a*). During the 1790s he completed a set of logarithmic and trigonometric tables that had been partly compiled by Borda with the help of Callet (see Mascart *1919a*, 580-593). Delambre's task was much smaller in scale than de Prony's (as his remark at the head of this sub-section hints): but he had no bureau to complete it, so that both projects were finished in 1801.

The major dissimilarity between the two projects concerned the number of decimal places taken. The Borda/Delambre set was prepared basically to seven, with indications of how to find the next three: thus the *Imprimerie de la République* was presented with a manageable printing job of a few hundred pages. (Nevertheless, according to their stock-list held at *AN*, F^{18} 31, they had over 1,100 copies still in stock a decade later.) By contrast, de Prony, desirous of preparing 'the most immense and the most important monument to calculation that would ever have been executed or even conceived' (*1801b*, 1)—whether or not the cadastral survey needed it—calculated values to between 14 and 29 decimal places. Each copy of his tables comprised 18 large folio volumes, together with another volume describing the mathematical procedures; and the firm of Firmin Didot were faced with printing 1,200 folio pages of tables, together with the introduction.

Negotiations and plans to print the tables went on for years. 80,000 *francs* were needed; apparently about 500 pages were on proof by 1802. But then serious financial problems in France caused the printing to be stopped (I have found only four of the pages). The printing was announced by Firmin Didot in catalogues of the 1810s, and the project was revived in 1819 with the encouragement of the British scientist Sir Charles Blagden, who hoped to obtain support from his government. Estimates were obtained, but the plan failed, apparently because the British wanted to convert the tables to the sexagesimal system. de Prony *1820a* reported this development, and in a companion notice *1820c* asked that this 'unique monument' be kept as it was. Later he reviewed the project in a public lecture *1824a* to the *Académie*, and printing soon started again; but it stopped once more, and in 1833 Firmin Didot returned the manuscripts to the *Bureau des Longitudes*. Even at the time of his death in 1839, hopes still were set for publication (according to Dupin *1840a*, 336); but they were not fulfilled.

One set of the tables was placed in the Paris *Observatoire*, and de Prony kept the other. This set was passed to his niece, Madame Corancez, as part of his *Nachlass*, but it did not go on later to the *Ecole des Ponts et Chaussées* (§ 2.5.3). It was traced in 1858 by Biot's grandson-in-law F. Lefort to Madame Corancez's daughter, who gave it to the *Académie*, to the pleasure of Biot and Leverrier and the hopes of Elie de Beaumont that the tables could at last be printed.[1])

de Prony claimed some novelty in the design of the project. Instead of calculating the entries directly by some mathematical means and using the resultant differences as checks on accuracy, he devised a scheme in which only some basic entries would be produced directly, the rest being calculated principally by members of the third *section* of the *Bureau*, from differences of various orders. The fundamental equation for the work was

$$u(x+nh) - u(x) = \sum_{r=1}^{n} n\,C_r\,\Delta^r u(x), \quad \text{where} \qquad (354.1)$$
$$\Delta u(x) := u(x+h) - u(x),$$

which we saw Euler put to other uses as (341.2). The second *section* calculated an initial entry up to (say) fifteen decimal places, and differences up to about the sixth, which could successively furnish the later places, usually two from each difference: de Prony indicated some of the interpolations involved in *1796a*, 555-556. These figures were laid out in columns under the headings

$$u(x) \quad \Delta u(x) \quad \Delta^2 u(x) \quad \ldots, \qquad (354.2)$$

with the later values of x set down the left hand side. The third *section* then filled in the page by additions and substractions as instructed.

The initial entries for logarithms were calculated from

$$\log(y+x) - \log(y-x) = 2M \sum_{r=0}^{\infty} [(x/y)^{2r+1}/(2r+1)], \quad M := \log e, \qquad (354.3)$$

---

[1]) See the report in de Prony *1858a'*: this set is now at *BI*, mss. 1496-1514. Documents concerning the plans to print the tables in 1802, 1819 and 1825 are held at *ENPC*, mss. 1181, 1183; *AS*, *dossier personnel* for de Prony; *BN*, ms. n.a. fr. 15778, fol. 79; and *AN*, F^{17} 13571. See also de Prony's handing over of the tables to the *BL*, and related matters early in 1801 in *BL*, *2*, meetings for 4 and 9 *pluviôse*, 9 *ventôse* and 22 *floréal an* 10; the volume containing the mathematical procedures about to be described in March 1805, in *3*, meeting for 15 *germinal an* 13; the unavailability of 72,000 *francs* in *5*, meeting for 10. 10. 1821 (see also 21. 7. 1824); on restarting the printing, several meetings between June and October 1826; and the return of the manuscript on 16. 10. 1833.

de Prony's and Delambre's projects would be worth a detailed study, both for the history of numerical analysis, and for aspects of state support and international collaboration. Not a word on either project appears in Goldstine *1977a*, but the context of de Prony's is provided in Konvitz *1987a*, 47-62.

with differences calculated by means of, say,

$$\log(y+x) - 2\log y + \log(y-x) = -2M\sum_{r=1}^{\infty}[(x/y)^{2r}/r]. \qquad (354.4)$$

(Delambre also used such relations in his work: see *1793a*, 147–148.) Logarithms of prime numbers led to logarithms of compound numbers by addition.

For the trigonometric functions the new decimal system of angles (§ 5.5.4) was used: 100 degrees per quadrant, 100 minutes per degree, 100 seconds per minute. de Prony's sine table was calculated for 10°, 20°, ...100° by standard power-series, with intermediate values obtained from

$$\sin(x+y) = 2\cos x \sin y + \sin(x-y). \qquad (354.5)$$

For differences Legendre suggested the recursion

$$\Delta^n \sin x = -(2\sin\tfrac{1}{2}\Delta x)^2 - (\Delta^{n-1}\sin x + \Delta^{n-2}\sin x): \qquad (354.6)$$

check formulae included Euler's

$$\sin x + \sin(40° - x) + \sin(80° + x) = \sin(40° + x) + \sin(80° - x). \qquad (354.7)$$

Delambre used formulae for $\log(\tan x/x)$ and $\log(\sin x/x)$ to get tables for $\sin x/x$ and $\tan x/x$, the so-called 'Delambre tables', important for calculating lengths of small arcs. Other functions, and difference formulae, were obtained from functions already dealt with and tables already prepared.

de Prony indicated various errors in previous tables (especially in *1804d*); but his own procedures were attacked by the Scottish mathematician Sang in *1874a*. For example, Sang pointed out that the rounding or truncating errors introduced in determining the highest-order difference could cumulate into serious errors when carried through the lower differences to the final values. In the logarithmic tables he criticised de Prony for not having used the equality of the mantissae of (for example) log 1001 and log 10010 either as a check relation or as a means of reducing the number of interpolations and the need for high-order differences. Lefort wrote a reply *1875a*, defending de Prony's method (and clarifying some aspects of the history), but Sang *1875a* expressed some of his criticisms in more detail.

Finally, in 1891 the *Service Géographique de l'Armée* published an eight-place "extract", in 640 large folio pages, of de Prony's tables (de Prony *1891a'*). They checked the tables and the proofs, finding few mistakes in the former. So presumably de Prony and his calculators and ex-hairdressers had conducted themselves with commensurate zeal.

The affair of the tables characterises de Prony well—his interests, his industry, his limitations. Above all, it reveals the fervour of his hope that mathematical techniques such as tables would bring the worlds of mathematics and physics into proper rapport. But in this case his judgement seems for once to have left him. After all, who actually needs 29 decimal places? And had the tables been published in full, how would the user have been able to work with them efficiently? Some element of notionality (§ 132.3) is evident here; or maybe the project was conceived on this vast scale so that the *Directeur* of the *Bureau de Cadastre* would have to be in Paris for a long time ...

3.6 Concluding comments: the aspirations of Lacroix

Despite its length, this chapter has been able to treat only some of the technical results and foundational questions handled in the late 18th century. Yet even with this severe selection, it is clear that the calculus had made enormous advances, especially with the development of its multi-variate side. The full significance of this still has to emerge, when the calculus of variations is introduced in § 4.5 and especially when some of the applications to mechanics are described from Chapter 5 onwards. Solutions to differential equations were the principal concern, both the general theory and forms appropriate to specific types of equations; and in the train of these studies a wide range of special series, functions and integrals had been presented, although for some of them the theory as such was still in a fragmentary state. As for foundational questions, the 18th century was less careless than its reputation allows, with a plurality of approaches adopted to the calculus and some sensitivity evident to questions such as the convergence of infinite series, the definability of functions and the existence of integrals.

Among the *savants*, from the 1780s Lagrange and Laplace had become the dominant pair of figures, with Legendre occupying a respectable third place and others like de Prony and Monge making special contributions of their own. In addition we see Lacroix, in his mid thirties at 1800, with the first edition of his large *Traité* recently fully published and a career of textbook publishing consciously lying before him. The *Traité* is by far the most comprehensive work of its kind for that time. The extent of its circulation is not known, and it may not have been very large: it is certainly harder to find now than is the second edition of the 1810s. But it was as well known as any other treatise of its time, and certainly more worth reading than any other, especially for the emerging new generation. Lacroix certainly had such an audience in mind, for near the end of his preface (*Calculus*, *1* (1797), xxviii) he wrote:

Animated solely by the desire to do a book [which will be] useful to young people, I have avoided all the pretensions of self-esteem. Among many things extracted from the works of the great geometers of our time, there can perhaps be found some details which will belong to me; but I shall not dispute on that point, and I shall content myself with what one would be willing to leave me.

Chapter 4

The Lagrangian tradition in the calculus, 1800–1815

4.1 Plan of the chapter

The plan follows roughly the structure of Chapters 2 and 3 together. After introducing some important new figures in §4.2, we examine a wide range of responses to Lagrange's approach to the calculus in §4.3. All algebraic in some form or another, they range from foundational questions through technical results to speculations of various kinds on the power and scope of operator algebras. Then in §4.4 some contributions to the theory of general and singular solutions of differential and difference equations are described.

The next three sections are devoted to subjects neighbouring on the calculus which are taken up here for the first time. §4.5 is based on Lagrange's treatment of the calculus of variations, and §4.6 on his approach to some aspects of the theory of equations; in each section others' work is also described. §4.7 takes up again from §3.2.5 the relationship between geometry and algebra in three cases: a theory of 'geometry of position' due to Carnot, the geometrical representation of complex numbers, and the growth of both analytic geometry and also differential geometry. Finally, in §4.8 the publication of two second editions (Lagrange's *Fonctions* and Lacroix's large *Traité*) is noted, and also the passing of Lagrange. The first of the interludes occurs round here also: 480.1, on the calculus outside France in the early 19th century.

4.2 New figures, and a new journal

Gathered together here is information on five new figures whose earliest work is pertinent to this chapter and whose other contributions will be of importance later on. All of them were connected with the *Ecole Polytechnique*, four initially as students. The sixth figure, Gergonne, stands out for the mathematical journal that he founded.

4.2.1 Biot: career scientist

> With regard to M. Biot, I had an opportunity of pretty fully appreciating his character when we were together in the Zetland [Shetland] Isles; and

> I do not hesitate to say that I never met with so strange a compound of vanity, impetuosity, fickleness, and national partiality, as is exhibited in his character.
>
> Olinthus Gregory, 1821[1])

Biot was a *polytechnicien par excellence*. Nobody else in our story shows the same degree of versatility—of *polycapacité*, as it were—in his career. When he died in his 88th year in 1862, he had been a member of the *classe* and the *Académie* for nearly sixty years: he also belonged to the *Académie Française* in recognition of his literary writings, and the *Académie des Inscriptions et Belles-lettres* for his historical researches.

Born in 1774 into a peasant family who had recently progressed to a modest level in the Parisian civil service, Jean Baptiste Biot rejected parental aspirations for a life of commerce. He studied mathematics with Cousin at the *Collège Royal* for a short time, joined the army in 1792 and then enrolled at the *Ecole des Ponts et Chaussées* in 1794. However, soon afterwards the *Ecole Polytechnique* started, and this was obviously the place for him to commence his higher education. Slightly older than many of the other students, his career there was distinguished, and on graduating in 1795 (in one year), and studying again for some months at the *Ecole des Ponts et Chaussées*, he was appointed to the *école centrale* at Beauvais, north of Paris. Two years later he was also appointed an admissions examiner for the *Ecole Polytechnique*, a post he held until 1807. He rapidly acquired other appointments: *associé non residant* of the *classe* in May 1800 to succeed Montucla (no *PV*), and full membership in April 1803 when he gained the place left vacant by Delambre's appointment as *secrétaire perpétuel* (*PV*, *2*, 642); a professorship in mathematical physics at the *Collège de France* in 1800 to succeed his old teacher Cousin (a post which brought him back to Paris, and which he held for the rest of his life); a place in the *Bureau des Longitudes*, in 1806; founder *Professeur* of astronomy at the new Paris *Faculté des Sciences* in 1809; ...[2])

---

[1]) Letter from Gregory to H. Schumacher, 3. 7. 1821, transcribed by Schumacher to Delambre, 5. 8. 1821 (*Nachlass* Schumacher, *Staatsbibliothek* (Berlin, German Democratic Republic). On Biot's expedition to the Shetland Islands, see §6.5.2.

[2]) The earliest useful writings on Biot are the bibliography *1862a* and obituary *1867a* by his grandson-in-law Lefort. Of the later surveys of his life, Picard *1928a*, and Frankel *1972a* and *1978a*, stand out. His *Nachlass* was left to Lefort, to be used sparingly, since 'posthumous works generally do more harm than good to the memory of their authors' (Biot *1862a*, vii: compare §4.8.1 for Lagrange). Thus few manuscripts survive, but some connected with his pendulum experiments (§6.5.2) are kept in *OB*, mss. E 14–56. He was quite a prolific correspondent, and so letters from him are locatable in various places: in particular, several sent to Lacroix are held at *Smith*, and a bound volume is at *Wellcome*. Letters to him are kept at *BI*, mss. 4895–4896. A file on his *PF* career is held at *AN*, F^{17} 20178.

Do we see here an example of *cumul*, the aggregation of job upon job (§ 2.6.2)? Indeed, we have an outstanding early case of career-building in the new (largely) state-supported career of science, of which cumulation was only a part. For the young Biot cultivated the right people around town and worked on the respectable problems with approved methods. His first scientific work was in differential and difference equations (§ 4.4), which Frankel has shown in *1978a* from his surviving letters to have been planned as a cultivation of Lacroix. But his main target was Laplace, whom he won over by his exegeses of Laplacian astronomy (§ 5.4.6) and his work on physics as a member of the Laplace/Berthollet *Société d'Arcueil* (§ 7.2-5). Articles on numerous topics poured from him: research papers on physics and astronomy; philosophical articles in the general journals and newspapers (his philosophy of science manifests itself in § 13.2.6 and elsewhere); frequent reports on manuscripts submitted to the *classe*; book reviews all over the place, and on virtually anything. The young man was enjoying himself. Absent from his output, however, after the first flourish, were research papers on mathematical topics. This area he seems to have handed over to a colleague with similar careerist ambitions.

4.2.2 Poisson: career mathematician

> His only passion has been science: he lived and is dead for it.
>
> Libri on Poisson (Libri *1840b*, 433)

Seven years younger than Biot, Siméon Denis Poisson was admitted top of the list to the *Ecole Polytechnique* in 1798. Like Biot, he was of fairly low social origins, being born to a soldier turned administrator in 1781 at Pithiviers, south of Paris. His mathematical ability showed itself while he studied at the *école centrale* at Fontainebleau, and still more at the *Ecole Polytechnique*; but it lay only in algebra and the calculus, for he was virtually incapable of doing geometry (including descriptive geometry). In terms of § 1.3.6, he exhibited an algebraic *Denkweise*; and he sustained it throughout his life, which ended in 1840.[1])

Poisson's early career shows other similarities with Biot's. He did not pass on

[1]) The better surveys of Poisson's life and career are Libri *1840b* (unreliable on some details), Parisot *1845a* and Arago *1854d*. A file on his *UN* career is held at *AN*, F^{17} 21521. His *Nachlass* seems to have vanished, although a few manuscripts and letters survive here and there. They reveal one of the vilest hands imaginable; possibly this is another aspect of his non-visual cognitive processes. There is no edition of his works, but a bibliography of his works—better than his own list *1851a*—is included in the bicentenary volume Poisson *1981a'*.

to an *école d'application* but was appointed at once as an *adjoint* at the *Ecole Polytechnique* in 1800, following Garnier in Fourier's position in 1802 (he had Fourier's rooms in the school) and taking over the appointment itself in 1806.[1]) He followed Biot on the *cumul* trail, with an appointment to the *Bureau des Longitudes* in 1808 (§ 6.3.2) followed the year after by an assistantship at the *Collège de France* (§ 10.1.1) and also the post of founder *Professeur* of mechanics in the new Paris *Faculté des Sciences*. But his election to the *classe* did not occur until 1812, when the premature death of Malus gave him a seat in the *section* for 'General Physics', a topic on which by then he had barely touched. This election is considered further in § 7.3.1.

Like Biot, Poisson consciously guided his career. Whereas Biot cultivated Lacroix and then gained the support of Laplace, Poisson cultivated Biot (in ways described in § 4.4, § 6.4.4 and § 7.2.5) and then reached Laplace. According to Libri (*1840b*, 422), Laplace described the young Poisson as a 'fine genius', and took him in to the *Société d'Arcueil*. The period 1800-1815 contains the rise to the peak of reputation in his career, with a continuous stream of papers, and textbooks, related to Lagrangian mathematics and Laplacian physics. After that, as we shall see, things did not go too well.

4.2.3 Poinsot: mathematician

Poinsot, like Biot, was a member of the founding class of the *Ecole Polytechnique*. There the similarities end. Whereas Biot pursued a frantically busy career, Poinsot lived a much quieter life, working in a few specific problems in mathematics and mechanics. Thus he differs in career from Poisson also; indeed, we shall note, in § 11.6.3 and § 17.4.1, some sharp disagreements between them. There is also another difference between Poinsot and both Poisson and Biot, of specific interest to the historian: Poinsot's *Nachlass* has been preserved.[2])

Born in (probably) Paris in 1777, Louis Poinsot seems to have lost his parents before he enrolled in 1795 in the founding *promotion* of the *Ecole Polytechnique*.

---

[1]) The successorship to Fourier, who was *Préfet* of the *Département* of Isère (§ 9.1.1) but hoped in vain to return to the *EP*, often recurred on meetings of the school's administration (see *EP*, *CoI*, *2-3 passim* for this period). Poisson was an admissions examiner, with Hachette, for 1806.

[2]) This chaotic collection is kept at *BI*, mss. 948-965; see also ms. 4738. *AN* has file F^{14} 2302^{1} on Poinsot's career at the *EPC*, and F^{17} 21518 on his longer association with the *UN*. Little work has been done on his life or achievements: the best guides are Bertrand *1872a* and *1890a*, and Bailhache *1975a*. There is no edition of his works.

Three years later he went on the *Ecole des Ponts et Chaussées*; but they found him 'not assiduous and occupy[ing] himself from morning to night only with mathematics and not lik[ing] representations'.[1]) He taught at a Paris *lycée* in the 1800s, and then in 1809 he was appointed *inspecteur* in the new *Université* (as was mentioned in §2.4.3) and also an extra *répétiteur* at the *Ecole Polytechnique* (where he had already worked as an admissions examiner). He held these posts until the reorganisation after the fall of Napoléon. His health was already poor while a student, and his teaching at the *Ecole Polytechnique* had to be taken over in some years. However, he lived until 1859; in fact, some of his most important research was published quite late in his life. Despite his early orphanhood, he appears to have had some financial means, and therefore was not forced to participate in the *cumul* system as much as the others. Unusually, he also seems to have lived much of his life north of the favoured parts of Paris illustrated in Figure 263.1: "away from it all", perhaps.

Poinsot worked mainly on the principles of mechanics and of the calculus, especially following Lagrangian methods. It was thus quite appropriate for him to be elected in May 1813 to Lagrange's place in the *classe* (§4.8.1). He also studied specific questions in astronomy, aspects of geometry and number theory. Although his mathematics shows a strongly algebraic *Denkweise*, he retained the guidance of geometry and/or mechanics in his calculations (§1.3.2), which may have caused some of his disputes with Poisson.

4.2.4 Binet: mathematician

Jacques Philippe Marie Binet is similar to Poinsot in some ways: a mathematician, working in specific topics in mathematics, publishing much late in life, taking various appointments, but not participating so seriously in the career competition. He was born in 1786 to an architect at Rennes in north-western France, and entered the *Ecole Polytechnique* in 1804, being appointed a *répétiteur* in 1807. This was the first of various jobs which he held at the school, of which the most important was *Directeur des études* from 1816 to 1830. His elder brother Paul René, a *polytechnicien* who entered the school in 1798 and occasionally submit-

[1]) See *ENPC*, ms. 2432: Poinsot was 'point assidu et ne s'occupe du matin au soir que mathématiques et n'aime point les représentations'. Thus he was refused admission into the *Corps PC* (*AN*, F^{14} 2302[1]).

ted papers to the *classe* or the *Académie* which were not even reported, also taught at the school in the 1810s and 1820s, which makes identification of their duties rather difficult. (To add to the fun, young brother Philippe Thomas enrolled at the school in 1809.) Among his other appointments, Binet worked for some years from 1808 at the *Ecole des Ponts et Chaussées*, and in 1823 succeeded Delambre in the chair of astronomy at the *Collège de France* (§ 13.6.2).

Binet's research papers began to be produced around 1812, and covered various topics: mechanics, perturbation theory, determinants, elasticity and calculus. His work would have stood out as excellent, except that he was working in Paris in the early 19th century. He was candidate for the *Académie* at various times, but he was elected only in 1843 (to succeed Lacroix); when he died in 1856 he rapidly passed into oblivion, despite being *Président* at the time. (However, he was the subject of the only obituary, *1856b*, that Cauchy ever wrote: apparently Cauchy was motivated by their common Bourbon fervour, which had caused Binet to lose most of his appointments after the 1830 revolution.) No *Nachlass* seems to have survived, and no substantial study exists of his life and work.[1]) He would merit some attention.

4.2.5 Ampère: philosopher

No-one moves more mysteriously through our pages than Ampère. One of the most brilliant minds that we meet, he is also among the most elusive, imbued with philosophical concerns to an extent, and of a kind, far different from those of his colleagues. Although he also joined in the *cumul* system, he seems always to have been an outsider. His contemporaries knew him as an absent-minded eccentric: stories enlivened café conversation of him greeting the trees on his way to the *Ecole Polytechnique*, or throwing his watch in the river and taking a piece of paper to a shop for mending. Remembered today mainly for his contribution to electrodynamics, this work was in fact only part of his total output, and was confined to a few years of his life. He is better seen as a polymath, of which that time boasted several other examples: Bolzano, Boscovich, Erasmus Darwin,

---

[1]) A large manuscript on a point in astronomy, which is described in § 6.4.1, is kept in *AS*, *dossier personnel*. There is also the autobiographical list *1828a* of Binet's works, prepared in the usual manner, in connection with a candidature at the *AS*. Some information is kept at *EP*, VI 2a2; and *AN*, F^{14} 2171^2 (for Paul, see F^{17} 20178). His neglect is rather extraordinary: he is omitted from May *1973a*, and has no article in the *Dictionary of scientific biography*.

Lambert. The only subject on which he published throughout his life was mathematics.[1])

Adrien Marie Ampère was born in 1775 at Lyon in eastern France to a merchant of independent means, who soon retired to the tiny village of Polymieux. His isolation began there; and so did his mathematical and philosophical interests. His father educated him following rather free-thinking views due to Rousseau, but his mother brought to him devout Catholicism: the fusion led him to metaphysics. He suffered personal catastrophes: the guillotining of his father in 1793, the sudden death of his beloved first wife in 1803, and a short unhappy second marriage a few years later.

By that time Ampère was in Paris and working at the *Ecole Polytechnique*. He never studied there; indeed, he seems to have been largely self-trained, especially in mathematics, teaching himself Latin in order to read Euler and Daniel Bernoulli. In 1802 he had been appointed to the *école centrale* at Bourg-en-Bresse, north of Lyon, where he wrote a monograph *1802a* on probability which attracted attention sufficient to gain him in 1804 a post as *répétiteur* in mechanics and analysis at the *Ecole Polytechnique*. In 1809 he was also made an *inspecteur* in the *Université*, like Poinsot; but six years later he was promoted to *inspecteur général* (§ 10.1.1), a post which he held for most of the rest of his life—indeed, he died in 1836 on a tour of inspection. He also took on various other appointments; for example, for some years he was involved in organising lectures and other activities of the *Athénée des Arts*, a Paris society for the encouragement of science and industry,[2]) and he also served on a *Bureau de Consultation des Arts et Manufactures*. Overall, his *cumul* is one of the most interesting. He was elected to

---

[1]) Although there is an (incomplete) edition of Ampère's *Correspondence*, and two biographies, the best recommendations are still the article *1843a* by Arago's brother Etienne, and the obituaries Arago *1854c* and Sainte-Beuve and Littré *1837a*. This latter was mostly reprinted in Ampère *Philosophy*, *2* (1843): his own preface to *1* (1834) is also worth reading. Taton *1978a* is a useful bibliographical guide to his publications. Little research has been carried out on his work outside electrodynamics; in particular, the absence of mathematics from the article in the *Dictionary of scientific biography*, and from the Ampère number of the *Rev. d'hist. sci.*, *30* (1977), is painful. L. Pearce Williams is preparing a biography.

A crucial source is Ampère's enormous *Nachlass* in *AS*, which is fairly well sorted into categories corresponding to his wide interests. I cite it in the style '*AS, Ampère*, ch. 318', where 'ch.' indicates a *chemise* number. An interesting autobiographical essay is kept at ch. 326: part is published in Blondel *1982a*, 171–173, although unfortunately the parts on his mathematical work are omitted and the wrong reference to ch. 314 is given. On his *UN* career, see Aulard *1911a passim*, and the file in *AN*, F[17] 20023; the dossier at *EP*, VI 1b2 is also of note. Relevant manuscripts by his son survive in *BI*, mss. 2379 and 4439. Other manuscripts are listed in *Correspondence*, 815–816. There are editions only of his electrodynamical papers.

[2]) Some manuscripts connected with this *Athénée* are held at American Philosophical Society (Philadelphia). They include quite a lot of letters and notes by Ampère.

the *classe* in November 1814 to succeed Bossut (*PV*, *5*, 434), on the strength of his mathematical writings. His later career is taken up in Chapter 14.

4.2.6 Gergonne and his *Annales*

Our last figure is of the generation of Fourier and Hachette, and features here not for the emergence of his research but for his launching of an important new journal. Joseph Diaz Gergonne was born in 1771 to a painter and architect at Nancy in eastern France. He lived until 1859, and passed his career entirely in the provinces. He taught at the *école centrale* in Nîmes in southern France from 1795 until 1816, when he moved to the nearby town of Montpellier to take the chair of astronomy in the local *Faculté* of the *Université*.

We noted in § 2.3.3 the paucity of journals in mathematics suitable for educational purposes, especially as the *Journal de l'Ecole Polytechnique* soon converted to a research periodical. Gergonne decided to fill this lacuna, and in 1810 he launched the *Annales de mathématiques pures et appliquées*. In a preface *1810a* to the first volume, written with his co-editor P. Thomas-Lavernède of the first two volumes, he stated that the 'exact sciences' had no journal of their own, and that this one would pay attention to pedagogical concerns as well as to research interests. By and large it fulfilled its aims during its 22-year run; but it also became something of a Journal of Non-Standard Mathematics, with articles on unusual applications and algebraic speculations along with more traditional topics, and with provincial authors more frequently represented than the Paris *savants*. Gergonne himself was by far the most frequent contributor, with articles signed with his own name or 'A subscriber' as well as anonymous pieces.[1]) He supplied a large proportion of the non-standard pieces: I cited an article of his on "analytic" and "synthetic" in § 3.2.5, and other contributions discussed (for example) definitions, the introduction of money into society, and numeration. His orthodox research interests centred on aspects of geometry and analysis. He also took part in others' articles, in the form of footnoted comments on their work. He is one of the wittiest writers in mathematics, as the quotations in § 3.2.5 showed.

Gergonne must have been efficient as an editor. He published his 400 pages each year in numbers which appeared fairly regularly (to judge by their dates of presentation to the *classe* and the *Académie*, recorded in the *Procès-verbaux*). He

---

[1]) According to Lafon *1860a*, xli, Gergonne's own copy of the *Annales* was kept in the public library at Nancy, with his own contributions marked '**' (listed on pp. xli-lxxiv). This little-known obituary is the best article on Gergonne; Stigler *1976a* has further useful references. The launching of the *Annales* was noted in various French journals and also abroad (for example, Buzengeiger *1812a*).

prepared the diagrams himself, and presumably also the tables of contents, where he adorned some articles with new titles. He sustained the journal through the transition to the Restoration, and also through some difficulties of the late 1820s (described in §17.1.2); but he stopped it in 1832, early on into volume 22, because in 1830 he had been appointed *Recteur* of the *Académie* of Montpellier in the *Université*. During his 14 years in this post, he wrote 44,000 letters (Lafon *1860a*, xxxviii). In 1844 he retired both as *Recteur* and as *Professeur* of physics in the *Faculté*, a post which he had taken in 1836 in succession to the chair in astronomy.

The print-run of the *Annales*, at least in its early years, was 500 copies; but it soon became rare, for it was in demand by booksellers already in the mid 19th century (see Friedländer *1861a*, 22), and is certainly hard to find now. But we cannot ignore it on these grounds, for it was available to our historical figures,[1]) who sometimes contributed to it. It contains some of the liveliest writing in mathematics of its time, and would repay a detailed study.

4.3 Consequences of Lagrange: functions and the new algebras

This section covers a range of developments which Lagrange's approach to the calculus received between 1800 and 1815. I start with certain treatments of the derived function and Taylor's and Lagrange's series (§4.3.2-4), move on to a generalised treatment of power series (§4.3.5) and conclude with some algebraic speculations, mainly in operator algebras and functional equations (§4.3.6-8). The sequence is a progression from orthodox to fringe mathematics—and the authors change from the new Parisians whose biographies have just been sketched to fringe figures, mostly provincials, usually contributing to Gergonne's *Annales*.

4.3.1 Lagrange's *Leçons sur le calcul des fonctions* of the 1800s

In preface to this sequence of studies, I note Lagrange's final treatise on the calculus: his *Leçons sur le calcul des fonctions*, delivered at the *Ecole Polytechnique* in 1799, his last year there.

---

[1]) See, for example, Liouville's notes on some volumes in *BI*, ms. 3617.1; and Poncelet's notes in a box in *EP*, IX (Poncelet), box 1. Friedländer *1861a*, just cited, is a sale catalogue of the libraries of the mathematicians Dirichlet, G. Piola and C. Ramus, one of whom owned a set of the *Annales*. The print-runs are not included in the declarations in *AN*, F^{18} 128 after 1817.

The lectures have a peculiar publication history. The first 20 were published in 1801 in the *Séances des écoles normales* (although they had not been delivered to the *Ecole Normale*) and reprinted, in slightly revised form, four years later in the *Journal de l'Ecole Polytechnique*. This edition, printed in 2,400 copies, also contained three small additions: Poisson *1804c* on singular solutions to differential equations, and Hachette *1804a* on a similar theme and *1804b* on Monge's characteristic (§ 3.3.7).[1])

Lagrange then prepared a book version which appeared, for some reason anonymously, from Courcier in 1806 (his 71st year). He removed lecture 14 on conditions for a differential to be exact, renumbered lectures 15–19 as 14–18, split lecture 20 into lectures 19 and 20, added lecture 21 embodying a revision of the material of the discarded 14th, and finished with a new lecture 22 on the calculus of variations. The two new lectures were also published in 1808 in the *Journal de l'Ecole Polytechnique*. Since the 1804 printing had contained the old lecture 14, the new material was in part repetitious; nevertheless, this *Journal* version of the 22 lectures was put out in 1808 by the *Imprimerie Impériale* as a book—different, therefore, from the Courcier edition of 1806. In 1810 a jury composed of Laplace, Monge and de Prony awarded the book an *Institut* prize as the most distinguished publication in geometry or mechanics during the decade (see *Institut de France 1810a*, 5–8); it could also have won a prize for versatility of form.

The Courcier edition must be regarded as definitive; it is the one that was republished in Lagrange's *Oeuvres*,[2]) and is the one that I shall cite (as *1806a*). However, I shall not discuss it in detail; for, as Lagrange explained in his preface, it was conceived 'to serve as commentary and as supplement' to that part of *Fonctions* dealing with the calculus.

4.3.2 The derived function and the difference quotient

In lecture 9 of *1806a* Lagrange returned to the theorem (322.10) of *Fonctions*:

$$\text{if } f(x+i) = f(x) + i(f'(x) + P(x, i)), \quad \text{then} \quad P(x, 0) = 0, \tag{432.1}$$

and to the construction (323.2) of intervals $\{[a + ri, a + (r+1)i]\}$ for each of

[1]) I recall from § 2.3.3 that Hachette and Poisson were *de facto* editors of the *JEP*. Materials on the printing of the *Leçons* are contained in *EP*, VIII 1, and *Col, 2*, fol. 309; and *AN*, F^{17} 1084. Poisson's and Hachette's contributions were to have been more substantial, comprising lectures 21 and 22 (see *Col, 4*, meeting of 29. 9. 1806).

[2]) The *Oeuvres* edition shows a typical (§ 1.4.1) piece of editorial carelessness; the editors referred Lagrange's references to *Fonctions* to the second (1813) edition, whereas Lagrange was, of course, citing the first (1797) edition.

which
$$f(a+(r+1)i)-f(a+ri)>0. \quad (432.2)$$

This time he was explicit about using positive values (though he did not distinguish strong from weak inequalities), and showed that

$$\begin{aligned} &0<|f(x+ni)-f(x)|<2niP, \\ &P:=\max\{|f'(x+ri)|, \quad 0\leq r\leq n-1\}. \end{aligned} \quad (432.3)$$

He used the same type of argument to show that

$$s_n(x)+i^n f^{(n)}(p)/n! \leq f(x+i) \leq s_n(x)+i^n f^{(n)}(q)/n!, \quad (432.4)$$

where $s_n(x)$ was the n-th partial sum of the Taylor series, and $f^{(n)}(p)$ and $f^{(n)}(q)$ were respectively the minimum and maximum values of $f^{(n)}(x)$ over $[x, x+i]$. (432.4) gave him estimates of the error involved in taking $s_n(x)$ for the Taylor expansion of $f(x+i)$ (compare (322.12)).

Ampère used these ideas to examine further the convergence of Taylor series and the notion of the derived function, in a paper *1806c* published in the *Journal de l'Ecole Polytechnique* and doubtless based on his teaching there. His paper is suggestive but very incoherent, especially in distinguishing apart assumption, theorem and proof; I shall indicate only some principal features. Since he was much concerned with properties of the difference quotient, I shall for convenience introduce the definition

$$\mathrm{DQ}(p, q) := (f(q)-f(p))/(q-p). \quad (432.5)$$

Ampère wanted to show that $\mathrm{DQ}(x, x+i)$ was neither zero nor infinite, except at occasional points, within a given interval $[a, k]$. From Lagrange's (432.4) with $n=1$,

$$f(x)+if'(p) \leq f(x+i)+f(x)+if'(q). \quad (432.6)$$

Ampère specified $f'(x)$ as the function satisfying the relation

$$f'(p) \leq \mathrm{DQ}(x, x+i) < f'(q) \quad (432.7)$$

for some two points p and q lying within $[x, x+i]$ and for all values of x and i. He hoped to conclude from the finitude of $\mathrm{DQ}(x, x+i)$ that $f'(x)$ took a determinate value (*1806c*, 150–156).

Suggestiveness and unclarity are both well exemplified here. (432.7) attempts to define $f'(x)$ in a new way; but it was specified over an interval whereas the property is definable only at a point. The definition used inequalities; but, as also with Lagrange, strong and weak inequalities were not distinguished in the discussion. The emphasis is laid on the function f rather than on Lagrange's faith in Taylor series; but Ampère seemed to be in the vague Lagrangian territory of "any" function rather than restricting the scope of functions discussed. Thus it is not obvious that (432.7) either defines a function f', or that f', if defined, is unique: Ampère did try to deal with this last question.

The same characteristics carry over into Ampère's reasoning. He took a uniform partition $\{x_j\}$ of $[a, k]$ and asserted that there would always be sub-intervals such that

$$\mathrm{DQ}(x_r, x_{r+1}) \geq \mathrm{DQ}(a, k) \geq \mathrm{DQ}(x_s, x_{s+1}). \tag{432.8}$$

Thus from Lagrange's (432.6) for sufficiently small i, there existed members p and q of $[a, k]$ such that

$$f'(p) \geq \mathrm{DQ}(a, k) \geq f'(q), \tag{432.9}$$

which was Ampère's own definition (432.7) (pp. 152–156). He also "deduced" on pp. 162–164 Lagrange's property (322.2) in the form:

$$\text{if } f(x+i) = f(x) + if'(x) + iI(x, i), \quad \text{then} \quad I(x, 0) = 0. \tag{432.10}$$

The essential argument here seems to be due to the elder Binet, Paul Réné, and Ampère wrote it up for him as a short paper for the *Société Philomatique*.[1]) P. Binet *1809a* tried to prove by *reductio ad absurdum* the conjecture that for "any" function f and $a \leq x \leq (a + b)$ apart from isolated exceptions, $\mathrm{DQ}(x, x + h)$ tended to a fixed finite value as h approached zero. Taking b sufficiently small so that f was monotonic over $[a, a + b]$, it followed that $\mathrm{DQ}(x, x + h)$ and $\mathrm{DQ}(a, a + b)$ took the same sign. If $\mathrm{DQ}(x, x + h)$ approached zero, then he could choose an h satisfying

$$\mathrm{DQ}(x, x+h) < \mathrm{DQ}(a, a+b). \tag{432.11}$$

Generalising by partitioning $[a, b]$ into n equal sub-intervals, such that $b < nh$, then

$$\mathrm{DQ}(a + (rb/n),\ a + (\overline{r+1}\, b/n)) < \mathrm{DQ}(a, a+b),\ 0 \leq r \leq n-1. \tag{432.12}$$

Summing over r, and using (432.5), he obtained the falsehood

$$n\,\mathrm{DQ}(a, a+b) < n\,\mathrm{DQ}(a, a+b). \tag{432.13}$$

If $\mathrm{DQ}(x, x + h)$ went to infinity, the reasoning proceeded as above, with the opposite sign to the inequality. Hence by *reductio*, the conjecture was proved—to Binet's satisfaction.

Ampère's and Binet's arguments attempted to break away from Lagrange's ap-

[1]) This paper is often attributed to the "well-known" Binet, although it describes the author as a *professeur* at Rennes; Lacroix is accurate in his account in *Calculus*$_2$, *1* (1810), 241–242. The manuscript, in Ampère's hand, is held in *AS, Ampère*, ch. 82; on its verso it is addressed as a letter for Daburon, a late-18th-century French fluxionalist who taught Ampère as a student at Lyon (see ch. 63).

proach (although they continued his tendency of offering proofs for "any" function). The change seemed to be towards the use of limits, but neither man offered a theory of limits (or used positive values of quantities at times when he should). Ampère's commitment to limits was more clearly indicated in some manuscripts (although unfortunately most of them are undated). In one of them, related to his lecture course of 1808-1809, he proved that

$$\text{if} \quad u_r < v_r < w_r \quad \text{and} \quad \lim_{r\to\infty} u_r = \lim_{r\to\infty} w_r = V, \quad \text{then} \quad \lim_{r\to\infty} v_r = V. \tag{432.14}$$

He even used Lhuilier's notation 'lim' (§ 3.2.6), which was not widely known then.[1])

However, the progress towards limits was not steady (or monotonic, one might say). In 1812 the *Conseil de Perfectionnement* of the *Ecole Polytechnique* reported that 'For the exposition of the differential calculus, for the method of limits has been substituted that of infinitesimals, which is easier and to which one is obliged to have recourse in mechanics' (*EP 1812* prog, 5). This change was stimulated by criticisms of the school's teaching, recently made by the *Ecole du Génie* at Metz. While being reciprocated, they were accepted, including by Laplace (see de Prony *1835a*); normally inclined to advocate modifications towards theoretical studies at the school, Laplace must have felt that the change had gone too far.

In the wake of this decision Poinsot soon gave a lecture course on the calculus, of which a version was published in the *Correspondance sur l'Ecole Polytechnique* as *1815a*. He regarded differentials as 'quantities next' to orthodox quantities, chosen such that their 'final ratios' took 'the ratio of equality' with a suitable orthodox quantity. Thus if y was a function of x, and Δy and Δx diminished together, then their ultimate ratio equalled that taken by $dy \div dx$ and yielded the 'differential function' (pp. 112-116). In other words, Poinsot taught an approach based on Euler's (§ 3.2.4), but bolstered with an appeal to the notion of limit. However, he did not expound any theory of limits; on the contrary, for example, he assumed that the ultimate ratio of $\Delta y \div \Delta x$ must always exist. He also drew on Lagrange's approach, as I shall explain in the next sub-section.[2])

---

[1]) The manuscript, from *AS, Ampère*, ch. 100, appeared in my *1981b*, 675-676; see also manuscripts in chs. 63-65 (from around 1790), 59-82 and 100-101 (some from the 1800s and others from the 1820s (§ 11.6.2), though many are undated); and at *BI*, ms. 955, fol. 11 (in the *Nachlass* of Poinsot). See also Pontécoulant's notes of Ampère's lectures of 1811 (*BI*, ms. 4068), where infinitesimals are somewhat more prominent than limits; and Olivier's notes of Ampère's lectures of 1811-1813 (*EP*, III 2a2).

[2]) Francoeur used Poinsot's lectures to develop a course, perhaps for use in *PF* (*Ecole ... des Beaux Arts*, ms. 401). Comte took notes of Poinsot's lectures in analysis in late 1814 (*Maison d'Auguste Comte*, student lecture notes file). Related notes by Poinsot himself are kept at *BI*, ms. 964.

The impression given by these texts is strengthened by inspection of some others of the time (for example, du Bourguet *1810a*, Gergonne *1815a* or Lacroix *Calculus*$_2$, *1* (1810)); that the situation was no more tidy around 1815 than it had been around 1800. Things, however, were to change; for in addition to Binet, Ampère's student audience at the *Ecole Polytechnique* included Cauchy (§ 11.3.3).

4.3.3 Taylor's series and some proofs

One respect in which Poinsot *1815a* showed Lagrange's approach to the calculus to the students at the *Ecole Polytechnique* was to obtain the derived functions of simple functions by 'purely algebraic means', especially their basic properties, and also functional equations. Among other contexts, Euler used these equations when deducing the binomial series (see, for example *1775a*, arts. 8–11; and Dhombres *1985a* for commentary), and Lacroix had included a clumsy but suggestive presentation in *Calculus*$_1$, *1* (1797), 20–26. Lagrange may have been inspired in part by these efforts to the account given in lecture 3 of the *Leçons*. Here is Poinsot's version of finding the derived function $\varphi(x, m)$ of the function x^m (*1815a*, 120–122).

From the chain rule of differentiation and the property

$$(ax)^m = a^m x^m, \quad \text{then} \quad \varphi(x, m)x^{1-m} = \varphi(ax, m)\,(ax)^{1-m}; \tag{433.1}$$

that is, each quotient of $(433.1)_2$ was independent of a.

$$\therefore \ \varphi(x) = f(m)\, x^{m-1}, \quad f \text{ unknown.} \tag{433.2}$$

From the additive property of powers,

$$x^{m+n} = x^m x^n, \quad \text{then} \quad f(m+n) = f(m) + f(n). \tag{433.3}$$

Hence, for an integral multiple e of m,

$$ef(m) = f(em); \quad \text{but then} \quad ef(m) = mf(e) \tag{433.4}$$

by symmetry between e and m. Taking the case $e = 1$, for which the derived function of x with respect to x was obviously 1, $(433.4)_2$ became

$$f(m) = mf(1) = m; \qquad \therefore \ \varphi(x) = mx^{m-1} \tag{433.5}$$

from (433.2), as required.

In lecture 3 of the *Leçons*, Lagrange regarded his own version of this argument as 'leav[ing] nothing to be desired for rigour and generality', and he had taken the approach somewhat further in lectures 4–7, with derivations of derived functions of various orders for further simple functions. Since Poisson was a student at the *Ecole Polytechnique* in 1799, he will have heard Lagrange deliver these lectures. Three early papers in the calculus relate closely to the considerations just recorded. They were all published in the *Correspondance sur l'Ecole Polytechnique* for pedagogical use, and one was first summarised in a textbook by Garnier (*1801a*, 201-203), whom Poisson was soon to succeed as *répétiteur* at the school. In that paper, Poisson *1810d* used Taylor's series to find series expansions of $(1+x)^n$, a^x, $\ln(1+x)$, $\cos x$ and $\sin x$. The work was very close to Lagrange's and included a repetition of the reasoning above. During his first weeks as a student Poisson sent a proof of the binomial theorem to Lagrange, who incorporated it into his lectures.[1]) In another paper, Poisson *1811e* pointed out that the expansion

$$(2\cos x)^m = \sum_{r=0}^{[m/2]} m\, C_r \cos(m-2r)\, x \tag{433.6}$$

was multi-valued if m was rational, and that the values should be calculated separately by putting x, $(x+2\pi)$, $(x+4\pi)$, ... into (433.6). The point may be obvious now, but it seems to have been novel then—for example, Lagrange did not consider rational values of m in of his *Leçons*—and it was incorporated into the teaching at the *Ecole Polytechnique*. (433.6) was to be discussed again, by Poisson and others, in the mid 1820s (§ 11.6.3).

The third of these papers, Poisson *1805a*, contained a proof of Taylor's theorem. He proceeded by assuming the more general expansion

$$f(x+h) = f(x) + h^a p(x) + h^b q(x) + \ldots, \qquad 0 < a < b < \ldots, \tag{433.7}$$

and comparing coefficients of $p(x)$ in

$$f(x+2h) = f(x) + (2h)^a p(x) + (2h)^b q(x) + \ldots \tag{433.8}$$

and $$f((x+h)+h) = f(x) + 2h^a p(x) + \ldots \tag{433.9}$$

to show that

$$(2h)^a p(x) = 2h^a p(x) \tag{433.10}$$

and hence that $a = 1$. As a particular case of (433.7), therefore,

$$(x+h)^m = x^m + mx^{m-1}h + \ldots, \tag{433.11}$$

where the further coefficient functions still had to be determined.

---

[1]) See *EP, Col, 2*, fols. 13, 18. The story was told in Fourcy *1828a*, 168. On the use of functional equations in mechanics, see § 5.3.2.

In order to find the coefficients, and the other indices, in (433.7), Poisson put

$$f((x+k)+h) = f(x+k) + hp(x+k) + h^b q(x+k) + \dots \qquad (433.12)$$

and $$f(x+(h+k)) = f(x) + (h+k)p(x) + (h+k)^b q(x) + \dots . \qquad (433.13)$$

Equating coefficients of powers of h yielded

$$f(x+k) = f(x) + kp(x) + \dots, \qquad (433.14)$$

$$p(x+k) = p(x) + kp'(x) + \dots, \qquad (433.15)$$

and so on, where in the equations of the type (433.15) the prime was used since $p(x+k)$ was a function of $(x+k)$ whose 'primitive function' was $p(x)$. As a sleight of mind to introduce the prime to relate to 'primitive functions', this move is admirable, like Lagrange's at (322.6), but it is not traceable in (433.11–13): it must not made *via* Lagrange's approach to the calculus, since a proof of Taylor's series is the task at hand. However, it led to a smooth completion of the proof; for substituting (433.11, 14, 15) and like equations into (433.12, 13) yielded two double power series in h and k, where equating coefficients provided the values $b = 2$, $c = 3$, ...; and reading the primes in terms of 'differential coefficients' (to quote Poisson's surprising use of the Eulerian term) furnished the coefficient functions needed.

In an addendum, Poisson added to the mystery. He claimed that his proof was valid if $f(x+h)$ were regarded as a function of two variables x and h; but if x were fixed, the expansion 'following the powers of h, would not have the general form that we have shown; from which one sees *a priori* that Taylor's formula would have to find itself in default for certain particular values of x' (*1805a*, 55). He obviously had in mind the possibility of infinite-valued functions, against which Lagrange had warned (§ 3.2.2); but I cannot relate it to this argument.

Taylor's series was studied more carefully by Ampère in his *1806c*, 162–167. He took the definition (432.5) of the difference quotient as an identity, in the form

$$f(z) = f(x) + (z-x)p(x,z). \qquad (433.16)$$

Holding z constant, he subtracted (433.16) from itself with $(x+h)$ substituted for x, used Lagrange's property (322.12) to replace terms in f by terms in f', and then put $h = 0$, thus yielding

$$0 = f'(x) + (z-x)p'(x) - p(x). \qquad (433.17)$$

He could have obtained (433.17) more easily by differentiating (433.16); but he probably avoided it because he was investigating the convergence of Taylor's series, with which differentiation was so closely linked. Repeating his procedure successively furnished

$$0 = f^{(n)}(x) + (z - x)p^{(n)}(x) - p^{(n-1)}(x), \quad n \geq 2. \tag{433.18}$$

Substituting for p in terms of p', p' in terms of p'', and so on produced Taylor's series, complete with remainder term in differential form:

$$f(x) = \sum_{r=0}^{n} [(z - x)^r f^{(r)}(x)/r!] + p^{(n)}(x)h^{n+1}/n!. \tag{433.19}$$

A straightforward delivery of Taylor's series is provided by repeated integration by parts. Series so produced were (and sometimes are) called 'Bernoulli series', after John: the method was used again in d'Alembert *World*, *1* (1754), 50. Fourier drew on it in his teaching at the *Ecole Polytechnique*, as was reported in Garnier's printed lecture notes (*1801b*, fol. 6*bis*). Perhaps independently, de Prony also employed it, and I shall follow his account from a pamphlet *1840b*. There it was applied to $\int_0 f(x - i)di$, and after iteration produced

$$f(x) = \sum_{i=0}^{n-1} (i^r f^{(r)}(x - i)/r!) + \int_0 f^{(n)}(x - i)d(i^n)/n!; \tag{433.20}$$

he then used geometrical considerations to convert the remainder term into the differential form

$$i^n f^{(n)}(x - \lambda i)/n!, \tag{433.21}$$

where λ was a 'positive fraction'.

We shall see Cauchy use this proof in his own textbooks much later (§ 11.5.4): appearing in these rather obscure publications in the 1800s gave it little impact. However, Cauchy was a student at the *Ecole Polytechnique* soon after Garnier's lectures were published, and at de Prony's *Ecole des Ponts et Chaussées* not long after his pamphlet appeared. Also, Laplace ended Book 1 of *Probability*$_1$ (1812) with this proof. Influences must have been accumulating on Cauchy.

4.3.4 Parseval on two series, 1806

So far we have seen the early work of some professionals. Our story is graced also by some fascinating fringe figures, and the first of them takes his place here.

Marc Antoine Parseval was born in 1755 into a family which was distinguished enough for his father to be guillotined in 1794. A younger brother, François Parseval-Grandmaison, was a poet of distinctive mediocrity. Parseval himself wrote poetry occasionally; some of it criticised Napoléon's regime, and he fled the country when his arrest was ordered. His brother was a member of Bonaparte's Egyptian campaign, and later was elected to the literary *classe* of the *Institut*; but Parseval himself was only a candidate for the scientific *classe* on a few occasions. The possibility of his election rested on five papers, submitted to that *classe* between 1798 and 1803 and all published in the 1806 volume of the *Savants étrangers*. He submitted other papers from time to time, but seems to have published nothing more before his death in 1836. His fame during his lifetime rested on his eccentricities.

Table 434.1 contains principal points or sources of data that I have been able to find about his life and career. It amounts only to a partial reply to the call for information uttered in Eneström *1892a*: more would certainly be welcome, for his work, which we now begin to consider, shows a considerable talent, about the development of which nothing seems to be known.

Parseval's name has survived until today mainly because of a paper submitted in May 1798 to the *classe* (*PV*, *1*, 397), in which he gave a formula for summing series. It was mentioned in § 3.3.4 that functional forms were preferred as solutions to differential equations, or failing them, power-series forms; Parseval's theorem gave a closed form to a series obtained from two power series. He let

$$U(x) := \sum_{r=1}^{\infty} a_r x^r \quad \text{and} \quad V(x) := \sum_{r=1}^{\infty} b_r x^{-r} \tag{434.1}$$

and put $x = e^{iu}$. Then by multiplying the series together and integrating over $[0, \pi]$—a procedure 'which is evident by itself', and relied on de Moivre's theorem—he obtained

$$\sum_{r=0}^{\infty} a_r b_r = \frac{1}{2\pi} \int_0^{\pi} [U(e^{iu})V(e^{iu}) + U(e^{-iu})V(e^{-iu})]du \tag{434.2}$$

(*1806e*, 638-640: compare § 10.2.1 on the use of i in evaluating integrals).

Parseval did not consider convergence but applied (434.2) to some series produced by Euler and Laplace. For example, starting from Laplace's results with his method of cascades (§ 3.3.5), he took the series

$$\exp(\sqrt{\alpha}\, x) = \sum_{r=0}^{\infty} (\sqrt{\alpha}\, x)^r / r!$$

and (434.3)

$$\exp(\sqrt{\alpha}/x) = \sum_{r=0}^{\infty} (\sqrt{\alpha}/x)^r / r!$$

Table 434.1 Principal known information on the scientific career of Parseval.

The information is laid out in roughly chronological order. Additions to the sources from which it is drawn are enclosed in square brackets. Most events relate to *CI* or *AS*.

Information	Source
Worked on de Prony's cadastral survey, 1790s (§ 3.5.4)	*ENPC*, mss.2629 *bis*, 2636; *AN*, F[14] 2146
Calculations on pendulum [for de Prony *1800c*?]	*ENPC*, ms.1357
Considered at *CI* as *associé non-résidant*, 14.2.1795	*PV*, *1*, 12
Letter on an experiment to exhibit Kepler's law, 1797	Parseval *1797a*
Submits *1806b* on partial differential equations, 5.5.1798 (§ 4.4.3)	*PV*, *1*, 387
Submits *1806e* on 'Parseval's formula', 20.5.1798; *CI* report Laplace *1799a* (§ 4.3.4)	*PV*, *1*, 397, 546-547
Submits [lost] manuscript on elasticity, 26.11.1798	*PV*, *1*, 496
Candidate for Borda's place, 5.4.1799; third to Lacroix	*PV*, *1*, 547
Submits *1806a* on sound, 5.7.1801; *CI* report Lacroix *1804b* (§ 7.2.7)	*PV*, *2*, 262-263, 373, 392
Has written paper on lunar theory; to publish a history of integral calculus	Lalande *1803a*, 829
Candidate for Delambre's place, 12.4.1803; third to Biot	*PV*, *2*, 640, 642
Submits *1806c* on fluid mechanics, 16.8.1803; *CI* report Lacroix *1804b* (§ 7.2.7)	*PV*, *2*, 692; *3*, 55-57
Submits *1806d* on Lagrange's series, 6.5.1804; *CI* report Lacroix *1805a* (§ 4.3.4)	*PV*, *3*, 96, 178-179
Manuscript of 1804 on sound and fluids	*AS*, *dossier personnel*
Manuscript [of 1804?] on differential equations	*AS*, *dossier personnel*
Candidate for Lagrange's place, 31.5.1813; joint bottom to Poinsot's election	*PV*, *5*, 217-218
Candidate for Bossut's place, 28.11.1814; no votes; Ampère elected	*PV*, *5*, 431, 434
Wants to copy out [lost] manuscript submitted on perturbations of Pallas, 12.6.1815	*PV*, *5*, 519
Complains on omission of his name in connection with researches on sound, 26.7.1819	*PV*, *6*, 469
Submits [lost] paper on sound, 18.12.1820	*PV*, *7*, 118
Submits [lost] paper on analytic functions, 28.2.1825	*PV*, *8*, 194
Submits [lost] paper on differential equations, 3.9.1827, 8.10.1827	*PV*, *8*, 588, 593, 604
Candidate for Laplace's place, 27.10.1828; no votes; Puissant elected	*PV*, *9*, 136
Dies in Paris, 16.8.36: described as mathematician and geographer, and as an eccentric	*MU*, (1836), 1770

Biographical articles:
Nouvelle biographie générale, *39* (1862), col. 253
Biographie universelle ancienne et moderne, nouvelle édition, *32* (1863?), 182
Dictionary of scientific biography, *10* (1974), 326-327
See also my and Ravetz's *1972a*, 239-241

to find, from (434.2),

$$\sum_{r=0}^{\infty} \alpha^r/(r!)^2 = \frac{1}{\pi}\int_0^{\pi} \exp(2\sqrt{\alpha}\cos u)du. \tag{434.4}$$

(pp. 644–646); a nice result for Bessel-function theory. Fourier and Laplace were both to find it in the 1800s, as will be noted at (928.23).

Equation (434.2) made some impact. Laplace praised it in his report *1799a* to the *classe*, Lacroix published it in his *Traité* (*Calculus*$_1$, *3* (1800), 377, 514), and de Prony discussed in his mechanics lectures at the *Ecole Polytechnique* (*1800a*, 344–346). Parseval used it himself in later papers (*1806a*, 390–393; *1806d*, 570–572). The interest lay entirely in the result as a summation formula, *not* in its implications for analysis, which have kept Parseval's name alive in more recent times. Indeed, (434.2) is not the 'Parseval's theorem' that we know, which is a property of trigonometric series. He produced this (to him) variant in a note appended to a later paper on fluid mechanics, as a corollary to the result that if

$$P(x) := \sum_{r=0}^{\infty} a_r \cos rx \quad \text{and} \quad Q(x) := \sum_{r=0}^{\infty} b_r \cos rx, \tag{434.5}$$

then (correcting a small error that he made)

$$2a_0b_0 + \sum_{r=1}^{\infty} a_r b_r = \frac{2}{\pi}\int_0^{\pi} P(x)Q(x)dx \tag{434.6}$$

(*1806c*, 544–545). Again, for Parseval this was a summation formula. Fourier found it himself, later (*1822a*, art. 235, para. 2), apparently independently; in his case it was just one of many properties of Fourier series. The modern interest in (434.6) dates only from the late 19th century; previously the phrase 'Parseval's theorem' referred to (434.2), not (434.6) (see, for example, Boole *Equations*$_4$ (1877), 477–478).

Parseval also submitted a very interesting paper on Lagrange's series (341.14) in May 1804 to the *classe* (*PV*, *3*, 96). In his report Lacroix *1805a* noticed a common theme with Parseval's earlier papers, including the one just discussed; the conversion of series to integrals. Parseval took Lagrange's equation (341.10) in the form

$$\alpha - x + \varphi(x) = 0, \tag{434.7}$$

which gave the required root p of x in the form

$$p = \alpha + \sum_{r=1}^{\infty} ((\varphi(\alpha)^r))^{(r-1)}/r! =: \alpha + U(\alpha), \text{ say}. \tag{434.8}$$

In order to find a closed form for $U(\alpha)$, he took the power series in (u/s)

$$V(\alpha) := \sum_{j=0}^{\infty} u^j (\varphi(\alpha))^{j+1}/s^j \tag{434.9}$$

(where u and s were new variables), formed the series

$$T(\alpha) := \sum_{j=0}^{\infty} s^j V^{(j)}(\alpha)/j!, \tag{434.10}$$

and noted that the terms in $T(\alpha)$ which possessed no powers of s formed a subseries similar to $U(\alpha)$, namely

$$\sum_{r=1}^{\infty} u^{r-1} \left((\varphi(\alpha))^r\right)^{(r-1)}/(r-1)!. \tag{434.11}$$

In fact, if $T(\alpha)$ were integrated over [0, 1] of u, $U(\alpha)$ would be obtained. Hence attention focussed on $T(\alpha)$ (*1806d*, 567–569).

Parseval noted that (434.9) was a geometric progression, so that

$$V(\alpha) = \varphi(\alpha) \Big/ \left(1 - \frac{u}{s}\,\varphi(\alpha)\right). \tag{434.12}$$

Hence, from (434.10),

$$T(\alpha) = \varphi(\alpha + s) \Big/ \left(1 - \frac{u}{s}\,\varphi(\alpha + s)\right), \tag{434.13}$$

and integrating over [0, 1] of u yielded

$$W(\alpha, s) := \int_0^1 T(u)\,du = -s \ln\left(1 - \frac{1}{s}\,\varphi(\alpha + s)\right). \tag{434.14}$$

To determine $U(\alpha)$, he used on $W(\alpha, s)$ a procedure similar to his derivation of the trigonometric form (434.6) of his summation formula. Putting

$$s = e^{iv} \quad \text{and} \quad s = e^{-iv} \tag{434.15}$$

into (434.14), integrating each formula over $[0, \pi]$ of v and adding, he obtained

$$U(\alpha) = \int_0^{\pi} [W(\alpha, e^{iv}) + W(\alpha, e^{-iv})]\,dv \tag{434.16}$$

$$= -\frac{1}{2\pi} \int_0^{\pi} e^{iv} [\ln(1 - e^{-iv}\varphi(\alpha + e^{iv})) + e^{-iv} \ln(1 + e^{iv}\varphi(\alpha + e^{-iv}))]\,dv. \tag{434.17}$$

He concluded by converting (434.17) into a trigonometric form free from i (pp. 569–576); alternatively, one could work directly on $W(\alpha, s)$ in (434.14) for a particular Lagrange equation (434.7).

Both these papers by Parseval received an unusual compliment in 1813, when Professor Doctor Jean Philippe Gruson presented two papers, written in German, to the Berlin Academy, which published them as *1816a* and *1816b*. The coincidence of titles, ideas, phraseology, notations and examples between these two "contributions to knowledge" and Parseval's papers is amazing. A delightful piece of cheek in the paper on Parseval's theorem occurs at the phrase '[...] Integration durch bestimmte Integral (*intégrale définie*)' (Gruson *1816a*, 24).

Nobody seems to have told the *classe* of Gruson's thieving, although Abel mentioned some of it in a letter of 1823 to Holmboe (Abel *Correspondance*, 6). Gruson had already published translations into German of the first two volumes of Lacroix's *Traité*, Lagrange's *Fonctions*, and some works by Euler; presumably he felt that with small fry like Parseval, he could help himself without acknowledgement.

4.3.5 Brisson: linear operators, 1808

The career of Barnabé Brisson was intertwined with Biot's in its early years. Born in Lyon in 1777, he also entered the *Ecole des Ponts et Chaussées* just before the *Ecole Polytechnique* was started but then transferred to the founding year of the new school. By then his family was living at Beauvais, where we saw in § 4.2.1 the young Biot taking a teaching post at the *école centrale*; Biot also married Brisson's sister there in 1797. After that their careers diverged, for Brisson entered the *Ecole des Ponts et Chaussées* and spent his career in the *Corps*; he also taught at that school from 1820. He died in 1828, when he was being considered as possible candidate for Laplace's place at the *Académie*.[1])

Brisson's research was largely devoted to engineering: two examples will be described in § 8.3.3 and § 17.7.4. But he also worked on differential equations, both in applications (he contributed to a paper on spheroids by Biot (§ 7.2.6)) and in general theory, which is our concern here. In view of his engineering interests,

[1]) According to the anonymous obituary Brisson *1828a'*. The most substantial obituary is Jullien *1828a*. He left a great impression both on his *Corps*, who struck a medal in his memory (*PV*, 9, 427), and on the students and the *EPC* (see Brisson *1830b'*). Student notes of his lectures of 1820–1821 at *EPC* on canals and rivers survive at *AN*, F^{14} 11057, and a large file of professional and career documents is held at 2182^1; on his role in the building of canals, see also 13859^2, and § 8.3.3. His wife, who was Monge's niece, was to survive him by sixty years (see Pothier *1887a*, 386).

On Brisson's paper (about to be described) see Petrova *1974a*; and on the tradition of symbolical methods in which he participated, see Lyusternik and Petrova *1977a*.

the paper to be described here is rather surprising, for it shows the spirit of algebraic speculation common to all works discussed in this section. Submitted to the *classe* in June 1803 (*PV*, *2*, 673), and again for him by Biot exactly a year later (*PV*, *3*, 103), it was never assessed, although Biot was one of the reporters appointed. Brisson added a section in July 1807, and the whole was published as *1808a* in the *Journal de l'Ecole Polytechnique*. He submitted a sequel paper to the *Académie* in 1823, which caused some disturbance (§ 11.5.6); but it is now lost.

Brisson was anxious to convert solutions of partial differential equations into closed forms, especially those containing integrals. For this purpose he began with a series expansion of a function of the product of two functions (*1808a*, 192–197). I shall not use his notations, since his symbols for "any" function included 'Π' and 'Γ'; nor shall I dwell on the proofs, but they of an algebraic cast which is discussed in the next two sub-sections. His first step was a rather vaguely argued claim for the relationship

$$V(f(u)g(u)) = g(u)V(f(u)) + M_{g(u)}(f(u), g(u)), \tag{435.1}$$

where M was a linear function of f and g which was zero if g were a constant function. Iterating (435.1) to generate a sequence of such functions, he eventually came to the expansion

$$V(f(u)g(u)) = \sum_{r=0}^{\infty} h_r(f(u))g^{(r)}(u)/r!, \tag{435.2}$$

where the $\{h_r\}$ were defined inductively as follows:

$$h_0(f(u)) := V(f(u)), \quad h_n(f(u)) := h_{n-1}(uf(u)) - h_{n-1}(f(u)). \tag{435.3}$$

Brisson's statement of this latter recursion is a good example of his algebraic style; to quote his notations for a moment, he actually expressed h_2 in the form

$$\text{`}\Gamma^2\left(\frac{V\cdot\Psi u}{u^2}\right) = \Gamma\left(\Gamma\left(\frac{V\Psi u}{u}\right)\Big/ u\right)\text{'} \tag{435.4}$$

(p. 196). He also took the normal view about questions of rigour: (435.2) was true for "any" function, apart from exceptional cases involving infinite values in the functions: convergence was not a problem. His result (435.2) makes an interesting comparison with Burmann's (343.3), as another extension of Taylor's series. Among other similarities with work already studied, at one point Brisson took the power series

$$g(x+s) := \sum_{r=0}^{\infty} s^r g^{(r)}(x)/r!, \tag{435.5}$$

and $$Q(\alpha) := -se^{\alpha s}\int W(\alpha)e^{-\alpha s}ds = \sum_{r=0}^{\infty} s^{-r}W^{(r)}(\alpha), \tag{435.6}$$

multiplied them together, put e^{iu} and also e^{-iu} for s, integrated over $[0, \pi]$ of u and came to a form of Parseval's theorem (434.2):

$$\sum_{r=0}^{\infty} W^{(r)}(\alpha)g^{(r)}(x)/r! = (e^{\alpha x}/\pi)\int_0^{\pi} [Q(e^{iu})g(x+e^{iu}) + Q(e^{-iu})g(x+e^{-iu})]du \tag{435.7}$$

(pp. 240-242). The result appears in the main body of his paper, written before the publication of Parseval *1806e*: he did not mention Parseval in his 1807 addition.

Thus Brisson's paper has various similarities with other work of the time. It also shared the common interest in power-series solutions of partial differential equations, for in the body of his paper he took forms such as $f(u)g(u)$ for the dependent variable z of the equation and used (435.2), and related equations and corollories such as (435.3, 7), to convert power-series solutions into closed forms. (435.7) itself, for example, led him to a double-integral solution of a class of differential equations in two independent variables (pp. 243-245). The novelties of the paper lay in the generality of the treatment, extended to first-degree equations of any order and in any number of variables; on the possibility of finding the 'complete integral' of such equations by using subsidiary equations to determine each of the 'particular values'; and especially in the use of the differential equation as a linear operator on the possible solution functions. He wrote the equation in the form '$\nabla z = 0$' (in those days '∇' had no particular connection with specific operators) and applied ∇ to the various solution functions. In a typical passage on p. 238, 'we represent the equation which concerns us by

$$\xi + \nabla\xi = M\text{'}, \tag{435.8}$$

which by iteration on itself led to

$$\xi = \sum_{r=1}^{\infty} (-1)^{r+1}\nabla^r M. \tag{435.9}$$

If the powers on differential operators were read as indices—that is, for example, if z_{xx} were taken to be $(z_x)^2$—then ∇M changed into another expression, written (confusingly) as $\nabla' M$, and (435.9) became

$$\xi = \sum_{r=1}^{\infty} (-1)^{r+1}\,\nabla'^r M = (1+\nabla' M)^{-1}. \tag{435.10}$$

Many of Brisson's results involved conversions from powers to indices, often effected by inserting a solution form $e^{\alpha x}V(y)$ into $\nabla M = 0$ (when it had two independent variables) and obtaining an ordinary differential equation $\nabla' V = 0$ by separation of variables. The rigour is, shall we say, optimistic, but the conversion of a differential equation into operator form, and the procedures that thereby followed, were ideas which flourished in others' hands, as we shall see in § 10.4.

4.3.6 Mathematics from Alsace: Arbogast, Jacques Français and the calculus of derivations

> Thereupon Arbogast gave D the new meaning of 'derivation'. It looked indeed as if the different mathematical architects engaged in erecting a proud mathematical structure found themselves confronted with the curse of having their sign language confounded so that they could less readily understand each other's speech.
>
> Cajori *Notations*, *2* (1929), 225–226

Like Parseval, Arbogast was an outsider, even outside Paris. Born in 1759 at Mutzig, near Strasbourg, he spent most of his life in Alsace, and died in 1803 at Strasbourg. His contacts with Paris seem to have been limited, and mostly concerned with some teaching at the *Collège Royal* and with his position as a *député* in the *Convention Nationale* from 1791; one aspect of those activities is noted in § 5.5.4. He was also only a part-time mathematician, making his living as a lawyer. Although he kept his involvement in the foundation of the *Ecole Polytechnique* to a minimum (§ 2.3.1), he was active in mathematical education in his own area, holding chairs and/or rectorships at various colleges and universities there. He was also very interested in the history of mathematics, especially concerning Marin Mersenne, and built up a large and valuable library. His contacts with other mathematicians seem to have been fairly slight, although we know of his extensive correspondence with Charpit (§ 3.3.4). He was elected a *correspondant* to the *Académie* in 1792, becoming an *associé non résidant* in 1796 soon after the *classe* was set up.[1])

Closely linked with Arbogast were François Joseph Français and his younger brother Jacques Frédéric, both Alsatian mathematicians strongly influenced by him and by Lagrange. François taught in Alsace, La Fère (near Reims) and Mainz, dying there in 1810 in his 43rd year. Jacques studied at the *Ecole Polytechnique* from 1797 to 1800 and then entered the *Ecole du Génie* at Metz: he was

---

[1]) A most useful guide to his life and career is Fréchet *1920a*. On the known extant parts of his *Nachlass*, see my and Engelsman's *1982a*.

to teach there from 1811 until his death in 1833, aged 58 years. Arbogast left his library to François, who in turn left it to his brother, who sold it in the 1820s (see J.F.Français *1823a*). Their research work is as similar at times as their names; authorship is welcomly unravelled in Taton *1972a*, which adds to the information given in Eneström *1903a*.

We saw in § 3.3.2 that Arbogast won a prize offered by the St. Petersburg Academy on solutions to partial differential equations with an essay *1791a* where a distinction between contiguity and discontiguity emerged somewhat uneasily out of an algebraic approach. Shortly before, he had submitted a manuscript *1789a* to the Paris *Académie* on the principles of the calculus (see Panza *1987a*); he followed Lagrange's approach and received Lagrange's praise in *Functions*$_1$ (1797), 5. This paper contained the seeds of his book *Du calcul des dérivations*, published in Strasbourg in 1800, which is our concern here.

The book is guided by two aims. Firstly, Arbogast put forward a generalisation by analogy, in which the basic notion was not the derived function but a 'derivation', which had the same kinds of properties and of which the derived function was a special case. Secondly, it was not clear from Lagrange's writings that "any" function could be expanded in a power series; so Arbogast gave rules for determining the coefficients of a power series of derivations from various standard forms of function then used. As he expressed it when sending a copy of the book to Lagrange in September 1800, 'I intended there to offer a type of calculus which embraces the theory of series and of which the differential calculus is a special case'.[1]) He asked for the book to be presented to the *classe*, and this was done at the end of the month (*PV*, *2*, 232). A month later Lacroix gave a verbal report, whereupon Biot, in typical manner, noted that a paper of his own on difference equations, which had already been approved by the *classe* but still awaited publication (§ 4.4.1), was 'relative to certain objects that citizen Arbogast has treated' (p. 264).

Lalande gave the book some early publicity (in Montucla *History*, *4* (1802), 659-661). But already François Français, who had helped Arbogast with the book (see Arbogast *1800a*, xiv), had published a long and admiring review *1800a* in the *Moniteur universel*, which was reprinted in two journals. In addition, he used the theory: a submission to the *classe* of November 1804 (*PV*, *3*, 159), which was recommended for the *Savants étrangers* in Biot's report *1805a* five months later, applied it to the motions of trajectories. But, as usual, François's paper never ap-

[1]) This letter, dated 22 *fructidor an* 8 (9.9.1800) is addressed to 'illustre confrère': the tone strongly suggests that Lagrange was the recipient. It is kept at *AS, dossier personnel* for Arbogast. The original passage translated reads: 'Je me suis proposé d'y offrir un genre de calcul qui embrasse la théorie des suites et dont le calcul différentiel est un cas particulier'.

peared there: it is now lost (§ 6.4.2). In fact, he published no papers, and from his *Nachlass* Jacques brought out only one paper on the calculus, a short essay *1811a* in Gergonne's *Annales* on obtaining the derived functions of the elementary functions using only the sum and product rules for differentiation.[1])

However, Jacques was also interested in Arbogast's methods, and operator algebras in general. He presented a paper himself to the *classe* in June 1803 (*PV*, *2*, 682) but withdrew it for lack of satisfaction with the method (see his *1813a*, 254). After his brother's death he presented another paper in November 1811 (*PV*, *4*, 554): it was ignored there but appeared in Gergonne's *Annales* as *1813a*. It contained an extensive vindication of the algebras of the D (written '∂'), Δ and E operators, converting various formulae into operator notation and obtaining other known results as corollaries. Then in *1815a*, which also appeared in the *Annales*, he summarised and simplified Arbogast's basic procedures. I shall make use of these papers as a guide to Arbogast's sometimes intriguing but often boring 404 pages, which, as Jacques put it, contained some 'truly remarkable results among a crowd of things which are, to put it thus, *hors d'oeuvre*' (*1815a*, 62).

The basic procedure of Arbogast's theory was to generate successively the coefficients in the power-series expansion of a function:

$$F(\alpha + x) = \sum_{r=0}^{\infty} a_r x^r / r! . \tag{436.1}$$

'Designating by D the operation that one must make on $F\alpha$ in order to deduce $[a_1]$ from it', he wrote

$$a_1 = \mathrm{D}F(\alpha) , \tag{436.2}$$

and thus by induction on the operator D,

$$a_2 = \mathrm{D}a_1 = \mathrm{D}^2 F(\alpha) , \; a_3 = \mathrm{D}^3 F(\alpha), \; \ldots . \tag{436.3}$$

Hence (436.1) became

$$F(\alpha + x) = \sum_{r=0}^{\infty} x^r \mathrm{D}^r F(\alpha) / r! \tag{436.4}$$

(*1800a*, 1-3).

Equation (436.4) reduced to Taylor's series when derivation was taken to be

---

[1]) A manuscript by François on the calculus of operations, which was supposed to have gone to the *Biblioteca Medicea-Laurenziana* in Florence in the late 19th century as part of the Ashburnham collection (compare § 3.3.4), was never received there.

differentiation, which Arbogast studied on pp. 305–324. He moved on to more general forms when considering the increment βx on α. He wrote that expansion in the form

$$F(\alpha+\beta x)=\sum_{r=0}^{\infty} x^r \mathrm{D}_c^r.\, F(\alpha), \tag{436.5}$$

where $\mathrm{D}_c^r.\, F(\alpha):=\mathrm{D}^r.\, F(\alpha)/r!\,.$ (436.6)

'D.' indicated the derivation given by the increment βx on α; 'D' corresponded to the case of $\beta=1$; 'D_c' absorbed the factorial in 'D.' (pp. 3–9 with p. 33).

The power of the method became more evident when functions of polynomials, and functions of functions, were considered; but then the laboriousness of Arbogast's procedures also strikes the eye. From (436.5),

$$\varphi(F(\alpha+\beta x))=\varphi\left(\sum_{r=0}^{\infty} x^r \mathrm{D}_c^r.\, F(\alpha)\right)=\sum_{r=0}^{\infty} x^r \mathrm{D}_c^r.\, \varphi(F(\alpha)), \tag{436.7}$$

and the calculation of $\mathrm{D}_c^r.\, \varphi(F(\alpha))$ involved some ponderous and partly repetitive work on the right hand side of (436.7) (pp. 3–13). Arbogast introduced a simplification by treating a function of a function as a function of a polynomial. Defining β as

$$\beta=\beta(x):=\sum_{r=1}^{\infty} a_r x^{r-1}, \tag{436.8}$$

if $\varphi\left(\sum_{r=0}^{\infty} a_r x^r\right)=\varphi(a_0+\beta x)=:\sum_{r=0}^{\infty} A_r x^r,$ (436.9)

then $A_n=\mathrm{D}_c^n.\, \varphi(a_0)=\sum_{r=1}^{n}(\mathrm{D}_c^r\varphi(a_0))\,(\mathrm{D}_c^{n-r}.\, a_1^r)$ (436.10)

(pp. 13–14). He followed his proof of (436.10) with various discussions and examples of successively calculating coefficients (pp. 14–38).

Jacques Français claimed that there was a basic conceptual distinction between differentials and derivations, in that differentials were concerned with variables but derivations, being a theory of operations, dealt with constants (*1815a*, 65). He also pointed out the underlying difference between D and D.: the definition of $\mathrm{D}^r\varphi(F(\alpha))$ was based on assuming that

$$\mathrm{D}F(\alpha)=1,\quad \mathrm{D}^r F(\alpha)=0 \quad \text{when} \quad r>1, \tag{436.11}$$

while $\mathrm{D}^r.\, \varphi(F(\alpha))$ did not make such assumptions but was a derivation of

$\varphi\left(\sum_{r=1}^{\infty} x^r \mathrm{D}_c^r F(\alpha)\right)$ (pp. 66-68). (In terms of an analogy with Euler's approach to the calculus (§ 3.2.4), D corresponded to a constant differential dx for which $d^r x = 0$ and thus x was an independent variable; D. corresponded to the case of non-zero higher-order differentials.) Thus the passage from D to D. was one from equal to non-equal derivations; no new principle after Taylor's series was involved (pp. 110-111).

Français also suggested a more convenient way of handling the derivations of $\varphi\left(\sum_{r=0}^{\infty} a_r x^r\right)$. Instead of Arbogast's "blockbuster" approach (436.9-10), he started from (436.7) in the form

$$\varphi(a + a_1 x) = \sum_{r=0}^{\infty} x^r \mathrm{D}_c^r.\, \varphi(a), \quad \text{where} \quad a := \varphi(\alpha). \tag{436.12}$$

From (436.11, 12_2), $(436.12)_1$ became

$$\varphi(a + a_1 x) = \sum_{r=0}^{\infty} (a_1 x)^r \mathrm{D}_c^r.\, \varphi(a). \tag{436.13}$$

Putting $(a_1 + a_2 x)$ for a_1 in (435.13), then $(a_2 + a_3 x)$ for a_2, and so on, then the expansion of $\varphi\left(\sum_{r=0}^{\infty} a_r x^r\right)$ could be built up (pp. 94-98).

Most of Arbogast's book was concerned with finding expansion formulae for sets of polynomials, multinomials, functions of several variables and various standard forms of function of the time (for example, inverse powers of polynomials). The applications covered a wide range of techniques in the calculus, including several intems discussed in this and the last chapters. As an example, take his treatment of Lagrange's series on p. 252. He aimed at the more general form (341.15) of the series rather than the case (341.14) which Parseval studied (§ 4.3.4): given

$$\alpha - t + y\varphi(t) = 0, \tag{436.14}$$

find a series expansion for $\psi(t)$. He had obtained on p. 236 an expansion for the form

$$\sum_{r=0}^{\infty} \left[a_r x^r \left(\sum_{s=0}^{\infty} c_s x^s \right)^{-mr} \right], \tag{436.15}$$

and he used (436.4) to convert (436.14) into the special case of (436.15) on p. 251:

$$y = x/\varphi(\alpha + x) = \left[\sum_{r=0}^{\infty} x^r \, \mathrm{D}_c^r \varphi(\alpha)\right]^{-1}, \quad \text{where} \quad x := t - \alpha. \tag{436.16}$$

The required function $\psi(t)$ became, from $(436.16)_2$,

$$\psi(t) = \psi(\alpha + x) = \sum_{r=0}^{\infty} x^r \mathrm{D}^r \psi(\alpha), \tag{436.17}$$

and comparing (436.17) with the expansion of (436.15) gave immediately

$$\psi(t) = \sum_{r=0}^{\infty} [y^r \mathrm{D}_c^{r-1} (\varphi^r(\alpha) \, \mathrm{D}\psi(\alpha))/r], \tag{436.18}$$

which was Lagrange's (341.15) when 'D' was taken as the differential operator.

Français gave various examples of operator algebras, especially in his first paper. In one example he took E as the forward operator, so that not only could Lagrange's (341.4) be written as

$$\varphi(x + h) = \mathrm{E}^h \varphi(x) = e^{h\mathrm{D}} \varphi(x) \tag{436.19}$$

but also the purely operational components be stated separately:

$$\mathrm{E} = e^{\mathrm{D}} \tag{436.20}$$

from (436.19) itself, and

$$\mathrm{D} = \ln(1 + \Delta) \tag{436.21}$$

from Euler's (341.1) (*1813a*, 250).

Arbogast's ideas, together with Français's stimulus, provoked a modest but continuing line of interest throughout the 19th century; for as Arbogast's own applications showed, the need to convert between a given form and a power-series expansion can occur in many areas of mathematics.[1]) The procedures not only fitted in well with the Lagrangian tradition of his own time, but also helped to encourage the development of new algebras of operators. However, over time Arbogast's aim became re-interpreted: *he* regarded his ideas as applicable to "all" functions, whereas they came to be seen as a restriction of functions expressible as combinations of operations. Further, the similarities of both terminology

[1]) For a useful survey of many papers see Tanner *1891a*. However, his account on pp. 83–87 of Arbogast's book is confusing, since he writes '*D*' for Arbogast's 'D_e.', 'Δ' for 'D_c', and 'δ' for '*D.*'.

and notations with those of the calculus were unfortunate, as my quotation from Cajori at the head of this sub-section indicates. Again, the rigour of the procedures was unclear: even Français prefaced his 'certain and rigorous' procedure (436.12–13) with the comment that Arbogast's methods of determining the coefficients 'are only conclusions of induction' and so 'can leave some doubt over the exactitude of the results that they furnish' (*1815a*, 94). This latter difficulty is still more evident in our next figure.

4.3.7 Servois: the algebra of functions

> The introduction of analogical reasoning in a branch of the pure sciences, is most decidedly objectionable, and it is used very freely in treating the doctrine of Functions. There is also a want of definitions, of axioms, and of principles, which give it a character not very unlike the researches in alchemy.
>
> Anonymous *1824a*

The career of François Joseph Servois is strangely entwined with those of the Français brothers. He had the same christian names of the elder brother and was born a year earlier, in 1767, also "into trade", at Mont-de-Laval in the Doubs area south of Alsace. Like François he began in the church, but like Jacques, he spent much of his time in the army. After entering the *Ecole d'Artillerie* at Chalons in 1794 (as a rather mature student), he taught at several such schools; briefly at the one in Besançon (on Legendre's recommendation) in 1801 before going back to Chalons, just in time for it to be moved into the *Ecole du Génie* at Metz (§ 2.4.4). He taught there until 1808, three years before Jacques arrived there, and moved to the *Ecole d'Artillerie* at La Fère, some time after François had left it. For a few months of 1816 he was back at Metz (with Jacques); but then he was called to become the *conservateur* of the *Musée d'Artillerie* in Paris. He stayed there until his retirement in 1827, when he returned to his home village, dying there in 1847. Like Jacques, he was unmarried; like François, he was an unsuccessful candidate as *correspondant* of the *classe* in March 1807 (*PV*, *3*, 504–505); like both of them, he was interested in Lagrange's and Arbogast's views on functions and operators, and published his ideas in Gergonne's *Annales*.[1])

[1]) The principal study of Servois's life is J. Boyer *1894a*. A file on his career is held at *VT, classement général alphabétique*, 1791–1848, where there is also a disappointing dossier on Jacques Français.

Servois first submitted a paper to the *classe* in February 1805 (*PV*, *3*, 179), and another one in October 1810 (*PV*, *4*, 383). Lacroix read a report in October 1812 on a paper by Servois which he described as an improved *résumé* of one submitted earlier and then withdrawn; perhaps the 1805 paper was the first version, and the 1810 paper this revision. He recommended that the two be merged and published in the *Savants étrangers* (Lacroix *1812a*, 100–101). Servois knew better than to wait for the *classe* to publish; but he seems to have followed Lacroix's advice, for in the end two papers, *1814a* and *1814b*, appeared consecutively in Gergonne's *Annales*, with a footnote at the beginning of the first indicating its origins in the earlier papers. He also reprinted the two papers together as a short book *1814c*: I shall cite the *Annales* version, as it is now rather less inaccessible.

§ 3.2.6 began with Servois's strong criticisms of infinitesimals (*1814b*, 143). He wanted instead to render the calculus as part of algebra, using the algebras of differences, functional equations and differential operators. He gave a survey of the algebraic properties of the difference operator Δ, the forward operator E, the integral $\int$ and the differential d, and ordinary functions. Some operators and functions possessed certain properties, others not; for example,

$$a\mathrm{E}z = \mathrm{E}az \quad \text{but} \quad \sin az \neq a \sin z. \tag{437.1}$$

To us (437.1) is concerned with commutativity; and so it was for Servois himself, for (437.1) followed his introduction of the words 'commutativity' and 'distributivity' into mathematics.[1] The passages are so admirably clear (*1814a*, 98) that they deserve quotation:

> 3. Let
>
> $$\varphi(x + y + \ldots) = \varphi x + \varphi y + \ldots \qquad [(437.2)]$$
>
> The functions which, like φ, are such that the function of the (algebraic) *sum* of any number of quantities is equal to the sum of corresponding functions of each of these quantities, will be called *distributive* [...]
>
> 4. Let
>
> $$\mathrm{f}fz = f\mathrm{f}z. \qquad [(437.3)]$$
>
> The functions which, like f and f, are such that they give identical results, whatever be the order in which one applies them to the subject, will be called *commutative between each other*.

[1]) In particular among earlier uses of these properties, Legendre had emphasized them, and also 'associativity' (his word), in the opening pages of his treatise *Numbers* on number theory, which appeared in editions of 1798, 1808 and 1830.

Among the influences on Servois's desire to emphasise these properties could have been Jacques Français's paper *1813a*, which contained not only many examples but also, for example, explicit remarks on p. 246 on the interchangability of operators and constants.

However, one must not expect to see an axiomatic approach attend the introduction of these terms. Like all figures discussed in this chapter, Servois was moving in the world of algebraic manipulations and inductions, not of algebraic structures, which were to gain prominence only towards the end of the century.[1]) For example, in *1814a*, 108 he took Euler's formula (341.1)

$$dz = \sum_{r=1}^{\infty} [(-1)^{r+1} (\Delta^r z)/r], \qquad (437.4)$$

which we saw Jacques Français express in operator form (436.21), as a definition of the differential dz (although it is not valid for periodic functions for which $\Delta z = 0z$): he used it rather than Lagrange's (341.4) as his means of obtaining relations between differentials and differences. He also produced the standard expansions of simple functions, relations such as Burmann's series (343.3), and expressions for total and partial differences and differentials.

Servois was quite aware of the use of analogy in forming definition (437.4). He described it there as '*infinitinome*', and compared it with the expansion of $\ln(1 + a)$ 'thus, by simple analogy, the differential is *like* the logarithm of the variable state $(z + \Delta z)$' (*1814b*, 142). In a succeeding paper *1817a* he produced similar algebraic expressions for numerical quadrature, including Lagrange's interpolation formula (352.14).

For Servois Lagrange's approach to the calculus was, of course, dominant. However, he seemed also to accept that 'the means of union between the calculus and geometry: necessarily the *method of limits*' (*1814b*, 143). Infinitesimals, as we have already noted, were for him a horror—as much as was his favour for analogy to his anonymous British contemporary quoted at the head of this subsection!

4.3.8 Hoëné Wroński as a critic

If Parseval, Brisson, Arbogast, the Français brothers and Servois were outsiders, then Józef Maria Hoëné Wroński was positively out of sight. Born in Poland in

---

[1]) See Wussing *1969a* for an excellent account of the emergence of group theory. Novy *1973a* gives some indications of the state of mid- and late-19th-century algebra.

1776, he followed a military career for a few years, then studied at German universities until 1800, when he came to France. He lived in Marseille for ten years, and then moved to Paris for the rest of his life, which ended in 1853.

Wroński is best remembered for his philosophy, which began as a divine revelation in 1803 and became an allegedly rational 'absolute philosophy' which would solve all problems. Its roots lay in Kant, but it was fed on the compost of Jacob Böhme and the cabalists. He charged his backer, a banker called Pierre Arson, 200,000 *francs* for knowledge of the Absolute; but Arson felt that the Absolute was not worth the money, and in consequence a libel suit took place in 1819, which Wroński lost absolutely.

Arson's support began in the 1800s, when Wroński's mathematical research also started. Wroński wrote criticisms of views of Lagrange and Laplace on mathematics, and advocated his own universal approach instead. Later he wrote extensively on astronomy, mechanics and machines, though in his later life explanation of the Absolute took up more of his time. He made a living from private teaching to those such as Arson who would pay, and had enough money to publish his works as books and pamphlets with various Paris publishers. He also submitted two of them to the *classe*, where they attracted some attention. It is easy to see him as Paris's resident lunatic fringe; but amidst the paranoia and the polemicising a perceptive mind peeps through from time to time, and the reactions by *classe* members casts useful light on their own positions.[1])

Wroński's first main mathematical work was his introduction *1811a* to the 'philosophy of mathematics, and technics [*technie*] of the algorithm'. The algorithm used functions which were distributive with respect to addition and multiplication (p. 10), to use Servois's (later) terminology. Wroński also manipulated differentials, integrals, differences and sums, although with much less care than Servois was to show (as Servois himself noted in *1814b*, 154–155). After dismissing Taylor's and Lagrange's series as only rather special (p. 209), he came to some primitive algorithms, and finally to the 'Absolute Law' (p. 252), which he would repeat again and again in later writings:

$$F(x) = \sum_{r=0}^{\infty} A_r \Omega_r(x), \tag{438.1}$$

where the $\{\Omega_r(x)\}$ were arbitrary 'generating functions'. They and their coefficients could be calculated from given relations by the algorithms; the details

[1]) Dickstein wrote several works on Wroński, especially a biography (with bibliography) *1896*a: *1892a* is a short survey of his mathematics. A few manuscripts are preserved in *AS*, *dossier personnel* for Wroński. The library of the *Institut Mittag-Leffler* (Stockholm) possesses many of his publications.

filled up many pages. After all that, the philosophical structure was summarised in a synoptic table.

Wroński never actually offered a proof of (438.1), but extolled its merits at length. Its complete generality and reliance only on algorithms to calculate values of $F(x)$ marked a change in the philosophy of mathematics (*1811a*, 256–257), which hitherto had used only inductions

> and, in consequence, had only a relative or apparent generality. Their intellectual origin, the source from which one would be able to derive their absolute generality, was unknown. Thus, the limits or the conditions of the application of these principles, could not be fixed; and in consequence [...] they forced one to declare that, in the question of mathematical principles, THE GREATEST GEOMETERS ARE OBLIGED TO CONFESS FRANKLY THEIR IGNORANCE.

The greatest geometers responded nobly. Wroński had submitted the manuscript of this work to the *classe* in August and September 1810 (*PV*, *4*, 373, 382), and in the report that Lacroix rapidly prepared for the *classe* (and which was published later in the *Moniteur universel*), he stated: 'But what has struck your commissioners in the Memoir of Mr. Wronsky [*sic*], it is that he obtains from his formula all those that one knows for the development of functions, and which are only special cases of it' (Lacroix *1810a*, 386: Lagrange, no less, was his co-signatory). Thus the aim of generality embodied in (458.1) seemed to them to more than compensate for the lack of proof (on which Lacroix had expressed reservations). The praise of Gergonne and Thomas-Lavernède was still greater: in 'an announcement' in their new *Annales* they regarded Wroński's work as among those which 'tends to produce some revolution in the exact sciences', and they proposed to 'translate into French' its doctrines in a series of articles, in order 'to make known its substance' (*1811a*, 66–67).

The geometers must have come to rue their praise. Lacroix told the *classe* in June 1811 that the printed version of the work 'has nothing in common' with the manuscript on which he had reported (*PV*, *4*, 489), but it was too late. Wroński quoted Lacroix's earlier praise many times in his later writings, beginning with his next submission to the *classe*, which came in during September 1811 (*PV*, *4*, 522). Here he "refuted" Lagrange's theory of functions (and similar theories, such as Arbogast's). This time Arago *1811c* reported, in November; but he found little to extol. For example, Wroński had objected to Lagrange's rejection of fractional powers of h in the series expansion of $F(x+h)$, on the feeble grounds that perhaps the different values would compensate when added up, so preserving a unique value for $F(x+h)$. We noted at (433.6) Poisson's attention to multivaluedness in his *1811e*; but when Wroński published his manuscript, he not only reprinted Arago's report and castigated it generally (*1812a*, 83–110) but also added a long section (pp. 41–81) on Poisson's proof *1805a* of Taylor's series (§ 4.3.3). All his criticisms were based in the same way on the inferior generality of the expansions as compared with his own absolute law (458.1), although Poisson's failure with Taylor's series was also due to his assuming that a function could be so expanded in the first place (p. 53). This interesting remark, prophetic of Cauchy (§ 11.4.3), was supplied with a remarkable explanation: Poisson's error had occurred because of the influence on him of materialistic philosophy (p. 54).

In his later publications Wroński became rather boring. A general resolution *1812b* of 'equations of all degrees', based on (458.1), was criticised in Gergonne *1812a* for clumsiness and in *1812b* for inapplicability beyond quartics (a view reinforced in Ruffini *1818a*). When Wroński *1812c* wrote in the *Moniteur universel* on the difference between the form and the nature of roots of equations, Gergonne *1813a* responded by saying that he did not understand the distinction, and invited Wroński to explain himself in the pages of the *Annales* (where Servois delivered lengthy philosophical criticisms of Wroński in his *1814b*). Instead Wroński published a philosophical book *1814a* on the foundations of the calculus; Carnot's views (§ 3.2.6) were castigated for begging the question, the second (1813) edition of Lagrange's *Fonctions* was found to be basically no better than the first, and Napoléon was invited to launch transcendental philosophy in Europe. A large three-part successor *1815–1817a* to *1811a* went over the ground in still more detail (or more words and symbols, anyway). Then Laplace's generating functions (352.1) were let off with a mere 'critique' *1819a*, concerned with the lack of generality and the use of induction. By 1819 Wroński was busy not obtaining 200,000 *francs* from Arson, and so his work subsided for a time, enough to allow us to call a halt.

What survives of value out of this flummery? The issue of generality and the status of inductive proofs show aspects of philosophical thinking in mathematics of the time. As for technicalities, the calculation of the coefficients in (458.1) contained useful passages; some power series were similar to Burmann's (343.3), and certain combinatorial sums led later to the name 'Wrońskian' being applied to a particular kind of determinant.[1]) The supreme law does encompass many ex-

---

[1]) For commentary on these 19th-century developments, see West *1886a* and Petrova and Romanovska *1979a*. A good summary of Wroński's method is given in the mathematical dictionary of his brother-in-law Montferrier (*1836a, 2*, 448–459, 544–551), and a modern appraisal is provided in Banach *1939a*. On Wroński's criticisms of Carnot, see Guerraggio and Panza *1985a*. His contributions to mechanics, engineering and probability are dismissed in Bertrand *1897a*.

In places Wroński's text looks uncannily like Cantor's set theory. For example, in *1811a*, 65 he used 'ℵ'-functions for (in effect) symmetric functions of roots of equations, and when working with infinite systems of equations he allowed on p. 90 different orders of the actual infinite. His account of the calculations in *1815–1817a* may be slightly less difficult to follow. Incidentally, he gave the letter

pansions known at the time; not only Lagrange's series and Laplace's generating function but also Burmann's series (343.3). Yet even as we give credit to Wroński, he himself tries to take it away; for in a letter to the *Moniteur universel*, in response to the publication there of Lacroix's report *1810a*, Wroński *1810a* explained that the algorithms were only examples of the absolute law (458.1), in which the generating functions were quite general.

The great geometers initially showed some appreciation of Wroński's efforts, as we have seen. In the autumn of 1811 he even gave a course of lectures on his ideas, with *Université* approval.[1]) But as the 1810s wore on, more and more noise came from him, and less and less new signal. Gergonne came to express the feelings of doubtless all geometers, both great and small (*1813a*, 209):

> If [...] Mr. Wroński persists in enveloping himself in obscurities [*ténèbres*], and only offers us promises; if he confines himself thus to what he has done up to now, to explain enigmas by other enigmas; if, in a word, he neglects to legitimate *his assertions by rigorous calculations*; I will be authorised to think that he writes views *completely foreign to science*, and consequently justified in no longer *giving any attention* in the future to his productions.

4.4 Differential and difference equations

The tradition of preferring power-series and functional solutions to (partial) differential equations was maintained; as some of the best contexts occur in physical interpretations, I postpone examples of their use until Chapter 5 onwards. In this section I have chosen a few papers by Biot, Poisson and Parseval, which are not widely known but which show well how some 18th-century procedures were modified and extended. Lacroix *Calculus*$_2$, *2* (1814) is the most detailed general source of information for the period.

4.4.1 Biot and Poisson on singular solutions to difference equations, 1802–1806

In his lectures on difference equations at the *Ecole Polytechnique*, de Prony devoted a short section to general and singular solutions of linear difference equa-

---

'ש' to Wronskians of any order.

In *1814a* Wroński reprinted most of the strange letter on Lagrange (§ 2.5.1) that had recently appeared in *MU*. He conjectured, without success, on its author (Maurice *1814a*).

[1]) See the announcement in *MU*, (1811), 878.

tions of degree above the first. 'Students who would like further details', he told them, 'will be able to consult several papers by Monge and by Charles, published in the collections of the *Académie des Sciences* from 1783 to 1788' (de Prony *1796a*, 502).

Biot began his scientific career by taking this advice, for in February 1797 he submitted a paper to the *classe* (*PV*, *1*, 174). de Prony himself eventually presented a lengthy report at the end of November 1799, mis-dating the date of submission by several months but recommending either the publication of the paper in the *Savants étrangers* or his own report in the *Mémoires*. The latter option was, of course, taken up, the report appearing as de Prony *1801a*. Biot submitted another (version of his?) paper in January 1800 (*PV*, *2*, 87), and either this or the original paper was published in the *Journal de l'Ecole Polytechnique* as *1802a*. Lacroix had already mentioned the paper in his *Traité* (*Calculus*$_1$, *3* (1800), 237), and gave a lengthy account of it in the second edition (*Calculus*$_2$, *3* (1819), 250–260). Both Lacroix and de Prony expanded on Biot's geometrical interpretation of his results; I shall confine myself to the results themselves.

Biot imagined that we have a relationship F between the variable x, the constant a, and the variable $y(x, a)$, and that an increment on x gave the value x_1 of x. Then

$$F(x, a, y(x, a)) = 0 \quad \text{and} \quad F(x_1, a, y(x_1, a)) = 0\,, \tag{441.1}$$

and eliminating a between these equations furnished the difference equation $Z = 0$. Equation (441.1) should hold if the parameter a were varied; thus

$$F(x_1, a_1, y(x_1, a_1)) = 0\,. \tag{441.2}$$

Now if (441.1_2, 2) were to hold simultaneously to yield $Z = 0$, then

$$y(x_1, a_1) = y(x_1, a)\,, \tag{441.3}$$

when (441.2) became

$$F(x_1, a_1, y(x_1, a)) = 0\,. \tag{441.4}$$

Eliminating $y(x_1, a)$ from (441.1_2, 4) produced a difference equation in a, whose solution showed the law of variation of a (Biot *1802a*, 182–185).

Hence the situation with difference equations was not the same as for differential equations. $(441.1)_1$ incorporated a relation between x and a expressed by the solution to a new difference equation, complete with constant of integration, and thus was a complete solution of $Z = 0$. By contrast, if $(441.1)_1$ were a solution to a *differential* equation, particular values of a gave particular integrals of it. When Charles had studied such questions, especially in his *1791a*, he had presumed that if he set $\Delta x = \Delta y = 0$ in a difference equation to obtain a differential equation, the plurality of solutions of which the former was susceptible became 'particular solutions' of the differential equations. However, there were more of them than had previously been known, including some which did not even satisfy the original differential equation.[1])

Biot attempted to find the sources of such paradoxes. From (441.3),

$$\Delta a\, g(x, a, \Delta a) = 0, \quad \text{where} \quad \Delta a := (a_1 - a), \tag{441.5}$$

where only $\Delta a = 0$ led to $(441.1)_1$; $g(x, a, \Delta a)$ furnished other possibilities. (Compare Lagrange's treatment of singular solutions to differential equations at (336.11): in lecture 18 of the *Leçons* he did not use this approach to solve Charles's difficulties.) In the example

$$\Delta x = x \quad \text{with} \quad y(x, a) = ax - a^2, \tag{441.6}$$

Biot eliminated a to obtain an equation

$$y = \Delta y - (\Delta y)^2/x^2, \tag{441.7}$$

rather similar to one studied by Charles. $(441.1_2, 4_2)$ yielded

$$y(x_1, a) = ax_1 - a^2 \quad \text{and} \quad y(x_1, a) = a_1 x_1 - a_1^2, \tag{441.8}$$

and eliminating $y(x_1, a)$ led in turn to

$$(a_1 - a)\,(x_1 - (a + a_1)) = 0. \tag{441.9}$$

$a_1 = a$ recovered $(441.6)_2$; but $(441.9, 6_1)$ offered the difference equation in a

$$\Delta a = 2(x - a), \tag{441.10}$$

whose solution

$$a = \tfrac{2}{3}x + b(-1)^{\ln x/\ln 2}, \tag{441.11}$$

containing the constant b of integration, could be inserted into $(441.6)_2$ to yield

$$y(x, b) = \tfrac{2}{9}x^3 - \tfrac{1}{3}bx(-1)^{\ln x/\ln 2} - b^2(-1)^{2\ln x/\ln 2}. \tag{441.12}$$

The determination of a and b from $(441.8_1, 12)$ and (441.3) with $a_1 = 1$ led to four solutions to the difference equation (441.7) (Biot *1802a*, 185–188). Similar considerations could explain the plurality of solutions to factorisable difference

[1]) On the identity of Charles, an obscure figure often confused with the physicist J. A. C. Charles, see Gough *1979a* and Hahn *1981a*. Note that Biot was using Laplace's terms (Table 337.1) for types of solution to differential (and difference) equations.

equations, such as

$$(\Delta y)^2 = c^2, \tag{441.13}$$

which had concerned Monge in *1786b*.

Biot was teaching at Beauvais when this paper, and also some of his other early work, was being prepared or submitted to Paris. His surviving letters to Lacroix show that he cultivated Lacroix as his initial supporter, and also gained the still more important favour of Laplace, thus gaining in 1800 the coveted situation of *classe* membership and a Paris chair (§ 4.2.1).[1]) In turn, Poisson used Biot as Biot used Lacroix, including writing papers on the same subjects. He had a similar reception to Biot's in some ways; the two papers we consider now were also recommended to appear in the *Savants étrangers*, but were placed (by Poisson himself: § 2.3.3) in the *Journal de l'Ecole Polytechnique*.

Poisson's first paper, submitted in December 1800 (his 20th year) to the *classe* (*PV*, *2*, 277) and approved by Legendre *1800a* a month later, appeared as Poisson *1802a* (immediately before Biot *1802a*). He used the idea of factorisation, as in Biot's (441.13), to the solution form $(441.1)_1$; but he sought particular values for a rather than let it become a variable. Assuming (without discussion) that F took roots $p, q, r, \ldots$ in a, he obtained an equation which I shall write as

$$F(x, a, y) = \prod_p (a - p) = 0. \tag{441.14}$$

Then, taking forward differences on x and y,

$$F(x_1, a, y) = \prod_{p_1} (a - p_1) = 0; \tag{441.15}$$

hence $$\prod_p F(x, p, y) = \prod_p \prod_{p_1} (p - p_1) = 0. \tag{441.16}$$

He now treated (441.16) in a manner similar to Biot's use of (441.9). For factors such as

$$p - p_1 = 0, \quad \text{then} \quad a = p, \tag{441.17}$$

and all such solutions lead to the complete integral; but the other factors produced quite different solutions, which correspond to Charles's.

---

[1]) The letters are mostly in *Smith*; one survives in *BI*, ms. 2397, and another in *Wellcome*, Biot volume, fol. 1 (see also the manuscript on singular solutions of differential equations at fols. 75–78, a letter to 'Monsieur' in fol. 92, and a seemingly unpublished review of Lacroix *Calculus*$_2$, *2* (1799) on fols. 68–70). § 20.1 contains a transcription of one of the letters in *Smith*: collectively they show Biot's awareness of the work of Parseval and Jacques Français discussed in § 4.3. Frankel *1978a* drew on them to demonstrate Biot cultivating his career.

On Poisson's work, about to be described, see Houtain *1854a*, *passim*.

Poisson's other paper, *1806a*, was submitted in May 1805 (*PV*, *3*, 211) and gained the approval of Lacroix *1806a* after eight months' wait. It dealt with 'particular solutions' to both differential and difference equations, and some applications. Poisson wished to use the result established by Laplace in his *1775a*, art. 6 and recently recalled by Lagrange in lecture 16 of his *Leçons*: a particular solution $Y = X(x)$ of

$$dy/dx = f(x, y) \tag{441.18}$$

had the property that

$$f(x, X(x)) = \infty\,. \tag{441.19}$$

Poisson's development of this idea involved in effect the use of power series on the form $\sum_{r=1}^{\infty} X_r(x)A^{\alpha_r}$—as at (433.7) in his proof of Taylor series. In particular, he introduced the criterion that a solution to a differential or difference equation was 'particular' or belonged to the 'complete solution' according as whether $\alpha_1 < 0$ or $\alpha_1 \geq 0$. His use of power series showed his continuation of the 18th-century tradition noted in § 3.3.4; and it led him to worthwhile comments on the use of differential equations in mechanics (as we shall see in § 6.2.3) and to the generality of solutions to the heat diffusion equation (§ 9.3.2). But the durability of his results here was fairly expressed by Boole (*1860a*, 149). 'Poisson says nothing about the geometrical character of these solutions', Boole remarked, pointing out Poisson's (typical) avoidance of geometrical considerations, 'But it is clear [...] that they are not in any peculiar sense the principal forms of those singular solutions of differential equations which are generally interpreted as envelopes. They are of extremely rare occurrence [...].'

4.4.2 Biot and Poisson on mixed equations, 1806

After his work on difference equations, Biot turned to differentio-differential equations, or 'mixed' equations, as the French called them. He submitted a paper in October 1799 (*PV*, *2*, 18) and three weeks (to be precise, two *décades*) later the *classe* received an excellent historical survey of the problem in Lacroix's recommendation *1799a* to the *Savants étrangers.* Obviously aware of the kiss of death that such a promise entailed, Biot published a short summary *1800a* in the *Bulletin* of the *Société Philomatique*; but, miraculously, his paper did appear in the designated journal, as *1806f.* Later Lacroix used the paper in his discussion of mixed equations in the second edition of his *Traité* (*Calculus*$_2$, *3* (1819), ch. 8).

As Lacroix pointed out in his report, the subject had been studied rather scrappily hitherto; it did not receive a mention in Lagrange's *Leçons* or in de Prony's lectures *1796a* at the *Ecole Polytechnique*. Confining himself to first-order (though not only first-degree) equations in two variables, Biot tried to build up a more general theory by distinguishing between two kinds of equations (*1806f*, 296–311).

'Equations of successive differences' were obtained by differentiating a difference, or differencing a differential. In the first case Biot took a difference equation containing a constant a:

$$V(x, y, x_1, y_1, a) = 0, \quad \text{with} \quad \Delta x = 1 \tag{442.1}$$

(where his notations of (441.1) apply), and eliminated a between

$$V = 0 \quad \text{and} \quad dV = 0 \tag{442.2}$$

to obtain the required equation

$$Z(x, y, x_1, y_1, p, p_1) = 0, \quad \text{where} \quad p := dy/dx, \quad p_1 := dy_1/dx_1. \tag{442.3}$$

In the second case he considered a differential equation containing a periodic function f:

$$V(x, y, p, f(x)) = 0, \tag{442.4}$$

and eliminated f between

$$V = 0 \quad \text{and} \quad \Delta V = 0 \tag{442.5}$$

to obtain an equation, again of the form (442.3).

In each case the method of solution, of course, was the reverse of the derivation; but it was important to observe the order of integration of (442.3). If the differential were dealt with first, then a arose, and led to no difficulties when handling the difference and then introducing f; but in the reverse case f, arising first, could be difficult to integrate, and should be determined from given conditions if they were available.

'Equations of mixed differences' proper did not have either of the iterative structures just described. Again, two cases arose. In the first Biot took an equation with two constants:

$$V(x, y, a, b) = 0, \tag{442.6}$$

and eliminated a and b between

$$V = 0, \quad \Delta V = 0 \quad \text{and} \quad dV = 0 \tag{442.7}$$

to obtain the mixed equation

$$Z(x, y, x_1, y_1, p) = 0. \tag{442.8}$$

In the second case V contained three constants:

$$V(x, y, a, b, c) = 0, \tag{442.9}$$

which he eliminated by means of

$$V = 0, \quad \Delta V = 0, \quad dV = 0 \quad \text{and} \quad d\Delta V = 0 \tag{442.10}$$

to obtain the mixed equation, again of the form of (442.3).

Like the difference equations, mixed equations admit of 'particular solutions', a plurality of solutions, and Charles's problems. To study them Biot appealed to his method of § 4.4.1, treating the constants as functions of x. Neither his account nor his examples (pp. 303-310) seems very clear; I shall try to capture the method as applied to mixed equations. He eliminated y, y_1 and p from (442.6, 8) and obtained

$$dV := V_x + pV_y + a'(x)V_a + b'(x)V_b = 0 \tag{442.11}$$

and

$$V(x_1, y_1, a_1, b_1) = 0, \tag{442.12}$$

where $a_1(x)$ and $b_1(x)$ were given by Taylor's series; for example,

$$a_1(x) = \sum_{r=0}^{\infty} a_1^{(r)}(0)x^r/r!. \tag{442.13}$$

The complicated equations that ensued were simplified in various ways. For example, he put

$$a'(x)V_a + b'(x)V_b = 0 \tag{442.14}$$

in $(442.11)_2$, since V_a and V_b were not involved in the determination of Z in (442.8). Again, conditions somewhat similar to those in (441.5), such as

$$\Delta a(x) = 0 \quad \text{or} \quad a'(x) = 0, \tag{442.15}$$

could be invoked.

Biot's analyses seem much less promising than those for difference equations. For example, the procedures just described led to a plethora of singular solutions. Again, despite his confidence expressed on p. 300, it may not be clear at first sight that a difference equation was an equation of successive differences; his definition was based on certain forms of solution to the equation, whereas in practise one usually had the equation first. In his summary paper he admitted that the distinction 'is without doubt too special, but it suffices for its purpose, which is to apply geometrical considerations to them' (*1800a*, 87). Yet even there the utility of the distinction seems dubious. For example, in both *1800a* and in *1806f*, 315-317, he took a problem in reciprocal trajectories from Euler *1766c*, which was expressed mathematically by the equations

$$y_1^2 = y^2(1+p^2) \quad \text{and} \quad x_1 = x + p. \tag{442.16}$$

Differentiating them with respect to x, he converted $(442.16)_2$ to

$$\Delta(yp) = 0, \tag{442.17}$$

which he used to solve $(442.16)_1$ as an equation of successive differences. He did not also treat it as a mixed equation (in his sense of the term), although it has the form of (442.3). Again, on pp. 321–327, a problem in geometrical optics, stated mathematically in the form

$$\Delta[(y - 2px - p^2y)/(p^2x - 2py - x)] = 0, \tag{442.18}$$

was said to be a mixed equation, but the solution was obtained without use of the considerations of (442.11–15).

Poisson wrote a paper on mixed equations which was closely linked with his second paper *1806a* on difference equations. It was submitted a month after that paper, in June 1805 (*PV*, *3*, 222); both papers were praised by Lacroix in reports *1806a* presented to the same January meeting of the *classe*; and it appeared as Poisson *1806b* in the *Journal de l'Ecole Polytechnique*, immediately after his first paper.

Poisson knew that Laplace's expansion (341.5) of $\Delta^n z$ as a series of derived functions could be used to convert mixed equations into differential equations, or Euler's expansion (341.2) of dz in a series of differences (and a companion set of expansions for higher-order differentials) could transform them into difference equations. However, perhaps with Biot's sortie (442.13) in mind, Poisson sensibly commented that 'these transformations do not facilitate at all the research for the integral', and he also pointed out that Euler's (341.1) was not valid anyway for periodic functions when $\Delta z = 0z$ (*1806b*, 128)—a point which Servois was to miss when offering (341.1) as his definition of (437.4) of the differential.

Poisson felt that 'it is under their own form that one must look to integrate equations of mixed differences' (p. 128). Thus he applied Laplace's method of cascades (335.1) to the first-order linear mixed equation, which he took in the form

$$y_1' + gy' + hy_1 + my = n, \quad \text{where} \quad \Delta x = 1 \tag{442.19}$$

and the coefficients were functions of x. (I write 'g' and 'h' instead of Poisson's unfortunate choice of 'p' and 'q'.) The transformation

$$z(x) := y_1 + gy \tag{442.20}$$

converted $(442.19)_1$ into the form

$$z' + hz = \alpha y + n, \quad \text{where} \quad \alpha := g' + gh - m. \tag{442.21}$$

When $\alpha = 0$, (442.21) was solved easily (in effect, as a Biotian equation of successive differences). Otherwise, y was eliminated between (442.20) and $(442.21)_1$ to produce an equation of the form of (442.19) in z. The process was repeated: a new $(442.21)_1$ obtained; perhaps the new $\alpha = 0$; and so on, as long as the process terminated. If not, then Poisson applied to (442.19) a transformation, which I write as

$$t(x) := y' + h_{-1}y, \tag{442.22}$$

where h_{-1} was the *backward* difference on h. If the new $\alpha = 0$, then he solved in the same way as before, except that difference and differential were now in the reverse order. If $\alpha \neq 0$, then he used the previous sequence of transformations; and so on again (pp. 129–135).

For many equations one or other of these procedures would indeed terminate: indeed, among his examples Poisson took Biot's non-linear (442.16) and solved it by means of a version of his first procedure (pp. 141–147). But obviously the scope of his suggestions was strictly governed by the coefficient functions admitted in (442.19); and the calculations could become very tedious if the procedure needed to be repeated many times. The same point applies to Laplace's method of cascades itself,[1]) to which I now turn.

4.4.3 Parseval on the scope of Laplace's method of cascades, 1806

Parseval submitted a manuscript in May 1798 (*PV*, *1*, 87). It was approved by Laplace in his report *1799a* (which was mainly concerned with Parseval's *1806e* on his theorem (434.2), submitted some days later), and appeared in the *Savants étrangers* as Parseval *1806b*. Poisson might have looked at the manuscript when writing his paper on mixed equations, for it touched on them.

It was mentioned in § 4.3.4 that Parseval applied his theorem in *1806e* to a result obtained by Laplace from his method of cascades. *1806b* treated Laplace's method itself, as developed in *1777a*, art. 7, as follows. Firstly, any second-order linear partial differential equation was converted by substitution into the form

$$z_{xy} + m_0 z_x + n_0 z_y + l_0 z = 0, \tag{443.1}$$

[1]) A propos of Laplace, Biot claimed in a later reminiscence *1850a* that it was the results in this paper *1806f* described here which Laplace had found long before him but had not then published, so that the young man could receive the credit. On Biot *1850a*, see his letter of 13.9.1850 to the printer (*Wellcome*, Biot volume, fols. 37–38).

where the coefficients were functions of x and y. Now (443.1) was the result of eliminating z_1 between

$$z_1 := z_y + m_0 z \quad \text{and} \quad (z_1)_x + n_0 z_1 + (l - (m_0)_y - m_0 n_0) z = 0 \tag{443.2}$$

(I recall from § 1.3.7 my use of the form of notation '$(z_1)_x$'.) Thus, if

$$\mu_0 := l - (m_0)_y - m_0 n_0 = 0\,, \tag{443.3}$$

then the integration of (443.1) via (443.2) was greatly simplified. If not, then Parseval differentiated $(443.2)_2$ partially with respect to y and eliminated z from this equation and $(443.2)_1$ to obtain a new equation of the form of (443.1) and a condition similar to (443.3). The process was iterable; it yielded for the r-th differential equation

$$(z_r)_{xy} + m_r(z_r)_x + n(z_r)_y + l_r z_r = 0\,, \qquad r \geq 1, \tag{443.4}$$

where
$$m_r := m_{r-1} - (\mu_{r-1})_y / \mu_{r-1}\,, \tag{443.5}$$

$$l_r := m_{r-1} - (n(\mu_{r-1})_y / \mu_{r-1}) + n_y - n m_r\,, \tag{443.6}$$

$$\mu_r := l_r - (m_r)_x - n m_r\,. \tag{443.7}$$

If $\mu_r = 0$, then the iteration could cease, for (443.4) would be integrable by means similar to that for $(443.2)_2$ by (443.3) (*1806b*, 478–482).

Parseval wanted to examine the form of μ_r, and its bearing on the solution of (443.1). Eliminating $\{l_r\}$ and $\{m_r\}$ from (443.7) by means of (443.5–6), he found the mixed equation

$$\Delta^2 \mu_r = (l_0 \mu_r)_{xy}\,. \tag{443.8}$$

He knew no means of solving (443.8) (and probably would have obtained little guidance from Biot's and Poisson's papers to come). But he derived the expression

$$\mu_r = \mu_0 + r((n_0)_x - (m_0)_y) + \sum_{j=0}^{r-1} (r-j)(l_0 \mu_j)_{xy} \tag{443.9}$$

from one of the equations obtained en route to (443.8), and drew from it a few conclusions and solved some examples (pp. 482–492). (443.9) was a compact statement of the process of iteration entailed by the use of the method of cascades; but it also indicated the cumbersomeness of the method, as with Poissons's procedures in § 4.4.2 for mixed equations.

Since Parseval's analysis had brought him to the mixed equation (443.8), it would be appropriate to end with a general comment on these equations. Lacroix described Parseval's work, and also some efforts due to François Français in a paper which is now lost (*Calculus*$_1$, *2* (1798), 595, *3* (1800), 543; and *Calculus*$_2$, *3* (1819), 596–599); and it seems clear that he found their study too difficult for the time. This judgement may seem surprising, for we know that the Laplace transform provides powerful means of solving mixed equations, and Laplace was still alive at this time. But the power of the Laplace transform was almost entirely missed in the early 19th century, as we note in Interlude 934.1.

4.4.4 Monge and Poisson on the general theory of partial equations

Monge still had things to say on differential equations. For example in August 1808 he read a paper to the *classe* on 'reciprocal surfaces' (*PV*, *4*, 93). Although it was never published, his result was stated the following year in the front matter of the latest edition of his *Application de l'analyse à la géométrie* (as his folioed lectures (§ 3.3.5) had now been entitled) and has become standard (*Analysis*$_4$ (1809), iv). Given a point (x, y, z) on a surface for which

$$dz = pdx + qdy, \tag{444.1}$$

the coordinates x', y', z' of its *reciprocal* point have for expressions

$$x' = p, \quad y' = q, \quad z' = px + qy - z. \qquad [(444.2)]$$

The location of all these reciprocal points is the *reciprocal* surface of the proposed surface. The reciprocity of these two surfaces consists in that the first surface is the location of the reciprocal points of the second, as the second is the location of the reciprocal points of the first.

To this edition Monge added an extensive essay on the integration of first-order equations in three independent variables (pp. 367–414). He extended his panoply of results on determining the characteristics of an equation, and their interpretation in the context of tangent planes and their envelopes (§ 3.3.7). Using Euler's notation, he took the equation in the form

$$F(x, y, z, p, q) = 0, \quad \text{where} \quad dz = pdx + qdy, \tag{444.3}$$

and differentiated it to produce, with the help of $(444.3)_2$,

$$0 = Pdp + Qdq + Xdx + Ydy + Zdz = Pdp + Qdq + (X + pZ)dx + (Y + pZ)dy. \tag{444.4}$$

On a characteristic pertaining to successive tangent planes, p and q held the same value; thus (444.4) split into

$$(X+pZ)dx+(Y+qZ)=0 \quad \text{and} \quad Pdp+Qdq=0, \tag{444.5}$$

in that $(444.5)_2$ applied to a surface which touched a 'trajectory' of the characteristic (that is, from p. 375, a curve cutting the sequence of characteristics). Further, if the surface were generated by the tangent planes in such a way as to hold dx, dy and dz constant, then, differentiating $(444.3)_2$ gave on p. 378

$$dp\,dx+dq\,dy=0. \tag{444.6}$$

The trio of equations (444.5-6), and related results, gave Monge an impressive range of possibilities with which to state properties of characteristics. In the second part of the essay he studied various methods of solving them, whether or not the original equation took the form (444.3).

Monge also made valuable contributions to the theory of second-order equations. Among other results, he found means of taking the equation

$$Rr+Ss+Tt+U(rt-s^2)=V(\text{or } 0), \quad \text{and via} \quad u=f(v) \tag{444.7}$$

making it susceptible to integration via intermediate integrals. In particular, it may have influenced a useful paper on singular solutions to second-order equations that Poisson published in the *Correspondance sur l'Ecole Polytechnique.*

Poisson *1813f* took the form

$$P(p,q,r,s,t)=(rt-s^2)^\alpha Q(x,y,z), \quad \text{where} \quad \alpha>0 \tag{444.8}$$

and P was homogeneous in r, s and t. If the substitution

$$q=f(p), \tag{444.9}$$

rather like Monge's $(444.7)_2$, were made, then $(rt-s^2)=0$, and (444.8) reduced to

$$P(p,f(p),r,rf'(p), \quad r(f'(p))^2)=0. \tag{444.10}$$

But homogeneity showed that some power r^n was a factor throughout (444.10) and so could be cancelled out, thus reducing (444.10) further to an *ordinary* differential equation in p, $f(p)$ and $f'(p)$. Its solution could then be substituted back into (444.9), which served as an intermediate integral en route to solving (444.8). The solution obtained, however, would not be general, since it contained only one arbitrary constant and one arbitrary function (*1813f*, 410-411).

Poisson gave two examples (pp. 411-414). The first,

$$t+2ps+(p^2-a^2)r=0 \quad \text{with} \quad a>0, \tag{444.11}$$

was homogeneous to first degree, and in it $\alpha=0$. He had already found some particular integrals for (444.7) in a paper on sound (*1807b*, 367; we consider this paper in § 7.2.6), but without the benefit of the procedures now known to him. The substitution of (444.9) into (444.11) and the elimination of r led to the ordinary differential equation

$$(f'(p))^2 + 2pf'(p) + (p^2 - a^2) = 0\,, \tag{444.12}$$

which factorised easily to produce the intermediate integral

$$\tfrac{1}{2}p^2 + q + ap = c, \quad \text{with } c \text{ constant}, \tag{444.13}$$

and thence by variation of parameters to the general integral

$$z = \alpha x + (c \mp a\alpha - \tfrac{1}{2}\alpha^2)y + \varphi(\alpha) \quad \text{and} \quad x = (a \pm \alpha)y - \varphi'(\alpha)\,, \tag{444.14}$$

where separate functions φ applied to the + and the − signs.

Poisson's other example was the equation

$$(1 + q^2)r - 2pqs + (1 + p^2)t = 0\,, \tag{444.15}$$

for which he found the solution

$$z = \varphi(x + ay) + y(-1-a^2)^{\frac{1}{2}}. \tag{444.16}$$

Now (444.15) was the equation (335.5) for minimal surfaces, as he well knew; for he both described it geometrically and stated that (444.16) was obtainable by known methods. It is rather surprising, then, or perhaps some sign of the times, that he never mentioned Monge once in the paper.

In a succeeding note in the *Correspondance* Poisson *1815e* took up another question which (without him saying so) related to Monge's methods: this time the conversion of linear forms of first-order first-degree partial differential equations by bringing in more independent variables. Naturally, more constants of integration thereby arose, and Poisson sought to simplify Lagrange's analysis, given in lecture 20 of the *Leçons* (*1806a*, 354-357), that these extra constants could be eliminated.

Poisson took the differential equation in the form

$$f(x, y, z, p, q) = 0\,, \tag{444.17}$$

but concentrated on the total differential of z and the integrability condition (334.1), that is,

$$dz = pdx + qdy \quad \text{and} \quad p_y + qp_z - q_x - pq_z = 0\,. \tag{444.18}$$

After invoking the auxiliary equations (334.3) to $(444.18)_2$ and writing the 'complete integral' of one of them as

$$q = \psi(x, y, z, a), \quad \text{where } a \text{ was constant}, \tag{444.19}$$

he substituted it in the other two solutions and those in turn into the 'general integral' of $(444.18)_2$ to obtain the relation

$$a = \Pi(\psi_1(x, y, z, a), \psi_2(x, y, z, a)). \tag{444.20}$$

He now took p from (444.17) and q from (444.19) and put them into $(444.18)_1$ to obtain, by holding a constant, a 'particular integral'

$$F(x, y, z, a) = k, \quad \text{where } k \text{ was constant.} \tag{444.21}$$

Were (444.21) to hold also for variable a (and thus k), then its a-derived function

$$F_a(x, y, z, a) = dk \quad (= \Pi_1\,(a, k), \text{ say}) \tag{444.22}$$

would have to be identical with (444.20), so that k was actually only a function of a. Thus (444.21) became the general integral of $(444.18)_1$:

$$F(x, y, z, a) = \varphi(a)\;(= k) \quad \text{and} \quad F_a(x, y, z, a) = \varphi'(a), \tag{444.23}$$

with only one constant a, and φ determined from (444.22) as

$$\varphi'(a) = \Pi_1(a, \varphi(a)). \tag{444.24}$$

Thus the solution of (444.17) was transferred to (444.18), and hinged on (444.19). While Poisson took for granted (in Lagrangian style) that the various algebraic processes involved could actually be effected, he provided a clear indication of a method used in the theory of non-linear equations, which was still little developed.

This paper appeared in the *Correspondance* as a reprint from its appearance in November 1815 with the *Société Philomatique*, with the addition of an example which Lagrange had given (*1806a*, 357–358). The new printing gave editor Hachette opportunity to point out that the method served 'as clarification of the method given by Mr. Monge, for the integration of the partial differential equations of curved surfaces'.

4.5 On the calculus of variations

A significant aspect of Lagrange's algebraic approach to the calculus was the role he assigned to the calculus of variations. Thus it requires some attention, partly in its own right and also for the role that it plays in his mechanics (§ 5.2). As various histories are available,[1]) I can confine myself to a brief summary of the subject as presented by Lagrange in his *Leçons* and to a paper by Ampère.

4.5.1 Lagrange's various treatments

Introduced at the end of the 17th century as an extension of the theory of maxima and minima into a study of isoperimetrical problems, the calculus of variations became part of Lagrange's conception in his *1762b*, when he introduced the letter 'δ' to indicate the operation of variation on the variable(s) of the analysis. In the opinion of Lacroix, this work was 'one of the first and the principal discoveries of Mr. Lagrange, and which rests on the highest degree of abstraction to which mathematics has been lifted up to now' (*Calculus*$_2$, *2* (1814), 721).

Lagrange restricted himself in *Fonctions* to a short account of the subject, as more complete versions were available in other textbooks (*Functions*$_1$ (1797), 200–223). He wrote 'ω' for the variation on y, ω' on y', and so on, but I shall use his δ-notation as it makes clearer the processes involved; for example, that it is a *theorem* that ω' is the derived function of ω, proved by means of the commutativity of δ and $'$:

$$\omega' := (\delta y)' = \delta y' = (\omega)', \quad \text{with} \quad \delta x = 0x. \tag{451.1}$$

Lagrange sought conditions for the extreme values of

$$I := \int_c f(x, y, y', y'', \ldots)\, dx, \quad c \text{ constant}, \tag{451.2}$$

by taking the variations on $y, y', \ldots$ and obtaining the expansion

$$\begin{aligned} &f(x, y+\delta y, y'+\delta y', \ldots) = f(x, y, y', \ldots) + \\ &(\delta y f_y + \delta y' f_{y'} + \ldots) + ((\delta y)^2 f_{yy} + 2\delta y \delta y' f_{yy'} + \ldots)/2! + \ldots. \end{aligned} \tag{451.3}$$

By reasoning which is familiar today and is essentially due to Euler (see his *1772a*, art. 15 for example), Lagrange found the relation

$$\sum_{r=1} (-1)^r (f_{y^{(r)}})^{(r)} = 0, \tag{451.4}$$

where the superscript indicates the number of differentiations with respect to x.

---

[1]) See *passim* in Todhunter *1861a*, Kneser *1900a* and Goldstine *1980a*; and Fraser *1985a*. The 18th-century pre-history is also treated in Woodhouse *1810a* and briefly in Lagrange's own books; Euler's contributions are discussed in detail in Carathéodory *1952a*. Lacroix gave the subject a long(-winded) chapter in both editions of his *Traité* (*Calculus*$_1$, *2* (1799), 655–724; *Calculus*$_2$, *2* (1814), 721–816).

(451.4) was known already as a condition for an ordinary differential equation to be exact, and was shown to be a necessary and sufficient condition for f to be a total differential. When only y and y' were present in (451.2), (451.4) reduced to

$$f_y - (f_{y'})' = 0. \tag{451.5}$$

Lagrange modelled his study of the second variation on his search for the second-order conditions (322.14) on $f(x, y)$. When only y and y' were varied, he determined new functions M, N and P from the condition that the second-order term

$$(\tfrac{1}{2}f_{yy} - M)(\delta y)^2 + (f_{yy'} - N)\,\delta y\delta y' + (\tfrac{1}{2}f_{y'y'} - P)(\delta y')^2 \tag{451.6}$$

took the form $(\delta y)^2(\nu(x) - \nu(a))$. He then showed that the extremal would be

$$\text{a maximum/minimum according as } P < / > 0,\ \ T := \left(M - \frac{N^2}{4P}\right) < / > 0, \tag{451.7}$$

unless P and/or T was always zero when $a \le x \le b$, or M, N or T were infinite-valued for some values of x. He also noted that ν could be determined by the condition $T = 0$, when the maximum/minimum would be given by $f_{y'y'} < / > 0$ (pp. 203-208). Legendre had offered this condition in his *1788a*; but Lagrange pointed out that ν could not necessarily be determined from $T = 0$, so that its finitude when $a \le x \le b$ could not be guaranteed. Some other problems in Legendre's paper were pointed out in Brunacci *1810a* (see Todhunter *1861a*, 229-242).

The rest of Lagrange's account was given over to some aspects of the general case (451.2), and to the situation where several variables were involved without or with a constraint. Here a Lagrange multiplier—which he unbelievably notated on p. 213 as 'Δ'!—was introduced.

Lagrange's account of the calculus of variations is rather partial, even given his restricted plan; (451.2) was the dominant tool, for reasons which will become clear in § 5.2.3 when we see his principles of mechanics. In the *Leçons* he did not much extend the scope of his account, but the approach was rather different. After deriving (451.4) in lecture 21 as a condition for integrability and pointing out its bearing on the calculus of variations, he began the last lecture with a typical claim that 'the method of variations [...] left nothing to be desired', but that it should be embedded in the theory of functions rather than allow it to stray towards the world of infinitesimals (*1806a*, 399).

Accordingly, Lagrange now considered the extreme values of a function $U(x, i, y)$, which was susceptible to power-series expansion in both the base variable x and the incremental variable i. Extreme values of U were given when its derived function V was zero; but the relationship $y = \varphi(x)$ between x and y was generally not known: 'Thus the difficulty consists in deducing these derived [functions] from the given V'. To find them, Lagrange varied i from an initial value of zero to produce the Taylorian expansion

$$V(x, i) := V(x + i) = \sum_{r=0} i^r V_{i^{(r)}}(x, 0)/r!, \tag{451.8}$$

in which not only x but also all its derived functions were held constant. 'We will call in general *variations* of the first order, of the second, etc. these derived [functions ...] relative to the quantity i' and 'to form these variations, one will follow the same rules as for ordinary derived functions', thus obtaining the properties for variations such as commutativity with differentials (pp. 402–403).

The previous theory went through much as before. If

$$V = f(x, y, y', \ldots), \tag{451.9}$$

then varying i with x held constant furnished

$$V_i = \sum_{r=1}^{\infty} y_i^{(r)} f'_{y^{(r)}}, \tag{451.10}$$

and since

$$V_x(x, i) = V_x(x + 0) = V_i(x, 0), \tag{451.11}$$

(451.10) yielded (451.4) once more as the first-order condition for extremity (pp. 403–404). If $\varphi(x)$ was known, the 'primitive' U of V could be determined. The extensions this time included the case where x and y were both independent variables, which corresponded to finding an extreme value for the double integral

$$U = \iint V(x, y, z, z_x, z_y, z_{xx}, \ldots)\, dx\, dy. \tag{451.12}$$

The first-order condition was now

$$f_z - (f_{z_x} + f_{z_y}) + (f_{z_{xx}} + f_{z_{xy}} + f_{z_{yy}}) - \ldots, \tag{451.13}$$

and by suitable manipulations of d and δ, δU reduced to a sum of single integrals (pp. 425–426); but their integrability was an awesome task, to be taken up by Poisson (§ 18.2.2).

The advantages of the new approach are difficult to appraise. The siting of the theory of variations in power series in i should free the process of variations from associations with infinitesimals, although of course to us (though not to Lagrange) other difficulties now come forth. Certainly the presentation, with primes, over-periods and commas used to distinguish the various operations of partial differentiation, is even more uninviting than its *Fonctions* predecessor. It is interesting to note that in the second edition of that work Lagrange reverted to a virtual reprint of the 1797 text (*Functions*$_2$ (1813), pt. 2, chs. 12–13), and passed over the intervening variant presentation.

4.5.2 Ampère on the philosophy of variational methods, 1806

Ampère's first submission to the *classe* was a two-part paper of May 1803 (*PV*, *2*, 662) on the calculus of variations. Characteristic of his individuality, he tackled not the technical aspects of the subject but its interpretations in geometry and in mechanics. Biot, member of the *classe* and at 29 years Ampère's senior by nine months, praised the paper in his report *1804a*, and it actually did appear in the *Savants étrangers*, as Ampère *1806a*.

Ampère wrote the paper while still teaching at the *lycée* at Lyon (§ 4.2.5). He sent Lagrange a grovelling letter about his 'feeble essay', written 'in reading those immortal works which have changed the face of mechanics and of mathematics'.[1]) The paper itself began in the same vein—'the immortal author of the *Mécanique analytique*', and so on—but soon became interesting. The basic insight was that in mechanics one had to allow for the continuity of the physical substances involved as the variations occur. In the equilibrium of an inextensible wire, for example,

$$\delta \int (dx^2 + dy^2 + dz^2)^{1/2} = 0 \tag{452.1}$$

had to be read in terms of the fact that neighbouring points of the wire were the same distance apart after variation as before it, and that corresponding points under the variation were the same point of the wire in two positions. Hence δx, δy and δz would not be arbitrary; so, in order to preserve the generality of (452.1), their coefficients must be zero. Since the number of resulting equations equalled the number of variables, no independent variables were left. These had to be provided by the Lagrange multipliers attached to the conditions of the problem, which meant that there must be enough such conditions to restore its generality.

[1]) The letter, dated 7 *floréal an* 12 (27.4.1804), is in *AS*, *Ampère*, ch. 345 ('faible essai ... en lisant ces ouvrages immortels qui ont changé la face de la mécanique et des mathématiques'). Pertinent to this paper are various manuscripts (chs. 103–109); a copy of the manuscript and of Biot's report (*ENPC*, ms. 1722); a letter from Delambre of 23 *frimaire an* 12 (15.12.1803) reporting that Biot was 'singulièrement occupé' (*AS*, *dossier personnel* for Delambre: published in Ampère *Correspondance*, 270–272); a letter to Lacroix of 20 *germinal an* 12 (10.4.1804), in which he still awaited Biot's report (*BI*, ms. 2396: the report was presented to the *CI* six days later); and some notes by Lagrange on the results of the paper (*BI*, ms. 904, fols. 133–138).

By contrast, when (452.1) arose in a geometrical context, such as the extreme positions of a curve of a given length, it said no more than that the variation in length was zero. Neighbouring points were not subject to the contraints indicated above, so that the variations were restricted only by (essentially) (452.1) itself and any specific conditions on the problem. Thus the number of arbitrary relations between the variations indicated the number of independent variables, and the setting to zero of the other variations gave the number of dependent variables (*1806a*, 492-501).

These considerations led Ampère to regard the calculus of variations as divided into two parts: 'the one where one can be content to vary only the variables which the nature of the question renders dependent upon the others, the other where one is obliged to allocate variations to all the variables' (p. 503). It also drove him to this criticism of the immortal author of the *Méchanique analitique* (pp. 503-504):

> [...] one must much regret that the celebrated mathematician whom I have already cited so many times, when he thought of stripping the calculus of variations of that supposition of the infinitesimals [...] which he has so happily banished from the differential calculus and the integral calculus, seems to have forgotten the application of this calculus to mechanics, which he was the first to attempt, and to which one must assign some part of his glory.

He then cited Lagrange's treatment of extrema in *Functions*$_1$ (1797), 200-223 as a case where only the variation of all variables was treated.

Ampère then took the case of the inextensible wire as an example of his approach (pp. 505-523). He took the equilibrium condition in the general form

$$\int \lambda \delta L(x, y, z) dx = 0 \tag{452.2}$$

and expanded the total variation δL and rearranged terms until reaching the form

$$P\delta x + Q\delta y + R\delta z = 0 . \tag{452.3}$$

Then, by his earlier considerations,

$$P = 0, \quad Q = 0 \quad \text{and} \quad R = 0 . \tag{452.4}$$

The first equation gave an expression for $d\lambda$, which could be substituted into the other two equations to yield the equation of the curve. The results were somewhat more general than those in Lagrange's *Mechanics*$_1$ (1788), 89-115, because Ampère's condition equation $\delta(Ldx) = 0$ embraced Lagrange's condition $\delta ds = 0$ (p. 90) as a special case.

The interesting philosophical points of the first half of Ampère's paper did not make much impact (Lacroix noted them briefly in *Calculus*$_2$, *2* (1814), 798), not even in its second half. There Ampère handled (452.3) in the way recommended for problems in mechanics, but not any corresponding geometrical problems. He did not even make use of a technical advantage that he had claimed for his approach over Lagrange's; that he did not have to make moves of the form

$$\int F\delta dx = \int Fd\delta x = F\delta x + \int dF\delta x \tag{452.5}$$

and thus use integration by parts. He promised in the paper to develop his ideas further, but he did not publish them; in particular, they are not found in his lectures on the calculus of variations given at the *Ecole Polytechnique* and published in Gergonne's *Annales* as Ampère *1825d*, which concentrated on basic principles and the condition (451.4). (In fact, the subject was dropped from the school's curriculum between 1811 and 1821.) Thus, regrettably, Todhunter's appraisal of the paper is fair: it 'contributes nothing to the theory of the Calculus of Variations; its only interest arises from its relation to mechanics' (*1861a*, 470).

4.6 Equations and their roots

4.6.1 Three types of problem

30 November 1812 was a very important day in the history of algebra in the period 1800–1815; for on that day Binet submitted to the *classe* a paper on determinants and Cauchy two on permutations (*PV*, *5*, 121). Poisson recommended them all for the *Savants étrangers*, in reports *1813a* and *1813b* presented to successive meetings of the *classe* the following April and published in the *Moniteur universel*; doubtless he was also responsible for the appearance of the papers in the *Journal de l'Ecole Polytechnique* as Binet *1813b* and Cauchy *1815a* and *1815b*. The papers have several mathematical features in common, especially concerning determinants, and their authors were good friends (§ 4.2.4); but each man stated in his paper that they were written independently.

Despite their importance I shall not describe these papers; on Cauchy's, see Dahan *1979a* and *1980a*. They are chiefly of importance for the histories of abstract and linear algebra, and their bearing on the theory of equations lies mainly on the resolubility of polynomials by radicals, and related topics. My concern with equations centres on two other problem areas, more closely related to the calculus: the detection of zeros of polynomials, especially by means of Descartes's rule of signs and related theorems; and the approximation to roots of equations (of any kind). (For convenience I speak of the roots of both the function f and of the equation $f(x) = 0$.) I have chosen some of the principal developments of the period 1800–1815, which built on the work of the principal figure—who was, of course, Lagrange.

4.6.2 Lagrange's 'treatise on algebra'

In 1798 Lagrange published with Duprat an edition of some of his earlier papers on equations in a book entitled *Traité de la résolution des équations numériques de tous les degrés*, adding a dozen new notes which far surpassed the original texts in length. Ten years later Courcier brought out the second edition, in which the principal change was two long additional notes. I shall be citing this edition, as *1808a*, for it seems to have made a much greater impact than its predecessor. In particular, Poinsot wrote a long review *1808a* in the *Magasin encyclopédique* which earned Lagrange's approval and was placed at the head of an 1826 re-issue by Courcier's successor Bachelier (§ 18.2.3). The editors of the *Oeuvres* chose to use the 1808 rather than the 1826 edition, so that Poinsot's essay is not well known; but as it is very useful, I can confine my remarks on Lagrange's book to a few points.[1])

For Lagrange the 'resolution of numerical equations' was concerned with finding explicit expressions for the roots of equations, or failing that (which was the case for polynomial equations above the quartic), devising methods of detecting and approximating to them—in other words, the two types of problem emphasised in § 4.6.1. (However, the book also contained various considerations of the kind on which Binet and Cauchy were to build.) In view of the emphasis of this book, and also the various algebraic experiments described earlier in this chapter, it is surprising for us to see Poinsot begin his review *1808a* by claiming that 'Algebra has been a little neglected' and welcoming Lagrange's book as a 'Treatise on Algebra'; 'for, as all algebra reduces, at root, to the analysis of equations [...]'. We shall see again in § 11.3 the same association of algebra with the theory of equations, and the role of equations in the emergence of mathematical analysis, in Cauchy's later work.

An example of the links between algebra and analysis occurred in the key theorem of Lagrange's first chapter:

---

[1]) Lagrange's book and Poinsot's essay are still among the best surveys of the state of the theory of equations around 1800. Hamburg *1976a* is a useful survey of Lagrange's earlier work, which also has significance for abstract algebra (see especially Wussing *1969a*, ch. 2 *passim*).

Theorem 462.1 Given a polynomial *f*, if $f(p) < 0$ and $f(q) > 0$, then $f(x) = 0$ took at least one real root within *(p, q)*.

Lagrange began his proof by factorising *f*:

$$f(x) = \prod_{r=1}^{m} (x - \alpha_r) \tag{462.1}$$

and pointing out that for at least one root (α_1, say) $(p - \alpha_1)$ and $(q - \alpha_1)$ took opposite signs; thus α_1 was the root desired. However, in his first note he recognised the difficulty cause by the possible presence of complex roots, so he offered an algebraic version of a geometrical argument. He set

$$f(x) = P(x) - Q(x), \tag{462.2}$$

where *P* and *Q* were respectively the terms of positive and negative sign. Assuming without loss of generality that $p > q > 0$, *P(x)* and *Q(x)* both increased with *x*; and if $P(p) \lesseqgtr Q(p)$ and $P(q) \gtreqless Q(q)$, $P(x)$ and $Q(x)$ would eventually equal each other as *x* increased 'by insensible degrees' to the value of the desired root. The argument recalls Lagrange's criticism from § 3.2.6 of the theory of limits—that one may pass through a limiting value and continue on the other side—and limits themselves recall the naive notion of continuity involved in the proof, a notion which Cauchy and others were to refine when proving more general theorems in mathematical analysis such as the intermediate value theorem (§ 11.3.5).

4.6.3 Descartes's rule of signs: the inductive proofs, and extensions

The primary theorem for the detection of roots of polynomials was Descartes's rule of signs: that the number of positive/negative zeros of the polynomial *f* was not more than the number of variations/permanences in the sequence of signs of its coefficients. Various proofs were offered in the 18th century, as Lagrange indicated in note 8 of his *1808a*: for him 'The simplest of these demonstrations' was the inductive proof in Segner *1758a*, where Segner multiplied $f(x)$ by $(x \pm a)$, $a > 0$, and showed, fairly clearly, that the number of variations/permanences increased by at least one. In fact, much of the substance of the argument can already be found in de Gua *1744a*.

Lagrange seems not to have taught the theory of equations much at the *Ecole Polytechnique*, although he discussed some aspects at the *Ecole Normale* (*1795a*, lectures 3 and 4). But the subject was dealt with by his assistant Fourier: indeed, Fourier seems to have discovered the inductive proof of Descartes's rule, and an extension, for himself while in his teens. He remained interested in the subject until his death, when he was preparing a book which was published posthumously as *1831d* under the editorship of Navier.[1]) His presentation of the inductive proof, in his teaching at the school was clearer than Segner's or de Gua's, and he also taught his extension, for a polynomial of degree m:

Theorem 463.1 If $f(a) \neq 0$ and $f(b) \neq 0$, the number of real roots of $f(x) = 0$ within (a, b) was not more than the number of variations in the sequences of signs for $x = a$ and $x = b$ given by the sequence of functions

$$f^{(m)}(x), f^{(m-1)}(x), \ldots f'(x), f(x). \tag{463.1}$$

The proof was based on the insight that (463.1) yielded a sequence of variations when $x = -\infty$ and a sequence of permanences when $x = +\infty$, and that variations passed into permanences as roots were crossed, because $f(x)$ changed sign through the root but $f'(x)$ preserved it. Multiple roots could be found, but pairs of complex roots were not directly detected; so that the theorem gave only an upper bound.

Fourier did not publish any of his results until the late 1810s, although he continued to work on equations. His silence led him into a running battle; for similar results were published by Ferdinand Budan de Boislaurent, *inspecteur* and later *inspecteur général* at the *Université* by profession, student of the theory of equations by passion, Royalist by conviction, and polemicist by temperament. He had submitted a paper in May 1803 to the same meeting of the *classe* which received also the manuscript of Ampère *1806a* on the calculus of variations (*PV*, *2*, 661). His principal result was:

Theorem 463.2 Let $f(x)$ be expanded in powers of $(x \pm p)$ as the polynomial $g(x \pm p)$. Then the number of roots of f within $[0, \pm p]$ was not more than the difference in the number of variations in $f(x)$ and in $g(x \pm p)$.

Biot was made reporter for this paper, as he was for Ampère *1806a*. He reported on Budan in a week, much more quickly than for Ampère; but he found less to praise, seeing nothing essentially new in Budan's paper, which was thus

[1]) On this aspect of Fourier's work see my and Ravetz's *1972a*, *passim*; Charbonneau *1976a*, *1*, ch. 3; and the relevant manuscripts (mainly in *BN*, mss. fr. 22502–22516). Fourier's book is described in § 17.3.2. The inductive proof is sometimes attributed to Gauss *1828a*, although the novelties there lay more in details rather than substance of presentation (as the commentaries Gergonne *1828a* and Cournot *1828d* noted).

fitted just 'to be usefully placed in the elements of algebra' (Biot *1803c*, 669). Presumably he had in mind Fourier's Theorem 463.1, which he will have heard Fourier teach; for Budan's result is essentially equivalent to it. Taylor's theorem for $f(x)$ could be written

$$f(x) = f(p + x - p) = \sum_{r=0}^{m} f^{(r)}(p)(x-p)^r/r! =: g(x-p), \qquad (463.2)$$

so that Fourier's sequence of signs from (463.1) was that of (essentially) Budan's (463.2). In addition, Budan's manuscript seems also to have lacked a proof of his theorem, although he gave rules for calculating the coefficients of the transformed polynomial $g(x \pm p)$.

Three weeks after Biot's report, Budan submitted a similar paper (*PV*, *2*, 675), which was reported in October by Lagrange *1803a.* Lagrange praised only the systematic calculation of the coefficients of the transformed polynomials and the bearing on the presence of real or complex roots of the changes of sign in successive polynomials for $(x - a)$ and $(x - a - 1)$.

Seeing no immediate prospect of publication by the *classe*, Budan published his work himself with Courcier as a booklet *1807a.* Then things moved quickly. Garnier, who had been Fourier's successor at the *Ecole Polytechnique*, praised the book quite highly in reviews *1807a* and *1807b* in the *Moniteur universel* and the *Journal de physique.* Poisson, Garnier's successor, wrote to Fourier in Grenoble, urging him to publish his results (Fourier *1831d*, xxi). And then Lagrange praised the booklet at the end of note 8 of his *1808a*: 'the author gives there a simple and elegant means of forming the coefficients of the transforms into $x - 1$, $x - 2$, etc.; [...] One can say that this Work leaves nothing to be desired on the resolution of numerical equations of which the roots are real, and it may, from this aspect, serve as supplement to the present Treatise'.

Could there be higher praise, and from Geometer Number One, who had been so cool in his 1803 report? However, as in 1803, Lagrange praised explicitly the relations between the coefficients of the polynomials. It was clear that Budan had not really proved Theorem 462.2; so in August 1811 (*PV*, *4*, 519) he submitted a proof. Legendre *1811a* reported in October at length, pointing out various difficulties, especially the need to add in Segner's proof of Descartes's rule (a theorem which should come out as a special case of the theorem) in the case of multiple roots. But he recommended publication in the *Savants étrangers*, along with his own report. Budan complained about the report the next week, but no special action was taken (*PV*, *4*, 548–550).

Budan must have been insufficiently well versed in the *classe* game to realise the long-term nature of the approbation that he had received. Fourier had been too busy in 1807 with heat diffusion and prefectural matters to take up Poisson's advice to publish; but when at last he published his Theorem 463.1, first in *1818b*, Budan went into a Wrońskian mode of vituperation. He published as *1822a* a revised edition of *1807a*, containing his 1811 paper, Legendre's report and his rejoinder, further comments on his formulae relating the coefficients of the different expansions of $f(x)$, and uncomplimentary comments on Fourier. During the 1820s he published various pamphlets (*1827a*, *1828a* and *1829a*), written with venom, repeating his ideas and complaining about not being read. In April 1830 he sent the *Académie* a pamphlet *1830a* on Fourier, in view of Sturm's recent submission of a paper on (to us) Sturm sequences (§ 17.3.5); he had heard of Fourier's forthcoming book on equations, and expected that its good bits would be his own theory and the rest would be wrong. Fourier, as *secrétaire perpétuel*, asked Poinsot to examine the pamphlet (*PV*, *9*, 427); and when Fourier died a few weeks later and his book appeared posthumously as *1831d*, Budan fired his last round of grapeshot *1832a*, rejecting the priority for Fourier that Navier had claimed from *pièces justificatives* included in his preface of the book. Yet neither Navier nor Budan seems to have noticed a passage of 1802 in Montucla's history of mathematics, suggesting Fourier's priority by reporting that Fourier had been concerned with Segner's proof and had given alternative proofs: 'His researches lead him to many other useful truths that it is just that he publish himself first' (Montucla *History*, *3* (1802), 39).

In later literature some writers have preferred Fourier's approach and some Budan's (see especially Darboux in Fourier *Works*, *2*, 311-314; and Runge *1898a*, 411-416). The calculations in Fourier's sequence (463.1) are fairly prodigious; but Budan's sequence requires some work too. The most straightforward approach to calculating the coefficients in $g(x-p)$ in (463.2) is to put $y := (x-p)$ and obtain from

$$g(x-p) = g(y) = \sum_{r=0}^{m} b_r y^r = f(y+p) = \sum_{r=0}^{m} a_r (y+p)^r \tag{463.3}$$

the relation

$$b_r = \sum_{j=r}^{m} r\, C_{j-r}\, a_r p^{j-r}; \tag{463.4}$$

but Budan usually expressed the relationships between the coefficients in a much more cumbersome way, stating the coefficients even for $(x-1)$ in terms of $\sum_{r=0}^{m} a_r$ and sums, sums of sums, ... of its partial sums, and appearing to imply that to reach $(x-p)$ one must effect such transformations p times. And still further calculations are needed if we follow Budan in using his methods to approximate to a root detected within $[a, a+1]$ of x by transforming from $(x-a)$ to $(x-a-0\cdot 1)$, $(x-a-0\cdot 2)$, and so on. Other methods of approximation are more efficient (though still cumbersome); Lagrange favoured Newton's method, certain techniques using continued fractions, and series with terms involving powers of $f(a)$ and $f'(a)$ (*1808a*, ch. 6 and notes 5, 8 and 11). Ruffini's procedure in *1804a* is more efficient than Budan's: known now as 'Horner's method', this point was made by Horner (*1819a*, 311) in his (independent) discovery.[1]) However, Budan's relations generated some interest,[2]) and has left his name, like Wroński's, in the mathematical literature.

4.6.4 Descartes's rule of signs: calculus proofs, and extensions

Of the various 18th-century proofs, Lagrange reported in detail (but without indicating the source) one using calculus techniques 'because it proceeds naturally from our analysis' (*1808a*, note 8, arts. 1-5). The polynomial F, of degree m, was expressed in the (monic) form

$$F(x) = \varphi(x) f(x), \quad \text{where} \quad \varphi(x) := \prod_r (x - p_r), \quad f(x) := \prod_r (x - q_r), \qquad (464.1)$$

$\{p_r\}$ were the real roots of F ordered by $>$, and $\{q_r\}$ the complex roots. The $\{q_r\}$ occurred in complex conjugate pairs, so that

$$f(p_r) > 0 \quad \text{and real; and } \varphi(p_r) = 0. \qquad (464.2)$$

Further, since

$$\varphi'(x)/\varphi(x) = \sum_r (x - p_r)^{-1}, \qquad (464.3)$$

---

[1]) On Ruffini's anticipation of Horner, see Cajori *1911a*. It is interesting to note that Horner presented his method as an offshoot of his treatment of Arbogast's expansion (436.4) in the calculus of derivations. Francoeur was presenting similar expansions at this time (see *Course*$_2$ (1819), *2*, 36; compare $_1$(1809), *2*, 30-33).

[2]) Among publications of the early 19th century, see Nicholson *1822a*; Ferroni *1828a*; J. R. Young *1835a*, ch. 4; and Christie *1842a*. Among manuscripts, see Lagrange's notes on Budan's 1811 paper (*BI*, ms. 908, fols. 107-125) and his complimentary letter to Legendre on Legendre's report *1811a* (*AS*, *dossier personnel* for Lagrange). Budan generalised his relationships between coefficients into a theory of linear recurrence relations, which he called 'syntagmatic sequences', in a paper which he submitted to the *CI* in May 1813 (*PV*, *5*, 218). It was not reported, but the ideas appeared in, for example, his *1822a*, 109-120.

then, by the ordering of the roots $\{p_r\}$,

$$\varphi'(p_1) > 0, \quad \varphi'(p_2) < 0, \quad \varphi'(p_3) > 0, \ldots . \tag{464.4}$$

Hence, since

$$F'(x) = \varphi'(x) f(x) + \varphi(x) f'(x), \tag{464.5}$$

it followed from (464.2, 4) that

$$F'(p_1) > 0, \quad F'(p_2) < 0, \quad F'(p_3) > 0, \ldots . \tag{464.6}$$

Thus, from Theorem 462.1, F' had at most one positive and one negative root less than F. Repeating the argument led finally to $F^{(m-1)}$, which took the form $(ax + b)$ and so contained one real root, which was positive/negative according as $F^{(m-1)}$ had no permanences/variations. Reversing the line of reasoning, and taking note also of multiple real roots of F, proved Descartes's rule.

Other uses of the calculus were outlined, in the course of historical and other considerations of Descartes's rule, by de Gua in *1744a*, 461-464. Having proved the rule by a form of the inductive proof discussed in the last sub-section, he cross-classified the maxima and minima of F into two other types of extrema, which I shall call 'concave' and 'convex': they were defined as the extrema which were concave or convex to the x-axis, or, alternatively, as the maxima or minima of the function $|F|$. Now, if $\{m_r\}$ were the real roots of F' and

if m_r were associated with a concave/convex extremum, then $F(m_r)\, F''(m_r) < / > 0$. (464.7)

de Gua used this result chiefly to try to detect complex roots of F by comparing their number with those for F' and the higher derived functions; but on pp. 470-472 he also applied it to determining the number of positive and negative roots. Basically, concave extrema were the only type to be found adjacent to intersections of $y = F(x)$ with the x-axis, although they occurred also in tandem with convex extrema when several extrema existed between neighbouring intersections. Consideration of this behaviour of F along the x-axis yielded the result

(number of intersections of $y = f(x)$ with the x-axis) $- 1 =$ (number of concave extrema) $-$ (number of convex extrema). (464.8)

This equation held also for intersections on the positive/negative parts of the x-axis, as long as the number of negative roots of F was even/odd (a feature revealed by the sign of the constant term in $F(x)$). Otherwise, the term '-1' in (464.8) had to be dropped.

The approach is interesting, but may suffer from limitations. For example, as with the proof of Descartes's rule reported by Lagrange, no account seemed to be taken of the multiplicity of a repeated root, when incidentally the intersection was also an extremum (but presumably not counted as one) or a horizontal point of inflexion (which de Gua regarded as one concave and one convex extremum). Again, it is necessary to resolve F' in order to be able to use (464.7). And the extension of (464.8) to cover positive and negative roots surely requires separate consideration of the numbers of concave and convex extrema between $x = 0$ and the nearest intersections on either side.

Nevertheless, the approach has some attractions; in particular, Lagrange praised it at the beginning of note 8 of his *1808a*, praise which Cauchy took as the inspiration of a paper submitted to the *classe* in May 1813 (*PV*, *5*, 214). Although Cauchy's mathematics is not studied in detail before Chapter 10, it is meet to take here this early piece, written in his 24th year.

Poisson *1813c* reported speedily on the paper and with approval, and Cauchy quickly published it (or part of it) with Courcier as the pamphlet *1813a*. When writing the paper he had known de Gua's procedure only from Lagrange's brief account, and he added a postscript, announcing that he had just read de Gua's paper itself. During the autumn he submitted two longer papers (*PV*, *5*, 253, 263). Poisson again promptly reported his approval in *1813j*; published with the *Société Philomatique* a short account *1814c* of de Gua's and Cauchy's results, based on his two reports; and (presumably) accepted for the *Journal de l'Ecole Polytechnique* Cauchy's final version *1815c* of his work, which appeared in the same volume as the earlier papers on permutations (§ 4.6.1). Poisson's interest in this research was explained *en passant* early in Cauchy's paper: 'Mr. Poisson having willingly indicated this subject of research [...]' (*1815c*, 171: see Belhoste *1982a*, *1*, 47–50).

Cauchy's basic insight was to extend de Gua's (464.7) and use *all* the roots $\{t_r\}$ of F. He formed two polynomials, each of degree $(n-1)$ (one less than that of F):

$$g(x) := \prod_{r=1}^{n-1} (x - F(t_r)\, F''(t_r)), \tag{464.9}$$

$$h(x) := \prod_{r=1}^{n-1} (x - t_r\, F(t_r)\, F''(t_r)). \tag{464.10}$$

He was able to formulate some of the properties of roots more elegantly than the geometrical considerations involved in de Gua's procedures; roughly speaking, g played the role of FF' in de Gua's argument, and the presence of t_r in h allowed discrimination between positive and negative roots. But there were various theoretical difficulties; complex roots of F' could nevertheless lead to real, but irrelevant, roots of g and/or h; special problems arose if g and/or h had 0 as a root; and so on.

To avoid some of these difficulties Cauchy developed various rules. In one of them he eliminated x between the two equations

$$F'(x) = 0 \quad \text{and} \quad x_1 + x\,F(x)\,F''(x) = 0 \tag{464.11}$$

to obtain a new polynomial F_1 of degree $(n-1)$ in the new variable x_1. Repeating the process in (464.11) on F_1 to yield $F_2(x_2)$, and so on, he produced a sequence of 'algebraic equations' (as polynomial equations were then sometimes called):

$$F_r(x_r) = \sum_{i=0}^{n-r} a_{ri}x^i = 0, \qquad 1 \le r \le (n-1), \tag{464.12}$$

where $r = 0$ was taken as the case $f(x)$, and obtained the result:

$$\begin{aligned}&(\text{number of positive roots of } F) - (\text{number of negative}\\ &\text{roots of } F) = \sum_{r=0}^{n} (\text{sign of } -a_{r1}a_{r0})\end{aligned} \tag{464.13}$$

(Cauchy *1815c*, 177).

The proof was elegant (it rested essentially on the role of h already mentioned and on the fact that $bx + c = 0$ took one positive/negative root if $-bc > / < 0$), but the procedure itself has limited practical potential. Eliminations can be difficult to execute; they also involve conceptual problems, such as the degree of the eliminant (as Poisson himself had found in *1802b*), and a sequence of eliminations is required to produce the $\{F_r\}$. Again, some rules require the resolution of g and h; but if this cannot be done, then we form two pairs of equations from the roots of g' and h' using definitions of the form (464.9–10), and if these equations be difficult to resolve, then we form .... Finally, the theory only extends as far as positive and negative real roots; there are no results analogous to Fourier's or Budan's on the number of roots within a given interval.

The key to Cauchy's interests is indicated in the opening lines of his paper: 'one resolves completely' (apparently) polynomials with numerical coefficients, but theory was still incomplete for 'literal equations' (that is, polynomials with algebraic coefficients). The main body of the paper, concerned with 'analytic developments', contained a wealth of properties of functions of the roots of polynomials, some produced with the help of the ideas developed in the earlier papers on permutations. Thus the paper may be better appreciated as concerned with the theory of functions rather than the theory of equations (as is argued in Bertrand *1857a*). In other words, it is an early indicator of the thrust of Cauchy's mathematical interests.

4.6.5 Complex roots and geometrical approaches

Complex roots of 'algebraic equations' posed particular difficulties, and all the methods discussed earlier were only partially successful in detecting them. Lagrange favoured using the equation of differences

$$w(x) := \prod_{r=1}^{n} \prod_{s \neq r}^{n} (x - (p_r - p_s)^2) = 0 \tag{465.1}$$

of F, where $\{p_r\}$ were its roots, for detection; if w had a negative root $-\alpha^2$, then F took the complex pair $(\alpha \pm i\beta)$ of roots (*1808a*, ch. 5). A fresh attack on this problem was mounted by another of our fascinating outsiders, Louis Corancez.

Two years Fourier's junior, son of a close friend and biographer of Rousseau, Corancez heard some of Lagrange's (and perhaps also Fourier's) lectures at the *Ecole Polytechnique* before embarking on one of the journeys that marked his career in diplomacy. He served in such a role during the Egyptian campaign, when both he and Fourier worked on the theory of equations; later he attested to Navier on Fourier's interest in the subject at that time (Navier *1831a*, xxi–xxii). His scientific publications were infrequent, but worthy of notice; they brought him candidature for Laplace's place in the *Académie* in 1828, although he received few votes. However, in 1811 his literary writings earned him election as a *correspondant* of the *classe* for *inscriptions et belles-lettres* of the *Institut*. During the 1810s he married a niece of de Prony, retired for health reasons, lived on de Prony's land at Asnières, and carried out his researches. de Prony wrote a notice *1836a* of him after his death in 1832; after de Prony's own death in 1839 the niece was made executor, as was noted in § 2.5.3.

Corancez submitted two papers to the *classe* on the roots of equations. The first was recommended to the *Savants étrangers*, and (therefore) was not published; the second was rather disparagingly regarded, but appeared in the *Journal de l'Ecole Polytechnique*. The first was submitted in September 1810 (*PV*, *4*, 377: see also p. 371). According to Lacroix's report *1811a*, Corancez sought means of distinguishing complex roots in the polynomial f of even degree by putting $y = f(x)$, setting $f'(x) = 0$ and eliminating x between these two equations to obtain $g(y) = 0$, where g was of degree one less than that of f. The roots of g would be the extrema of f; so that if f took no real roots, g had no negative ones (presumably because Corancez took f in monic form). Since the degree of f was even, that of g was odd, so that g had at least one real root; and if this root were positive, then f took a complex pair of roots.

There seem to have been various other considerations in Corancez's paper, especially concerning the number of complex pairs of roots. Apparently his method was an alternative to the use of the equation of differences (465.1), and had advanced up to the sextic, although it involved prodigious calculations of its own. It cannot have entailed the resolubility of the quintic; presumably he did not assert that it did.

Corancez's second paper went to the *classe* in June 1813 (*PV*, *5*, 221), shortly after Lagrange's death. It is a rather strange affair, offering geometrical constructions in terms of evolutes of standard curves such as the circle to represent algebraic expressions for real roots. But the constructions themselves were also expressed algebraically, in terms of Arbogast's calculus of derivations (§4.3.6), no less:

$$\text{`}x = \text{Arc. dév. } (M + \text{Arc. dév. } (N + \ldots)\ldots)\text{'}, \qquad (465.2)$$

where x was the root, M, N, ... relevant radicals, and 'Arc. dév.' referred to the operation of forming 'developpants' (or evolutes) (*1815a*, 218-220).

The applications of this idea led to formulae as forbidding as are the diagrams. One can understand Poinsot's puzzlement in his report *1813a* to the *classe*: he praised the speculations *per se* but found them irrelevant to the principles of the theory and ponderous in their procedures. It is also surprising that the paper was published in the *Journal de l'Ecole Polytechnique*; it seems too geometrical for Poisson, and too abstract for Hachette.

Perhaps de Prony exerted influence on Corancez's behalf to get his paper published. de Prony was interested in the bearing of the theory of equations on geometrical problems; one of his first publications had been a pamphlet *1791a*, for use at the *Ecole des Ponts et Chaussées*, on interpreting quadratics in the context of conic sections. But he offered *specific* results, of practical use for engineers. So did Monge, in a note *1817a* in the *Correspondance sur l'Ecole Polytechnique*, which gave a graphical solution of the cubic

$$x^3 - px - q = 0 \qquad (465.3)$$

in terms of the intersection of a straight line with a cubical parabola engraved on a board, a scheme prepared for engineering use during the Egyptian campaign. By contrast, Corancez was (somewhere) in general theory, like the more theoretical mathematicians of his time. His mixture of algebra and geometry is highly characteristic of not only his own mathematical thought but also that of several other contemporaries, as we shall now see further.

4.7 Case studies in the relationship between geometry and algebra

4.7.1 Carnot and the 'geometry of position'

> As long as algebra and geometry have been separated, their progress have been slow and their uses limited; but when these two sciences have been united, they have lent each other mutual forces, and have marched together with a rapid step towards perfection.
>
> Lagrange *1795a*, 271

In § 3.2.5 I mentioned the uneasy relationship between algebra and geometry, the alleged philosophical distinction between them based on the distinction between 'analysis' and 'synthesis', and the inadequacy of the term 'analytic geometry' to describe the subject denoted. These difficulties are marked in Carnot's 'geometry of position', which also tempers Lagrange's optimistic view cited above.

Carnot was concerned with some non-metrical properties in geometry, especially those which were preserved while a geometric configuration was continually changed. To take one of his simple examples, in a triangle ABC the foot D of the perpendicular from A fell between B and C as long as $\angle \mathrm{ABC} < \tau$ $(=90°$ (§ 1.3.7)), although it lay outside BC when the angle surpassed τ.

Carnot expressed his theory as far as possible in algebraic terms, both the algebra and trigonometry of directed line-segments (where he used the form of notation '$\overline{\mathrm{BC}}$') and also ordinary letters. But he gave geometry predominance over algebra, and he held that negative numbers should be avoided when they represent objects, or operations on objects, which (for him) could not be effected. For example,

$$x - a = \sqrt{b} \tag{471.1}$$

caused no trouble (when $x > a$); the minus sign merely told us that a must be subtracted from x. Thus (471.1) was an 'explicit formula'. By contrast,

$$x - a = -\sqrt{b} \tag{471.2}$$

was an 'implicit formula': quite satisfactory as a formula, it contained an operation $-\sqrt{b}$ which could not be effected.

From such considerations Carnot came to a fresh distinction between analysis and synthesis. 'Analysis' could embrace both explicit and implicit formulae, and so was closely associated with algebra; but 'synthesis' could use only explicit formulae, and in this way was associated with geometry (*1801a*, 142–145; *1803b*, 7–11). The occurrence of negative numbers in a formula was an indication of some kind of anomaly, or impossibility of correlation between configurations. In the earlier example of ΔABC, if D fell outside BC, then he wrote '$-\overline{\text{CD}}$', and not '$\overline{\text{DC}}$', for the reversed directed line-segment.

Carnot's opposition to negative numbers carried over to complex numbers; but he had to admit the need for 'complex correlations', between, for example, the configurations given by

$$a^2 = x^2 + y^2 \quad \text{and} \quad a^2 = x^2 - y^2. \tag{471.3}$$

He used proportions to express relations such as these: $(471.3)_1$ would be written

$$a + x : y :: y : a - x \tag{471.4}$$

(*1801a*, 177–188).

Carnot's position was opposed to that of the *idéologue* philosophers such as Condillac, whose support of negative numbers was noted in § 3.2.5. The results that he obtained had some significance in the development of both analytic and synthetic geometry (see Simon *1906a passim*, and Gillispie *1971a*, ch. 5). His preference for geometry was expressed in various contexts. For example, in a pamphlet *1800a* addressed to Bossut, which Bossut published in his *Cours de mathématiques* of that year, Carnot presented basic trigonometry and the geometry of the triangle in both ordinary algebra and the algebra of directed line-segments; but he did not attempt to algebrise geometry any further.

Carnot's approach stands in contrast with that of Corancez, who in a pamphlet *1798a* which Carnot seems not to have known, had attempted to algebrise the basic notions of geometry. Probably influenced by some sketched ideas of Legendre (*Geometry*$_1$ (1794), note 4: renumbered in later editions), Corancez developed the geometry of the triangle and of conic sections from these basic relations for the triangle (pp. 11–13):

$$A = f(B, C), \quad \text{where} \quad f(B, C) = f(C, B); \tag{471.5}$$

$$c/a = \varphi(b/a, A, B). \tag{471.6}$$

Corancez's interesting suggestions unfortunately aroused no interest, no doubt due to the obscurity of the author and the publication. Carnot's ideas produced what might politely be termed as expectancy. For example, Servois made some use of them in a short book *1803a* proving various theorems on the geometry of the plane and applying them to a selection of problems in practical geometry; and in a review Bonaffos *1803a* praised him for making Carnot's ideas seem interesting!

Carnot found Servois's book useful, for he drew on it when he prepared a short monograph 'on the relation which exists between the respective distances of any five points taken in space; followed by an essay on the theory of transversals' (Carnot *1806a*). Among other things, he essayed a theory of oblique coordinates; and in an appended 'digression on the nature of quantities called negative', he returned to his old theme by proposing to distinguish between 'negative quantities' (as truly affected by a minus sign) and 'inverse quantities' (taking a minus sign in order to render true the formulae in which they so occur).[1])

Carnot's work was coolly noted by those who took an interest in it; for example, in 1813 Jacques Français stated that 'the first principles of the *Géométrie de position* are not yet completely established' (*1813c*, 63). And at this time a *polytechnicien* lay in a Russian prison working out ideas which were to make a major contribution to projective geometry (§ 11.2.1). His name was Jean Victor Poncelet.

4.7.2 Argand and the geometry of complex numbers

In the paper just mentioned, Français tried to use Carnot's theory to develop a geometrical representation of complex numbers. The ideas were not his own; he had found them in a letter from Legendre to his now dead brother François. They were not Legendre's either, but had 'been communicated to him [Legendre] and as [an] object of pure curiosity'. Français hoped 'that the publicity that I am giving to the result to which I have come may determine the first author of these ideas to make himself known, and to bring to light the work that he has done himself on this subject' (*1813c*, 74). He published his paper in Gergonne's *Annales*, and Argand responded by identifying himself there as the missing link.

As the introduction of the 'Argand diagram' is now one of the best-known events in 19th-century mathematics, I can refer to suitable secondary literature,[2]) and restrict myself to a few comments on the French response. Its most notable

---

[1]) Carnot's ideas exercised an influence on Canard *1808a*, an 'elementary treatise on inequations' by a *lycée professeur*, in which he tried to restate the known theory of equations in terms of a calculus of inequalities. This work bore no anticipation of Fourier's conception of linear programming of the *1820s* (§ 17.3.2). Independently of Carnot, Jacques Français outlined a theory of oblique coordinates in two papers *1808a* and *1808b*.

[2]) Most of the relevant primary literature is published in Argand *1874a'*, a so-called 'second edition' of Argand's original pamphlet *1806a*. Among the older secondary literature, see Cartan *1908a*, arts. 1-3; and Cajori *1912a* on geometrical representations of i before the late 18th century. Among more recent literature, Crowe *1967a*, ch. 1 is a useful guide, including to other secondary literature; the little-known S. Bachelard *1966a* is worth adding. On Wessel's interesting anticipation of the geometrical representation, see Brun *1959a*, and especially Andersen *1979a*, and *1980a*, 327-338. Buée *1806a*, contemporary with Argand, was published in England.

feature is silence; there was little interest at the time, and Legendre's assessment of Argand's suggestion as 'an object of pure curiosity' is typical. The main reason for the uninterest was that the *algebraic* theory of complex numbers had been exhaustively developed, so that there seemed to be no need for alternatives.

The state of the algebraic theory around 1800 can be best appraised from a treatise by Suremain de Missery, another of our outsiders who was to achieve only candidature as a *correspondant* to the *Académie*. His interests showed an interesting link from music through acoustics to logarithms—and thus to complex numbers, for the definition of $\ln(-x)$ was one of the principal problems in complex analysis of the time. He presented a copy of his treatise *1801a* on a 'purely algebraic theory of imaginary quantities and of functions which result from them' to the *classe* in June 1801: as it was published, Lacroix gave only a verbal report.[1])

Suremain's 300 pages of continuous symbolism is not riveting reading, but it shows well his confidence in the algebraic approach. The logarithmic function, the binomial expansion $(a + ib)^m$, the complex roots of unity, trigonometric functions, and so on; all are developed as algebraic manipulations, especially concerning the functions f and g in the power-series expansion of e^{iz} given at the start (*1801a*, 13):

$$\exp(x + iy) = e^x f(y) + ie^x g(y). \tag{472.1}$$

Confusingly, later in the book he wrote (472.1) in the form

$$e^{iz} = f(z) + if'(z) \tag{472.2}$$

(p. 50), in which f and f' were two different functions (to be determined), *not* that f' was the derived function of f. This is an example of the drawbacks inherent in Lagrange's notations warned in § 1.3.7.

The impression that complex numbers were not problematic is strengthened by other literature; for example, Lagrange's 'treatise on algebra' *1808a* often referred to complex roots of equations, but made no special issue of complex numbers themselves. Hence there was no reason why Argand's pamphlet *1806a* should make an impact, quite apart from the obscurity of its author (he is said to have been a Paris bookseller, although he is not named in Delalain *1899a*).

Argand indicated no background to his work, but he may have been influenced by Carnot; for he, too, used directed line-segments and wrote them in the

---

[1]) See *PV*, *2*, 359, 377; a letter from Suremain to Lacroix of this period on algebraic constructions survives in *BI*, ms. 2396. Maybe the manuscript of this book was the 'treatise on pure mathematics' that he submitted to *CI* in June 1797 (*PV*, *1*, 225), and on which Lagrange and Legendre failed to report.

notation '$\overline{\mathrm{CD}}$', and worked with proportions like Carnot's (471.4) for 'complex configurations'. Indeed, he defined i as 'the quantity x which satisfies the proportion

$$+1:+x::+x:-1\text{'}; \qquad (472.3)$$

and since -1 lay in the reverse direction to $+1$, i had to be half way round (Argand *1806a*, 6). The rest of his pamphlet contained treatments of various theorems in the geometry of the plane and their algebraic analogues; for example, the expansion (433.6) of $(2\cos a)^n$ for integral n (pp. 44–45). The proofs were largely carried out in the geometry, so that the essay belongs more to the algebra of directed line-segments than to orthodox algebra. The most noteworthy passage occurs on pp. 58–59, where Argand tried to prove the fundamental theorem of algebra. In effect he expanded the polynomial $f(p+\varrho q)$, where p and q (which he actually wrote as 'i'!) were complex numbers and ϱ some complex root of 1, in a power series in ϱq which took the form

$$f(p+\varrho q)=f(p)+i\varrho Q(q) \qquad (472.4)$$

when q was an infinitesimal. He chose ϱ such that $f(p+\varrho q)$ and $i\varrho Q(q)$ were directed line-segments in exactly opposite directions, thus obtaining from (472.3)

$$|f(p+\varrho q)|<|f(p)|. \qquad (472.5)$$

By repetition of the procedure he obtained a decreasing sequence of moduli, which eventually would become zero. Hence the desired root was found. While the proof is hardly free from faults (as Argand admitted), and not up to the standards which Gauss was bringing to the problem (see Ostrowski *1920a* and Fränkel *1922a*), it does not compare too badly with "the state of the art" in France, as reported in notes 9 and 10 of Lagrange *1808a*.

In addition to Argand responding to Français's 1813 call for identification, there were various papers and letters for the next two years in Gergonne's *Annales*; by Gergonne himself, Français again, Lacroix and Servois. (All the papers are reproduced in Argand *1874a'*.) Argand reported that his pamphlet had sold badly, and so he repeated many of its principal points. The most interesting mathematical question discussed was the possible extension of the representation to three dimensions. Ignorant of the fact that Euler *1751a*, 222 had shown that

$$\ln i=(-1)^n(2n+1)\,i\tau, \quad n \text{ integral, and thus that } i^i=e^{-\tau} \qquad (472.6)$$

for the principal value, given by $n=0$ (see Eneström *1899a*), Argand had suggested that i^i needed a third dimension (*1813a*, 92–93). Servois pointed out Euler's result (*1814d*, 105), and both he and Français discussed ways in which the extensions might be achieved. Since all standard functions of complex numbers could be reduced to the form $(p + iq)$ (Lagrange *1808a*, note 9: compare Gergonne *1813b*), the task was difficult, though the discussion included some interesting *tentatives* towards hyper-complex numbers. Now aware of Euler's $(472.6)_2$, Argand *1815a* wondered how numbers of the form i^i could be defined in the first place, and recalled his alternative interpretation.

So the geometrical representation gained some currency in the early 1810s. Indeed, apparently it also exercised the mind of one H.-D. Truel in the 1780s, although we know this only from Cauchy's later testimony, collected during his days as an *ingénieur des ponts et chaussées* (see *1849a*, and the beginning of *1849b*). It was mentioned in §4.2.5 that Gergonne's *Annales* may not have been widely known; in this case, at any rate, the papers published there did not prevent another sequence of publications on the geometrical representation by Gauss and others around 1830![1])

4.7.3 On the establishment of 'analytic geometry' and 'analysis applied to geometry'

Perhaps the most substantial development of this period in the relationship between algebra and geometry concerned the establishment of analytical geometry as an organised body of knowledge on the application of algebraic methods to the study of curves, surfaces and bodies. It is an extensive story in its own right, and only partially related to ours,[2]) but its bearing is worth indicating in general terms. In this sub-section some general features and results are recorded: the next one is devoted to important properties discovered by Fourier and Lancret.

The subject in fact became known as this time as 'analytic geometry'. The phrase was even in the title of Biot's textbook *Essai de géométrie analytique*, written while he taught at the *école centrale* at Beauvais and first published in 1802 (Biot *Geometry*). It proved to be a popular work, an eighth edition appearing in 1834; it grew in size over the years, starting out at 320 pages but ending up at 474. As the rest of the title indicated, it was devoted to 'curves and surfaces of

[1]) In a sequence of letters to Airy of November 1857, de Morgan recalled that in lectures at Cambridge (around 1820, presumably), which de Morgan had attended as a student, Airy had claimed that he found the geometrical interpretation of i in his own student days (in the late 1810s) (Airy Papers, Royal Greenwich Observatory (Herstmonceux), ms. 433, fols. 199–205).

[2]) Of literature of the time, Hachette *1815b* provided a selection of results recently proved by Dupin, Fourier and Lancret, and in *1831a* he furnished some historical details. Among the modern histories, C. Boyer *1956a* is the most detailed: see also Taton *1951a*, ch. 3; and Kline *1972a*, ch. 23.

the second degree', and naturally gave much attention to the algebra/geometry of conic sections.

Analytic geometry was a useful background for workers in mechanics and mathematical physics, and sometimes played a larger role—for instance, in Malus's and Fresnel's treatments of double refraction (§ 7.4.5, § 13.4.3). In view of the difficulty noted in § 3.2.5 over the words 'analytic' and 'synthetic', the use here of the former needs explanation. Biot provided the following in the preface of his book, only slightly changing the wording in the various editions: 'I intend by this denomination the manner of applying algebra to geometry, not with the aid of particular constructions which must be varied for all the cases, but in employing the general methods that Messrs. Lagrange and Monge first made known in their works'. To this somewhat dubious history, he added a comment on the organisation of the book, noting that 'I have sought to follow the example of Euler, who is, in all his works, a model of clarity'.

Lacroix wrote two series of textbooks in this area, and with links to descriptive geometry: *Curves*, mainly for descriptive geometry (19 editions, 1799–1874); and *Geometry*, in analytical geometry (seven editions, 1795–1840). We take up the further stages in § 8.3, which is devoted to descriptive geometry as presented at the *Ecole Polytechnique* by Monge and his followers.

When the calculus was involved, the topic became what we call 'differential geometry' but then carried names such as 'analysis applied to geometry'. Indeed, this was part of the title of Monge's textbook at the *Ecole Polytechnique*, which first appeared in 1795, with three later editions to 1809, the latter two incorporating material written with his disciple Hachette (Monge *Analysis*). We have met this work already in § 3.3.5, 7 in connection with types of solution to differential equations and their interpretations as surfaces.[1])

Otherwise, and sometimes in harness, the theory was much concerned with tangent planes, developable surfaces, curvatures, arc lengths, and the like, including aspects of the calculus of variations such as Lagrange's equation (335.5) of minimal surfaces. The form of the equation quoted there was in fact proved in Meusnier *1785a*, 504, an important paper on the curvature of surfaces, building on various results given in Euler *1767d*. One of those had stated that if the principal radii of curvature of the surface $z = f(x, y)$ were R_1 and R_2, then the radius R_3 for the section making respective angles ω and $(\tau - \omega)$ with the principal sec-

[1]) I cited in § 3.3 the second (1801) edition of Monge *Analysis*, as it was reasonably well distributed (and is closer to the nominal starting date of 1800 for this book). By contrast, the first edition (1795) is extremely rare. It was printed on very large sheets, called 'Planches' and then 'Feuilles'. The first two editions do not much differ in content. Liouville brought out an extended and posthumous edition in 1850.

tions was given by

$$1/R_3 = (\cos^2 \omega)/R_1 + (\sin^2 \omega)/R_2 \tag{473.1}$$

(to quote the form first given in Dupin *1813a*, 109). Monge *1784a* also developed Euler's ideas: in particular, he showed in art. 25 that R_1 and R_2 were the roots of the quadratic equation

$$\begin{aligned}(z_{xx} z_{yy} - z_{xy}^2) R^2 - [(1 + z_x^2) z_{yy} - 2 z_{xy} z_x z_y + (1 + z_y^2) z_{xx}] R \\ + (1 + z_x^2 + z_y^2)^2 = 0.\end{aligned} \tag{473.2}$$

Theorems such as these were to serve as bases for important work done later: for example, at (648.1) for de Prony's study of latitude and longitude, (722.7) in Laplace's analysis of capillarity, or Poisson at (732.19) and Navier at (1525.3) in their examinations of elasticity.

The theory seems to have been reasonably thoroughly taught at the *Ecole Polytechnique* from various points of view, not only Monge's. For example, Lagrange treated certain aspects in his *Théorie des fonctions analytiques*, together with maxima and minima and some isoperimetry, relying as usual fairly heavily on Taylor-series expansions (*Functions*$_1$ (1797), 117–187). It was quite an active research area for *polytechniciens*, with papers appearing in both the *Journal* and the *Correspondance*. One sees the editorial encouragement of Hachette here, no doubt (§ 2.3.3). Gergonne's *Annales* was another ready source for papers; and here the enthusiasm of the editor (§ 4.2.6) was quite explicit! Other significant figures included Bobillier, Chasles, Dupin and Poncelet, and there were several more; but I stress from above that only fragments of their work appear in this book.

Naturally all these topics related closely to each other. A good example is provided by Puissant, a distinguished geodeter and cartographer who comes into our tale first in § 5.5.8. In 1801 he published a collection of 'geometrical propositions resolved or demonstrated by algebraic analysis' (*Geometry*). This book, especially in its revised editions of 1821 and 1824, provided a good representative coverage of the interactions that geometry, the calculus and algebra could achieve when applied together to curves and surfaces.

4.7.4 Lancret on differential geometry

Here we disclose important contributions to differential geometry, made by a figure new to our story: Michel-Ange Lancret, another member of the founding year at the *Ecole Polytechnique*, then to the *Ecole des Ponts et Chaussées* before joining the Egyptian campaign (§ 2.6.2). Upon his return he joined the commission charged with publishing the *Description de l'Egypte*, becoming its secretary.

He found a good friend in Fourier, who pointed out to him a property of curves in space which seemed to stimulate his researches. The first paper was submitted in April 1802 (*PV*, *2*, 496): Lacroix *1802b* recommended it three months later to the *Savants étrangers*, where it appeared as Lancret *1806a*.[1])

Curves in space were then called 'curves of double curvature' because research (especially in Monge *1785a*) had made clear that two parameters were needed to specify their path: the 'first flexion', marking the rate of change of the angle between consecutive tangents with arc length; and the 'second flexion', the similar rate of change of the consecutive normals to the osculating planes. (The terms 'contingence' and 'torsion' became fairly standard later for these notions.) At a point P of the curve C could be defined the osculating plane O and also the plane N of normals; and as P moved along C, N enveloped a developable surface D (called by Lancret its '*développée*'). Consecutive planes intersected along their 'characteristic' L (to use Monge's term from § 3.3.7: 'polar line' is often used now); and as P moved along C, the ensemble of lines L generated a curve A which Monge had called the '*arête de rebroussement*' ('edge of regression', in modern parlance), the curve which split L into its two sheets.

Lancret started out from the following unpublished theorem of Fourier. Firstly, since N was tangent to D, then the first flexion (contingence) of C was equal to the second flexion (torsion) of A. Secondly, since O was parallel to the plane of normals to A, then the second flexion of C was equal to the first flexion of A (Lancret *1806a*, 419–420).

Given this start, Lancret worked out a most elegant theory of the developables of curves, and their osculating planes, edges of regression, and so on. He also studied the 'rectifying surface' of C, the surface R containing C such that if R were stretched to a rectifying plane, then C would become a straight line. He showed that R was the developable of the surface generated by the osculating plane, so that the normals to this surface were tangents to R (pp. 420–425). Sometimes he argued geometrically, in terms of properties of planes and normals (although he furnished no diagrams); otherwise he followed Monge's use of the calculus which we have seen already in § 3.3.5, taking the equations of tangents and normal planes in terms of a parametric variable α for P, α-differentiating and -eliminating to find envelope curves and surfaces, and so on.

Lancret continued these interesting curves of thought in a submission to the *classe* in December 1806 (*PV*, *3*, 472). As last time, Lacroix *1807a* recommended

---

[1]) Hachette *1807b* comprises a short obituary notice for Lancret. No *Nachlass* survives, but letters from his Egypt colleague Malus are held in Malus's *dossier personnel* at *AS*. His work is noted briefly in various places, especially Lacroix *Calculus*$_2$, *1* (1810), 582–651 *passim*. On Fourier's contribution, about to be described, see the pertinent manuscripts in *BN*, ms. n.a. fr. 22519, fols. 1–22.

it to the *Savants étrangers*, where it appeared as Lancret *1811a*: in addition, a prosodic summary *1807a* appeared first with the *Société Philomatique*, while some of the sums made a rather tardy re-appearance as *1815a* in the *Correspondance sur l'Ecole Polytechnique*. This time he extended his theory to 'developoids' of plane curves (*1811a*, 2), surfaces which were generated by lines making an acute angle ω with the given (plane) curve C (so that developables were the special case defined by $\omega = \tau$). Again he found equations for these surfaces and related features by calculus methods: in addition, since for any C and ω two lines were definable (one for ω forward along C, the other back), two osculating centres P and Q applied, which acted as foci for an 'osculating ellipse' (pp. 10–14, fig. 3). He pursued these ideas for planar curves and curves in space, and announced at the end the intention to study the developoids of curves in space and of developable surfaces. Unfortunately his death at the end of 1807, at the age of 33, terminated these plans; otherwise no doubt another distinguished geometer in the tradition of Monge could have been added to the list of major Parisian talents in mathematics.

4.7.5 Lagrange and the geometry of Euclid, 1806

In this section I have considered aspects of the geometry of the plane, and thus of Euclidean geometry. I conclude with this postscript, recording a strange and little-known event, which occurred at the meeting of the *classe* on 3 February 1806 (*PV*, *3*, 314). Biot recalled it as follows (*1837a*, 84):

> Then one day Lagrange took out of his pocket a paper [*1806b*] which he read at the *Académie* [*sic*], and which contained a demonstration of the famous *Postulatum* of Euclid, relative to the theory of parallels. This demonstration rested on an obvious paralogism, which appeared as such to everybody; and probably Lagrange also recognised it as such during his lecture. For, when he had finished, he put the paper back in his pocket, and spoke no more of it. A moment of universal silence followed, and one passed immediately to other concerns.

4.8 Concluding comments

Chapter 3 was substantially built on the existence of various treatises written around 1800 and summarising the state of development in the calculus. It is appropriate to end its successor with a brief note of some works of a similar nature which appeared around 1815. For reasons of space, I shall do little more than note their existence, although I have already made use of them and shall continue to do so. In addition, Interlude 480.1 looks rapidly over the state of the calculus in some other European countries.

Interlude 480.1 The calculus outside France

This Interlude is restricted to five countries, giving a fair survey of knowledge and ignorance. On Belgium, see Bockstaele *1966a*; on Ireland, my *1988a* and Interlude 660.1.

1. *England* English mathematics was dominated by the Newtonian fluxional calculus, although ignorance of Continental mathematics was not as great as is sometimes thought. The Cambridge mathematician Woodhouse, for example, showed good knowledge of it in his history *1810a* of the calculus of variations—better than in his calculus textbook *1803a*, in fact. He may have written (some of) the reviews of Continental mathematics in the *Monthly review*: in particular, I attribute to him the long piece *1799a* on Lagrange's *Functions*$_1$ (1797). More important for reform than Cambridge, however, were the adoptions of Continental calculi by Gregory, Barlow and Hutton at the Royal Military College, Woolwich, and Ivory, Leybourn and Wallace at the Royal Military Academy (founded in 1799 at Great Malvern and moved in 1816 to Sandhurst), before the Cambridge reforms of the mid 1810s (see my *1986b* and Panteki *1987a*). A significant motivation seems to have come from Laplace's *Mécanique céleste*, when it began to appear in 1799 (§ 5.4.3).

The inferiority of the fluxional calculus does not seem to me to lie in its form (which appears to be the claim of Cajori *1917a* and *1919a*, and of Dubbey *1978a*, ch. 1), but in the limited range of problems that were tackled in the first place, a restriction that was largely imposed by the (surprising) failure of the fluxionalists to develop equivalents to multi-variate calculus and the calculus of variations. This view is the burden of Toplis's criticisms *1804a* of English ignorance. There were also influential opponents of mathematics in England; for example, Gregory was to opine of Sir

Joseph Banks that 'He had a great dislike to the *mathematical* sciences and did all in his power to suppress them; a natural consequence of which is, that mathematical knowledge has greatly depreciated in England during the last 40 years' (letter of 27. 6. 1820 to H. Schumacher, *Nachlass* Schumacher, *Staatsbibliothek*, East Berlin: quoted in my *1985b*, 87, which covers the early 19th century in England). This situation is discussed again in Interlude 1170.1, when further transitions in English calculus are given a non-standard interpretation.

Guicciardini *1987a* is a fine history of the fluxional calculus. On the organisation of science in England, see Cardwell *1972a*, chs. 1 and 2.

2. *Scotland* The Scots were more interested in geometry than in the calculus, in line with their philosophical traditions (see Gibson *1927a* and Olsen *1970a*). Playfair complained about the ignorance of the calculus in his review *1808a* of the *Mécanique céleste* (compare supra); an annotated French translation Playfair *1809a* of this review in the *Bibliothèque britannique* motivated Biot *1809b* to compare Britain with France.

3. *Germany* The Germans were more aware than the British of French developments; for example, they translated some of the major French works. Mathematical research was led largely by the combinatorial school of Hindenburg, whose work was similar to that of Arbogast and his followers (§ 4.3.6: the Alsatian Kramp belonged to both groups). Otherwise the work was done by respectable but isolated figures such as Trembley, and by a variety of *Mathematiklehrer* or professors at none too prestigious universities (see Mehrtens *1981a*). Some publications were literally encyclopaedic in the hands of Kaestner, Klügel, Reuss and Rosenthal, and so tended to report many details on all aspects of the calculus rather than choose any particular approach.

Separate from these movements is Gauss, who seems removed even from the tradition of 18th-century Göttingen (on which see Müller *1903a*). An important paper relative to the content of this chapter is his essay *1813a* on the hypergeometric series, which far surpassed any French writing of the period on convergence and related questions. However, in general his mathematical interests tended not to fall within the main thrusts of this chapter.

4. *Italy* With Italy we find the most remarkable part of our story outside France: the existence of a tradition of research into both the calculus and mechanics from the late 18th century right through to the rise of Italian mathematics in the 1860s after the *risorgimento*. There was even an organisation: the *Società Italiana delle Scienze*, which from 1782 published every

two years a volume of *Memorie di matematica e di fisica* (see my *1986c* for a survey up to 1815: and for related information, Malfatti *1982a′ passim*). No work has been done on these Italians as a group: my own impression is that they made no basic contributions but several useful additions from time to time, which the French did know. Current surveys of Italian science in general from 1816 were provided by the journal *Biblioteca italiana*, which played something of the "comparative" role followed by the Swiss *Bibliothèque britannique* (§2.2.4).

With regard to the calculus, Caluso *1788a* favoured fluxions among the various approaches. Brunacci and Paoli were perhaps the most noteworthy contributors to the technicalities of the calculus. But Ruffini was probably the best known abroad, for his work on equations, of which we briefly noted some morsels in §4.3.8 and §4.6.3.

5. *Russia* In Russia the principal figure by far was Euler, who died in 1783 but was still publishing papers in the *Mémoires* of the St. Petersburg Academy until the volume for 1830, many of them in the calculus and mechanics. (The *Royal Society catalogue* conveniently supplies the references.) It would be interesting to study the reception of these papers by early-19th-century mathematicians; they seem not to have been widely noted, although in §5.2.7 we note an exception. Work by others in Russia was strongly Eulerian in character—that of N. Fuss is the most notable. On the period see Yushkevich *1968a*, esp. chs. 6–12; and Ozhigova *1980a*. For a general survey of Euler's achievements, see the volumes *1983a′* and *1985a′* celebrating the bicentenary of his death.

4.8.1 Impressions of some surveys

Lacroix wrote a manuscript *1801a* on the recent history of mathematics, perhaps in connection with his work on Montucla's history (§3.2.8). The essay presumably served him in good stead a few years later, when in 1808 Napoléon ordered the *secrétaires perpétuels* to tell him how science had gone since the dawn of civilisation in 1789. Lacroix helped Delambre with the mathematical sections of his report on the "mathematical" sciences (§2.2.1), of which part was published as Delambre *1809b* in the 'Histoire' of the *Mémoires*, and the complete version as a book *1810a*, in quarto and octavo editions. The topics discussed in this chapter were placed in the section entitled 'Algebra', which is itself a fair comment. Foreign work was duly noted, though naturally French achievements dominated.

The impression given is the same as that which I hope this chapter has left—that the traditions of 1800 were maintained and extended, rather than changed in any substantial way.

The same view applies to the second edition of the *Traité*, which Lacroix published in three volumes in 1810, 1814 and 1819. New material, recording many of the advances made during the new century, were introduced throughout the text, which was rounded off by a long list of 'Corrections and additions' and a splendid 'Table of contents' (*Calculus*$_2$, *3* (1819), 601-772). In addition, the structure of the work was changed somewhat, especially in the third volume on series and differences. But the general impression is still that the main streams and directions of the calculus had been amplified and enriched, rather than changed in any substantial way.

If Lacroix's *Traité* revealed mostly amplifications and enrichment of the Lagrangian traditions on the calculus, Lagrange's own *Théorie des fonctions analytiques* showed few changes at all in its second edition of 1813 from the content of its 1797 predecessor, especially in the sections on the basic principles and applications to mechanics, which are of chief interest here.[1]) The only literature by others which he cited was a bare mention of Ampère *1806c* on the derived function (§4.3.3) on p. 85. However, the reference followed a rather more purposeful argument towards the differential form (433.21) of the remainder term of Taylor's series.

Lagrange died in April 1813, aged 77 years, two months after this second edition was published. Successors to his offices were elected with some difficulty: Poinsot obtained his place in the *classe* in May, on the third poll (*PV*, *5*, 217-218), while Legendre succeeded him at the *Bureau des Longitudes* after various disputes, in which the nominations of the young Cauchy, Mathieu and de Prony had been cancelled (see Bigourdan *Bureau*, pt. 1 (1928), A30-A31, A36). In 1815, apparently at the instigation of Carnot, the government purchased Lagrange's manuscripts and gave them to the *classe*, who set up a commission composed of Lacroix, Legendre, Poisson and de Prony to study their suitability for publication (*PV*, *5*, 513). The commission met on more than a dozen occasions, passing manuscripts to each other in the interim, so that each member had read (more or less) every manuscript; they also enlisted the advice of Lagrange's special friend, the Swiss mathematician Maurice (§2.5.1). Lacroix presented a re-

[1]) The principal additions, mentioned in the preface and reported in Poisson's notice *1813g*, concerned the solidity and surface of bodies in pt. 2, ch. 14, and the proof of the principle of virtual velocities (§5.2.1, §6.2.2) in pt. 3, ch. 5. In addition, there was considerable re-writing and some re-organisation in pt. 1, chs. 12-14 on aspects of multi-variate analysis. Among Lagrange's manuscripts pertinent to the revisions, see especially *BI*, ms. 907, fols. 105-126. 1,500 copies were printed.

port *1817a* to the *classe* (*PV*, *6*, 233), recommending the publication of only a very few pieces but the preservation of the manuscripts.[1]) Lagrange would have been relieved at their decision; for, according to Biot's later recollection (*1828a*, 201), he was burning some of his manuscripts on one occasion, and said to Biot: 'I do not want to make posthumous works'.

4.8.2 A modern impression

When compared with the inheritance outlined in Chapter 3, it is clear that developments in the early 19th century were largely conservative, adding details—often interesting ones—to the algebraic tradition which Lagrange had made prominent. New speculations were offered in the tradition of operator algebras, and in solution methods to differential equations, although not much was added to the knowledge of general and singular solutions.

However, in various ways, Lagrange's tradition was not holding so firm. We have seen that differentials and (to a lesser extent) limits were in competition with power series, while some aspects of the theory of equations drew on the geometry of curves, and Argand explicitly preferred the geometry of the complex numbers. As we look at concurrent developments in mechanics in the following chapter, we shall find Lagrange's approach again playing a major role; but we shall also see that when "in the field" these geometers often had to allow themselves considerable geometrical rein as they used their mathematics in contexts requiring a physical interpretation.

---

[1]) They are now kept at *BI*, mss. 901-916; the signatures of the members of the commission are to be found on most of them. The commission's own file is to be found at *AS*, *dossier personnel* for Lagrange, and other relevant papers at *AN*, F^{17} 3604.

Chapter 5

The 18th-century heritage: mechanics around 1800

5.1 Prefaces

5.1.1 Mechanics in the 18th century

After an attempt in the last two chapters to survey the vast amount of work done in and around the calculus, I come now to the even vaster and more complicated developments in mechanics. This subject was very much at the heart of 18th-century mathematics, and the quantity of the research is as awe-inspiring as its mathematical and historical complications. For example, the quantity and range of Euler's work alone is bewildering; not only the central questions of general principles but also a spectrum of special cases from sounding bells to swinging chains, the motion of elastic lines and membranes, the design of ships and of spectacles, the performance of machines and microscopes, the bending of rods and the buckling of columns, and so on and on—right across the five divisions of mechanics indicated in Figure 121.1. Further, mechanics provided the imperative for the mathematics: in Euler's case again, partial differential equations were of importance to him because of the vibrating string and other problems from the 1740s; his first paper "purely" concerned with solving such equations was *1766b* (although much of his textbook *Integral* (1768–1770) was written by then).

It would be quite foolish to attempt even a summary of 18th-century mechanics, especially as the secondary literature available is still rather scanty.[1]) Instead I adopt the view that was often held around 1800: that Lagrange's *Méchanique analitique* (1788) contained the basic principles and main lines of thought in mechanics, and also the definitive outline of the history of the subject. But I must also remark that the view was quite mistaken; for Lagrange had narrowed down the rich and varied heritage of 18th-century mechanics, with the result that some

[1]) Probably the best sources are the lengthy editorial introductions to certain volumes of the second and third series of Euler's *Opera omnia*: distinguished examples include Fleckenstein *1957a*, Habicht *1973a* and *1978a*, and Truesdell *1954a*, *1955a* and *1960a*. Some articles in the *Encyklopädie der mathematischen Wissenschaften* can be consulted *passim*: see especially Prange *1933a*, Stäckel *1905a*, Voss *1901a* and Wangerin *1907a*. Among general histories of mechanics, see *passim* in Dugas *1950a*, Duhem *1903a*, Mach *Mechanics*, Truesdell *1968a* and Szabo *1977a*; and Dühring *1873a*, pt. 3. Many of these accounts were influenced by Lagrange's own historical *résumés* in the *Méchanique analitique*, and are unsatisfactory both for that reason and for the defects described in § 1.4.1; the marked lack of either mathematical details and/or historiographical sensitivity. Truesdell's writings are free from dependence on Lagrange, but his rendering of the mathematics is very anachronistic. At least these writers tried to tackle the subject, in contrast to the extraordinary silence resounding from Cantor *1908a* and Dieudonné *1978a*. For a pleasant brief introduction to 18th-century mechanics and its historiography, see Bos *1980a*.

of it was forgotten (for a time) and some re-discovered. Euler suffered most of all, partly for the very modest place that his works received in Lagrange's historical accounts, and partly because of the long delay in publishing some of his posthumous papers, but mainly, I feel, due to his very prolificity; it must have been so easy to overlook a piece of his work when there was so much of it, and when, unlike the bibliographical assiduity of Lacroix and the German encyclopaedists (Interlude 480.1), the historical accounts then available were rather meagre, much less than those for the calculus (parts of Montucla *History*, *3-4* (1802) were the best source).

5.1.2 Some preliminary remarks on mechanics and its mathematicisation

Today we are accustomed to distinguish between pure and applied mathematics; and as the title of Gergonne's *Annales des mathématiques pures et appliquées* shows, the terminology was known then also. But it is not helpful; *any* piece of mathematics is applic*able*, so that its uses in (say) hydrodynamics or number theory are not distinguished by the presence or absence of application *per se*. The distinctions made in § 1.3.2 between the actual concern, potential concern, or lack of concern, with physical interpretations, are more useful. They are closer to distinctions made at the time, which were expressed by Montucla, for example (*History* , *1* (1799), 3-4). To him mathematics was

> the science *of the ratios of size or of number, which can have between them all the things that are susceptible of augmentation or of diminution.* [...] Mathematics divides up naturally in two classes; the one comprises those which one calls pure and abstract; the other those which one calls mixed, or more ordinarily physico-mathematical.

Thus mathematics was partitioned by the criterion of whether or not physical objects were among the referents of the mathematical variables and constants. It is interesting to note Montucla's confinements of 'pure' mathematics to number and magnitude (a view which Lagrange's influence was making out-dated already by 1799, as we saw in Chapter 3), and 'mixed' mathematics to physics (also an out-dated view, but correctly indicating the emphases of the time).

There was a philosophical connection between the post-Lagrangian algebras and certain attitudes to physics. As was noted in § 3.2.5, Condillac's work enjoyed popularity around 1800, and its advocacy of the clarity of signs in language (which we recognise today as semiotics) was sometimes conjoined with the primacy of algebras (as in § 4.3, for example). This view in turn lived alongside some of the empiricist (the word that I shall use) and/or phenomenological approaches taken to physics in general and mechanics in particular. Thus the mixed mathematics of such thinkers could ally a clarity of mathematical expression with an empiricist interpretation of its referents.

The recipes for making the mixtures with which I am concerned often used differential equations. Four distinct, though mutually interacting, stages were involved: forming the equations to represent the physical phenomenon; solving the equations (with or without boundary conditions) by some mathematical method; relating the solution to the phenomena (in terms of generality, basic properties, and so on); and deducing specific physical situations (including calculating values of variables from given numerical data). Variants on this procedure included using difference, or mixed, equations rather than differential equations (examples were given at (353.2) and (442.18)); obtaining expressions to represent a certain effect, rather than an equation to describe a situation; and using a method now often called an 'n-body model', where a continuous physical situation was replaced by an n-body analogue which was solved by some mathematical means to obtain an equation $E(n)$ and then n taken to infinity to yield the (alleged) solution $E(\infty)$ of the intended problem. If a solution could not be found, then as much information as possible was extracted from the mathematical means available.

5.1.3 Francoeur: reporter

Louis Benjamin Francoeur was born in 1773 in Paris to the superintendant of the King's Music, who was therefore imprisoned during the Revolution. Nevertheless, the young man was able to join the *Ecole Polytechnique* in the founding year, after which he studied and taught at the *Ecole de Géographes*; then in 1799 he joined the teaching staff of his old school as a *répétiteur* to assist de Prony until 1804, and also served as an admissions examiner until 1810. He was appointed to assist Lacroix when the Paris *Faculté* opened in 1809; but he suffered at the Restoration, for this was the only appointment that he retained. He was elected to the *Académie* relatively late in life; in April 1842, seven years before his death.

Francoeur's late election was probably due to his rather modest contributions to research. But his capacity to produce textbooks was prodigious. He is comparable to Lacroix; he also produced a *Cours complet de mathématiques*, though rather than Lacroix's eight-volume ensemble (§ 2.5.4) this was an integrated two-volume work, which appeared in five editions between 1809 and 1828. Unlike

Lacroix, he also wrote textbooks in mechanics, usually at the second of the three levels described in § 2.2.4. An elementary treatise on mechanics was published in five editions between 1800 and 1825. There were also textbooks on statics (1810), astronomy (called 'Uranography': six editions, 1812–1853), technology (two editions, 1833 and 1842), geodesy (eight editions, 1835–1895) and other subjects. His interests in technology were very wide-ranging; he contributed to a multi-volumed *Dictionnaire technologique* in the 1820s (and to an abridged version of the 1830s), and wrote over 300 reports for the *Société d'Encouragement pour l'Industrie Nationale* (§ 16.2.1). He was gifted in languages, knowledgeable about calendars, and a passionate student of botany, he reviewed books on all sorts of subjects for the *Revue encyclopédique*, and on his death he apparently left behind book-length manuscripts on algebra, celestial mechanics and the history of mathematics, works which were not published.[1])

Since his contributions to French science were rarely original, and since he entered the *Académie* only in 1842, Francoeur does not feature prominently in the story. However, his role deserves this advertisement, which is placed here since his writings covered many of the topics covered in this and several later chapters (as well as several which are not). In particular, his interest in geodesy is a sign of the rise of planetary mechanics at this time, a feature which this chapter duly takes into account.

5.1.4 Plan of the chapter

§ 5.2 conveys something of the principles and scope of mechanics in the late 18th century. Structure-similarity with the calculus described in Chapter 3 is very evident; for again Lagrange advocated a very prominent view but Euler had his supporters, and the French taught mixtures of the two. There was considerable interest in the proofs of basic results, as is illustrated in § 5.3 with two important examples: the principle of virtual velocities, and the composition of forces.

The rest of the chapter is devoted to celestial and planetary topics. Laplace now takes over prime position from Lagrange, with his *Mécanique céleste*; several themes from its volume 1 (1799) occupy § 5.4. Volume 2, also published in 1799,

[1]) These items are reported in the obituary I. Francoeur *1853a* written by one of his sons, who taught at the *Ecole Nationale Supérieure des Beaux Arts* in Paris and deposited the *Nachlass* there in 1884 (mss. 397–438). Half the collection is devoted to botany: the mathematical half largely consists of the lecture notes of his *PF* teaching or of certain editions of his books, but does *not* include a history of mathematics. Another useful notice is Jomard *1851a*; but since then essentially nothing has been done on his life or work. Several interesting letters are kept at *BN*, ms. n.a. fr. 1304, nos. 88–93.

guides § 5.5, which is devoted to various matters in planetary science: potential theory, cartography, the shape of the earth, tides, and the rotation of the earth.

Just as some aspects of the calculus were picked up only in Chapter 4, several significant parts of mechanics are postponed until the next two chapters. Thus elasticity, optics and sound do not appear until Chapter 7, fluid flow and hydrodynamics until § 8.5 and § 10.3 respectively, and machine theory until Chapter 16.

5.2 Principles and scope of mechanics in the late 18th century

5.2.1 The publication of Lagrange's *Méchanique analitique* in 1788

> Among the philosophers who are occupied with research into the laws of motion, some make of mechanics an experimental science, the others, a purely rational science; that is to say, the former in comparing the phenomena of nature break them up [...] in order to know what they have in common, and thereby reduce them to a small number of principal facts [...]; the others begin with hypotheses, then reasoning in consequence upon their suppositions, come to discover the laws which bodies follow in their motions, if these hypotheses conform to nature, then comparing the results with the phenomena, and finding them to be in accord, conclude from this that their hypothesis is exact.
>
> Carnot *1783a*, 120–121

Carnot's distinction, while perhaps optimistically sharp, neatly expresses the different approaches to mechanics at that time. In this chapter we are largely concerned with those who treated mechanics as 'a purely rational science', and who enlisted mathematics to aid the reasoning from hypotheses which they concluded from experimental confirmations to be 'exact'. Carnot doubtless had Lagrange in mind as such a *philosophe*, as evidenced by his various published papers which Carnot must have seen.

At the time when Carnot wrote, Lagrange, still at the Berlin Academy, was completing a treatise on mechanics. He wrote to Laplace in September 1782 that the prospects of its publication were not clear (Lagrange *Works*, *14*, 116). Eventually it was published in Paris in 1788 (the year after his move there), following the recommendation of an *Académie* commission composed of Laplace, Cousin, Legendre and Condorcet. Legendre took over, or at least helped with, the proof-reading. Lagrange (or his publisher, Desaint) rather quaintly spelt its title '*Méchanique analitique*'. At this time Lagrange was in his early fifties.

The book began with a famous preface, which is often quoted: here are a few main phrases. 'I have intended to reduce the theory of this Science, & the art, of resolving the problems which relate it to general formulae'; 'this Work will have besides another use; it will reunite and present the different Principles found up to now under one same point of view, in order to facilitate the solution of ques-

Table 521.1 Outline of the contents of Lagrange's *Méchanique analitique* (1788).

Section	Topic	Page
	Part 1: Statics	
1	'On the different Principles of Statics'	1
2	Equilibrium of any system of forces	12
3	'General properties of equilibrium' deduced from section 2	24
4	Equilibrium of system of masses (or mass-points)	44
5	Solutions of statics problems (composition of forces; forces acting on linked system of masses; equilibrium of a line, and of a body, under a system of forces)	58
6	'On the Principles of Hydrostatics'	122
7	'On the equilibrium of incompressible fluids'	130
8	'On the equilibrium of compressible & elastic fluids'	155
	Part 2: Dynamics	
1	'On the different Principles of Dynamics'	158
2	Motion of 'a system of bodies, animated by any forces'	189
3	'General properties of motion' deduced from section 2	198
4	Equations of motion of any system of bodies animated by forces	216
5	Solution of dynamical problems (small oscillations of 'any system of bodies'; motion of a body 'attracted to one or more centres'; motion of several mutually interacting bodies)	233
6	Rotation of bodies (general formulae; rotation of a body under any forces; motion of a heavy body of any shape)	337
7	'On the Principles of Hydrodynamics'	428
8	'On the Motion of incompressible Fluids' (general equations; motion in receptacles or canals of "any" shape)	437
9	'On the motion of compressible & elastic Fluids'	492
	End	512

tions of Mechanics'; and, above all, 'One will not find Figures in this Work. The methods that I expound require neither constructions, nor geometrical or mechanical arguments, but only algebraic operations, subject to a regular & uniform course'.

The mechanical principles on which the book was based had also been introduced and used in earlier papers (see Truesdell *1968a*, 246–247), while the mathematical methods were those which he was to develop in his *Fonctions*: mechanics was (allegedly) reduced to geometry, then down to analysis, and then to algebra. Thus we have a mechanics where there was no dependence on 'mechanical arguments'!

Table 521.1 indicates the contents of the treatise by section, partly to indicate its range and balance and partly because it is not widely known (the second edition of 1811–1815 is the one published in Lagrange's *Oeuvres*). It reveals an elegant similarity of structure between the two parts; basic principles for bodies, chief properties, solutions of special cases, and applications to fluids. I follow its structure by devoting a sub-section to some basic features of each part—including the notations.

5.2.2 Lagrange's *Méchanique*: aspects of its statics

'Statics is the science of the equilibrium of forces', Lagrange began, thus laying at once the emphasis on equilibrium which is to characterise most mechanics of this period. A difficulty arises with the term 'equilibrium'. Although it was usually taken to cover both the condition of rest and states of uniform velocity (as in Newton's first law, for example), often the emphasis fell only on the first case, so that equilibrium became very closely associated with statics. This point will be opened up in § 16.3–4.

The first section of Lagrange's treatise was largely historical (and influential in itself: many later historians seem not to have looked further back). In it, he discussed three basic principles for equilibrium: the lever principle, 'composition of motion', and 'virtual velocities' (p. 2). Since the relationship between them was a major concern around 1800, I defer consideration of them until § 5.3.1.

Lagrange had little to contribute to the philosophy of physics. 'One understands in general by *force* or *power*', for example, using the two words as synonyms (a fairly common usage then, though surprising to us now), 'the cause, whatever it may be, which impresses or tends to impress on the body to which one supposes it [to be] applied' (p. 1). Again, mass was introduced much later, and *en passant*: gravitational forces 'will be, as one knows [*sic*], proportional to the masses of bodies' (p. 37). His philosophy of mechanics, however, was clearer;

the mathematics was based on variational methods, and the physics on the principle of virtual velocities, preferred precisely because of its expressibility in variational terms. If forces *P, Q,* ... acted in directions *p, q,* ... at certain fixed points, and if *dp, dq,* ... denoted 'the variations, or differences of these lines' (p. 13) caused by an infinitesimal disturbance of the system from equilibrium, then the principle yielded the fundamental equation

$$\sum Pdp = 0\,, \tag{522.1}$$

where the summation ranged over the forces in the situation. 'We will call [...] *Pdp*, the *moment* of the force *P*, in taking the word moment in the sense which Galileo gave it' (pp. 15–16). Perhaps the reference to Galileo clarified the word 'moment' and thus that *dp* was actually the projection of displacement of the point in the direction of *P*. But the reader was on his own to note that '*p*' was not Euler's differential coefficient z_x, and thus that *dp* was not *rdx*. Since Lagrange did express *dp* in terms of the direction cosines of *P* and the differentials *dx, dy* and *dz*, his use of '*p*' (and also the letters following) was unfortunate, as was also his employment of primes to play the role of subscripts rather than to indicate derived functions, and of '*d*' to represent variations ('δ' was brought in only on p. 51).

From (522.1) or its trigonometric analogues, Lagrange obtained other basic results. For example, (grandly) assuming that

$$\sum Pdp = d\varphi\ (=0) \tag{522.2}$$

and thus threatening the primitivity of the concept of force in his system,[1]) his theory of extrema supplied him with a proof that equilibrium corresponded to extreme values of φ; second-order conditions on the expansion of φ showed whether the equilibrium was stable or unstable (pp. 36–44). Again, if the system were under 'indeterminate' constraints, expressed by the equations $\{L = 0\}$, then the Lagrange multipliers (as we now say) were introduced to modify (522.1) for mass-points to

$$\sum P\delta p + \sum \lambda\delta L = 0\,, \tag{522.3}$$

---

[1]) We would regard φ as the potential work function for a system of conservative forces; but Lagrange did not so name it. We have seen already that (522.1) was a consequence of the 'principle of virtual *velocities*'; the word 'work' did not gain general currency until the 1820s, succeeding a chaos of terms (§ 16.4). A good study of the development of Lagrange's foundations of mechanics is given in Fraser *1983a*.

Just as Arbogast came to a view of the calculus similar to Lagrange's (§ 4.3.6), he advocated a form of mechanics like that of Lagrange's, with main place given to virtual velocities. His manuscript *1786a* was crowned by the Mantua Academy but was not published.

since $\sum \lambda\delta L$ could be shown to be equivalent to a set of extra forces acting on the system. For finite bodies, of masses $\{m\}$, it was necessary to allow that P acted only on an 'element' dm of m; to admit 'determinate' constraints $\{A = 0\}$ describing conditions applicable to given 'points' ('elements'?) of the body; and to include also terms of the form $P'dp'$ to allow for 'all the quantities which relate to determinate points of the mass', each point being identified by the number of its distinguishing primes. This line of reasoning led to the most general equation of equilibrium:

$$\mathrm{S}\left(\sum P\delta p\right) dm + \mathrm{S}\left(\sum \lambda\delta L\right) + \sum_{\text{primes}} P^{(r)} dp^{(r)} + \sum \alpha\delta A = 0\,, \qquad (522.4)$$

where 'S' was a special integral sign applied both to dm and to the 'indeterminate' constraints (because the functions $\{L\}$ might involve differentials $\{dx\}$ as well as variables $\{x\}$). The calculus of variations, especially the commutativity of d and δ (§ 4.5.1), generated developments of (522.4) (pp. 44–58).

After this display of algebra and generality, Lagrange showed his reader how four cases could be treated. I summarise the last, which concerned a non-deforming body of any shape (pp. 116–121). The main constraint was the constancy of distance between any two of its 'elements', which I write as

$$\delta \sum_n (x_{n+1} - x_n)^2 = 0\,, \quad \text{where} \quad x_1 := x\,, \quad x_{n+1} := x_n + dx_n\,. \qquad (522.5)$$

Expansion of these expressions, and the commutativity of d and δ, furnished

$$\sum_x d^r x d^r \delta x = 0\,, \quad r \geq 1\,. \qquad (522.6)$$

Now x could be taken as the independent variable, so that $d^2x = 0$. Hence for $r = 2$ and $r = 3$, (522.6) reduced to

$$d^2y d^2\delta y + d^2z d^2\delta z = 0 \quad \text{and} \quad d^3y d^3\delta y + d^3z d^3\delta z = 0\,. \qquad (522.7)$$

Straightforward manipulation of these equations (for example, eliminating $d^3\delta y$ between $(522.7)_2$ and the differential of $(522.7)_1$ with respect to y) and their integration produced

$$\delta x = \delta\lambda - y\delta N + z\delta M\,, \qquad (522.8)$$

and two similar equations for δy and δz, where $\delta\lambda$, δM and its companions were constants of integration. Hence their substitution in the form

$$\mathrm{S}\left(\sum_x X\delta x\right) dm = 0 \qquad (522.9)$$

of (522.1), where X, Y and Z were the appropriate components of the force acting on *dm*, yielded the equation of equilibrium

$$\delta\lambda \mathrm{S} X dm + \delta\mu \mathrm{S} Y dm + \delta\nu \mathrm{S} Z dm + \delta N \mathrm{S}(Yx - Xy)\, dm + \delta M \mathrm{S}(Xz - Zx)\, dm + \delta L \mathrm{S}(Zy - Yz)\, dm = 0\,. \tag{522.10}$$

For Lagrange it was self-evident that if the variations were arbitrary, then each of their coefficients had to be zero. This move produced six equations from (522.10), asserting the sum to zero of the component forces, and of torque, with respect to the axes. Again, if the body were fixed at the point with coordinates (*a, b, c*), then (522.8) converted to

$$\delta\lambda = b\delta N - c\delta M \tag{522.11}$$

and two similar equations, which when substituted in (522.10) yielded an equation in δL, dM and δN, all of whose coefficients were therefore zero. Hence he obtained three equations of the form

$$\mathrm{S}(Y(x-a) - X(y-b))\, dm = 0\,. \tag{522.12}$$

In several ways these arguments are typical of late-18th-century mechanics. The results themselves, of course, were not new. The manipulation of variations, especially the equating to zero of their coefficients, was a common way of forming the basic equations; today we would obtain them as the scalar components of a vectorial treatment. The use of differentials, and the assignment of *dx* as constant, was standard Eulerian practise (§ 3.2.4), although it is surprising to see it in Lagrange. And yet, while not new, the results were open to fresh interpretations, as we shall see in § 6.2.1 when Poinsot began to develop his theory of couples.

5.2.3 Lagrange's *Méchanique*: aspects of its dynamics

Anyone expecting to see Lagrange begin his treatment of dynamics with Newton's laws of motion will be very surprised, for the laws are not easy to find. They were often not used before the 1740s, and thereafter there was much controversy of whether they or some other principles should be taken as fundamental: for example, whether the second law was a statement about dynamics or a definition of force. Further, the laws were not readily applicable to continuum mechanics anyway; an important generalisation in their use was Euler's suggestion in *1752a* that Newton's second law should be stated separately for each direction of a chosen system of Cartesian coordinates instead of the common practise of applying the laws along the normal and tangent (or in the tangent plane). For some reason Lagrange attributed this suggestion to MacLaurin *1742a* (*Mechanics*$_1$ (1788), 165: see also Montucla *History*, *3* (1802), 613).

Some of the difficulties of thought (and of terminology) that attended many writing on mechanics of the late 18th century may be judged from Lagrange's discussion on p. 168 of various things called 'force', following a description of the phenomenon of impact:

> Just as the product of mass & velocity express the finite force of a body in motion, so the product of mass & accelerative force[,] which we have seen to be represented by the element of velocity divided by the element of time, will express the elementary or nascent force: & this quantity, if one considers it as the measure of the effort that the body can make in virtue of the elementary velocity which it has taken, or which it tends to take, constitutes what one calls *pressure* [*pression*]; but if one regards it as the measure of the force or power necessary to impress this same velocity, then it is what one calls *motive force.*

So we take the body's mass and multiply it by its velocity to obtain its 'finite force'; and by its 'accelerative force' (defined as the quotient of infinitesimal increments on velocity and time, apparently) to obtain its 'elementary' or 'nascent force', which may be regarded as its 'pressure' if related to the body's own motion, or as 'motive force' if interpreted in terms of the 'force or power' (§ 5.2.2) to cause the motion. The definition of 'accelerating force' was d'Alembert's (see, for example, his *1758a*, 25); he stood in opposition to Euler's position that force was to be defined in terms of the impenetrability of matter (see Euler *1752b*, among other places; and Gaukroger *1982a* for commentary). The interpretation of 'force or power' as 'pressure' or as 'motive force' stemmed in part from differing views on the *mode* (if any) of definition; they go back at least to Hermann *1716a*, 2–5, a work which exercised quite an influence during the century.

One of the key-stones of Lagrange's treatment of dynamics was d'Alembert's principle, which not only related to the principle of virtual velocities but also for Lagrange (p. 179) offered

> a direct & general method of resolving, or at least of casting into equations[,] all the problems of Dynamics that one can imagine. This method reduces all the laws of the motion of bodies to that of their equilibrium, and thus takes Dynamics back to Statics.

d'Alembert himself had made no such claim, but then he had neither stated his principle clearly or fully exploited it (see Truesdell *1960a*, 186–191). Lagrange stated it as follows (*Mechanics*$_1$ (1788), 180):

> If several bodies tend to move with velocities & directions, which they are forced to change because of their mutual action, one can regard these motions as composed of those which the bodies will really take, & other motions which are destroyed, from which it follows that these latter must be such that the bodies[, when] animated by these motions alone[,] would make equilibrium.

It is worth noting that, like d'Alembert, Lagrange formulated the principle in terms of (neo-)vectorial notions, not as forces (of which d'Alembert was distrustful, preferring to have them defined via Newton's second law). They may have conceived the principle as a sort of generalisation of Newton's third law.

Given this principle, Lagrange could found dynamics on his equilibrium equation (522.1). Body m at (x, y, z) was moving at time t under the effect of forces $\{P\}$ acting at points distant $\{p\}$ from m. Hence equilibrium for a collection of such bodies was given by

$$\mathrm{S}m(\ddot{x}\delta x + \ddot{y}\delta y + \ddot{z}\delta z + \sum P\delta p) = 0 \tag{523.1}$$

(where 'S' was the summation over bodies), which could be converted, by straightforward transformations of the forces $\{P\}$ into components (X, Y, Z) along the axis directions, to

$$\mathrm{S}m(\ddot{x} + X)\delta x + \mathrm{S}m(\ddot{y} + Y)\delta y + \mathrm{S}m(\ddot{z} + Z)\delta z = 0\,. \tag{523.2}$$

Since the variations were arbitrary, then their coefficients had to be zero, relations which yielded at once the differential equations of motion (pp. 189–201).

Lagrange now derived 'the different principles of Dynamics' from (523.1). He proved without much difficulty the '*conservation of the motion of the centre of gravity*' (or linear momentum) and that of angular momentum (or moment of momentum), together with its related 'principle of areas' (pp. 198–206). However, he had to put more effort into the '*conservation of living forces*' (or energy, although 'living forces' (note 'force' being used again) normally referred only to (twice the) kinetic energy). Writing d for δ in (523.1) and assuming again the exact integrability (522.2) of $\sum Pdp$, he converted (523.1) to

$$\mathrm{S}\left(m \sum_x \ddot{x}dx + d\varphi\right) = 0\,. \tag{523.3}$$

Integration converted this to the required

$$\mathrm{S}\left(m \sum_x \frac{1}{2}\dot{x}^2 + \varphi\right) = F \quad \text{(a constant)} \tag{523.4}$$

(pp. 206–209; compare pp. 37–38). Finally, the '*principle of the least quantity of action*' was obtained as follows. A (rather laborious) conversion of (523.1) into

$$\mathrm{S}\left[m\left(\sum_x \ddot{x}\delta x - u\delta u\right)\right] = 0, \quad \text{where} \quad u^2 := \sum \dot{x}^2, \tag{523.5}$$

was in turn transformed via

$$\sum d^2x\delta x = d\left(\sum_x dx\delta x\right) - \frac{1}{2}\,\delta(ds^2) \quad \text{and} \quad udt = ds \tag{523.6}$$

into

$$\mathrm{S}m\left(d\left(\sum_x dx\delta x\right)/dt - \delta(uds)\right) = 0 \tag{523.7}$$

(compare (522.4–6) for method). Integration and rearrangements yielded

$$\mathrm{S}\left(m\sum_x dx\delta x\right)/dt - \delta(\mathrm{S}m\int uds) = \text{constant}. \tag{523.8}$$

If the variations $\{\delta x\}$ were each equal to zero at the end points of the integral, then the constant was zero, so (523.7) reduced to the required principle of least action (pp. 209–216):

$$\delta\mathrm{S}m\int uds = 0. \tag{523.9}$$

While the performance appears impressive, some doubts come to mind, especially with regard to generality. Firstly, the treatment covers only a finite number of mass-points, so that its extension to continuous bodies by treating them as an infinity of such points (explained on pp. 50–52 for statics) appears to be only a speculative induction. (On the relation between mass-points and continuous bodies in 18th-century mechanics, see Körner *1904a*.) Secondly, the constraints must be holonomic, or independent of time (as Lagrange admitted on p. 199), and the forces conservative, as he assumed when obtaining (523.8) from (523.7) and the φ adopted in (523.3). Thirdly, the proof of the conservation of angular momentum assumes that the inter-attractive forces between the masses add no torque to the system. Truesdell has shown that without this assumption the principle is *independent* of Newton's laws or equivalent principles (*1968a*, ch. 5), and that Lagrange was not fully aware of the point; perhaps under the influence of thought generated by d'Alembert's principle, his treatment of angular momentum covered only some special cases where zero torque obtained and so independence would not arise. Fourthly, the existence of φ, required in (523.4), was lamely argued on p. 207 from a few cases rather than issued as a general lemma which such a basic principle as energy conservation requires. Finally, the use of infinitesimals, both in these and other parts of the treatise (such as at (522.12)), seems contrary to Lagrange's search for certainty. Presumably in response to comments on this point, he inserted the following remark in the introduction to the second edition of the treatise which runs quite counter to the spirit of his own *Fonctions*:

> When one has well understood the spirit of this system, and one has been convinced of the exactitude of its results by the geometrical method of the first and last ratios, or by the analytical method of derived functions, one can employ the infinitesimals as a sure and commodious instrument to abridge and simplify the proofs.

My final remarks concern the principle of least action. Although it was the most intractable to derive from Lagrange's principles, it held an especial attraction for him, from his early *1762c*, for reasons that are worth noting.[1])

The principle was a physical law expressed in terms of extrema (shortest path, least resistance, and so on). In the mid 18th century Maupertuis argued for it as a branch of teleology, regarding it as a manifestation of God, or at least nature. A controversy developed over these teleological aspects; Poincaré later put the point beautifully (*1903a*, 128):

> The molecule seems to know the point to which we want to take it, to foresee the time that it will take it to reach it by such a path, and then to know how to choose the most convenient path. The enunciation of the principle presents it to us, so to speak, as a living and free entity.

There were also disputes on the relationship between (least) action and *forces vives* (the contemporary term, twice in value of our kinetic energy).

Euler came independently to a version of the same principle, considering the forms in which action should be defined for various statical and dynamical situations, and the bearing of the conservation of energy on its generality. 'Mr. Euler recognised [it] only in the motion of solid bodies', stated Lagrange (*Mechanics*$_1$ (1788), 189) after describing only one of Euler's relevant writings,

---

[1]) Lagrange's emphasis on it has led to various historical studies. See especially Montucla *History*, *3* (1802), 643-658; Jourdain *1912a*; Dugas *1955a*, 254-275; Fleckenstein *1957a*; Szabo *1977a*, ch. 2 *passim*; and above all Pulte *1987a*.

> I have extended it since to the motion of bodies which act one upon the other in any manner, & this new Principle has resulted from it, that the sum of the products of the masses and the integrals of the velocities multiplied by the elements of the spaces traversed, is constantly a maximum or a minimum.
>
> Such is the Principle to which I give here, although improperly, the name of *least action*, & that I regard not a metaphysical principle, but as a simple & general result of the laws of Mechanics.

The attraction of the principle for Lagrange lay in its expressibility in variational form (p. 189):

> This principle[,] combined with that of the conservation of *forces vives*, & developed following the rules of the calculus of variations, gives directly all the equations necessary for the solution of each problem; [...] this method is itself only a corollary of that which is the object of the second Part of this Work, & which at the same time has the advantage of being drawn from the first Principles of Mechanics.

Further, for all its supporters it had a quantitative side, in measuring the minimum.

Mention of variations naturally brings to mind Lagrange's mathematics, and his *Fonctions.* He devoted the last fifty pages of the 1797 edition to mechanics, and followed an approach rather different from that in the *Méchanique.* 'Here the functions relate essentially to time' was the opening gambit (*Functions*$_1$ (1797), 223), so that only dynamics was considered. Derived functions were defined in terms of time; consideration of uniform velocities yielded the composition of velocities, and of uniform accelerations the composition of forces. The terms of Taylor's series were interpreted in terms of uniform motion, uniformly accelerated motion, and so on (following MacLaurin *1742a*, art. 255, which he did not cite). '[O]ne can regard [force] as an exterior cause acting continually on the body' (p. 224); attractive forces such as gravity 'act equally on all the material parts of the body', but the others, such as pressure and tension, 'act only on the exterior' (p. 237). Mass was mass (*ibid.*). Before any general principles were discussed, the passage of a mass-point through a resistant medium was analysed, including Newton's supposed error committed in the *Principia*$_1$ (1687), Bk. 2, prop. 10 (pp. 241–253). Among the various principles, conservation of linear and angular momentum, and of *forces vives*, were proved for a finite system of mass-points acting under no external forces (pp. 256–267). Conservation of energy was demonstrated in roughly the same way as in (523.3–4) (pp. 268–274); but there was no discussion of the principle of least action, or variational techniques, and d'Alembert's principle was used only implicitly. The process of reduction pursued so relentlessly in the *Méchanique analitique* was relaxed here by one stage.

5.2.4 Lagrange's derivation of the 'Lagrange equations'

> For since the fabric of the universe is most perfect and the work of a most wise Creator, nothing at all takes place in the universe in which some rule of the maximum or minimum does not appear [...]
>
> Euler *1744a*, 1st addition, art. 1

Since one of the best-known features of Lagrange's contributions to mechanics is the 'Lagrange equations' for the motions of a dynamical system, I shall briefly record their presentation in the *Méchanique*, pp. 216–232. His notations are too atrocious to follow—for example, differentials are differentials but primes play the role of sub-scripts—so I shall write $\{\xi_j\}$ for the set of variables (possibly including time) into which the spatial variables x, y and z were transformed.

Lagrange's method was to consider separately each term in the equilibrium equation (523.1) under the transformation. He gave two forms of the transformations of the acceleration term. One was based on transforming and differentiating $(523.6)_1$, while in the other he used variational principles to obtain two relationships between the differentials (of any order) with respect to both sets of variation of "any" function α, and then took the particular case

$$2\alpha := (dx)^2 + (dy)^2 + (dz)^2 \;(= (ds)^2)\,. \tag{524.1}$$

In both cases the following result obtained (pp. 220, 223):

$$\sum_x d^2x\delta x = -\sum_j \left(d\left(\frac{\delta\alpha}{\delta d\xi_j}\right) - \frac{\delta\alpha}{\delta\xi_j}\right)\delta\xi_j\,. \tag{524.2}$$

He then defined

$$T(\{\xi_j\}, \{d\xi_j\}) := \mathrm{S}\,\frac{1}{2}\,m\sum_x \dot{x}^2: \tag{524.3}$$

δ-variation of (523.4) yielded, from (523.3),

$$\mathrm{S}m\left(\sum_x \ddot{x}\delta x\right) = \sum_j \left(d\left(\frac{\delta T}{\delta d\xi_j}\right) - \frac{\delta T}{d\xi_j}\right)\delta\xi_j\,. \tag{524.4}$$

For the "work" function Lagrange assumed (522.2) again, and from its form

$$d\Pi = \sum Pdp \tag{524.5}$$

he showed that

$$\mathrm{S}mP\delta p \quad = \mathrm{S}m\delta\Pi = \delta(\mathrm{S}m\Pi) =: \delta V, \text{ say.} \tag{524.6}$$

$$\therefore \mathrm{S}mP\delta p = \sum_j (\delta V/\delta\xi_j)\,\delta\xi_j. \tag{524.7}$$

(523.4, 7) provided the required alternative expressions for the two terms in the equilibrium equation (523.1), so that substitution yielded

$$\sum_j \left(d\left(\frac{\delta T}{\delta d\xi_j}\right) - \frac{\delta T}{\delta\xi_j} + \frac{\delta V}{\delta\xi_j}\right)\delta\xi_j = 0\,. \tag{524.8}$$

As in (523.2), the last step was to note that the variations $\{\delta\xi_j\}$ were indeterminate, so that each of their coefficients in (524.8) was zero, thus producing the 'Lagrange equations' for a system without constraints. If such constraints did apply, then (522.3) rather than (522.1) was the basic equation, and the Lagrange equations took the form

$$d\left(\frac{\delta T}{\delta d\xi_j}\right) - \frac{\delta T}{\delta\xi_j} + \frac{\delta V}{\delta\xi_j} + \sum_s \lambda_s \frac{\delta L_s}{\delta\xi_j} = 0\,. \tag{524.9}$$

If 'the proposed system is composed of an *infinity* of particles or elements of which the assemblage forms a finite mass of variable figure' (p. 229, my italics) then the reasoning had to be modified for S to sum over the mass Dm of each 'particle' of the continuous body.

Readers familiar with the derivation of Lagrange's equations from a (relatively) modern textbook may be surprised by the elaborate nature of Lagrange's own procedure, especially if they also read the background mathematics to (524.2) on pp. 217–224. The abstract nature of some of the steps may also cause concern: for example, the definition of T in (524.3) is mathematically rather slick, while that of V in $(524.6)_4$ is wanting in physical intuition. In fact, the equations themselves lack contact with the basic concepts of dynamics with which the 'different principles of mechanics' are concerned. And the generality of applicability of the equations is in doubt; the extension to continuous bodies is suspect, for the same reasons as were indicated just after (523.9).

The subsequent sections (pp. 233–428) contained extensive treatments of various special cases (see Table 521.1), but some of the omissions are a little surprising. For example, Lagrange did not take the opportunity to apply his solution of a system of linear difference equations (§ 3.5.1) to a system of collinear masses. Again, the famous vibrating string problem (§ 3.3.2–3) received only a paragraph, with the functional solution (332.2) just written down (p. 334), although the passage was proceeded by a lengthy discussion of the corresponding *n*-body model (pp. 300–329, where he worked out from the equation for the conservation of *forces vives*). Further, contrary to his claim that his equations provided 'the simplest method' of determining motions (p. 216), his treatment of several cases was not only non-intuitive but even rather laborious. For example, he analysed the rotation of a continuous body, largely by means equivalent to the manipulation of third-order vectors, matrices and determinants, in terms of setting up a rectangular coordinate system and determining the axis of instantaneous rotation (pp. 338–371); but he paid no attention to Euler's equations or moments of inertia. He defined these moments only on p. 379 (relative to the coordinate system), in the course of expressing the *forces vives* function *T* relative to the instantaneous axis. His purpose was to state the Lagrange equations in spherical polar coordinates, in order to study specific examples, such as the libration of the moon. Later figures, such as Poinsot and Cauchy, were to ponder on the quadratic form *T* from the point of view of the momental ellipsoid (§ 17.4.2, § 18.2.4): geometrically formed considerations, rather than the 'regular march' of algebra. Let us continue this theme with a quotation for our next sub-section.

5.2.5 On Euler's approach to mechanics

> [...] the illustrious *Lagrange* took up again this famous problem [of the rotation of a continuous body], in order to deepen and develop it in his manner, I want to say by a sequence of formulae and of analytical transformations which present much order and symmetry. But one must admit that, in all the solutions, one hardly sees only calculations, without any nice image of the rotation of the body.
>
> Poinsot *1834a*, 486

Mention of non-Lagrangian approaches to mechanics will from now on become more frequent. Thus it is appropriate to consider here the principal rival tradition; that of Euler. The view of Euler as mainly clear and elegant was fairly standard for decades into the 19th century; thus, for example, Liouville could write 'After d'Alembert, Euler, and with him elegance' (*1858a*, 1: see also § 19.4.4).

I mentioned in § 5.1.1 that in fact the Eulerian tradition of mechanics was broader in scope than that of (d'Alembert and) Lagrange; but I also disavowed any intention of trying even to summarise Euler's achievements. Here I shall confine myself to one general remark, and to two examples of his work which have been mentioned before.

In contrast to Lagrange's desire to reduce dynamics to statics, Euler kept the subjects separate, devoting most of his attention to dynamics. His only extended attempt to deal with statics is a manuscript, apparently written in his early years but never completed and published only posthumously in 1862. The key principle was that two equal and opposite forces were in equilibrium (*Statics*, art. 28), and the account was assisted by diagrams all over the place. His *Mechanica* (*1736a*), which in fact only covered dynamics, dealt successively with various cases: motion of a mass-point in a resistant medium, motion under centres of attraction along given paths, rigid body dynamics, and so on. Differential models were the order of the day, and differential equations—that is, equations relating differentials, as emphasised in § 1.3.7—the usual outcome. The presentation was rather Euclidean—definition, corollary, theorem and so on—and thus much tighter than Lagrange's.

An instructive and important example of Euler's use of a differential model is finished by his derivations of the wave equation, whose solutions were discussed in § 3.3.2-3. Figure 525.1 follows the kind of figure that he used (though not his

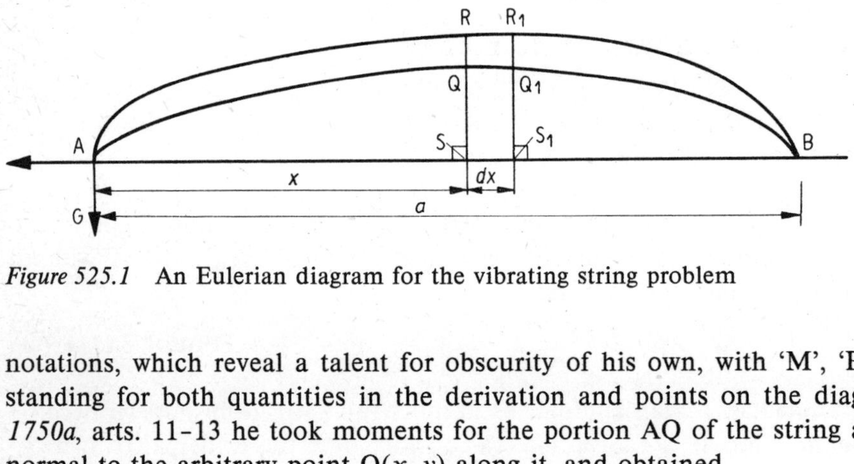

Figure 525.1 An Eulerian diagram for the vibrating string problem

notations, which reveal a talent for obscurity of his own, with 'M', 'P' and 'n' standing for both quantities in the derivation and points on the diagram). In *1750a*, arts. 11-13 he took moments for the portion AQ of the string about the normal to the arbitrary point Q(x, y) along it, and obtained

$$Fy - Gx + \int dx \int P(Mdx/a) = 0, \quad \text{with} \quad d^2x = 0, \tag{525.1}$$

where F and G were components of the reaction at the end A, P the 'accelerative force' at Q along QR, and M the mass of the string. Since the transverse vibrations were small, he assumed that F and G were constant, and differentiating twice with respect to x yielded

$$Fy_{xx} + MP/a = 0. \tag{525.2}$$

Now 'mechanical principles' applied to Q showed that

$$P = -y_{tt}, \tag{525.3}$$

so that (525.3) became

$$y_{xx} = (M/Fa)y_{tt}. \tag{525.4}$$

Euler gave an alternative derivation in *1755a*, arts. 13–19, which has remained more durable. Applying Newton's second law to the differential segment QQ_1 in the direction QR, he found that

$$Fy_x - F(y_x + dy_x) = -(Mdx/a)y_{tt}, \tag{525.5}$$

where F was now the tension of the string (assumed constant along it, and thus presumably equal to the F of (525.4)), which manipulation of the differentials converted to a form which I shall write as

$$Fy_{xx} = \varrho y_{tt}, \quad \text{where line density} \quad \varrho := M/a. \tag{525.6}$$

The examples are instructive not only for the place assigned to intuition and experience in the model but also for the escapades that handling differentials can entail (see Interlude 525.1).

Interlude 525.1 Notes on handling differential models in mathematical physics

The constancy of F in both of Euler's derivations leads to some conceptual problems which are typical of the difficulties involved in differential models. Firstly, the constancy of the component F of reaction at end A in (525.4) is strictly true only if the string is stationary. Secondly, its appearance in (525.4) assumes that it is also the component of reaction at the end B; but this in turn suggests that the motion of the string is symmetrical, which is obviously not the case, on either experiential grounds or from considering the solution forms (332.2) or (333.1). In the second derivation, where F is now the tension of the string, the first of these difficulties still applies.

A preferable version of the second derivation could proceed as follows.

Using ϱ from (525.6) for convenience, replace (525.5) by

$$(F + dF) \sin (\psi + d\psi) - F \sin \psi = \varrho ds y_{tt}, \quad \text{where} \quad \tan \psi := y_x. \tag{1}$$

Neglecting second-order small terms, $(1)_1$ becomes

$$F d\psi \cos \psi + dF \sin \psi = \varrho ds y_{tt}. \tag{2}$$

Meanwhile, $(1)_2$ yields on differentiation

$$\psi_x = \cos^3 \psi y_{xx}. \tag{3}$$

Hence (2) becomes

$$F y_{xx} \cos^4 \psi + dF \sin \psi = \varrho y_{tt}, \tag{4}$$

which reduces to the wave equation (525.6) because the small vibrations allows us to replace $\cos^4 \psi$ by 1 and ignore $dF \sin \psi$. Modern textbooks usually follow Euler's assumption of the constancy of F; and so, like him, they beg one of the questions.

Euler's second derivation came to him doubtless as a consequence of his 'Discovery of a new principle of mechanics' in *1752a*, which I described in § 5.2.3 as an important extension of Newton's second law. The essence of that paper, including the role of the differential model, was expressed in art. 20:

> Let there be an infinitely small body, or [one] of which the mass be reunited in a single point, this mass being $= M$; that this body have received any motion, and that it be affected by any forces. In order to determine the motion of this body, one has to have regard only of the distance of this body from a fixed and immobile plane; let the distance of the body from this plane at the present instant $= x$; [let] one decompose all the forces which act on the body, along directions which are either parallel or perpendicular to the plane, and let P be the force which results from this [de]composition along the direction perpendicular to the plane and which in consequence will endeavour either to move away or to approach the body from the plane. After the element of time *dt*, let $x + dx$ be the distance from the body to the plane and[,] taking this element *dt* for constant, it will be [that]
>
> $$2 M ddx = \pm P dt^2 \qquad [(525.7)]$$
>
> according as the force P tends either to move away from or to approach

> the body from the plane. And it is this sole formula, which contains all the principles of Mechanics.

In fact, the principle is not quite as general as it would have been, had variable mass been allowed for:

$$2d(Mdx) = \pm P dt^2 . \tag{525.8}$$

Even then it cannot cover 'all the principles of Mechanics' but only dynamics, which was obviously so dominant in his thought. The 2 in (525.4) occurred because of his practice of measuring mass in terms of the weight M at the earth's surface and assigning gravity the value $\frac{1}{2}$. The use of special units such as these is manifest in many of Euler's (and others') writings,[1]) but it was to decline substantially in popularity by the end of the century.

Another important feature of Euler's *1752a* was his derivation of the equations of rotation of a rigid body around its centre of gravity. He fixed the centre at O and set up moving rectangular axes. If the point P(x, y, z) moved with angular velocity $(\omega_1, \omega_2, \omega_3)$, then

$$\dot{x} = y\omega_3 - z\omega_2 . \tag{525.9}$$

(I have changed Euler's notations.) Differentiating and using the corresponding equations for $\dot{y}$ and $\dot{z}$ produced

$$\ddot{x} = (y\dot{\omega}_3 - z\dot{\omega}_2) + (z\omega_1\omega_3 + y\omega_1\omega_2) - x(\omega_2^2 + \omega_3^2). \tag{525.10}$$

Similarly,

$$\ddot{y} = (z\dot{\omega}_1 - x\dot{\omega}_3) + (x\omega_2\omega_1 + z\omega_2\omega_3) - y(\omega_3^2 + \omega_1^2) \tag{525.11}$$

(and also for $\ddot{z}$: arts. 40–44). From the new principle of mechanics (525.7) the accelerative force on the differential element of mass dM at P in the direction of Ox was $2dM\ddot{x}$ (complete with the 2 of (525.7)). Hence, the torque T_3 about Oz was given by

$$dT_3 = 2dM(\ddot{x}y - y\ddot{x}) \tag{525.12}$$

$$= 2dM[(x^2 + y^2)\dot{\omega}_3 - (yz\dot{\omega}_2 + xz\dot{\omega}_1) + (yz\omega_1\omega_3 - xz\omega_1\omega_2) + (y^2 - x^2)\omega_1\omega_2 - xy(\omega_2^2 - \omega_3^2)] , \tag{525.13}$$

---

[1]) The practise is described in Truesdell *1954a*, xliii-xliv, and discussed in Ravetz *1961a* and Hankins *1970a*, ch. 10. It occurred, for example, in Euler's derivations of the wave equation just described, but for the sake of clarity it was not used in (525.1).

On the influence of d'Alembert upon Euler *1752a*, see Wilson *1987a*. Some 18th-century problems in dynamics are noted in Cayley *1862a passim.*

from (525.10–11). Summing (525.13) over all elements of the body introduced moments I_1, I_2 and I_3 and products J_1, J_2 and J_3 of inertia with respect to the three axes:

$$\begin{aligned}\tfrac{1}{2}T_3 = I_3\dot{\omega}_3 - (J_1\dot{\omega}_2 + J_2\dot{\omega}_1)\\ + (J_1\omega_1\omega_3 - J_2\omega_1\omega_2) + (I_1 - I_2)\omega_1\omega_2 - J_3(\omega_2^2 - \omega_3^2)\end{aligned} \quad (525.14)$$

(arts. 44–49). But J_3 was determinable *independently* of this analysis, as the torque about Oz of the impressed forces. With this information, (525.14) and its two companions served as Euler's equations (arts. 54–55).

As is so often the case, the original formulation of a well-known result differs from its later standard forms. Firstly, Euler expressed the equations for any set of moving axes; not the set of principal axes, for which the products of inertia are zero and (525.14) reduces to

$$\tfrac{1}{2}T_3 = I_3\dot{\omega}_3 + (I_1 - I_2)\omega_2\omega_3 . \quad (525.15)$$

Secondly, he simplified (525.14) and its companions not in the way of (525.15) but by assuming that the axis of instantaneous rotation lay along ω_1 (so that $\omega_2 = \omega_3 = 0$) and solving the resulting equations for $(\dot{\omega}_1, \dot{\omega}_2, \dot{\omega}_3)$ by the long-hand equivalent of inverting a third-order matrix. It was *these* enormous expressions, and not (525.14) and its companions, which for Euler were the 'three formulae, which contain the new principles of Mechanics' (art. 58). Finally, although he mentioned that the assumption that O was the centre of gravity would entail

$$\int x dM = \int y dM = \int z dM = 0 \quad (525.16)$$

(art. 45), he could not use these results in the derivation of (525.14); but then the assumption can be discarded. In other words, in a paper on general principles of mechanics one set of main results was more general than it was stated to be!

The quotation at the head of the previous sub-section shows that Euler saw optimisation as a general principle; and he pioneered variational methods in mechanics (see, for example, *1744a* on elastic curves, or *1753c* on lunar theory). But variations never *dominated* his mechanics as they were to do in Lagrange's; nor was he so concerned with reducing the subject to a minimal number of principles.

5.2.6 Lazare Carnot on the principles of mechanics, 1803

I noted in § 3.2.6 Carnot's defence of infinitesimals in the different editions of his *Réflexions* on the calculus. He reprinted the first, 1797, edition that same year in Basel (whither he fled after losing office), in a volume called *Oeuvres mathématiques*. The other work in the volume was a reprint of his *Essai sur les machines en général* (*1783a*): a revised version *1803a* came out in his 50th year from Bachelier, under the title *Principes fondamentaux de l'équilibre et du mouvement*. The 1783 version seems to have been little studied, despite its three printings (it also appeared in 1786) and its intrinsic interest. On his views on mechanics, including some manuscripts, see Gillispie *1971a*, ch. 3; and also Fraser *1981a*, pt. 2.

We have already heard from Carnot's *Essai*; for some philosophical remarks were quoted as the second gobbet of § 1.1 and at the beginning of § 5.2.1. His concern with philosophy is typical of both books. The *Essai* contained a first part on the principles of mechanics (arts. 1–30), while a second part pondered 'On machines properly called' (arts. 31–63). The *Principes* was differently organised: the first part was a purportedly 'experimental' (p. ii) account of the principles of mechanics (pp. 1–106), while the second part, dealing with the algebraic expression of these principles and their applications to machines (pp. 107–262), 'commences at the point where I considered the science as ceasing to be experimental in order to become entirely rational' (p. xvii). The revision also shifted the emphasis from machines to mechanics in general, as the change in title suggests.

The algebra which Carnot used in the *Principes* was that of his geometry of position (§ 4.7.1), where this time the directed line-segments represented velocities. He wrote the projection $\overline{\mathrm{Aa}'}$ of the velocity $\overline{\mathrm{Aa}}$ in the direction $\overline{\mathrm{AB}}$ in the notation

$$\overline{\mathrm{Aa}'} = \overline{\mathrm{Aa}} \cos \overline{\mathrm{Aa}} \wedge \overline{\mathrm{AB}} \tag{526.1}$$

(*1803a*, 16–17). However, he was not interested in reducing mechanics to algebra, *à la* Lagrange; for example, both of his books contained diagrams. On the other hand, he did not subscribe to Eulerism either. For example, from his distinction between the experiential and the rational he asserted, against Euler, that mechanics was based on (empirical) notions rather than on the (rational and metaphysical) concepts of force and cause (pp. xi–xiv). Thus 'force' was understood merely as a way of speaking—and in consequence of this view he gave priority to dynamics over statics (p. xv), contrary to normal practise. This and related aspects of his theory will be taken up in Chapter 16.

In order to clarify the plethora of terms in dynamics, Carnot listed his principal terms and their dimensions (see Table 526.1). His lexicon was not complete, even for his time—for example, he had no concepts corresponding to (our notions of) power or impulse—and he was sometimes a little inconsistent in his usages. Nevertheless, a list such as this would have graced Lagrange's discussion of 'force' quoted in § 5.2.3, and many other sources of that time.

Carnot introduced the term 'force of inertia', with acknowledgement to Euler

Table 526.1 Carnot's explanation of his main notions of dynamics in *1803a*.

His initial list was given on pp. 12–13. He represented dimensions by m (mass), e (distance) and t (time). '/' connect his synonyms; '*or*' divides our distinctions.

Carnot's term	Dimension	Page	Modern equivalent
velocity	e/t	12	velocity
quantity of motion/	$m(e/t)$	12	momentum
force of percussion		23 }	change in momentum at impact
force of inertia	me/t	68 }	
accelerative } force	$(e/t)/t$	12	{ acceleration
decelerative }			deceleration
motive force/force of pressure/dead force	$m(e/t^2)$	{12, 22, 28}	force
force/power	me/t or me/t^2	12	momentum *or* force
live force	$m(e/t)^2$	12	(twice) kinetic energy
moment of motive force	$m(e/t)(e/t)$	12	
moment of activity/ latent live force	$me(e/t^2)$	{12, 37}	work
moment of quantity of motion/	$m(e/t)\,e$	12	(units of angular momentum)
quantity of action	$m(e/t)^2\,t$		
moving force(s)	me/t^2	33	force in context of machines

on p. 73, in the general context of Newton's first law, which he stated on p. 49 (without so naming it) and called a 'hypothesis'. His use of the word 'hypothesis' is unusually frequent for that time. Newton's law was the first of seven hypotheses, which were listed on pp. 49–51 and prefaced with this explanation of their status: 'next we compare the consequences which result from them, with the phenomena, and if we find that they agree, we conclude that we can consider these hypotheses as the veritable laws of nature'. This passage, which is in accord with the quotation from him at the head of § 5.2.1, is one of the senses of 'hypothesis' described in § 1.3.5, where an example from Carnot's contemporary Laplace was quoted.

Carnot's hypotheses were all drawn from the classical principles of mechanics: forces in equilibrium took a zero resultant in any direction; impact depended on relative, not absolute, velocity, and so on. He followed his list with a lengthy justification of these hypotheses (pp. 51-67); and much of the second, 'rational' part of the book was taken up with theorems on conservation or loss of *forces vives*, momentum and angular momentum (especially in the context of impact), on the equilibrium or disequilibrium of systems of forces, and proofs of other principles such as least action. The book was very notable for its concern with impact and percussion, and for properties of hard and elastic bodies (see Scott *1970a*, chs. 5-6), as we shall see in § 16.3. The presentation was Euler-tight in its structure; but proofs proceeded by means of a mixture of algebraic statements, *à la* (526.1), of geometrical configurations and a dash of differential and variational calculi. The impression left by the book is ambiguous: a mixture of Euler and Lagrange, with Euler perhaps dominant; in level sometimes reading like a treatise, at other times like a textbook. It enjoyed less reaction than the other works discussed in this section; but, as we shall see in Chapter 16, it was to gain some acclaim during the 1820s.

5.2.7 The mechanics teaching of de Prony around 1800

I noted in § 5.2.4 the brief treatment of mechanics in Lagrange's *Fonctions*, due doubtless to the desire not to repeat large parts of his *Méchanique* and the pressures on his time. In the first years of the *Ecole Polytechnique* the lack of a suitable textbook for mechanics was felt, and de Prony, *Professeur* of mechanics there, was urged to convert his lectures into such a book. He had presented an outline *1798a* of his lecture course in the school's *Journal*, and his *Mécanique philosophique* appeared as *1800a*, in his 46th year. It was published both as a book and as *cahier* 8*bis* of the *Journal de l'Ecole Polytechnique*, although it was incomplete.[1]) It was not his first book on mechanics, for he had published a treatise on engineering during the 1790s, as we shall see in § 8.1.1.

'Why philosophical?' asked the young Biot of Lacroix concerning de Prony's title during their correspondence around 1800 (see § 20.1). Lacroix's reply has

---

[1]) de Prony's full text was designated as *cahiers* 7-10 of the *JEP*; but in the end 7 and 8 came out in 1812 with a reprint of Lagrange's and Laplace's lectures at the 1795 *EN*; copies of Lagrange *Functions*$_1$ (1797) were issued at some time as *cahier* 9 (and *Functions*$_2$ took this same *cahier* number in the year (1813) of its publications); and *cahier* 10 made a tardy appearance in 1810 as a normal collection of papers. Fourcy says that de Prony's book was withdrawn after publication (*1828a*, 290).

Delambre *1800a* described the book to the *CI* in July (*PV, 2*, 182). Manuscripts related to this and de Prony's later textbooks (§ 6.6.1), or to his *EP* teaching, are held at *EP*, III 3b (box 1); *AS, Ampère*, ch. 136; *ENPC*, ms. 2 771; and *AN*, F^{17} 1084 and F^{14} 11057.

not survived, but doubtless he explained the significance of de Prony's reference in the preface to Fourcroy's *Philosophie chimique*, which had first appeared as an article in the *Encyclopédie méthodique* as an attempt to classify chemistry, in the *encyclopédiste* tradition. de Prony wished to do the same for mechanics, presenting the principal definitions and theorems as a 'reasoned analysis of the parts of the science of equilibrium and of motion', to quote the sub-title of his book.

The author promised more than he delivered in his preface, for there a five-part work was outlined, whereas only the first three parts were published. After an opening part on basic conceptions, which took up pp. 1-93 of the text, the second part covered statics and dynamics (pp. 94-203) and the third hydrostatics and hydrodynamics (pp. 204-477). 'The FOURTH PART is entirely devoted to objects of practice and of everyday utility' such as machines and friction, while the last was to be devoted to 'transcendental mechanics' (that is, the basic principles of mechanics). Although these last two parts were not written, we shall see in this chapter, § 6.6, § 8.4-5, and § 16.3.4 that de Prony dealt with these matters elsewhere.

Two general features of the book stand out. One is the layout; the narrative is usually confined to the left-hand pages, and its main points stated and/or adumbrated in the right hand pages in four columns of 'Notation', 'Definitions', 'Theorems' and 'Problems'. (The presence of problems is a pleasant novelty in an educational work of this period.) The second feature is *Denkweise*; the algebraic style that we saw already in de Prony's calculus (§ 3.5.2-4) is equally marked in his mechanics, for this book contained no diagrams. One recalls Lagrange's treatises, perhaps with surprise, given the vast intellectual distance between Lagrange's (overly?) theoretical concerns and de Prony's desire for practicality (§ 3.5.2); but it was stressed in § 1.3.6 that *Denkweisen* are concerned with the style and not the content of the science, and I gave there this common feature of Lagrange and de Prony as an example.

As fluid mechanics is taken up in § 8.5 and § 10.3, remarks are confined here to a few observations on the first two parts of de Prony's book. The opening was devoted to questions of cognition and metaphysics, together with an *encyclopédiste* synoptic table of the subject matter (Table 527.1). Time and force 'are borrowed from *metaphysics* or *ideology* [*sic*], which itself has deduced them from the first products of *sentiment*; the six other [categories] are taken from *geometry* and from *physics*, which draws them immediately from *metaphysics*' (*1800a*, 3). The distinction is not easy to understand, but it seems that time and force were placed on the same metaphysical level, with space located at a more material level. 'The nature of this cause of motion, called *force* or *power*, is completely unknown to us', he added, 'But we have no need, in mechanics, to know [its] nature' (p. 18). However, he also felt that 'the law of *continuity* is constantly observed';

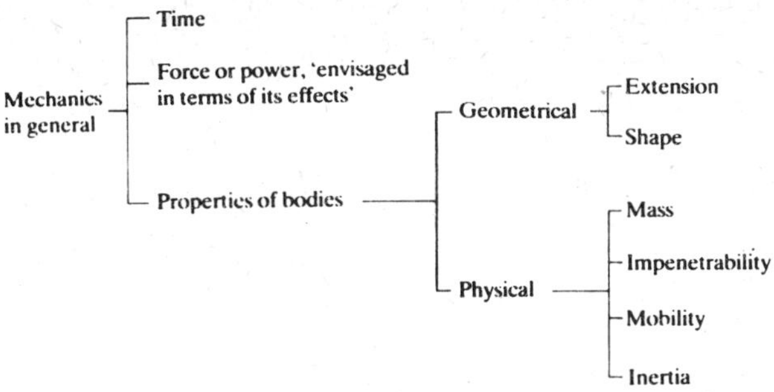

Table 527.1
de Prony's synoptic table of mechanics (*1798a*, 3; *1800a*, 3; *1801d*, 5)

even in the case of impact 'a body does not pass from one velocity to another, without having had all the intermediate velocities' (p. 20). So mobility was apparently not without its metaphysical aspects.

The influence of Lagrange is detectable not only in terminology ('force or power', for example; and also 'pressure' (§ 5.2.3) elsewhere) and vile, prime-strewn, notations, but also in presentation; the more elementary account of *Fonctions* seems to have been preferred to the sophistications of *Méchanique*. For example, de Prony began his dynamics with uniform acceleration and velocity, and he assumed principles, such as areas or virtual velocities, without proofs (which would have been given in the fifth part). Proofs of other results (the composition of velocities, for example) were cast in trigonometrical terms.

However, de Prony was also influenced by Euler. For example, he revived for the students Euler's sensible but neglected suggestion that the centre of gravity be renamed 'centre of inertia' (Euler's term), 'because the position of this point not at all depending upon powers that one supposes to be acting on the body, its name must not recall any' (p. 116). Again, he used only Euler's calculus when such techniques were called for (which again recalls Lagrange's *Fonctions* rather than *Méchanique*). With regard to specific theorems in mechanics, he described as 'one of the most beautiful of mechanics' (p. 110) Euler's result that if the torques of a system of forces with respect to the axes of a rectangular coordinate system were respectively P, Q and R, then the torque M relative to the axis through the origin with direction cosines $(\cos\alpha, \cos\beta, \cos\gamma)$ was given by

$$M = P\cos\alpha + Q\cos\beta + R\cos\gamma, \tag{527.1}$$

so that the torque was maximal for the axis with direction ratios (P, Q, R).[1]) Again, his lengthy account of moments of inertia and their bearing on the mo-

[1]) Euler's theorem had recently been published posthumously as *1793b*. de Prony cited Laplace's proof *1798b* in the *JEP* (§ 5.4.4), and also his own in *cahier* 9 of the journal; but I can find his neither there (it contains only Lagrange's *Fonctions*) nor elsewhere in the journal, although he gave a trigonometric proof in a pamphlet *1804a* published by his *EPC*.

tion of a rigid body (pp. 142–202), while algebraic in form (mainly "mixed" (§ 5.1.2) trigonometry, in fact) was in content largely a version of Euler's treatment, as he acknowledged on pp. 178–180. He also brought his own mathematical concerns (§ 3.5.3) into play by using differences and differential equations on occasion, especially on pp. 124–136, where he presented an algebraic rendering of (even so geometrical technique as) the polygonal representation of systems of forces, known then by the rather delightful name of 'funicular polygons' (see Hennenberg *1903a*, 349–351).

de Prony's attachment to Eulerian mechanics was underlined in a pair of lengthy papers *1799b* and *1802f* in the *Journal de l'Ecole Polytechnique* on the dynamics of a rigid body; for the papers built both on Euler's *Theoria motus corporum rigidorum* (*1765a*) and on sections of his own *Mécanique philosophique* (which was published before the second paper but after the first). He ran through his own presentations of moments of inertia and their properties, the axis of instantaneous rotation, and so on; in particular, he derived Euler's equations (525.7) by Euler's method of taking moments, and ignored Lagrange's equations (524.8–9) (*1799b*, 326–336: compare *1800a*, 170–194). His treatment of the problems, in fact, does not seem to add much to Euler's; but his papers were of educational merit, appearing in a publication and language convenient for the students, and containing several large tableaux which either explained details of calculation or presented "philosophical" classifications of the various cases of motion. His papers displayed both educational concern and sympathy with Eulerism in another way, which is of especial interest as it marked a departure from *Mécanique philosophique*: they contained a few diagrams. Indeed, he related the diagrams to other parts of the students' course by claiming that they 'facilitate extremely the intelligence of the analysis to those [students] who have the first notions of descriptive geometry' (*1802f*, 88).

de Prony's approach to mechanics is conveyed most concisely and comprehensively in another book *1802a*, published by Courcier, which he modestly called a 'Rational plan' of his teaching of 'equilibrium and motion of bodies'. In 224 pages, aided by diagrams, he covered the subject more briefly than in his *Mécanique philosophique*, but more completely in that he included sections on 'machines and motors' and 'general considerations on equations and motion' that the earlier work had lacked. He did not introduce variational methods until pp. 213–215, and then he used them only to solve a few isoperimetrical problems rather than develop a general treatment of mechanics. 'Philosophical mechanics' for de Prony seems to have been largely Eulerian in its approach to mechanics, but, especially in the *Mécanique philosophique*, Lagrangian in its *Denkweise*. In other words, like Carnot, de Prony was choosing his own mixture of the principal traditions.

5.2.8 Two elementary presentations of mechanics: Monge and Francoeur

de Prony's works were not the only teaching materials available around 1800. I shall briefly notice two rather more elementary textbooks which were not only used at the *Ecole Polytechnique* and elsewhere at that time, but also became popular and appeared in many editions during the century.

Monge's *Traité élémentaire de statique* (*Statics*) first appeared in 1788 (his 43rd year), before the Revolution, 'for the use of marine schools' of which he was an *inspecteur*. The next editions were published in 1795, 1799 and 1801, and two more under Hachette's editorship came out in 1810 and 1826. The basic structure was not radically altered; after opening chapters on fundamental aspects, based on the assumption that two equal and opposing forces are in equilibrium, he described several engineering applications, such as pulleys, cogs, screws and wedges. He confined himself to 'statics properly called': hydrostatics was omitted. As might be expected, the text was plentifully assisted by diagrams; for example, the 1826 edition had 100 of them for a text of 215 octavo pages. The discussion—the proofs of theorems, for example—usually drew on the diagrams. An interesting algebraic feature of these proofs is that he followed an 18th-century tradition of arguing by proportions of quantities of the same type; for example,

$$P : Q :: \mathrm{AB} : \mathrm{AC}, \tag{528.1}$$

where P and Q were forces and AB and AC were lengths. This approach was an alternative to defining some quantities in terms of others, which we saw in Euler when interpreting (525.4).

Monge's book was one of the few elementary works available around 1800, and its range is limited. de Prony's works, despite his educational imperative, were formidable challenges for the students, especially the *Mécanique philosophique*. Francoeur, one of the founding-year students at the *Ecole Polytechnique*, noted the lack of more elementary mechanics texts, and so wrote a *Traité de mécanique élémentaire*, 'for the use of the students of the *Ecole Polytechnique*, written following the methods of R. Prony [...] and intended to serve as introduction to his *Mécanique philosophique*'. Its first edition appeared in 1800, by which time he was a *répétiteur* at the school. Its level corresponds to the more elementary works of Lacroix, whose help he acknowledged in the preface. The book is cited as Francoeur *Mechanics*.

Francoeur's book is not as close to de Prony's *Mécanique philosophique* as its title suggests. There were 153 diagrams, occuring with a frequency more akin to Euler or Monge. Again, 'In order not to complicate the considerations' of de

Prony's synoptic Table 527.1 of categories, which he reproduced (*Mechanics*$_1$ (1800), xi), he began his treatment of mechanics in terms of mobility and forces, and then added shape when handling statics, time for dynamics of mass points, and mass and extension for rigid bodies. In contrast to de Prony's extensive account of the dynamics of rigid bodies, Francoeur's was brief (pp. 315–330), while he included a substantial section on machines (pp. 112–194), a topic which de Prony had not written up. He also used sources, including himself, other than de Prony for some principal results: examples will be given in § 5.3.1.[1])

Both in the (many) ways in which Francoeur followed de Prony and in some of the ways in which he differed, especially the use of diagrams, he seemed to suggest a return to Euler's methods, or at least away from Lagrange's. Take for example, de Prony's treatment of equilibrium of a system of forces (*1800a*, 42–48, including a reference to Francoeur's interest) which Francoeur followed in his own book (*Mechanics*$_1$ (1800), 26–64): a system of forces was resolved into components of the rectangular coordinate system, and equilibrium asserted if the sums of the forces and of their moments were zero. Francoeur appraised this *standard* technique of the past as follows (pp. 64–65):

> We have thus taken back these conditions of equilibrium between all the forces of a system, so as to depend only on this principle, *two equal forces, and directly opposed, destroy each other* [...] It is to cit[izen] *Prony* that this happy simplification is due. The interesting researches which he has done on mechanics, have led him to detach the theory of equilibrium from all idea of motion. A long experience has convinced him that nothing was more distanced from the simplicity necessary to the first principles of mechanics, and less proper to give clear notions of the equations of equilibrium, than to combine thus heterogeneous elements. Those who have already known the procedures employed before him to prove the preceding theorems, will sense the value of this luminous path.

Connoisseurs of veiled comments will rate this gobbet highly. What can be the system receiving 'happy simplification' from de Prony but that of the first geome-

---

[1]) Francoeur also introduced a systematic method of lettering equations which I have not seen before, using different founts, cases and alphabets to distinguish equations in different parts of the book, and different numbers of primes to indicate the chapter of a part. This last feature, in a work already spattered with Lagrangian primes serving various purposes in both equations and text, probably saw to the demise of his method.

Francoeur's manuscript summary of de Prony's *EP* lectures are held at *Ecole ... des Beaux Arts*, mss. 408–409. See also his notes based on the writings of Lagrange and Laplace, and Biot's teaching of physics at *CF*, at ms. 400.

ter of the Republic, 'reducing dynamics to statics' in his *Méchanique analitique* by the principle of virtual velocities? Was this lowly young *répétiteur* of the *Ecole Polytechnique*, deeply concerned with the practical side of mechanics, associating his own ideology with that of a distinguished colleague, whose book he is basically following? Was it to imitate the politics of his situation, as well as to honour another major figure, that Francoeur dedicated his book to Laplace?

5.2.9 Concluding remark

The survey of approaches to mechanics provided in this section is the analogue to the review of the calculus given in § 3.2. A picture emerges of competing traditions, as was the case also in § 3.2; indeed, we have here an example of structure-similarity (§ 1.3.1). The correspondence extends to issues involved in the competition: algebra versus geometry, empirical versus rational in the sense of Carnot's distinction (§ 5.2.6), and so on.

However, there are differences in degree between the calculus and mechanics. Firstly, the status of empiricism(s) just mentioned is more important for mechanics, since it relates both to the experientiability of concepts in physical theories (§ 1.3.3) and the degree of guidance of mathematics by physical considerations (§ 1.3.2). Lagrange's mechanics was not 'notional' in the sense of § 132.3; but (for example) it introduced the impressed forces on a rigid body at (522.8) as constants of integration, which is hardly evidential either. Secondly, there were differences in scope between these two approaches to mechanics, especially concerning the status of the principle of the conservation of angular momentum, differences which do not quite obtain in the calculus. Finally, while there was not much objection to Lagrange's assumption of Taylor's series, his views on the principle of virtual velocities provoked much reaction. This matter is one of the topics treated in the next section.

5.3 Two main results, and their various proofs

My task here is to exemplify the discussions of proof-methods that obtained in mechanics around 1800. In § 5.3.1 I fulfil the promise of § 5.2.2 of discussing the principle of virtual velocities, which Lagrange had elevated to high status in his *Méchanique*. The other result discussed is the composition of forces, which received a fascinatingly wide range of proofs.

5.3.1 The principle of virtual velocities: axiom or theorem?

Lindt has published an excellent history *1904a* of this principle from its origins in the 17th century up to the time of his writing. One of his early points is that '"Proof" and "principle" contradict each other' (p. 150); but by intellectual inertia the 'principle' retained this designation even after proofs had been offered. Of course not all proofs were valid, or perhaps sufficiently general, as the discussion here will demonstrate.

The principle is concerned with the state of equilibrium of a mechanical system. Lagrange's explanation of equilibrium for statics at the beginning of the *Méchanique*—'In the state of equilibrium force has no actual exercise: it produces only only a simple tendency to motion' (p. 2)—neither made clear whether equilibrium was a property of the forces or the bodies involved, nor did it clarify the 'simple tendency to motion' that an equilibriate system possessed. (This is the problem of interpreting the word 'equilibrium' which was mentioned in § 5.2.2.) The importance of equilibrium was partly caused by the principle of virtual velocities, which he expressed 'very generally in this manner' thus (pp. 10–11):

> *If any system of as many bodies or points that one wants taken, each one held by any forces* [puissances], *is in equilibrium, & one gives to this system any small motion, in virtue of which each point would travel an infinitely small space which will express its virtual velocity; the sum of the forces, each one multiplied by the space that the point where it is applied, travels along the direction of this same force, will always be equal to zero, regarding as positive the little spaces travelled in the sense of the forces, & as negative the spaces travelled in an opposite sense.*

Note that Lagrange's formulation was similar to d'Alembert's statement of his principle in § 5.2.3: motions and distances more than forces as such, and little emphasis on a work concept (which in sum was zero anyway).

Lagrange claimed that the principle could be interpreted 'as a type of axiom of Mechanics' (p. 12), and so he offered no proof of it. However, it had not always been granted such status, and was not so treated by Lagrange and others in the 1790s. In particular, *cahier* 5 (1798) of the *Journal de l'Ecole Polytechnique* contained three relevant papers: by Lagrange, Fourier and de Prony.

Lagrange's paper *1798a* introduced the principle of pulleys. He considered a system $\{A_r\}$ of K mass-points (or of individual points of a system of finite bodies) remaining in equilibrium under the action of forces $\{F_r\}$. He assumed that the forces were commensurable; that is,

$$F_r = n_r P, \qquad n_r \text{ an integer.} \tag{531.1}$$

He replaced F_r by a pair of small pulley-blocks, one sited at A_r and the other located in space at B_r such that A_rB_r lay along the direction of F_r. A smooth flexible thread joined the blocks, starting at B_r, passing n_r times over the smooth pulleys in each block, and ending with a weight $\frac{1}{2}P$ suspended from B_r. In the same way he replaced the system $\{F_r\}$ of forces with a single light inextensible thread, starting at B_1, passing around each pair A_rB_r of blocks n_r times and ending with a weight $\frac{1}{2}P$ suspended over the last block B_k. If the length of suspended thread were u, then the total length l of thread was given by

$$l = u + \sum_{r=1}^{k} 2n_r A_r B_r + \sum_{r=1}^{k} B_r B_{r+1}. \tag{531.2}$$

Now in equilibrium $du = 0$, which from (531.2) yielded

$$du = \sum_{r=1}^{k} 2n_r d(A_r B_r) = 0. \tag{531.3}$$

On multiplying by $\frac{1}{2}P$ and using (531.1), (531.3) became

$$\sum_{r=1}^{k} F_r d(A_r B_r) = 0, \tag{531.4}$$

which was an alternative to (522.1) for expressing algebraically the required principle. The argument was extendable to incommensurable forces, for which (531.1) was not applicable, 'by known arguments' (presumably *reductio ad absurdum*, since in § 3.2.4 we saw his criticisms of the theory of limits). The case where $du = 0$ led to a minimum rather than a maximum value for u (corresponding to the desired *lowest* position of the weight) could be dealt with by taking $-F_r$ for F_r throughout.

Lagrange the engineer, providing his students with an argument for the truth of the 'axiom of mechanics'. Probably few of them were convinced that certainty was furnished by this notional (§ 132.3) small-smooth-pulleys and light-inextensible-thread mechanics. Among other points, they may have been uncertain about generality, especially if they knew of Lagrange's casual remark following (524.9) that a continuous body was composed of an infinity of mass-points.

The students may have found Fourier's treatment in *1798a* more impressive, though also tainted with some artificiality. Beginning with an Aristotelian tag, 'geometrae est probare', to show where his mathematical tendencies lay, Fourier based his account of equilibrium on the balance of moments. The first part of his paper (arts. 1-16) was grounded on a (well-known) argument of a system of forces $\{F_r\}$ acting at a point O. If F_r made an angle u_r with an arbitrary line OH, which was displaced when the equilibrium of the system was disturbed, then from the equilibrium condition

$$\sum_r F_r \cos u_r = 0 \tag{531.5}$$

the principle followed in the form

$$\sum_r F_r(dp \cos u_r) = 0\,, \tag{531.6}$$

where dp was the displacement of OH (consistent with the constraints on the situation, presumably), and $dp \cos u_r$ expressed the notion, not clearly defined by Fourier, of virtual velocity (art. 3). He extended the argument to 'any material system', which for him were systems of mass-points, rigid bodies, 'flexible bodies' (that is, rigid bodies linked by inextensible threads) and incompressible fluids.

Presumably Fourier did not think that he established the principle with sufficient generality, for in the second part of his paper he attempted such a proof, which was geometrically oriented (arts. 17-19). Figure 531.1 (of which he surpri-

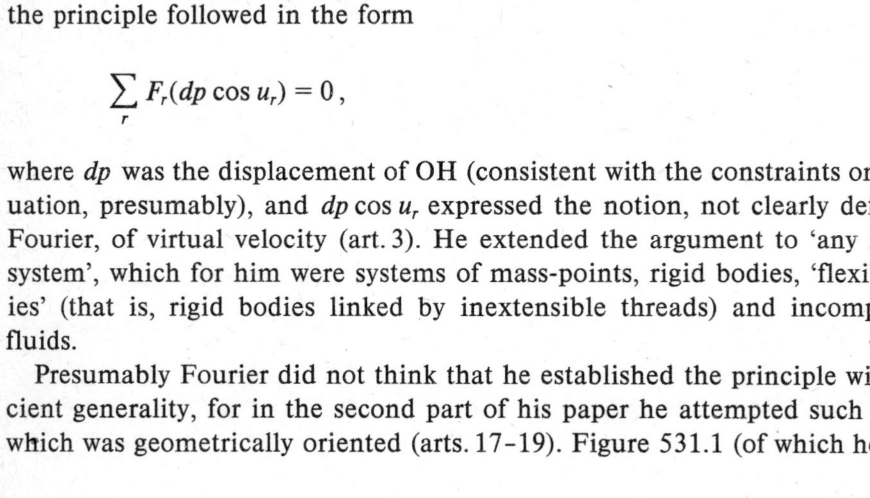

Figure 531.1 Illustration of Fourier's second proof of the principle of virtual velocities
The lettering has been changed:
Here: A_1 B_1 A_2 B_2 OC OA_1 OL A_2L
Fourier: p p′ q q′ d h h′ h″

singly did not furnish a version) shows the displacement of the force F_1 in the direction A_1B_1 and of F_2 along A_2B_2. OC was the line of intersection of the two planes α_1 and α_2 for which A_1B_1 and A_2B_2 were normals, and O itself was the foot of the perpendicular from A_1; OL was the perpendicular to OC in α_2, and L the foot of the perpendicular in α_2 from A_2. Now OA_1 and OL formed a lever with OC as axis; and LA_2 could be interpreted as a lever about *any* axis EG in α_2.

Thus the displacement of A_1 to B_1 was transmitted through O, L and E to move A_2; and the position of F on A_2L was determined such that A_2 moved to B_2. (Fourier assumed a continuity situation here; the effect of this lever increased gradually from nothing as E moved (either way) along A_2L from A_2.) The argument was generalised to the system of forces $\{F_r\}$ by taking them successively in pairs (presumably retaining this mechanism for F_2 when linking it to F_3, and so on: he was not clear on the point).

Fourier assumed that the forces $\{F_r\}$ exerted zero moment on displacement; that is,

$$\sum_r F_r(A_rB_r) = 0. \tag{531.7}$$

It then 'results quite clearly from the principle of the lever and that of the composition of forces' that the system of levers would remain in equilibrium. If now the points $\{A_r\}$ were regarded as also points of the original physical situation, then the conclusion just drawn showed that its equilibrium would not be disturbed by the situation described in (531.7), which established the principle (art. 19).

Fourier's final proof, which was given in art. 23, used the lever and also a scheme like Lagrange's pulleys. He replaced each force F_r with a lever, where a suitable heavy weight W_r hung from one end to cause the tension F_r in a thread joining the other end to A_r in the direction of F_r. The displacement of A_r to B_r induced the moment $F_r \times (A_rB_r)$, which the lever principle showed to be balanced by the moment of W_r. If the lengths of the arms of the lever were arranged so that the velocities of the weights were equal, then (531.4), which was presumed to be satisfied, showed that the combined weights which ascended, equalled those which descended. Hence equilibrium was preserved.

This important paper is rich in other details.[1] For example, in arts. 20-21 Fourier explored the possibility that the equality in (531.7) could be replaced by '$\geq$', on the grounds that the argument of the lever showed that admissible motions of the members of the system had to admit of a non-negative moment. This was an excursion into inequalities which he was to resume in the 1820s (§ 17.3.2). He also recalled his use of the principles of lever and of composition of forces in his proofs, and claiming rather optimistically that 'the one reduces easily to the

[1]) Aspects of the paper are discussed in, for example, Lindt *1904a*, 166-169; Dugas *1950a*, 361-366; Costabel *1972a*; and Bailhache *1975a*, 47-52. Fourier's connections between composition and lever principles may show some influence of d'Alembert *1773a*. His manuscripts on mechanics are assembled at *BN*, ms. fr. 22520.

other', concentrated on proving the composition of forces. However, the proof was interesting.

Firstly, Fourier proved in art. 32:

Theorem 531.1 In Figure 531.2, the downward forces F at A and B were in equilibrium with the upward force $2F$ at the midpoint E of AB.

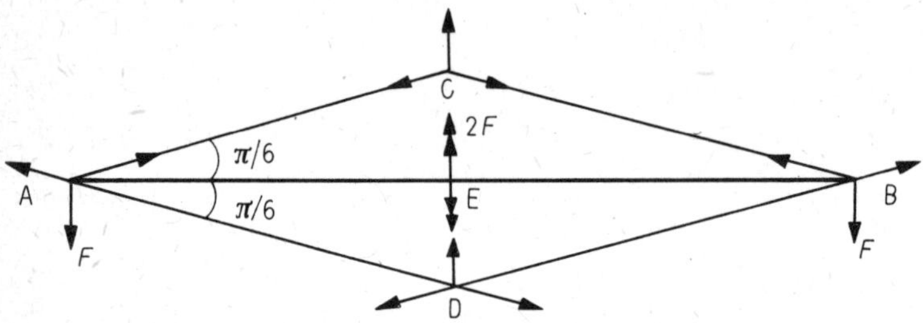

Figure 531.2 Fourier's diagram for theorem 531.1

In the rhombus ABCD, the trios of forces F at the corners were 'manifestly in equilibrium', as was the pair $2F$ at E; and cancelling equal and opposite forces yielded the theorem. The proof, and the diagram, were presented later by Monge (*Statics*$_5$ (1810), fig. 3).

The lever principle was now proved in art. 33 from Figure 531.3 (which Fourier

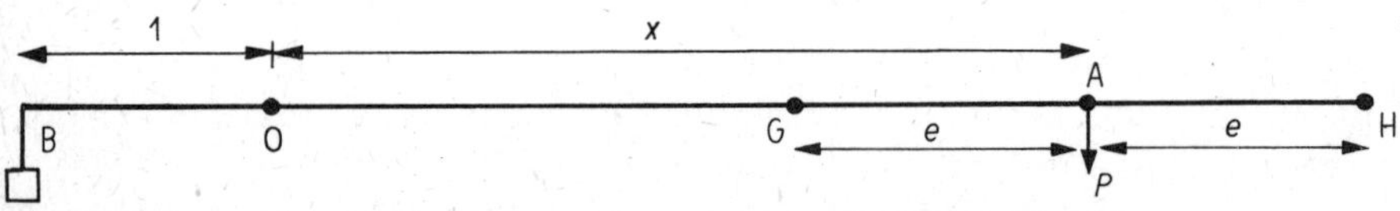

Figure 531.3 Diagram for Fourier's proof of the lever principle

did not supply). The weight at B balanced the force P acting at A distant x from the pivot O of the lever; the moment $Q(P, x)$ was to be determined.

If P increased by dP, then

$$\text{increase in the moment} = \text{moment due to extra force } dP; \tag{531.8}$$

that is,

$$dQ = Q_p(P, x)\, dP = Q(dP, x)\,. \tag{531.9}$$

Hence, since $(531.9)_3$ did not contain P, differentiation produced

$$Q_{pp}(P, x) = 0\,, \tag{531.10}$$

whose solution, given that $Q(0, x) = 0$, could be written

$$Q(P, x) = P\psi(x) \quad \text{with } \psi \text{ unknown.} \tag{531.11}$$

Fourier now used Theorem 531.1 to replace P at A with $\frac{1}{2}P$ at G and H, and thus convert (531.11) into the functional equation

$$2\psi(x) = \psi(x - e) + \psi(x + e)\,. \tag{531.12}$$

Differentiation, and the conditions

$$\psi(0) = 0 \quad \text{and} \quad Q(P, 1) = P, \tag{531.13}$$

showed that $\psi(x) = x$, and hence that (531.11) became

$$Q = Px\,, \tag{531.14}$$

from which the lever principle followed.

'The demonstrations that c[itizen]s *Lagrange* and *Fourier* have given of the principle of virtual velocities, proves incontestably that the formula that one deduces from it applies, without exception, to all cases of equilibrium', began de Prony in his contribution (*1798c*, 191); but he warned the students that they would find *general* proofs difficult to follow. His own proofs, both in this paper and elsewhere, were similar to the argument used by Fourier in (531.5-6) but expressed in the trigonometrical terms which saw him favouring in § 5.2.7 and laced with a touch of difference mathematics (§ 3.5.3). Thus from the equation for zero moment

$$\sum_r F_r p \cos u_r = 0 \tag{531.15}$$

'one infers the two noteworthy equations'

$$\sum_r F_r \Delta p_r = 0 \quad (\text{with } \Delta p_r := p \cos u_r) \quad \text{and} \quad \sum_r F_r dp_r = 0\,, \tag{531.16}$$

where the difference or differential case applied according as the constant length p in the given direction was finite or infinitesimal (*1798c*, 195-198; *1800a*, 36-49; *1802a*, 205-207). As a proof it was, of course, only an educationally suggestive sketch; but it spared the students Lagrange's smooth wheels and Fourier's rigid levers.

An interesting point in Fourier's paper was his reference to Carnot's 1783 *Es-*

sai in the course of his final proof (*1798a*, art. 23)—one of the very few citations of that book by a contemporary author of which I am aware. Carnot regarded the theory of machines as concerned with virtual velocities in systems of bodies (*1783a*, art. 60), and both there in the succeeding 1803 *Principes* he offered unusual proofs, argued from impact theory (see Bailhache *1975a*, 81-90). I shall briefly note the main steps of an argument in the later work (*1803a*, 179-181), which started out from consideration of 'a system of hard bodies free or applied to any machine without spring[s]'. A typical body m had velocity V and 'motive force' (acceleration) P at time t, and would take velocity u if its actual motion were suddenly stopped. During an instant dt, m would gain velocity $P \cos u \wedge P dt$ in the direction of u, while the component of that velocity was already $V \cos V \wedge u$. Thus, 'in virtue of the reciprocal action of bodies' and earlier theorems,

$$\mathrm{S} mu (P \cos u \wedge P) dt = \mathrm{S} mud (V \cos V \wedge u) . \tag{531.17}$$

Now in equilibrium the right hand side of (531.17) was zero. Further, mP could be taken as any 'moving force' F, and u any 'geometrical motion' (that is (p. 108), a motion which did not affect any forces or motions in the given situation) and thus the virtual velocity for F. Hence (531.17) became

$$\mathrm{S} Fu \cos u \wedge F = 0 , \tag{531.18}$$

which was 'visibly the expression of the famous principle of virtual velocities' (*1803a*, 181). Like de Prony's proof, it is free from contrivancies; and like Fourier's it has some hope of generality, although it is compromised by the restrictions of its premises, which the later discussion (pp. 194-195) did not entirely remove.

Two features are absent from all these proofs. Firstly, nobody seemed to offer necessary and sufficient conditions for the principle in terms of the zero 'moment' effected by the impressed forces. The other is a proof based on argument from equations of the form of Lagrange's (522.4):

$$\sum_j P_j \delta s_j + \sum_j R_j \delta r_j = 0 , \tag{531.19}$$

where impressed forces $\{P_j\}$ are displaced $\{\delta s_j\}$ and reactive forces $\{R_j\}$ are constrained to move $\{\delta r_j\}$ along the curves or surfaces of the problem. In fact, in his *Mécanique céleste* Laplace did try to develop a proof from (531.19) by decomposing P_j into three component forces defined relative to these curves and surfaces (*1*(1799), 72-82). The argument is hard to follow; and in fact Poinsot *1838a* was to show that it essentially begged the question (see Bailhache *1975a*, 70-80). So an "obvious" move turned out not to be such a good idea after all.

5.3.2 Proofs of the composition of forces

We saw Monge's Theorem 531.1 on parallel forces reported by Fourier with the help of Figure 531.2. For the general planar case Monge used Figure 532.1, in

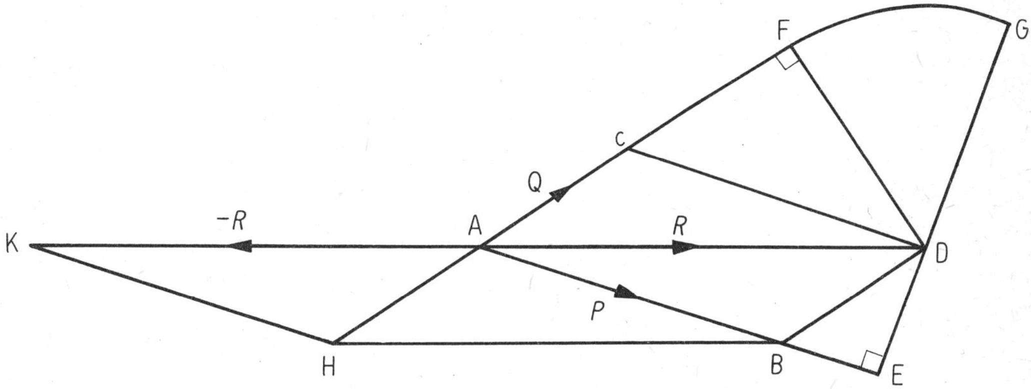

Figure 532.1 Mongean diagram for proving the composition of forces

which AB and AC represented the forces *P* and *Q* in magnitude and direction. He set up the arc FG of a circle centred at D as shown, and at G he lay a force of magnitude *P* along the tangent. He also moved *P* to act at E, and used earlier theorems on parallel forces to show that their resultant passed through D, and thus the resultant *R* of the original *P* and *Q* did so as well. Now *R* obviously acted at A, and hence lay along AD. In order to ascertain its length along that direction, Monge noted that *P* and *Q* were in equilibrium with $-R$ set along AK. Hence by examining the parallelograms ABHK and ADBH and using the result just proved, he showed that AD gave the length and direction of *R* (Monge *Statics*, ch. 1; for example, $_6$(1826), 26–34 and figs. 10 and 11, where hands grasp lines such as AB and AC).

As we saw in § 5.2.7, de Prony's approach to mechanics was close to Euler's (and Monge's) in content, though Lagrangian in its avoidance of diagrams and stress on algebra. Both features were prominent in his treatment of this theorem. 'The beginning of statics has been the generalisation of the principle of the composition and of the decomposition of forces', he explained (de Prony *1798a*, 7) and in his *Mécanique philosophique* he left the proof as a problem for the students (*1800a*, 29). It is not clear *how* he expected them to prove it, for he did not introduce the cosine law for another three pages; his proof of the principle of virtual velocities, in the form (531.16), also assumed composition (*1798c*, 197). Three of his students took up the question, and produced two algebraic proofs which have become well known.

Charles Duchayla (or Blanquet-Duchayla, to use his full surname) studied at the *Ecole Polytechnique*, but only for part of the academic year 1795-1796. He passed much of his later career in the *Université*, teaching in the *académie* in Turin, and then at Montpellier from 1815 to 1822, where he was *Recteur*. He then acted as *Recteur* at Aix until 1825, when he was promoted to *inspecteur général*. We shall see him in § 7.4.7 and § 9.3.4 as an acute observer of Parisian rivalries. His brief paper *1804a* on the composition of forces was, it seems, his only publication in science; but it left his name in several 19th-century textbooks in mechanics, which presented, with attribution, his proof. It was inductive, for he assumed that the theorem was true for parallelograms ACFG and GFDB of Figure 532.2, and showed it to be true for parallelogram ACDB also. The resultant r_1 of

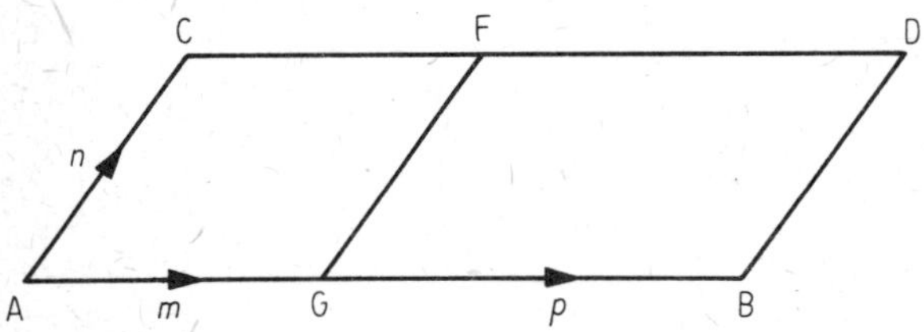

Figure 532.2 Diagram for Duchayla's proof of the composition of forces

forces n at AC and m at AG lay along AF. Now r_1 could be decomposed at F into m along FD and n along GF; and the resultant r_2 of this n and the third force p along GB lay along GD. Hence the resultant of r_2 and m along FD passed through D; but r_2 was the resultant of n along AC and $(m + p)$ along AB, so that the theorem was proved for the parallelogram ACDB, as required. To start the induction, it was clear from symmetry that the theorem was true for $m = n = 1$; thus, by the construction just described, it also held for any pair of rational values m and n. 'One will then show', he concluded, 'by the ordinary argument of reduction to the absurd, that the position includes also the case of two incommensurable forces' (that is, in irrational proportion). This remark recalls from § 5.3.1 Lagrange's unspecified appeal to 'known arguments' in a similar context: Duchayla's preference for a cumbersome *reductio* argument corroborates the judgement of § 3.2.6 concerning the low prestige given to the theory of limits around 1800.

Francoeur gave two proofs of composition in his textbook on mechanics. In the main text he reported Laplace's proof from the *Mécanique céleste* (see Figure

532.3$_1$), in which two concurrent forces P and Q were taken at right angles and resolved along and perpendicular to their resultant R. Arguing from proportions, Laplace had found that the magnitude of R could be represented by the diagonal

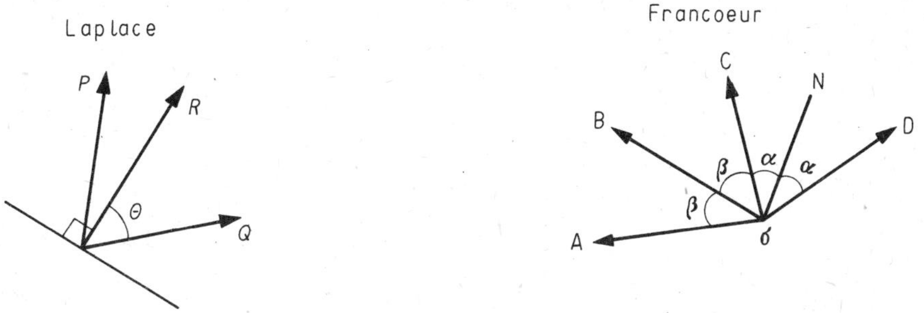

Figure 532.3 Diagrams for Laplace's and Francoeur's proofs of the composition of forces

of the rectangle defined by the sides P and Q. To find the function of the angle Θ which gave the decomposition, he added differential increments on to the components of P and Q which lay along and perpendicular to R, and came to the differential equation

$$(PdQ - QdP)/(P^2 + Q^2) = kd\Theta, \quad k \text{ constant}, \tag{532.1}$$

and the solution

$$P = R\cos(k\Theta + \varrho), \quad \varrho \text{ constant}, \tag{532.2}$$

in which special cases showed that $k = 1$ and $\varrho = 0$, as required.[1])

On p. 417 of his notes to the book, Francoeur presented Fourier's proof of the equilibrium of parallel forces based on the functional equation (531.12); and perhaps under its inspiration he found his own proof of the composition of forces (pp. 411–416). The component of any force P at angular distance α was put as $Pf(\alpha)$, and four unit forces acted from O as shown in Figure 532.3$_2$. By symmetry the resultant R of OA and OC lay along OB, and had magnitude $2f(\beta)$. Therefore, the component of R along ON equalled the sum of the components of the original forces in that direction, a relationship which the diagram showed to

---

[1]) Laplace *Mechanics, 1* (1799), 2–6; Francoeur *Mechanics*$_1$ (1800), 19–22. In his proof Laplace implicitly drew on the functional equation

$$(f(\theta))^2 + (f(\tau - \theta))^2 = 1\,, \tag{1}$$

as Babbage noted in *1827a*, 327.

be expressible mathematically by the functional equation

$$2f(\alpha)f(\beta) = f(\alpha + \beta) + f(\alpha - \beta). \tag{532.3}$$

Differentiating twice with respect to α and with respect to β yielded

$$f''(\alpha)f(\beta) = f(\alpha)f''(\beta), \tag{532.4}$$

with solution

$$f(\alpha) = Ae^{\alpha\sqrt{c}} + Be^{-\alpha\sqrt{c}}, \quad c \text{ constant.} \tag{532.5}$$

Substituting (532.5) back into (532.3) showed that $A = B = \frac{1}{2}$, and the special case of $f(\alpha) = 0$ when $\alpha = \tau$ yielded $\sqrt{c} = i$. Hence (532.5) reduced to the cosine law, as required.

(532.3) has become known as 'Poisson's functional equation', because of his methods of solving it; and Poisson's context was none other than this proof of the composition of forces, as presented in his textbook on mechanics.[1]) One might imagine a spot of borrowing; after all, he studied at the *Ecole Polytechnique* during Francoeur's first years of teaching there, and was on the staff himself in 1800 when Francoeur's textbook was published. However, when Hachette accepted Petit's write-up of Poisson's version of the proof in a paper (Poisson *1808d*) for his *Correspondance sur l'Ecole Polytechnique,* he attested that 'This demonstration had been communicated to me before Mr. Francoeur would have made it known in his *Mécanique*'. Indeed, the proof was first published in d'Alembert *1772a*, 278-280 (perhaps by generalising on the special cases proved geometrically in Daniel Bernoulli *1726a*; so to d'Alembert the priority of invention seems truly to be due.

5.4 On celestial mechanics

5.4.1 Introduction: the *Bureau des Longitudes*

Most of the major developments in this history occurred in connection with terrestrial mechanics; but celestial mechanics played a role, which we now begin to

[1]) See Poisson *Mechanics*$_1$ (1811), *1*, 11-15 (where he solved (532.3) by his preferred (§ 4.3.3) method of power series); and *Mechanics*$_2$ (1833), *1*, 43-55 (where he showed by substitution that cos α solved (532.3), and tried to argue for its uniqueness as a solution by induction on the number of satisfied values of α). For contemporary interest in equations such as (532.3) see Ampère *1825b* and Gergonne *1824a*; and for a later treatment of (523.3) see, for example, Andrade *1900a*.

consider. The astronomical problems usually concerned the paths and perturbations of the heavenly bodies, outstanding specific questions such as the three-body problem and the stability of the planetary system, and the fitting of data to theory. The mathematics was normally based on differential equations and their solutions, and so interacted with the mathematics of terrestrial mechanics. The topics discussed in this book are but a fragment of an enormous amount of work which exercised important influence on both the intellectual and social developments of science; and I cannot cite authoritative historical sources, since unfortunately historians of astronomy appear to have paid least attention to the mathematical details attending their subject.[1])

It is necessary to begin with the *Bureau des Longitudes*, founded in 1795 as an institution to develop mathematical and theoretical astronomy, including the preparation of annual tables for the use of navigators; observational astronomy was largely carried out at the Paris *Observatoire* (and also in observatories elsewhere). The *Bureau* published an *Annuaire* on its business (and some general articles); but the principal publication was the *Connaissance des tem(p)s*, also an annual serial, which contained not only these tables but also technical articles (especially from the 1810s onwards).

The tables were compiled by calculators; as was mentioned in § 3.5.4, some transferred to the *Bureau* when de Prony's cadastral survey tables were completed. As in de Prony's empire, their work was conducted under the supervision of *savants*, for whom membership of the *Bureau* was one of the posts which they would accumulate. Of figures with whom this history is concerned, its founding members, who initially received 8,000 *francs* each year, included two 'geometers' (Lagrange and Laplace) and four astronomers (among them Lalande and Delambre), while Borda specialised in navigation. In 1801 de Prony was added as a *surnumeraire*, at 6,000 *francs* per year.[2])

---

[1]) A comprehensive history of mathematical astronomy for the late 18th century is sadly lacking; Gautier *1817a* and Wilson *1980a* provide valuable information on certain aspects. Some sources are disappointing; for example, the articles on astronomy, geodesy and geophysics in the *Encyklopädie der mathematischen Wissenschaften, 6*, pt. 2 (1905-1923) are very patchy on this period, while Montucla *History*, *4* (1802), devoted to astronomy, is very slight on mathematical details. This volume was completed by Lalande (§ 3.2.8), whose own *1803a* provided bibliography to 1803 and on pp. 661-880 non-mathematical history of the last 20 years. Delambre presented largely anecdotal biographies of 18th-century astronomers in his *1827a*, although sometimes he provided useful technical information. Arago *1844a* gave his own view of Laplace's achievements. Grant *1852a* contains a fine account of much of the physical astronomy pertinent to this study.

[2]) Bigourdan *Bureau* (1928-1931) describes many aspects of the history of the *BL*. The minute-books are still held in its offices; they are cited by volume number, followed by the date of the meeting. For other manuscripts, see the archives of the *Observatoire* (square 8A of Figure 263.1); and *AN*, F^{14} 3702-3717 and 13569-13571.

As usual in this chapter, I use the French literature published around 1800 as the main guide. There is a dominant figure again—but this time it is Laplace.

5.4.2 Laplace's *Exposition du système du monde*, 1799

> If man were restricted to collecting facts; the sciences were only a sterile nomenclature, and he would never have known the great laws of nature. It is in comparing phenomena between each other, in seeking to grasp their relations; that he is led to discover these laws [which are] always impressed in their most varied efforts.
>
> Laplace *Exposition*$_2$ (1799), 94

Laplace wrote a non-mathematical introduction to astronomy called *Exposition du système du monde*. The first edition appeared in 1796; the sixth, in preparation at his death in 1827, was published posthumously in 1835 (and also appears in both editions of his *Oeuvres*); I am concerned here with the second edition, which was published by Duprat in 1799. Some later editions will be noted from time to time, especially in § 13.6.1: see also § 19.3.3.

The quotation above exemplifies Laplace's philosophy of physics (and also his often eccentric punctuation). The discovery of laws required more than just the accumulation of facts. As he said on p. 171 of the volume, in a gobbet quoted in § 1.1, the correspondence between the 'phenomena of nature' and the 'intellectual truths of nature' was 'one of the greatest charms attached to mathematical speculations'. When correspondence occurred, such speculations, whether mathematical or prosodic, became 'veritable hypotheses', as he stated in the passage on pp. 248-249 which was quoted in § 1.3.5. The simpler the theory the better, in some probabilistic sense: 'One can increase the probability of a theory, either by diminishing the number of hypotheses upon which it rests, or in augmenting the number of phenomena which it explains' (p. 282). The thrust of his contributions to probability theory and to statistics will be indicated in § 6.4.7.

Determinism was a significant guide to the philosophy outlined in Laplace's book. Apparently all features of the physical world were determined—even 'the curve described by the light atom which the winds seem to carry along by hazard, is regulated in a manner [which is] as certain, as the planetary orbits' (p. 134)—and so were susceptible to confirmation by veritable hypotheses. The generality of scope of this view will emerge during the rest of this chapter, and in the next one. However, he did not see the need to carry this confidence down to the ontological level. For him, 'the first causes of the intimate nature of beings, will be unknown to us for ever' (p. 5).

The *Exposition* was divided into five Books, a scheme which Laplace preserved in all editions. A discussion 'On the apparent motions of celestial bodies' (pp. 2-93), which also included the motion of the sea and atmosphere refraction, was followed by a companion account 'On the real motions of celestial bodies' (pp. 94-133), which covered not only the planetary, satellitary and cometary paths but also the effects caused by the rotation of the earth. The third Book, 'On the laws of motion' (pp. 134-177), concerned with 'force' (momentum in this case: p. 140), equilibrium and motion, was succeeded by a companion Book, 'On the theory of universal gravity' (pp. 178-287), consisting of an account of perturbation theory, the motion of the sea again, and the shape of the earth. The final Book, a 'Precis of the history of astronomy' (pp. 288-351), read rather like Lagrange's historical comments in his treatises: a genuine interest was shown, but a limited capacity for historical research. It ended with the best-known, if least argued, part of the whole work: his so-called 'nebular hypothesis', Buffon's suggestion that the planets and comets were the product of coalescence of matter after the collision of a comet with the sun (see Jaki *1976a*).

Like Lagrange's *Méchanique* (§ 5.2.1), and indeed much mechanics of the late 18th century, Laplace laid great emphasis on stability and equilibrium. However, in an interesting passage on p. 208 he briefly allowed for the possibility that the earth had not always been a safe ride:

> However, the small probability of a similar encounter [of the earth with a comet], can become very great in adding up over a long sequence of centuries. It is easy to picture to oneself the effects of this impact upon the earth. The axis and the motion of rotation changed; the seas abandoning their old position to throw themselves towards the new equator; a large part of men and animals drowned in this universal deluge, or destroyed by the violent tremor imparted to the terrestrial globe [...]

Soon after this 1799 edition of Laplace's book was published, his newly-formed brainchild (§ 2.3.1), the *Conseil de Perfectionnement* of the *Ecole Polytechnique*, decided in 1802 to introduce a course in astronomy, using the book as its basis. The physics *Professeur* Hassenfratz took the course, and published his lectures as a textbook *1803a*. In addition, introductory though the *Exposition* was intended to be, it received a summary of its own in the pamphlet de Prony *1801c*, an expanded version of articles in the *Journal de Paris* written by him as part of one of the reports on the work of the *classe* which he had prepared as one of the current *secrétaires* (§ 2.2.2). It presaged the summarising, that several people were to attempt of Laplace's *Mécanique céleste*, to which we now turn.

5.4.3 Laplace's *Mécanique céleste*, 1799: the Bowditch translation

1799 also saw the appearance of the first two volumes of Laplace's *Traité de mécanique céleste.* The publisher was again Duprat; the print-run was 1,500 copies for each volume. In an announcement in volume 2, Laplace stated that the third volume 'will be put to press immediately' and that 'The two remaining volumes will follow at similar intervals'. In fact, volumes 3 and 4 did duly appear, in 1802 and 1805 respectively (§ 6.3.1); but volume 5 was not published until the 1820s, when it came out in parts (§ 12.6), and probably was not the tome which he had had in mind here.

By the 1820s the American astronomer and navigator Nathaniel Bowditch had completed his translation, with commentary, of the first four volumes, and was trying to accumulate enough money to pay for its printing. Eventually it was published in Boston in four volumes between 1829 and 1839, the last volume in the year after his death. Although he had planned the translation of the fifth volume, it was never executed.

Bowditch sent complimentary copies of his volumes to prominent scientists around the world, and their reactions provide insight into the current perception of of Laplace's original version.[1]) There was universal gratitude for his clarification in the notes on difficult passages (and also for his provision of cross-references and descriptions of much later work). He probably spoke for many when he said: 'Whenever I meet in La Place with the words "Thus it plainly appears", I am sure that hours, and perhaps days, of hard study will alone enable me to discover how it plainly appears' (H. Bowditch *1839a*, 62). His explanations included statements of basic calculus and details of easy passage work, which are excessively pedantic; and sometimes the commentary is myopic, explaining small details and passing over large features. But the majority of his contributions are very useful; and as the versions of the *Mécanique* published in both editions of Laplace's *Oeuvres* are merely reprints of the original, I shall cite the Bowditch edition throughout this book, while attaching Laplace's dates of publication to the volumes. Thus '*Mechanics, 1* (1799), 265' cites the first volume, of 1799, but from p. 265 of Bowditch's edition. I quote his translation, even when it is rather free; but I do not follow his enthusiasm for italicising Laplace's original text.

---

[1]) A few of the letters to Bowditch are quoted on pp. 63-65 of the obituary *1839a* by his son Henry, printed in the fourth volume of the translation and moved to the first volume of the 1966 reprint. The originals of these and other letters, and some notebooks, are kept in Bowditch's *Nachlass* in Boston Public Library: this collection is now microfilmed (see Montgomery and Monti *1983a*). The names of recipients of the translation are given on pp. 166-168 of the obituary.

5.4.4 Laplace's principal mathematical procedures

> There was moreover a difference between their geniuses which would be noticed by all those who studied their works: whether it was the libration of the Moon or a problem about numbers, Lagrange seemed most often to see, in the questions which he dealt with, only the mathematics to which they gave rise; and from that comes the high price he placed on elegance of formulas and the generality of methods; for Laplace, by contrast, mathematical analysis was an instrument that he bent to the most diverse applications, but always subordinating the particular subject matter to the business [*fonds*] of each question. Perhaps posterity will judge that one was a great geometer, and the other a great philosopher [...]
>
> Poisson *1827c*, 20 (or Laplace *1827c'*, 307-308)

Volumes 1 and 2 of Laplace's *Mécanique*, called the 'first part', more or less comprised a general theoretical basis for mathematical astronomy; the next two volumes, which constituted the 'second part' and are introduced in § 6.3.1, dealt with many of the details and special cases. Volume 2 was largely concerned with planetary mechanics, and is our main guide in § 5.5; the rest of this section is devoted to volume 1, which contained two Books, one 'On the general laws of equilibrium and motion' (pp. 1-238), covering systems of bodies (and also fluids), and the other 'On the law of universal gravitation, and the motions of the centres of gravity of the heavenly bodies' (pp. 239-725). As ever, I must be *very* selective over the topics discussed here; some others will be taken up later.

As with Lagrange's *Méchanique*, Laplace's treatise drew substantially on earlier work, both by the author and on others. (Indeed, Laplace's title was of course an imitation of Lagrange's.) Unlike Lagrange's little historical essays, however, Laplace gave virtually no references, which makes attribution and influence particularly difficult to appraise. Todhunter expressed the problem in a delicious phrase, in the context of Laplace's treatise on probability but applicable to all of his writings: 'we conclude that he supposed the erudition of his contemporaries would be sufficient to prevent them from ascribing to himself more than was justly due' (*1865a*, x-xi).

Laplace was in fact noticeably, if silently, influenced by Lagrange (and, through him, by d'Alembert) in various ways. There were virtually no diagrams in his treatise, even when, as Bowditch's figures show, the configurations were complicated. Variational methods were introduced at once, in a rather careless passage where 'variation' and 'differential' were taken as synonyms (*1* (1799), 8).

Equilibrium was emphasised, with a prominent place assigned to the principle of virtual velocities (as was noted in § 5.3.1); the general equations of motion were grounded in the principles of least action and conservation of 'living force' (pp. 131–136). Building on his *1793a*, art. 3, he also included on pp. 137–143 a discussion of the possibility that 'force' denote "any" function of velocity, rather redolent of d'Alembert and Daniel Bernoulli *1726a* (see Vuillemin *1958a*).

However, as the quotation from Poisson above indicated, Laplace gave intuition a much greater role than Lagrange allowed. For example, we saw at (532.1) that he began his text with the composition of forces. Again, he expressed the rotation of a rigid body in terms of a form of Euler's equations (525.14), and ignored Lagrange's equations (524.9) (pp. 146–158, including some rather pedantic help from Bowditch). His treatment of the motion of fluids was similarly Eulerian, in that he imagined (and Bowditch drew) differential parallelepipeds of fluid as his means of deriving the equations of fluid flow (pp. 194–210). His theorem on the existence of the invariable plane, repeated on pp. 106–122 from the articles *1798b* and *1799c* in the *Journal de l'Ecole Polytechnique*, was proved as follows. From the principle of areas, for a system of bodies $\{m_j\}$ moving under no impressed forces and located at $\{(x_j, y_j, z_j)\}$ at time t, the quantities

$$c_1 := \sum_j m_j(y_j\dot{z}_j - \dot{y}_j z_j), \quad c_2 := \sum_j m_j(z_j\dot{x}_j - \dot{z}_j x_j), \quad c_3 := \sum_j m_j(x_j\dot{y}_j - \dot{x}_j y_j) \quad (544.1)$$

were constant. Further, from Euler's torque theorem (527.1) the angular momentum about the normal to the 'invariable plane'

$$c_1 x + c_2 y + c_3 z = 0 \quad (544.2)$$

was maximal; and, to simplify calculations, the axes could be adjusted so that two of the constants in (544.1) were zero (*1* (1799), 106–122). 'It is chiefly in the theory of the solar system, that the investigation of this plane is important', explained Laplace in a later discussion (p. 628), 'on account of the proper motion of the stars, and of the ecliptic, which makes it very difficult for astronomers to determine with precision the motions of the heavenly bodies'. In fact, Lagrange had given a similar argument in, for example, an analysis of the three-body problem (*1777c*, art. 17). The invariability of Laplace's plane was to be questioned in 1828 by Poinsot (§ 17.4.1).

Laplace was content to study the initial approximations to the motions in these Books; more detailed accounts were provided in the later volumes. He began with the two-body problem, where a small body m moved in elliptical orbit with mean angular velocity n about a stationary large one M distant r away. Setting polar coordinates about M, with radius r and eccentric and true anomalies u and v respectively, he found on p. 356 the usual equations for motion in an ellipse of semi-major axis a and eccentricity E:

$$\begin{aligned} nt &= u - E \sin u, \quad r = a(1 - E \cos u) \quad \text{and} \\ \tan \tfrac{1}{2} v &= \sqrt{(1+E)/(1-E)}\, \tan \tfrac{1}{2} u . \end{aligned} \tag{544.3}$$

(Laplace wrote 'e', and called 'eccentricity' the distance aE between the centre and a focus of the ellipse (p. 325); I shall follow his use of the word.) $(544.3)_1$ (which is often called 'Kepler's equation') took the form of Lagrange's equation (341.10), so that "any" function ψ of u could be expanded in a Lagrange series (341.15) (p. 369):

$$\psi(u) = \psi(nt) + \sum_{j=1}^{\infty} [E^j(\psi'(nt) \sin^j nt)^{(j-1)}/n^{j-1} j!] .^{1)} \tag{544.4}$$

Laplace presented (544.4) as part of an account of his theory of generating functions (352.1) (pp. 357–368). The purpose of such expansions was to produce explicit forms for the variables of (544.3): the mean angular velocity n, the true anomaly v, the eccentric anomaly u, and the radial distance r between m and M. Thus we see again the prominent place assigned to power series in the mathematics around 1800, with convergence surely guaranteed by the very small value of E. To take another example, *we* see some of his results derived from (544.4) as expansions in series of Bessel functions $\{J_j(x)\}$ (pp. 369–383, presumably following Lagrange *1771a*); but that theory did not exist as such around 1800 (§ 3.4.4), so Laplace simplified these series by using (433.12) and other corollaries of de

---

1) Laplace's units are difficult to follow. On p. 33 he stated Newton's second law in Euler's form (525.7)

$$\ddot{x} = P, \quad \ddot{y} = Q, \quad \ddot{z} = R, \tag{1}$$

without mention of the gravitational constant G. He carried the form (1) through the equations of motion of a system of bodies (pp. 268, 304, 313, 316) to the equations from which he deduced those for the elliptic orbit (544.3) (p. 322). He also defined

$$\mu := M + m \tag{2}$$

on p. 321, again without reference to G, and was to lead to dimensionally chaotic definitions such as

$$a := h^2/(\mu(1-e^2)) \tag{3}$$

on p. 325, where a was the semi-major axis of the ellipse, and h a constant whose units ($L^2 T^{-1}$) are not clearly determinable from the defining equation. I preserve the chaos in my account below.

The discussion is not helped by some notations. 'μ' was used not only in (2) but also, from a definition on p. 287, as $\cos \theta$; and M' was introduced as an acceleration term on p. 307. Bowditch seems to have been happy with the procedures discussed in this footnote: his notes do not illuminate their foundations.

Brinkley *1803a* contains an historical survey of methods of deriving and treating $(544.3)_1$.

Moivre's theorem to convert from $\sum_j a_j E^j \sin^j nt$ to $\sum_j b_j E^j \sin jnt$—and thus exhibit *power-series* expansions rather than *Bessel-function* expansions. In the 1820s he was to study the convergence of such series, as we shall see in § 11.6.4.

Laplace's first desimplification of the two-body problem was to consider, as his 'second approximation', the perturbing forces caused by the attractions from other bodies. Retaining M as the centre of coordinates, the motion of body m_j, located at $P_j(x_j, y_j, z_j)$ of a system $\{m_j\}$ of N bodies, was given by the system of ordinary differential equations

$$0 = \ddot{x}_j + \left(\frac{M + m_j}{r_j^3}\right) x_j + R_{x_j}, \qquad 1 \le j \le N, \tag{544.5}$$

and two other equations for y_j and z_j, where R was the perturbing function

$$R := \sum_{\substack{k=1\\k \neq j}}^{N} m_k \left(\frac{x_k x_j + y_k y_j + z_k z_j}{r_k^3}\right) - \frac{1}{m_j} \sum_{\substack{i,k=1\\i \neq k}}^{N} \left(\frac{m_i m_k}{r_{ik}}\right); \quad r_k := MP_k, \quad r_{ik} := P_i P_k \tag{544.6}$$

(pp. 268, 504–505: this function is discussed in § 5.5.2). He performed various manipulations and integrations of these equations. For example, integrating (544.5) and its companions and adding them together yielded

$$0 = \dot{r}_j^2 - (M + m)\left(\frac{2}{r_j} - \frac{1}{a_j}\right) + 2\int (R_{x_j} dx_j + R_{y_j} dy_j + R_{z_j} dz_j), \tag{544.7}$$

where a_j was the major semi-axis of the ellipse of m_j around M. Again, multiplying (544.5) by x and the others by y and z respectively and adding to (544.7) produced on pp. 505–506

$$0 = \tfrac{1}{2}\, \ddot{\overline{r_j^2}} - (M + m)\left(\frac{1}{r_j} - \frac{1}{a_j}\right) + Q, \tag{544.8}$$

$$\text{where} \quad Q := 2\int (R_{x_j} dx_j + R_{y_j} dy_j + R_{z_j} dz_j) + R_{x_j} x_j + R_{y_j} y_j + R_{z_j} z_j. \tag{544.9}$$

Laplace solved such equations in two steps: the determination of the elements of the orbit of each planet, and the forms of solution for a planetary system. For one planet (544.7) reduced to

$$\dot{r}^2 = (M + m)\left(\frac{2}{r} - \frac{1}{a}\right). \tag{544.10}$$

He saw that $M + m = U^2$, where U was the mean velocity of the planet m if it followed a circular orbit of unit radius, and so he regarded the semi-major axis as determinable from the initial position r_0 and velocity V_0 and the modified (544.10) as

$$V_0^2 = U^2\left(\frac{2}{r_0} - \frac{1}{a}\right), \tag{544.11}$$

and derived conditions for the orbit to be elliptical, parabolic and hyperbolic (pp. 393–394). He carried out passage work on the equations to find formulae for the eccentricity aE, the true anomaly v, and (via $(544.3)_3$) the eccentric anomaly u. He also discussed other features of the equations; for example, that (544.10) furnished the velocity without knowledge of the eccentricity (pp. 394–407). He was also aware of the weaknesses of the approach; the impossibility of knowing the values of U, V_0 and r_0, which would have to be determined, or replaced, by calculations from data-points (pp. 407–417). This brought him to error theory, which I shall discuss in § 5.5.5.

Laplace's solution of equations such as (544.4–5) for a planetary system used harmonic analysis. It was obtained from his method of successive approximations, where a system (335.11) of nth-order differential equations

$$0 = y_j^{(n)} + P_j + \alpha Q_j, \quad \alpha \text{ small}, \tag{544.12}$$

was solved by developing a sequence of (assumedly) successively approximating solutions, based on the known closed-form solution to (544.12) when $Q_j = 0$; for the smallness of α in (544.12) corresponded to the smallness of R in (544.5) and Q in (544.8). Thus, in a simple case he took the equation

$$0 = \ddot{y} + a^2 y + \alpha Q, \tag{544.13}$$

and obtained from the solution form (335.20) of (335.18):

$$\begin{aligned} y = (c_1/a)\sin at + (c_2/a)\cos at - (\alpha/a)\sin at \int Q(t)\cos at\, dt \\ + (\alpha/a)\int Q(t)\sin at\, dt \end{aligned} \tag{544.14}$$

(pp. 482–483). Now if Q (or R) could be converted into polar coordinates, it would be expanded in a series of terms of the form

$$K_j \frac{\sin}{\cos}(k_j t + \varepsilon_j) \tag{544.15}$$

(as proved on pp. 522–530), so that the solution (544.14) would gain a series of terms

$$\left(aK_j \frac{\sin}{\cos}(k_j t + \varepsilon_j)\right) \Big/ (k_j^2 - a^2) \quad \text{if} \quad k_j \neq a, \tag{544.16}$$

or

$$\pm(\alpha Kt/2a)\frac{\sin}{\cos}(at + \varepsilon) \qquad \text{if} \quad k_j = a. \tag{544.17}$$

When these terms were introduced into the solution (544.14), he had the next approximation to the value of Q in (544.15). The process could be repeated to produce further approximate solutions for y. The technique could apply to systems of equations, and to equations of higher order than (544.13) (pp. 483-503). The degree of approximation depended on the numerical accuracy required (by observational data, for example); the rate of approximation depended on the position within the formulae of powers of small quantities such as the eccentricities of the planetary orbits and the inclinations to the ecliptic.

Laplace then presented his 'second approximation' to the solutions. Confined to elliptical orbit solutions, the variation of their 'elements' were found by taking their first-order total differentials with respect to constants of integration and other parameters (*1* (1799), 504-515). The method was in line with Lagrange's variational techniques, and Laplace made good use of it; but it is worth noting that he gave it a very limited brief.

From mathematical reasoning and astronomical considerations such as these emerged one of the most distinctive features of French mathematical astronomy of this period and later: differential equations of the form of (544.5) with (544.15), and solutions of the form (544.14) with (544.16-17), in both cases with queues of trigonometric terms on the right hand sides. The estimation of size of such terms, especially after integration, was a major question: 'one is never sure that the neglected quantities are very small', Laplace warned in a contemporary paper (*1798c*, 192). A decade later Poisson was to look into this matter, as we shall see in § 6.3.2.

5.4.5 Laplace's principal astronomical concerns

> Laplace had a marvellous feeling for grasping[,] amidst all the inequalities of which the multiplicity forms the principal difficulty of the theory of planetary motions[,] those which must acquire sensible values and those which will always remain inappreciable [...]
>
> Pontécoulant *1835a*, 395

The mathematics just described was basically Eulerian: literally differential equations were formed and solved, and discussed in terms of spatial motions. Laplace ended volume 1 with a 'second method of approximation' (pp. 634-725). He took the first integrals of (544.5) and then formulated expressions involving differentials of the constants of integration now treated as variables. Since these constants were interpretable in astronomical terms, their differentials could indicate variations in value due to perturbation. For example, he showed that c_3 in

(544.1), a coefficient in the equation for the invariable plane, was subject to a variation given by

$$dc_3 = (zR_y - yR_z)\,dt, \tag{545.1}$$

where R was defined in (544.5); similar expressions pertained for dc_1 and dc_2 (p. 639).

Laplace did not merely repeat himself when developing this method; he also used it as a setting for presenting his (and others') earlier results. I have space only for a few major ones, which were of interest to later workers.

Euler and Lagrange studied in detail the secular inequalities of the planets, especially Jupiter and Saturn; and yet there remained discrepancies between theory and observation. Laplace recalled his own solution to the problem in his *Exposition*$_2$ (1799), 197-198:

> Struck by these differences, I examined again this matter, bringing the greatest care to the discussion, and I arrived at the veritable analytical expression for the secular inequality of the mean motion of the planets. On substituting there the numerical values relative to Jupiter and to Saturn, I was surprised to see that they became nul. I suspected that that was not peculiar to those planets, and that if one put this expression, into the simplest form [...] all the terms would cancel each other out. Calculation confirmed this suspicion, and informed me that in general, the mean motions of the planets and their mean distances from the sun, are invariable, at least when one neglects the fourth powers of the eccentricities and of the inclinations of the orbits, and the squares of the perturbing masses, which is more than sufficient for the present needs of astronomy.

Laplace was recalling (for the erudite reader, anyway) part of an intensive and overlapping quantity of work by himself and Lagrange on perturbation theory in papers published between 1775 and 1790 (see Wilson *1980a*, 207-257; and *1985a* on the Saturn/Jupiter inequality). In the *Mécanique céleste* he proceeded by expanding the perturbation function R of (544.6) for two bodies in power series of eccentricities and related small quantities (*1* (1799), 522-531) and then evaluating the perturbations of their orbits (pp. 646-665). These studies showed the importance of nearly rational proportions in the theory; for in the case of the three bodies sun/Saturn (♄)/Jupiter (♃) the mean motions satisfied the relations

$$p := 5n_{♃} - 2n_{♄} = n_{♄}/74 \tag{545.2}$$

(p. 647). Further, since the analysis involved the mean motion of a planet expressed in terms of the double integral $\iint n\,dt\,dR$, then, when the expansion (544.15) was invoked, expressions arose which took the form

$$(K/p^2)\sin(pt + A), \tag{545.3}$$

and thus became 'very sensible' when such small denominators as those afforded by (545.2) were involved, even though they were 'of the third order relative to the eccentricities and inclinations of the orbits' (pp. 647–648). Hence astronomers had to pay special attention to commensurability when studying perturbations, and for many purposes they needed to analyse only terms of the form of (545.3), 'which ought to be the most considerable' (p. 648). Laplace mentioned another example of commensurability a little further on, when asserting the relation

$$n_1 - 3n_2 + 2n_3 = 0 \tag{545.4}$$

between the mean motions of three of the four (then) known satellites of Jupiter, with a similar relation between their mean longitudes (pp. 669–674). His optimistic view of accuracy is well exemplified by (545.4), which for him was not an approximation like (545.3) but a relation which 'is accurately and invariably equal to nothing', even though he immediately allowed for an error term (p. 671).

Laplace was surprisingly quiet about the case $n = 3$ of (544.5): the general three-body problem. Lagrange and Euler shared an *Académie* prize in 1772 on it, and Lagrange's paper *1777c* seems to have given the standard formulation for the time (see Gautier *1817a*, 79–82, 121–123). It appeared in Laplace's work on lunar theory from time to time, including the contemporary papers *1798c* and *1798d*. As the *Mécanique céleste* is our principal current concern, I shall postpone the problem until § 6.4.3; its absence in this volume may be due to the fact that he did not want to commit himself to a method of solution in his "definitive" treatise when the theory was still so incomplete. However, he did denote space to another major problem, to which we now turn.

5.4.6 Laplace (with Lagrange) on the stability of the planetary system

A principal drive behind Laplace's studies of perturbations and inequalities was his desire to prove the stability of the planetary system. This hope, held also by Lagrange, marked a change in the tradition from Newton to Euler, where the possibility of instability was admitted. Even then, the studies of Saturn and Jupiter suggested that the former was slowing down in its orbit while the latter was accelerating, so that a good deal of galactic discomfort might occur before the (very long) period of the associated inequalities could be completed. Stability laid emphasis again on equilibrium, as was noted in § 5.4.2 in Laplace's *Exposition*. Mathematically, it was required to show that the various parameters associated with perturbations always remained bounded. It was in the course of such studies that Laplace made his advances in Saturn/Jupiter theory (see Wilson *1980a*, 248–249) and examined the mean motion of Jupiter's satellites (see Gillispie *1978a*, 325–327); his hope was that all perturbations would show the same periodic, and thus bounded, character. His argument to this end involved an important example of Lagrange's influence upon him.

Lagrange's proof, given in a paper *1778a* published by the *Académie*, started out from Newton's second law cast in Euler's form (525.7) for a Cartesian coordinate system. (However, in proving stability at all, the influence on Lagrange from Newton and Euler was *negative*.) His analysis drew on a transformation of dependent variables into ones now often named after him. Laplace cast them as follows:

$$h := E \sin \omega \quad \text{and} \quad l := E \cos \omega, \tag{546.1}$$

where E was the (variable) eccentricity of the orbit of a planet and ω the longitude of its perihelion; a companion pair of variables related to the inclination of the planet, instead of E, was also used. Laplace saw the value of (546.1); indeed, he helped to publicise it by ending his own *1775a* with a letter that Lagrange (*1775b*) sent him about it. He then used both it and (apparently) Lagrange's *1778a* as the basis of his own analysis of secular inequalities (Laplace *1776a*, arts. 3–5),[1]) and from this study he argued for the stability of the planetary system (*1787a*, art. 13). These pieces of work appeared together in the *Mécanique céleste* (*1* (1799), 583–609). Lagrange already had given a version of his procedure in his own *Méchanique* (*1* (1788), 241–262), in a study of the small oscillations of a system of point masses; he had started out from his equations (524.9) of motion.

For a system $\{P_i\}$ of K planets of masses $\{m_i\}$ Laplace found a system of linear ordinary differential equations, which I write as

$$\dot{h}_1 = \left(\sum_{j=1}^{K} A_{ij}\right) l_i - \left(\sum_{j=1}^{K} B_{ij}\, l_j\right), \quad 1 \leq i \leq K, \quad i + j \ \text{(here and below)}, \tag{546.2}$$

$$-\dot{l}_i = \left(\sum_{j=1}^{K} A_{ij}\right) h_i - \left(\sum_{j=1}^{K} B_{ij}\, h_j\right), \tag{546.3}$$

---

[1]) My nervous '(apparently)' is caused by a difficult of chronology typical of Laplace/Lagrange studies in the late 18th century. Lagrange *1778a* was presented to the *AS* in August 1774, although not published until 1778. Laplace *1776a* was published earlier, but seems to have been presented, in part, to the *AS* in December 1774 (Gillispie *1978a*, 391, 396). But Laplace acknowledged Lagrange's manuscript in arts. 3 and 5 of his paper, and so was presumably working under its conscious influence.

where A_{ij} and B_{ij} were given, via the expansion

$$(a_i^2 - 2a_i a_j \cos \Theta + a_j^2)^{1/2} = \sum_{s=0}^{\infty} b_s^{ij} \cos s\Theta \tag{546.4}$$

of the distance function for the i-th and j-th planets, in various forms, including

$$\begin{aligned} A_{ij} &= -\frac{3n_i a_i^2 m_j a_j b_1^{ij}}{4(a_j^2 - a_i^2)^2}, \\ B_{ij} &= -\frac{3n_i a_i m_j[(a_i^2 + a_j^2) b_1^{ij} + a_i a_j b_0^{ij}]}{2(a_j^2 - a_i^2)^2} \end{aligned} \tag{546.5}$$

(pp. 585-591).[1]) The linearity of the differential equations precluded the need for the method of successive approximations; Laplace just assumed the simple-state solutions

$$h_j = N_j \sin(gt + \beta) \quad \text{and} \quad l_j = N_j \cos(gt + \beta). \tag{546.6}$$

The task centred on showing that the $\{g, \beta\}$ all took real values, and then that the series expansions were bounded. Today this is a standard case-study in latent roots; but for Laplace, living in a time when linear algebra was virtually non-existent, he had to proceed on his mathematical wits. His method (that is, Lagrange's) was to define a quadratic form φ in the variables $\{N_j\}$ by the property that

$$\varphi_{N_j} = 0 \tag{546.7}$$

was the set of linear equations obtained by substituting (546.6) into (546.2) or (546.3) and cancelling off the trigonometric terms: namely,

$$\varphi := \sum_{j=1}^{K} \left[N_j^2 m_j \sqrt{a_j} \left(g - \sum_{i=1}^{K} A_{ij} \right) \right] + 2 \sum_{j=1}^{K} \left[N_j m_j \sqrt{a_j} \left(\sum_{i=1}^{K} B_{ij} N_j \right) \right]. \tag{546.8}$$

[1]) Laplace seems to have advanced beyond Lagrange *1778a*, art. 16 in taking different values for A_{ij} and B_{ij}. On their papers see Wilson *1980a*, 212-240; and Gautier *1817a*, 166-177. Properties of the expansion (546.4), and of more general forms for (distance)^n, were widely studied in the 18th and 19th centuries (see Burkhardt *1908a*, 62-92; and von Zeipel *1912a*).

The dimensional chaos evident in these equations was caused by Laplace's omission of the gravitational constant (mentioned at (544.4)) and also by his decision on p. 547 to put $\mu = 1$. This led to δr being linear with m^2 (pp. 550, 554 and especially p. 560, where h and l were used); this term was then added on p. 570 to a term which had been shown on p. 376 to be within $O(E^2)$ of r. (546.2-3) were basically derived for two planets on p. 572, and explicitly in this form on p. 584.

By a property of homogeneous functions, and (546.7),

$$\sum_{j=1}^{N} N_j \varphi_{N_j} = 2\varphi = 0. \tag{546.9}$$

Now (546.7) yielded just one solution, say q_1, for N_j; substituting it in φ gave

$$\varphi(q_1, N_2, \dots, N_K) = 0. \tag{546.10}$$

Repeating the process produced the form

$$\varphi(q_1, \dots q_{K-1}, N_K) = N_K^2 f(g), \tag{546.11}$$

where f was a Kth-degree polynomial in g which reduced to the characteristic equation

$$f(g) = 0 \tag{546.12}$$

after the final differentiation. The zeros $\{g_j\}$ of f were the latent roots of the problem; working the process backwards furnished the solutions, which I write after (546.6) in the form

$$\left.\begin{matrix} h_j \\ l_j \end{matrix}\right\} = \sum_{i=1}^{K} \left(N_{ij} \begin{matrix} \sin \\ \cos \end{matrix} (g_i t + \beta_i) \right), \tag{546.13}$$

where the constants $\{N_{ij}\}$ and $\{\beta_i\}$ were determinable from the relations obtained by the reversed process and from initial observational data.[1])

---

[1]) In terms of linear algebra, (546.2-3) are

$$\dot{h} = C\, l \quad \text{and} \quad -\dot{l} = C\, h, \tag{1}$$

where the matrix C is defined by

$$C_{ii} := \sum_{j=1}^{K} A_{ij}, \qquad C_{ij} := -B_{ij} \qquad \text{when} \qquad i \neq j. \tag{2}$$

From (1), both h and l obviously satisfy

$$\ddot{p} + C^2 p = 0, \tag{3}$$

and so are susceptible to simple-state analysis and thus the solution (546.13). Unable to find the latent roots from the secular determinant

$$|C^2 - gI_K| = 0, \tag{4}$$

Laplace had in effect to manipulate the quadratic form

$$\Phi := N^T (C^2 - gI_K)\, N \tag{5}$$

to obtain (546.12). From (546.7) the value 0 of Φ will be an extremum, as he noted (assuming a maximum, in fact); but the property is not relevant.

I am not following Laplace's unhelpful notations, where primes are rivalled in ubiquity by round brackets.

In order to deduce the required stability, Laplace argued from (546.13, 1) that

$$\text{if}\quad N_{ij} > 0, \quad \text{then}\quad E_j^2 < \left(\sum_{i=1}^{K} N_{ij}\right)^2, \qquad 1 \le j \le K \tag{546.14}$$

and

$$\tan(\omega_j - g_j t - \beta_j) = \left[\sum_{i=1}^{K} N_{ij} \sin \gamma_{ij}\right] \bigg/ \left[\sum_{i=1}^{K} N_{ij} \cos \gamma_{ij}\right], \tag{546.15}$$

where

$$\gamma_{ij} := (g_i - g_j)\, t + (\beta_i - \beta_j). \tag{546.16}$$

$(546.14)_2$ showed that E_j was bounded, while from (546.15) its positive denominator implied that $\tan(\omega_j - g_j t - \beta_j)$ would remain finite-valued, so that its angle would never reach a right angle (and thus the mean motion of the perihelion of m_j equal $g_j t$). Hence the stability of the planetary system was assured; as the $\{N_{ij}\}$ were small, so were the $\{E_j\}$ and the mean motion always remained uniform (pp. 599–603).

A crucial matter for Laplace was to determine the reality and the inequality of the latent roots $\{g_j\}$ in (546.12). He argued for their reality by multiplying each equation in (546.2) by $m_j \sqrt{a_j}\, h_j$ and in (546.3) by $m_j \sqrt{a_j}\, l_j$, adding, and using (546.1) and known properties (including symmetry) of $\{A_{ij}\}$ and $\{B_{ij}\}$, to obtain

$$\sum_{j=1}^{K} m_j \sqrt{a_j}\, E_j\, dE_j = 0 \quad \text{and thus} \quad \sum_{j=1}^{K} m_j \sqrt{a_j}\, E_j^2 = \text{constant}. \tag{546.17}$$

Previous work showed that, since the planets rotated in the same direction (a feature of astronomy for which neither Newton nor the followers of his theory provided any explanation) the positive values of $\{\sqrt{a_j}\}$ could be used in $(546.17)_2$, which was thus a sum of positive terms (p. 591); and since the sum itself was small, it always was and always will be. Hence all the latent roots were real, for otherwise unbounded exponential terms would replace the trigonometry in (546.13) (pp. 603–607). Further, they were all unequal, since otherwise the corresponding terms in (546.13) would convert to the form $t^p \sin(gt + \beta)$, where again boundedness would be infringed (pp. 607–609). Thus, to Laplace's satisfaction 'The system of the orbits [of the planets $\{p_j\}$] is therefore perfectly stable, as it respects the eccentricities. The ellipticities of the orbits oscillate about their mean values, from which they vary but little, while the transverse axes remain invariable' (p. 609), showing the strict sense of stability with which he worked. A similar analysis of inclinations of the orbits showed the impossibility of flip-over of the system (pp. 612–618).

Laplace's reasoning shows astronomical faith brought in to bolster mathematical arguments which are wanting for a lack of linear algebra (see Hawkins *1975a*, 3–15). He made only partial use (at (546.17)) of symmetry properties associated with the matrices involved in (546.2, 3); he did not know that, even with multiple latent roots, the solutions would still be bounded. (By a counter-example, Clausen *1845a* refuted Laplace's belief that unbounded terms would arise.) Thus he felt compelled to prove the inequality (and reality) of the roots via uncomfortable assumptions such as the positivity of the $\{N_j\}$ in $(546.14)_1$, and mathematically unrigorous claims about the effect on (546.17) of multiple and/or complex roots. Further, the calculation of the latent roots from (546.7) and the coefficients in (546.13) involved calculations which were not only laborious but might be difficult to compute with sufficient numerical accuracy. The astronomical faith made its presence felt in the differential equations (546.2-3) themselves, which were valid only to the first order of masses. Laplace did not discuss this point, but it emerged in a later discussion of secular and periodic inequalities, where he took the differential equations in more extended forms such as

$$\dot{h}_j = \left(\sum_{i=1}^{K} A_{ij}\right) l_j + \left(\sum_{i=1}^{K} B_{ij}\, l_i\right) + Y_j, \qquad 1 \le i \le K, \tag{546.18}$$

where Y_i was a series of perturbation functions (pp. 690-693). He discarded terms of $O(m_j^2)$ and so reduced (546.18) to

$$\dot{h}_j = Y_j. \tag{546.19}$$

However, the discarded terms were those which for Laplace guaranteed stability, so that those left in (546.19) were presumably larger; so what hope for stability now? It is not surprising that Poisson was to look again at the stability problem and its differential equations, as we shall see in § 6.3.2.

5.4.7 Trends in astronomy in the *Mécanique céleste*

It was mentioned in § 5.4.2 that Bowditch's edition did the "proper job" on Laplace's *Mécanique céleste* that was so badly needed; namely, it provided a readable version of the text. Prior to its appearance from the late 1820s on (and indeed afterwards), various simpler versions were written by many authors, both in France and elsewhere. First off the mark, predictably, was Biot, who helped his career along by publicising volumes 1 and 2 of the treatise. He reviewed them for the *Société Philomatique* in *1799a*, and also edited and summarised parts of their contents in a three-part article *1799-1800a* which took occupied over 80 pages of the *Magasin encyclopédique* (and also appeared in booklet form). Then he published a non-technical account *1801d* of 140 pages in an edition of the *Séances des écoles normales*. (Of course he will not have taught the material at the *Ecole Normale*, for it closed in 1795, while he was still a student at the *Ecole Polytech-*

nique.) He gave attention to the other volumes as they appeared, and then began to publish his own *Traité d'astronomie* from the mid 1800s on, as we shall note in § 6.4.1.

The thrust of Laplace's work, and its influence on Biot, is of course to the theoretical. Laplace left the same stamp on the *Bureau des Longitudes*, with consequences over which Delambre sorrowed in an autobiographical essay.[1]) His remarks make an interesting contrast with those of the theoretician Poisson at the head of § 5.4.4:

> [...] never should one put a geometer at the head of an observatory; he will neglect all the observations except those needed for his formulas. [...] One can reproach him with the fact that in more than 20 years of existence the *Bureau des Longitudes* has not determined the position of a single star, or undertaken the preparation of the smallest catalogue.

5.5 On planetary mechanics

5.5.1 Introduction: the *Dépôt Général de la Guerre*

'Planetary science' was not a term used in the early 19th century; indeed, even 'planetary physics' only gained some currency in recent decades (see Brush *1978a*). But 'planetary mechanics' usefully describes a group of topics whose late-18th-century developments will be described in this section: topics where some aspect of a planet (usually the earth) as a whole is considered, in contrast to both the corporeal mechanics discussed in § 5.2-3, and to the celestial mechanics of § 5.4.

An important part of this activity dealt with geodesy and cartography. A corps of *ingénieurs géographes* had been formed in France in the mid 17th century, belonging to the army. During the 1760s it became part of the *Dépôt (Général) de la Guerre*, which continued until the Revolution, when it was suppressed in 1791. Re-established in the following year, it received several reforms until the late 1800s. For our purposes it may be regarded as an organisation analogous to the *Bureau des Longitudes* (§ 5.4.1), and there were various connections between

[1]) This manuscript is kept in *BI*, ms. 2041, fols. 408-420. The appropriate part was published in Bigourdan *Bureau*, pt. 4 (1931), A 106-A 108: I quote from the translation in Wilson *1980a*, 284.

them; for example, *Bureau* members Delambre and Laplace (and also Callet of § 3.4.3) belonged to a group of *savants* to guide the *Dépôt*'s activities.[1])

One of the component institutions of the *Dépôt Général* was the *Dépôt des Cartes et Plans*, which was housed in the same buildings as the *Ecole des Ponts et Chaussées* in the reorganisations of 1794 and 1795. Connection with de Prony became still closer in 1795 when (as was mentioned in § 2.4.5) he was made director of the little *Ecole de Géographes*, which was founded at Meudon near Paris. His directorship lapsed with its closure in 1803, but it was re-opened a few years later as part of the *Dépôt Général*. These moves had an effect at the *Ecole Polytechnique*, where in 1811 a course in construction was replaced by one on geodesy (including elementary astronomy).[2])

de Prony was also a member of a 20-man commission which planned to regularise the preparation of maps and charts: their report was published as Sanson *1803a* in the *Mémorial topographique et militaire*, a journal launched by the *Dépôt Général* in 1802. From the point of view of this study, map-making is a particularly important branch of mathematics: it occurs in § 5.5.3 below and again in the closing § 5.5.8, which is devoted to Puissant, the member of the *Dépôt* most closely examined in these pages.

The main guide to this section is again Laplace's *Mécanique céleste*; but this time it is volume 2. It began with Book 3, 'On the figures of the heavenly bodies' (pp. 1-525), largely concerned with the theory of spheroids in general, and the earth in particular; continued with Book 4, 'On the oscillations of the sea and atmosphere' (pp. 526-801); and finished with Book 5, 'On the motions of the heavenly bodies about their own centres of gravity' (pp. 802-990). We shall do the same.

5.5.2 Potential theory and Legendre functions

Neither of these topics has now a special association with planetary mechanics; but their early histories, in the 18th century, showed strong connections with a problem then called 'attraction of spheroids': to express mathematically the

[1]) On the history of the *Dépôt Général* up to 1800 see Anonymous *1802a*. Berthaut gives a rather chaotic account of our period in his *1902a*, esp. *1*, 126-248; a briefer but perhaps clearer account appears in his *Maps, 1* (1898), 71-82.

[2]) On the teaching at the *Ecole de Géographes*, which has been almost entirely forgotten, see the synoptic table of its teaching in *EP, CoP, 1*, fol. 52, and Sanson's report of 1802 at fols. 148v-151. We recall from § 5.1.3 that Francoeur was also on its staff. On the changes at the *EP*, see the criticisms by J. Sganzin (the teacher of construction) of 3.12.1811 in *EP*, III 3a; *EP, CoI, 4*, meeting of 7.6.1811; and *EP 1801* prog, 72-74 and *1810* prog, 46-48. Konvitz *1987a* has valuable information on cartography education.

Newtonian theory of attractions between spheroidal objects such as planets. These developments have been fairly well studied;[1]) I restrict myself to some comments on the potential function and the Legendre functions (and so omit, for example, the treatments via elliptic integrals in papers such as Lagrange *1798c*).

Laplace contributed substantially to both these theories; but great erudition is demanded of the reader (§ 5.4.4) to determine the extent to which Lagrange is present in the potential theory, and Legendre in the treatment of Legendre functions, that Laplace laid out in the *Mécanique céleste*. Lagrange had some velocity potential theory implicit in the hydrodynamics of his *1762c*, arts. 42–43 (§ 10.3.1), and he used the potential function V for astronomy in *1775a*. The perturbation function R of (544.6), which he introduced in *1776c*, art. 10, was an off-shoot, as he made clear in various places; for example, when he remarked in *1779b*, art. 1 that the components of attractive force of a mass m_j on the other members of a system $\{m_j\}$ of masses were given by

$$(V_{x_i}, V_{y_i}, V_{z_i})/m_i, \quad \text{where} \quad V := \sum_{j \neq k} m_j m_k / r_{jk}. \tag{552.1}$$

In these early treatments of potential theory, the function was defined for a (finite?) sum of discrete masses; but of course the full theory needed to be developed for continuous masses, and the work of Lagrange, Laplace and Legendre was involved in handling the corresponding integrals. Laplace briefly treated the potential function for point-masses and continuous bodies in volume 1 of the *Mécanique céleste* (pp. 274–303), and began volume 2 with a detailed account of expressions for the attractive force (552.1) of an ellipsoid on internal and external (but not surface) points in various coordinate systems. Among the various properties the most famous is 'Laplace's equation' (known before Laplace):

$$V_{xx} + V_{yy} + V_{zz} = 0\,, \tag{552.2}$$

and also transformed into other coordinate systems (pp. 63–71: compare *1*, 280–288). He committed a serious oversight in not noticing that the equation applied only for attraction to external points; in § 6.5.1 we shall see Poisson sieze on this point.

When cast in spherical polar coordinates (r, ω, Θ), (552.2) took the form

$$((1-\mu^2)V_\mu)_\mu + V_{\omega\omega}/(1-\mu^2) + r(rV)_{rr} = 0\,, \quad \text{where} \quad \mu := \cos\Theta, \tag{552.3}$$

[1]) Todhunter *1873a* covers the history from Newton until about the 1830s; the end of the first volume and the beginning of the second are relevant here. His account of Legendre functions is more satisfactory than that of potential theory, which is treated rather better in Bacharach *1883a*, and Sologub *1975a*, ch. 1. Wangerin *1904a* gives a useful survey of Legendre and other functions, though it is rather brief on the early history. Various textbooks on both topics have brief historical or expository versions of Laplace's work: Pratt *1871a* is a good example of the latter kind.

from which Legendre's equation in μ would come by separation of variables. (If ω were absent and r and Θ separated, elliptic integrals in r arose.) Laplace continued to think in terms of power series, and assumed for an external/internal point the expansion

$$V = \sum_{j=1}^{\infty} U_j/r^j \quad \text{or} \quad V = \sum_{j=1}^{\infty} v_j r^j \tag{552.4}$$

and examined U_j or v_j as a 'rational and integral function of μ, $\sqrt{1-\mu^2}\sin\omega$, $\sqrt{1-\mu^2}\cos\omega$, depending on the nature of the spheroid' (p. 74). Thus the theory of Legendre functions (as we see it) was embedded in the surface harmonics $\{P_j(\cos\Theta)\,{}^{\sin}_{\cos}\,\omega\}$. In addition, the special condition of spheroidicity was borne in mind—which was the context in which Legendre functions had been developed in the late 18th century.

Laplace's use of the functions in the *Mécanique céleste* led Whewell *1830a* to introduce the name 'Laplace's coefficients', a term which lasted until the late 19th century; but then terms like 'spherical functions' and 'zonal harmonics' came in, and also 'Legendre functions'. If an individual's name must be attached to the function, Legendre's is perhaps the best choice, for in his own papers, especially *1785a* and *1787a* (with its successor *1793a*), he succeeded more than Laplace in isolating the function itself from its surrounding mathematics. Both men, naturally proceeding from equations such as (552.3), defined $P_n(x)$ as the coefficient function in the generating function

$$(1 - 2\alpha x + \alpha^2)^{-1/2} = \sum_{j=0}^{\infty} P_j(x)\alpha^j \tag{552.5}$$

(compare (546.4)). Most of the now standard power-series and (real-variable) integral forms for $P_n(x)$, together with recurrence relations and orthogonality, were known for integer values of n; theory was much more scrappy for non-integral n, for associated functions $P_n^m(x)$, and for second-kind functions. Orthogonality and uniqueness of expansion were known properties; but convergence was usually taken for granted, in the spirit described in § 3.4.

Legendre started out from MacLaurin's theorem (*1742a*, arts. 648-650) that the attractions of two confocal ellipsoids of revolution at an external point lying along a principal axis were in the proportion of their masses, of which an "analytical" proof for all ellipsoids had been given in Lagrange *1777b*. Legendre gave a proof himself (*1785a*, 413-415) and then introduced Legendre functions to determine, as a power series, the attraction of an ellipsoid of revolution to any (exterior) point (pp. 418-420). He submitted a version of this paper to the *Académie* in January 1783, when he was 30 years old. Laplace, three years his senior and newly promoted to *associé*, presented in March a praising report (recently published, with valuable commentary, in Gillispie *1980a*, 262-264); within days Legendre had been elected to Laplace's place as *adjoint* in the *Académie*. From then on their work was closely intertwined. Laplace *1785b* reduced to single integrals Legendre's triple-integral formulae for the attraction of a body to an external point, and in *1784a*, 76-86 he extended Legendre's result in attraction to all ellipsoids. All these results and procedures, but no references, occurred in the *Mécanique céleste* (*2* (1799), 1-51).

Laplace treated spheroidicity by a technique akin to his method of successive approximations (§ 3.3.5); he put the radius in the form

$$r = a(1 + \alpha f(\mu, \omega)) \qquad (552.6)$$

(where a was the semi-major axis) in which the small quantity α could be ignored from α^2 up (p. 94). The potential function, including its Legendre-function expansions, was modified to include terms linear in α (pp. 91-153).

The other major extension to be noted here was the assumption that the spheroid was covered by a fluid stratum of variable density. This was the best practicable approximation to the presence of the sea: Laplace's theory of sea-flow is the subject of § 5.5.6, while some general results will be described now.

Laplace began with a study of a rotating homogeneous fluid body. He assumed, as an 'hypothesis', that the mass was an ellipsoid of revolution rotating about its axis of revolution (p. 210), and argued that the equation of equilibrium could be expressed in the form of the equation of an ellipsoid (p. 214). His argument rested on the assumption that $\lambda^2 > -1$, where λ was given by

$$\lambda(9 + 2q\lambda^2) = 3(3 + \lambda^2)\tan^{-1}\lambda, \quad q := 3g/4\pi\varrho, \quad \varrho \text{ density} \qquad (552.7)$$

(p. 214). He put $(552.7)_1$ into his calculations and used data on the rotation of the earth to produce estimates on the allowable values of λ and q (p. 232). This work made his hypothesis 'legitimate' (p. 210), presumably a category lower in truth-status than 'veritable' (§ 5.4.2), in view of the assumption about admissible values of λ and the inadequacy of current knowledge in the theory of equations (§ 4.6) to determine analytically the roots of $(552.7)_1$.

Laplace next considered a homogeneous fluid stratum covering a solid spheroid (pp. 241-290). He maintained the approximation (552.6), but confessed that 'Notwithstanding this simplification, the investigation of the figures of the planets is very complicated' (p. 241). His basic move was the same as Lagrange's at (522.2); to assume that in the equation of fluid equilibrium

$$d\Pi/\varrho = \sum_j F_j df_j, \qquad (552.8)$$

where Π was pressure and $\{F_j\}$ the system of applied forces acting in directions $\{f_j\}$, the right hand side was an exact differential $d\varphi$, so that

$$d\varphi := \sum_j F_j df_j = 0 = d\Pi/\varrho. \tag{552.9}$$

The integral of this equation,

$$\text{constant} = \int \sum_j F_j df_j, \tag{552.10}$$

gave the equation of a 'level surface' (where he used the term introduced in MacLaurin *1742a*, art. 640), including the surface of the stratum itself (pp. 241–244).

The rest of this study was based on expressing (552.10) in forms suitable for a spheroid; (552.6) allowed Laplace to limit his expansions to terms of the order of α. I write one of them in the form

$$\begin{aligned}\text{constant} = \frac{4\pi a^2}{3} - \frac{8\alpha\pi a^2}{3}\sum_{k=2}^{\infty}\left[\frac{k-1}{2k+1}P_k(\cos\Theta)\cos(\omega+K)\right]\\ + \frac{a^2 g}{3}(1 - P_2(\cos\Theta))\\ + \pi a^2 \sum_{k=3}^{\infty}\left[\frac{8(k-1)}{3(2k+1)}\sum_j\left(\frac{m_j}{s_j^k}P_k(\cos\Theta_j)\cos(\omega_j+K_j)\right)\right]\end{aligned} \tag{552.11}$$

(p. 255), where the first two terms provided terms for the centrifugal force (p. 97) and the other terms attractions due to (point-)masses $\{m_j\}$ located at polar coordinates $\{(s_j, \Theta_j, \omega_j)\}$ from the centre of gravity of the spheroid (p. 253) (the large values of $\{s_j\}$ provided the smallness required by (552.6), and the absence of the gravitational constant (furnished) the dimensional chaos); K and $\{K_j\}$ were constants. (552.11) could be related to their generating functions involving perturbation terms (pp. 255–257).

Laplace's procedure for an heterogeneous fluid stratum (pp. 296–350) also started from (552.8); but, since the form of the density function ϱ was not known, he had to clutter his analysis with integrals whose integrands contained ϱ, and so he could not proceed very far. He assumed ('We shall observe' was the optimistic turn of phrase on p. 296) from equilibrium conditions that the fluid lay in sub-strata of densities decreasing with distance from the centre of gravity, a property expressed rather vaguely by the power series

$$\text{`}\varrho = \beta - \gamma . a^n - \&c\text{'}. \tag{552.12}$$

This move helped him to convert some of the differential equations into types studied already and thus discover a few properties of the fluid, although he still often left ϱ unaltered inside integrands. For example, on p. 308 he found a second-order ordinary differential equation relating (essentially) the eccentricity of the ellipsoid of revolution to ϱ; but 'This equation is not integrable by any known method, except in some particular hypotheses, relative to the density ϱ'. He was to return to this topic in the late 1810s (§ 12.4.2). He also found that gravity varies in proportion to the square of the sine of the latitude' (p. 350); erudite readers know this law as Clairaut's (*Earth*$_1$ (1743), 248–250).

Laplace concluded his discussion with a consideration of terms of order α^2 and higher (pp. 351–357). He replaced (552.6) by the next approximation

$$r = a(1 + \alpha f(\mu, \omega) + \alpha^2 f_1(\mu, \omega)) \tag{552.13}$$

and sketched out the modifications to previous analysis to allow for the new term. '[T]here is no other difficulty than that arising from the length of the calculations', he decided, and concluded that 'we can always find a series of figures, which, by substitution in the equation of equilibrium, will leave remainders that become successively smaller, and will finally be less than any assignable quantity' (p. 357). Thus he bore upon the problem the same line of thought as underlay his method of successive approximations (§ 3.3.5).

After conditions such as (552.11), and guesses such as (552.12) and the tendency towards notional theoretical astronomy (in the sense of § 132.3) that (552.13) and its successors entail, comparison of this theory with observations could launch a perilous course. Nevertheless, Laplace devoted a chapter (pp. 358–491) to the matter. Many of his considerations were based on determination of the shape of the earth by pendulum measurements, a topic to be taken up in § 6.5.2; others concerned the determination of the meridianal arc, to which we now turn.

5.5.3 Cartography: a choice of projections

The problem of projecting the details of a surface in space onto a plane is of considerable mathematical interest. Euler wrote an important paper, *1778a*, which he began by showing that a spherical surface was not congruent with the plane. The argument was a clever use of *reductio*. If congruence were possible, then there would be coordinates (t, u) on the sphere dependent on (x, y) on the plane for which

$$dx^2 + dy^2 = du^2 + \cos^2 u \, dt^2 \tag{553.1}$$

was an identity; but he showed that φ, as defined from

$$\tan \varphi := y_u / x_u, \tag{553.2}$$

was both dependent on, and independent of, u (arts. 1-9). He then went on to study some (for him, 'all') of the defining properties from which cartography could proceed.

A widely used projection for charts was Mercator's, in which the parallels were taken into parallel lines and the meridian made equally-spaced intersections perpendicular to the projection of the equator. Its conformality (to use a term introduced by Euler's colleague Schubert in *1789a* and popularised later by Gauss) had the important consequence that a curve which cut the meridian at a constant angle (a 'loxodromic curve') projected into a straight line (the 'rhumb line'), thus simplifying for navigators the taking of bearings even though the course sailed was not geodesic. Euler found the equation of the loxodromic curve (*1778a*, arts. 18-20), and also the (already known) expression

$$y = \int \sec u \, du = \ln \tan \tfrac{1}{2} (\tau + u) \tag{553.3}$$

for the distance of the parallel with latitude u from the equator (the second art. 16).

Conformality (or, at least, isogonality) can itself be taken as the defining property of a projection, as Euler discussed in arts. 24-51. His analyses are a fine example of his methods of solving differential equations, described in § 3.3.4; the search for exact differentials; the determination of 'particular solutions' (Table 337.1), here by separation of variables; and especially the 'general solution', which he found by complex-variable methods, which were then barely known. He explored these forms of solution further in a sequel paper *1778b*, starting from the forms

$$\left.\begin{matrix} x \\ iy \end{matrix}\right\} = F(e^{it} \cot u) \pm F(e^{-it} \cot u) \,. \tag{553.4}$$

Lagrange had been alerted to the mathematics of cartography by Lambert's major study *1772a*, and after Euler's *1778b* (in which Lambert was not mentioned) he wrote an extensive paper on certain forms of conformal projection also using complex variables. In his case he progressed from a version of (553.1) to the differential equation

$$dx + idy = (\alpha + i\beta)\,(du \pm idt) \tag{553.5}$$

and followed d'Alembert *1747a* to obtain the functional solution

$$\left.\begin{matrix} x \\ iy \end{matrix}\right\} = f(u + it) \pm F(u - it) \tag{553.6}$$

(Lagrange *1781b*, arts. 1-3).

Conformal mappings, complex variables; we seem to be in the area of the Cauchy-Riemann equations. Indeed, we shall see at (1023.16) that Euler was to obtain these equations, as a mathematical result, soon after his work on cartography. The closest approach here seems to occur in the last part of his first paper, where he considered another defining property: areal invariance (or 'equivalence'). This property was popular in the design of maps, and since the perpendicularity of parallel and meridian would be a great help (though not a necessary condition) in securing it, Euler stipulated that

$$y_t/x_t = -x_u/y_u \tag{553.7}$$

(*1778b*, art. 52). The Cauchy-Riemann equations, conjugate functions, potential functions satisfying Laplace's equation in two variables, are within a whisker; but they were not Euler's concern here. He explored the relationship between (x, y) and (t, u) given (553.7) and equivalence, and found interesting things; for example, in art. 53 a pair of differential equations inaccessible to general solution because of the intervening presence of the non-linear equation

$$\varphi_{tt} + (1/\varphi)_{vv} = 0, \quad \text{where} \quad v := \sin u. \tag{553.8}$$

For their national map the Russian government was using Delisle's projection, where two parallels were mapped into concentric arcs of true lengths and meridians taken to radii of these arcs; so Euler wrote a paper *1778c* on this procedure also. Meanwhile, the map of France was being prepared following the Cassini projection, a modified cylindrical projection, where a chosen central meridian was rectified and geodesic perpendiculars to it from all points plotted in true length, with the Paris *Observatoire* as the origin of coordinates. All parallels and non-central meridians went into curved lines, the meridians being too long. Its mathematics, mostly spherical trigonometry, caused some difficulties (see Puissant *1812a*). During the early 19th century a new map of France was being planned as the old one was being finished; and the *Dépôt Général de la Guerre* changed loyalty from Cassini to the Bonne projection, similar in type to the Delisle but working from one meridian in true length rather than two. An equivalent projection, it was a suitable choice for a country with small latitudinal extent; we shall note some features of its preparation in § 6.4.9.[1]) Among other projections used in France at that time, de Prony employed a central projection for the cad-

[1]) Although there are many histories of cartography, many avoid the sordid mathematical details. Lacroix *Maps*$_1$ (1802) is useful but very brief, as are Wangerin's notes in his editions of classic 18th-century papers (Euler *1898a'*, Lagrange *1894a'* and Lambert *1894a'*). For surveys of French cartography, see Berthaut *Maps*, *1* (1898) and Konvitz *1987a*. A good review of 19th-century projections is supplied in Bourgeois and Furtwängler *1909a*, and historical information (including a synoptic table) in d'Avezac *1863a*, arts. 23-30.

In a special case of the equator as chosen parallel, the Bonne projection is known as the 'Sanson projection'; but it is named after a 17th-century French cartographer, not the *Directeur* of the *Dépôt Général de la Guerre* in the 1800s. Bonne and Delisle were prominent 18th-century cartographers.

astral survey in the 1790s, where the sphere was projected from its centre onto a tangent plane. This projection gave good representations of both geodesics and areas—preferable to the distortions arising from Lambert's cylindrical equivalent projection, where the sphere was unfolded onto an enveloping cylinder and both parallels and meridians were projected into straight lines. de Prony carried out this survey as the *Directeur* of the *Bureau du Cadastre*, for which the large tables were being prepared (§ 3.5.4); it was also linked with the *Ecole de Géographes* which he directed from 1795 to 1803. In this case, he ran his projects together.[1])

5.5.4 The shape of the earth: the metre and the meridian

The mathematics of cartography just described is based on mapping a sphere onto a plane; it becomes much more complicated when the spheroidicity of the earth is considered. I now note two related projects of the 1790s concerned with the shape of the earth and with cartography: the definition of the metre, and the determination of an arc of a meridian.

The reform of weights and measures during the 1790s has been extensively studied by historians;[2]) I am concerned only with the metre. Under the stimulus of Prieur, who was now a member, the *Assemblée Nationale* instructed the *Académie* to study the reform in 1790; by October a committee including Borda, Lagrange and Condorcet had decided in favour of decimalisation. The following March a fresh committee, of which Laplace and Monge were also members, preferred to determine the unit of length by measuring the length of the meridian passing between Dunkerque and Mont Jouy, near Barcelona (and through Paris, of course) rather than using a seconds pendulum; it assigned the surveying work to Delambre and P.F. Méchain. The committee also reported on the progress made since 1788 by Cassini IV, Legendre and Méchain on the triangulation survey between the Paris and Greenwich observatories (see Legendre *1789c* and *1791b*: the committee's reports are listed as *Académie des Sciences 1791a*). Arbogast, as a member of the *Convention Nationale* (which had replaced the *Assem-*

---

[1]) Files on the *Ecole des Ingénieurs Géographes* and the *Bureau* are held at *AN*, F^{14} 2146 and F^{17} 1393. de Prony's projection is briefly noted in Lacroix *Maps*$_2$ (1824), 14.

[2]) Bigourdan *1901a*, 12–199, 424–435 remains a standard source; a useful introduction in English is provided by Hallock and Wade *1906a*, ch. 2. Valuable information on various aspects is furnished in Wolf *1882a*, Birembaut *1959a* and Crosland (in Bugge *1969a*, ch. 7), while Hellman *1936a* clarifies the role of Legendre. Several reports are reprinted in Mascart *1919a*, 670–680, 702–704, and the story is summarised on pp. 488–533. Crosland *1972a* provides useful background on the 18th-century conception of measurement.

blée), presented there a decree *1793a* on the reforms, together with a report prepared by Borda, Lagrange and Monge.

But from then on politics hampered progress. Prieur suspended Delambre and others at the end of 1793 from the *Commission Temporaire* which was in charge of the reforms. An *Agence Temporaire* took over in 1795; Delambre and Méchain were instructed to carry on with the survey under the aegis of the *Dépôt Général de la Guerre*, and Prieur published a survey *1795b* of the proposed new systems; but the *Agence* ceased to exist in February 1796. Delambre and Méchain managed to carry on, but completion of their measurements delayed until December 1798 an international conference where the *Commission des Poids et Mesures* checked all the work and determined the new standards. Delambre wrote or edited the material connected with the survey and the conference in his three-volume *Base du système métrique décimal*, published between 1806 and 1810, around his 60th year.[1])

The project was full of political, mathematical and technical complications: I mention only certain principal features of the mathematics. Legendre had already described some in his paper *1789c* on the triangulation between the Paris and Greenwich observatories. Firstly, the stations of the triangulation would be at different altitudes, so that the azimuthal angles would need to be converted to the horizontal; he showed that if two distant points took declinations α and β with respect to the point of observation and a sweep angle γ between them, then the correction $+x$ to γ was given by

$$x = \tfrac{1}{4}(\alpha + \beta)^2 \tan^2 \tfrac{1}{2}\gamma - \tfrac{1}{4}(\alpha - \beta)^2 \cot^2 \tfrac{1}{2}\gamma + O(\alpha^3). \qquad (554.1)$$

Secondly, the excess ω of the sum of these angles $\{\gamma + x\}$ in a spherical triangle of area T over π radians was given by

$$\omega = T/r^2, \qquad (554.2)$$

where r was the radius of the sphere on which the triangle was cast. Thirdly, this spherical triangle had to be mapped onto the plane; and for this Legendre provided a theorem which is now well known:

Theorem 554.1 To a small triangle $\triangle$ABC on the sphere there corresponds a

---

[1]) This work is listed as Delambre *Metrics*: see also his posthumous *1912a*. Among relevant manuscripts, see the notes of around 1794 in *ENPC*, ms. 726; the observations and correspondence in *OB*, ms. E2, 1-21; the coloured plates of 1810 for *Metrics* by Collin in *BI*, ms. 3807; Delambre's letters of 22.4.1793 to Lalande, and of 12 *vendémiaire an* 12 (5.10.1803) to (Sanson), the *Directeur* of the *Dépôt Général de la Guerre*, both in *Smith*; and the manuscripts of Lagrange in *BI*, ms. 910, fols. 2-52. An eyewitness account of the conference is provided in Bugge *1969a*, ch. 7; on the importance of the conference as an international gathering see Crosland *1969a*.

planar triangle ΔA′B′C′ with sides of the same length and angles given by

$$A' := A - \tfrac{1}{3}\,\omega, \quad B' := B - \tfrac{1}{3}\,\omega, \quad C' := C - \tfrac{1}{3}\,\omega. \tag{554.3}$$

The proofs of these results were not difficult, as he showed in a later paper (*1799a*, 11–14).

This paper was published as the beginning of a short book *1799a* by Delambre, in which was described in detail the mathematical formulae used in the determination of the meridian from Dunkerque to Barcelona. The procedure was to measure an initial base length along the meridian as accurately as possible, and to prolong the base by means of a sequence of observations from station to station near to the meridian with the measured angles modified by the procedures just outlined (see Figure 554.1). Legendre also furnished equations relating the coordinates of a station and of its perpendicular to the meridian (L and X in the Figure, for example: *1799a*, 14–16).

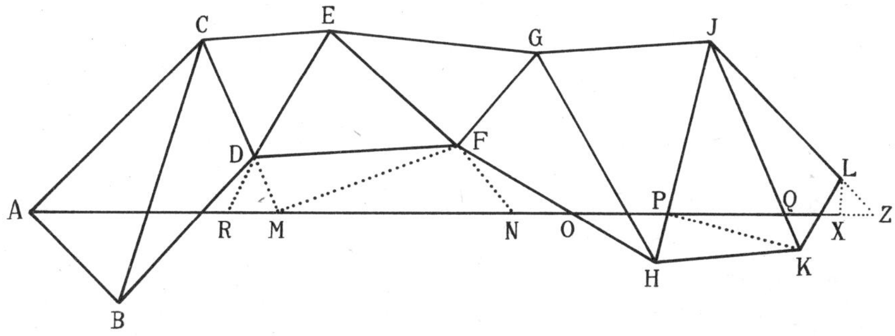

Figure 554.1 Layout of a chain of stations A, B, C, ..., L to determine the length of the meridian AMNX (Legendre *1799a*, 3)
The continuous lines indicate measured lines between stations. The dotted lines mark extrapolations of these lines, or additional lines, introduced for purposes of the calculations. The actual triangulation used in the Méchain-Delambre survey is shown in Delambre *Metrics, 3* (1810), Plates 2–6.

This calculation involved the principal desimplification of the theory not mentioned so far: the spheroidicity of the earth. Legendre drew on his work on potential theory and attractions (§ 5.5.2). He recalled from *1793a*, 394, 418 his formulae for the radial distance $v(\psi)$ from the centre of the earth to a point with latitude $(\tau - \psi)$:

$$v(\psi) = b(1 + m\sin^2\psi + Km^2\sin^4\psi), \tag{554.4}$$

where b was the length $v(0)$ of the polar semi-axis, m a small quantity of the order of flattening of the poles, and K a finite constant to be determined. He expressed the flattening factor α via the equation

$$b = a(1+\alpha), \quad \text{where} \quad \alpha > 0 \text{ and small} \tag{554.5}$$

and a was the equatorial semi-axis $v(\tau)$. Setting $\psi = \tau$ in (554.4) yielded

$$\alpha = m + Km^2. \tag{554.6}$$

Working to $O(\alpha^2)$, Legendre showed that if $K = \frac{3}{2}$, then the meridian would lie along an ellipse; otherwise the shape would be close to it. By straightforward manipulation of the equation

$$v^2 = a^2b^2/(a^2\cos^2\psi + b^2\sin^2\psi), \tag{554.7}$$

he found for the length of the meridian between latitudes L and L' the expression

$$f(L, L') = M\left[\frac{2(L'-L)}{\pi} - \frac{3}{4\pi}(2m + (2K-1)m^2)(\sin 2L' - \sin 2L) + \frac{15\,m^2}{16\pi}(2-K)(\sin 4L' - \sin 4L)\right] + O(m^3), \tag{554.8}$$

where M was one-quarter of the meridian (when $L' = \tau$ and $L = 0$):

$$M = \frac{b\pi}{16}(8 + 4m + (3+K)m^2) + O(m^3). \tag{554.9}$$

For Legendre the work centred on (554.8) with (554.9); making approximations to (554.8) (for example, neglecting its last term), calculating the value of $f(L, L')$ for successive stations on the triangulation, and so on (Legendre *1799a*, 6-11: (554.5) was mis-stated).

Delambre's essay was largely taken up with these and other questions. He began with an extra, special, difficulty: he had to convert between the normal 360° angular division and the decimal 400° division of the circle (100° per quadrant, 100 minutes per degree, and so on) which had been introduced. (However, since the determination of the metre was one of the tasks at hand, he used the *toise* as his unit of length.) Among the various factors with which he had to deal, there was the accuracy of the measurements themselves, errors in the lenses of the sighting telescopes (pp. 18–20) and the altitude of the sextant (pp. 52–56), and atmospheric refraction (pp. 105-111: on this see § 7.4.1).

The survey team used the 'repeating circle', an instrument which had been developed by Borda in the 1760s and 1770s and proved during the 1780s in studies of the positions of the Paris and Greenwich observatories. It was a form of theodolite with two sighting telescopes (just mentioned) on which successive measurements could be recorded, thus avoiding errors from intermediate readings in the sequence. It excited much interest: Borda himself wrote a short book *Circle* on its use, which enjoyed four printings between 1787 and 1816; Delambre described its use on the meridian project (*Metrics*, *2* (1807), 160–169: see also Mascart *1919a*, 270–290, 488–489); Biot *1808a* appraised its advantages over other instruments; and, later, de Prony *1822b* and Puissant *1823a* published pamphlets outlining the mathematics that attended its use.

A major part of the project was calculating numerically the various angles and lengths involved in the survey. This aspect may have motivated Delambre to find a formula different from Legendre's (554.8) for the length of the meridian. He calculated the lengths of the sides of the rectilinear triangle with vertices at the two ends A and B of the (elliptic) meridian and at the intersection M of the polar axis and the normal at A, and from these sides he could obtain $\angle$AMB and thus the arc AB desired (Delambre *1799a*, 77–83). The formulae were exceedingly complicated, impossible to summarise here, but they contained many products and quotients of (usually) trigonometric terms and so were amenable to calculation by logarithms—perhaps using the tables prepared by Borda which he was just completing for publication (§ 3.5.4). Delambre criticised Legendre's procedure on the grounds that checks would be rarely possible and difficult to execute, so that an early error could be compounded without control as calculations along the triangulation proceeded (p. 85). He also felt that Legendre's (554.1) could admit unacceptably large errors (pp. 36–39).

Legendre wrote a brief reply *1799b* as an appendix to Delambre's book. He defended his (554.1) on the grounds that the assumed smallness had been legitimate for Delambre's project, and rejected the assertion that his method lacked simple checks, indicating some obvious examples on his Figure 554.1. Indeed, with regard to lack of checks, it is Delambre (and Méchain) who are suspect, for they measured their initial base-lines only once, which surprised Bugge (*1969a*, 205).

Legendre let Delambre's equations off rather lightly, merely suggesting an alternative way of calculating with spherical triangles which contained very small angles, and offering a simpler way to calculate the errors due to a small error in an angular measurement. He could have been worried by the number of calculations that Delambre's required (especially those with the tiny quantity $\sin 1''$ in the denominator), and surprised both by Delambre's assumption that the meridian was elliptic and also by the effective identification of the arc with the circular arc AB (because of the use of the *rectilinear* $\triangle$ABC) rather than the great circle which he himself had determined in (554.7). Nevertheless, similar kinds of formulae were used in the survey, and written up in Delambre's *Metrics* (see especially *1* (1806), 97–166; and *3* (1810), 1–140 *passim*). Meanwhile in February 1806 Legendre read a sequel paper to his *1799a* at a meeting of the *classe* (*PV*, *3*, 317), which was quickly published in the *Mémoires* as his *1806a*. Here he extended Theorem 554.1 to nearly spherical spheroids of revolution, and determined triangles of certain kinds (for example, with two meridians and the geodesic joining given points).

Méchain and Delambre used their survey and Delambre's methods to calculate the length of the meridian between Dunkerque and Barcelona, and the latitudes of these two places (which is of the order of 9½°), and thence to compute the length of one quarter of the meridian. Then the metre, defined as 10^{-7} of this length by the 1793 temporary commission, was determined from this computation at the international conference of 1798–1799. Despite the confidence and mutual congratulations that were offered at the time—as in, for example, van Swinden's gushing review *1800a* of the Delambre/Legendre volume (a review which appropriately (§ 5.4.7) was republished as an appendix to Biot's summary *1799–1800a* of the *Mécanique céleste*)—it is no surprise that the calculation was to be criticised by Puissant, then a young member of the *Dépôt Général de la Guerre* but already on to the scent of difficulties in his study *1803a* of Delambre's and Legendre's work. Soon he was to raise the standards of study in French geodesy, as we shall see in § 6.4.9.

Another calculation was made: the measure α in (554.6) of the flattening of the poles. This was the central question in the problem of the shape of the earth—a problem which brings us back to Laplace.

5.5.5 The shape of the earth: Laplace's error theory

> Everybody believes firmly in [the law of errors,] because the mathematicians imagine that it is a fact of observation, and the observers that it is a theorem of mathematics.
>
> J. G. Lippmann, quoted (without attribution) in Poincaré *1892a*, v

It may be rather difficult today to realise why the problem of the shape of the earth attracted *so* much attention in the 18th century. Certainly the vindication in the 1730s of the prediction from Newton's theory of the oblateness of the spheroid was an important test for the continued triumph of the theory itself; and details of the shape would be important for cartography and aspects of astronomy. But it seemed to touch the scientific consciousness at a deeper level; perhaps as a wonder that man can calculate something about the whole earth, just as today we can record eruptions on Io.

Various methods were used to calculate the shape. We have just seen the direct geodetic approach in the meridian project; there was also the use of potential theory (§ 5.5.2), on which Legendre drew on his proposed equation (554.5) for the length of the radius vector; there were calculations drawn from lunar theory, which we pick up in § 6.4.3; and there were experiments by pendulum, which are related in § 6.5.2 in connection with Biot's and Arago's continuation of the meridian project. The answers provided by these methods showed significant variance, around the value of $\frac{1}{300}$, for oblateness: Laplace examined the question in volume 2 of his *Mécanique céleste*. As the various theories of mechanics have been noticed already or are to appear in Chapter 6, I shall describe here his views on comparing these theories with observations—that is, his theory of error.[1])

Among the few contributors to the theory of error before Laplace, Boscovich developed some ideas, in connection with the shape of the earth. Unable to find the same value for the (presumed) ellipticity E of the earth from different but apparently reliable determinations of meridianal arcs, he took the required law

$$l(\Theta) = 3l(0)E\sin^2\Theta + l(0) \tag{555.1}$$

for the length $l(\Theta)$ of a degree of the meridian at latitude Θ, and determined the coefficients A and B in

$$l(\Theta) = A\sin^2\Theta + B \tag{555.2}$$

by defining the 'corrections'

$$\varepsilon_j := l(\Theta_j) - 3l(0)E\sin^2\Theta_j - l(0) \tag{555.3}$$

and requiring that

$$\sum_j \varepsilon_j = 0, \quad \text{and also that } \sum_j |\varepsilon_j| \text{ was minimal.} \tag{555.4}$$

In other words, he attempted a linear regression using the criterion of absolute deviation. The details of his procedures are not of concern here: they were presented geometrically, rather than in analysis as given here; some of them amount to exercises in linear programming.

Most of Boscovich's principal writings in this area appeared in a French translation and edition *1770a* of his and Maire's account of their geodetic survey of the Papal states, which had been published in Latin in 1755. It seems likely that

---

[1]) This topic has been much neglected; useful guides from the mid 18th century to Laplace's death are provided in Tilling *1973a*. Laplace's and Boscovich's theories are described in Eisenhart *1961a*, Sheynin *1977a*, and Stigler *1986a*, 39–55. I change Laplace's notations below.

Laplace will have known this work before writing his first paper *1786a* on using error theory to determine the shape of the earth; as usual, Laplace gave no references, and the reader must appeal to his own erudition.

Instead of applying to (555.3) a best-fit condition such as (555.4), Laplace *1786a*, art. 3 tried the minimax conditions that

$$\sum_j \varepsilon_j = 0, \quad \text{and} \quad \left|\max_j \{\varepsilon_j\}\right| \quad \text{was minimal;} \tag{555.5}$$

but the technique was clumsy to use. So he tried a variant upon it which was rather more systematic in its search processes, and also Boscovich's method, in a sequel paper (*1793a*, arts. 8–14, where Boscovich's ideas were 'uselessly complicated from the consideration of diagrams'!) and in the *Mécanique céleste* (*2* (1799), 417–434, where Boscovich was not mentioned at all).

The second method based upon (555.5) allowed Laplace 'to ascertain whether the elliptical figure is within the limits of the errors of observation': he now used Boscovich's conditions (555.4) to determine 'the figure which is most probable' (p. 434) in the same best-fit spirit with which Boscovich had addressed his work. He modified it in an interesting way, for he applied it to the determination of whole meridianal arcs and not just a degree of the meridian and thus gave to each degree of the arc 'so much more influence, in the computation of the ellipticity of the earth, as the corresponding arc is of greater extent, which ought to be the case' (pp. 436–437). In other words, he weighted each arc in proportion to its latitudinal extent, so that if ε_j was the error associated with calculating the length of a degree from measuring an arc of n_j degrees, then Boscovich's condition (555.4) should be modified to a weighted form:

$$\sum_j n_j\varepsilon_j = 0, \quad \text{and also} \quad \sum_j |n_j\varepsilon_j| \quad \text{was minimal.} \tag{555.6}$$

He then, broadly speaking, followed an analytical version of Boscovich's treatment (pp. 437–442).

'We shall apply this method to the degrees of the terrestrial meridian which have already been measured', Laplace continued, choosing data from seven past surveys, including the Lapland expeditions of the 1730s and 1740s when the oblateness of the earth's spheroidicity had been calculated and the Méchain-Delambre project; the 'most probable' oblateness turned out to be $\frac{1}{312}$ (pp. 443–450). He then applied his own minimax method to five stations from the Méchain-Delambre data—and obtained $\frac{1}{150\cdot 6}$, which did not conform with other information (for example, as drawn from astronomical calculations: pp. 442–459). So now he was drawn into a lengthy discussion of ellipticity, the length of the Dunkerque-Barcelona arc, data from pendulum experiments, and the interpretation of the values which his error theory had furnished (pp. 451–486). He emphasised any conformities of calculation which arose, but he avoided talk of 'veritable hypotheses' which we noted in § 4.5.2 when he was writing his armchair mechanics. Since Bowditch found one of the calculations of ellipticity to be incorrect (p. 447), it was just as well.

5.5.6 Laplace's theory of sea-flow

> It would be useless to offer this theory in the same shape in which Laplace has given it; for the part of the *Mécanique Céleste*, which contains the Theory of Tides, is perhaps on the whole more obscure than any other part of the same extent in that work.
>
> Airy *1845a*, 259*

The theory of the tides, and the motion of the sea in general, was of considerable importance for both planetary mechanics and navigation. 18th-century studies (by Euler, MacLaurin, Daniel Bernoulli and others) tended to work in hydrostatical terms; but Laplace initiated a dynamical theory of the tides based on finding the differential equations for tidal flow. He wrote up his ideas, in rather high-powered style, in *1778–1779a*; but he gave a more luminous account, with some mathematical changes and attention to numerical data, in a paper presented to the *Académie* in December 1790 but not published until *1797a*. He largely followed this version in the *Mécanique céleste* (*1* (1799), 218–227; *2* (1799), 526–792). The calculations are formidably complicated, and so is the history if taken way back; I shall point out only a few features.[1)]

Laplace followed Lagrange in establishing the basic differential equations of the motions of a homogeneous incompressible non-viscous fluid from the equations for equilibrium, although he employed his characteristic mixture of Eulerian differentials and Lagrangian variations (*1* (1799), 90–95, 194–210: this topic is taken up in § 10.3). He considered fluids next when determining the shape of the earth when regarded as a solid spheroid covered by a fluid in equilibrium (*1* (1799), 210–218; *2* (1799), 241–357). He drew of course, on Legendre functions in the course of this work, and maintained the connection when studying the sea.

[1)] Laplace gave his own historical survey in *Mechanics, 5*, Bk. 13 (1824), ch. 1. On his work and/or its background, see Aiton *1953a*; Gillispie *1978a*, 295–301, 339–341; and Darwin and Hough *1908a*, arts. 11–17.

Laplace used spherical polar coordinates, with the polar semi-axis (of length 1), and took the coordinates of a particle dm of sea at latitude $(\tau - \Theta)$ and longitude ω at time t in the form

$$(r_0 + \alpha s, \Theta + \alpha u, \quad \omega_0 + nt + \alpha v), \tag{556.1}$$

where α was a tiny quantity and n the mean angular velocity of (544.3). The equation representing the motion of dm was

$$(u_{tt} - 2nv_t \sin\Theta\cos\Theta)d\Theta + (v_{tt}\sin^2\Theta + 2nu_t\sin\Theta\cos\Theta)d\omega = \alpha dV' - gdy, \tag{556.2}$$

where $\alpha dV'$ represented the forces acting on dm which was located αy above the surface of equilibrium. (They include components of the 'Coriolis force', as we call it (§ 18.2.4); it was introduced in the theory of tides in MacLaurin *1752a*.) The continuity condition for the sea furnished

$$y + (\gamma u)_\Theta + (\gamma v)_\omega + \gamma u \cot\Theta = 0, \tag{556.3}$$

where $\gamma(\Theta, \omega)$ was the depth of the sea, regarded as small relative to the polar semi-axis (p. 527). The mathematical task centred on (556.2-3); too difficult to solve in closed form, special cases and approximations had to be taken.

Laplace introduced the Legendre functions in the simple case of no earthly rotation ($n = 0$) and uniform depth γ. V' then reduced to a form of their generating function (552.5), and so they appeared as $\{P_j(\cos\Theta)\}$ in its power-series expansion (pp. 527-542). Next he introduced rotation, took γ as a function only of $\cos\Theta$, and obtained insoluble forms of the equations. As in his astronomy (§ 5.4.5), he assumed that periodic forces produced periodic effects, and so expanded the dependent variables in trigonometric series such

$$y_i = F_i(\{P_j(\cos\Theta)\})\cos(it + N\omega + \varepsilon), \quad N \text{ an integer}, \tag{556.4}$$

where F_i was some function of the Legendre functions, to be determined (pp. 543-549). He then classified the various sources of tidal flow into three types of oscillation: long-term, diurnal, and semi-diurnal (see Table 556.1). The core of his work lay in the detailed analyses of these oscillations, the assessment of their relative importance, and their combination to produce tidal effects.

Laplace's studies are noteworthy for their use of hypotheses. He assumed, reasonably, that the earth was an ellipsoid of revolution (p. 552), and deduced that the equilibrium of the sea was stable, because its mean density was less than that of the spheroidal core (pp. 623-628: the reasoning had certain similarities with that for the stability of the planetary system, described in § 5.4.6). By contrast, when he assumed that the depth γ of the sea was nearly constant, he found that the second-kind oscillations vanished (p. 578) and the other two kinds yielded timings for high and low water which were quite at variance with the known data

Table 556.1 Basic types of motion of the sea in Laplace's *Mécanique céleste* (from *2* (1799), 551-552, 558, 567, 581).

Oscillation	Period(s)	Principally caused by [a]	Series expansion of $\alpha V'$ had terms of form
'First'	Monthly or yearly (mostly)	Orbit of earth	$\alpha f_i(\cos\vartheta)\cos(it + A)$
'Second'	Diurnal	Rotation of earth	$\alpha f_i(\cos\vartheta)\cos(it + \omega + A),\ i \triangleq n$
'Third'	Semi-diurnal	Moon (mostly)	$\alpha f_i(\cos\vartheta)\cos(it + 2\omega + A),\ i \triangleq 2n$

[a] Laplace did not always consider the forces due to the attraction of the sea (introduced on pp. 529-530).

for Brest, so that 'the present hypothesis is very far from representing the observations in this particular' (p. 609).

The Brest figures had been collected by Cassini II in the early 18th century, found by his grandson (Cassini IV) much later, and published in Lalande *Astronomy*$_2$, *4* (1781), 161-231. Laplace made only a brief allusion to their origins (p. 688), but he used them for his extensive series of comparisons of theory with observations (pp. 668-792). Perhaps surprisingly, he did not employ his minimax method to fit his posited linear relations to the figures (§ 5.5.5); instead he usually employed, without explanation, methods of averages to reduce the number of equations to the number of unknown constants. 'We have treated fully on the ebb and flow of the sea', he concluded (p. 792), and called for figures to be prepared for other ports in order to fit theory to observations in still more detail. In fact, data had long been collected (though not usually to the quality of the Brest information); and Laplace's theory itself was to receive modification during the 19th century from foreigners, as the quotation from Airy at the head of this sub-section hints.

5.5.7 Laplace on the rotation of the earth

Laplace's volume 2 comes to an end with 'the motions of the heavenly bodies about their own centres of gravity'. He presented a paper on this subject in January 1796 to one of the early meetings of the *classe* (*PV*, *1*, 6) and revised it for publication as *1798a* in the first volume of the *Mémoires*; a similar version of the material appeared a year later in the *Mécanique céleste*. He dealt with the earth, the moon (§ 6.4.3) and Saturn's rings; I shall consider only the earth, which he regarded as the most important.

Laplace's principal concern was to calculate the precession and nutation of the earth's polar axis. Figure 557.1 contains two diagrams to indicate these effects

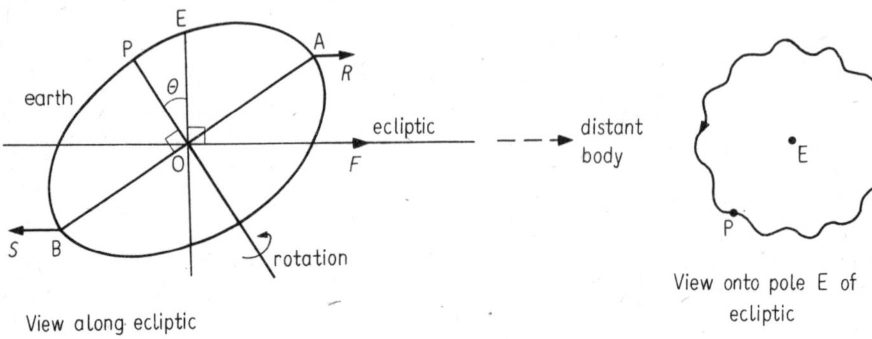

Figure 557.1 Illustrations of the principal effects of precession and nutation

when caused by one distant body. By the inverse square law, the force on the earth will be greatest in the area of A, the point of the equator nearest to it, and weakest at the antipode B. Thus if a mean force F, determined as the value at the centre O of the earth, were taken as applying to all of its points, the variation between A and B would have an effect like that of a couple given by forces R and S as shown. This couple tends to jerk the axis OP of rotation towards the pole E of the ecliptic; but since P is rotating out of the page already, the mean result is to send P into a perturbed circle around E. The forces which are not dependent on the longitude of the distant body cause P to *precess* and thus to cause such effects to other astronomical points (the equinoxes are usually studied) in circles around E. In addition, there are perturbing *nutational* effects on the circles, caused largely by the variation of the position of the ecliptic, itself due to the actions of the planet. The perturbations involved are far smaller than Figure 557.1 suggests; they are also much more complicated than described here, since the influence of both sun and moon must be considered, and also the equatorial bulge of the earth and the ebb and flow of the sea.

Laplace naturally started from Euler's equations (525.15) for the rotation of a solid body, which he had already derived (in Eulerian differential style) in *1* (1799), 144–171. Taking principal axes for the earth and assuming it to be a *solid* spheroid of revolution, he expressed the principal moments of inertia and the perturbing force of the distant body in various forms, all involving surface harmonics (*2* (1799), 802–824). He then rotated his axes about O from the system (x, y, z) given by the polar axis OE and two axes in the plane of the equator (see Figure 577.1) to (X, Y, Z) given by OE and two axes in the ecliptic, Ox lying along the (moving) vernal equinox, relative to which longitude φ was measured; the other angle employed was the obliquity Θ ($\angle$EOP) of the ecliptic. With (p, q, r) as the rotation vector relative to the original axis system, one Euler equation took the form

$$r = 0, \tag{557.1}$$

which stated that the rotation of the polar axis was constant; the other two were of the same kind, one being

$$Adq + (C - A)rpdt = [3L(C - A)f(X, Y, Z, \Theta, \varphi)/r_1^5]dt, \tag{557.2}$$

where A and C were the moments of inertia of the earth, L the mass of the distant body, r_1 its distance from O, and $f(X, Y, Z, \Theta, \varphi)$ a quadratic form in X, Y and Z resulting from the transformation of coordinates, with coefficients given by certain trigonometric functions of Θ and φ (pp. 824-828, with homework from *1* (1799), 106-114, 144-171).

Laplace assessed precession and nutation largely by solving the Euler equations and finding explicit expressions for Θ and related angles. To integrate the equations he followed his normal practise (544.15) of expanding the right hand side of (557.2) (and its companion) in trigonometric series and proceeding from there. He normally used only first-order terms in the perturbing functions, and deduced (among other things) that the rotation of the earth was uniform (p. 861); that precession and nutation 'are exactly the same as if the sea form a solid mass with the spheroid which it covers' (p. 899), since the depth γ of the sea, assumed to be small and constant, furnished only terms of the order of $\alpha\gamma$, where α was the small quantity of (556.2); and, in a splendid piece of mathematical optimism, that the principle of areas (§ 5.2.3) 'proves' that atmospheric effects such as winds 'must be insensible' (p. 812). He moved to second-order approximations in some formulae when comparing them with observations (p. 925), and calculated a few things, such as bounds of $\frac{1}{340}$ and $\frac{1}{508}$ for the ellipticity of the earth (p. 928), consistent with (though partly dependent on) his theory of the tides (§ 5.5.6). But he did not discuss the nutation found by Euler (*1765a*, arts. 841-849), where the complementary function solutions of the Euler equations (525.15) produced a period of oscillation of the earth axis of around ten months (depending on the value taken for the ratio of the moments of inertia). These topics were resumed later, as we shall see in § 6.3.5, § 12.4.2 and § 17.6.4.

5.5.8 Geodesy and topography: the aspirations of Puissant

All the topics in this section have dealt with planetary mechanics, especially concerning the earth. The growth of knowledge in this field, and its use in civil and military contexts, required expression in new textbooks and treatises, and these were provided by a new actor in the story, cited in passing in § 5.5.2-4: Louis Puissant.

Born in 1769 near Chatalet-en-Brie (south-east of Paris) to farming parents who died soon afterwards, Puissant drew much of his upbringing from an engineer called Lomet, ten years his senior. Puissant joined the *Dépôt Général de la Guerre* in 1796, and sat in on classes at the short-run *Ecole Normale* and maybe also at the *Ecole Polytechnique*, where Lomet was in charge of models. After five years teaching at the *école centrale* at Agen near Toulouse, where he wrote the book on the algebraic solution of geometrical problems (*Geometry*$_1$ (1800)) noted in § 4.7.3, he returned to the *Dépôt* as an *ingénieur géographe*, and carried out surveys of Elba and Lombardy before appointment in 1803 to the newly formed *Ecole Militaire* at Fontainebleau, near to his birthplace. (This school, and his teaching there, are noted in § 8.1.2.) During this time he wrote two treatises: a *Traité de géodésie*, which first appeared in 1805 (to a warm reception from Biot *1806c* in the *Moniteur universel*); and a *Traité de topographie*, which was published first in 1807 (when Hachette *1808a* welcomed it there in review). From 1809 until retiring in 1833 he directed an *Ecole de Géographes* within the *Dépôt*. In addition, he produced a steady stream of articles on these topics, publishing some in the *Mémorial* of the *Dépôt* but many in the *Correspondance sur l'Ecole Polytechnique*, with the *Société Philomatique*, and elsewhere.[1])

The book on geodesy showed well the range of material which Puissant regarded as relevant. A substantial quarto volume of 318 pages published by Courcier (Puissant *Geodesy*$_1$ (1805)), it fell into five Books, covering respectively the motion of heavenly bodies, plane and spherical trigonometry, geodetic operations, some aspects of positional astronomy, and 'astronomical observations' (including the use of instruments). After a short appendix, an extensive suite of tables was provided. The third Book, on geodesy, was by far the largest (pp. 59-251): the material included not only some aspects of cartography but also triangulation, the calculation of latitudes, longitudes and azimuths, the shape of the earth, barometric correction, atmospheric refraction (§ 8.4.1) and

[1]) Little attention has been paid to Puissant (Konvitz *1987a*, a good history of French cartography, omits him entirely!), beyond the usual tributes on his death in 1843 and dictionary articles; but Elie de Beaumont *1870a* provides an excellent survey of his life and work. No *Nachlass* seems to have survived, though one manuscript is noted in § 17.7.2; and his *dossier personnel* at *VT*, 43105 (1st ser.) is very interesting.

various matters of technique such as the use of certain instruments and signals. Many results of Delambre, Laplace and Legendre were rehearsed. The students of geodesy would have received a thorough grounding from the book, and doubtless Puissant's colleagues kept it to hand too. An expanded second edition appeared in 1819 in two volumes, also from Courcier.

The treatise of 1807 on topography 'is, to speak properly, the second part' of the geodesy book, as Puissant announced in its preface: he took his topics further rather than introduce new ones. Over 98 of Courcier's pages for his first Book, he summarised again the principal formulae from spheroidal trigonometry and the curvature of surfaces, and then dealt in turn with trigonometric and barometric methods of levelling, determination of the shape of the earth by means of the simple pendulum, and a variety of techniques in geodesy (including Laplace's error theory of § 5.5.5). The second Book (pp. 99–150) described some main features of map-making, with examples from certain projections (§ 6.4.8). Book 3 (pp. 152–207) furnished details on surveying techniques, including the evaluation of surface areas, and the measurement of polygonal boundaries; the next Book (pp. 208–270) dealt similarly with levelling, including allowing for atmospheric refraction and estimating the volume of a solid. Finally, in Book 5 (pp. 271–295) he described the use of instruments necessary for the drawing up of designs and maps. As in the earlier work, this one was completed by a few appendices, an index (for the two books together) and tables. A supplement to Book 2 of 110 pages, published in 1810, described in more detail the making of maps: the significance of this interest will be brought out in § 6.4.8. This treatise also took a second edition, in 1820.

In both volumes the mathematics used was 'the two trigonometries' (as Puissant called it: planar and spheroidal), the geometry and calculus of curves and surfaces, some interpolation and quadrature techniques, and dashes of other theories (such as Legendre functions) here and there. Geodesy and topography were shown to have become a mathematical discipline in their own right.

5.6 Concluding comment: the aspirations of Laplace

In this chapter we have seen a gradual swing of emphasis from Lagrange to Laplace. The order in which I have taken the topic from basic matters to large but specific areas of research, reflects both a natural sequence for the material and the relative positions of the two giants. Lagrange appeared as rather the imperious statesman of mathematics, laying down the laws: Laplace acted as the politician, directing much of the teaching and research in the field.

The difference between the men carries over from personality and philosophy to scientific strategy and aim. While Lagrange made many major contributions to research, the thrust of his main works, especially the *Méchanique analitique*, was to systematise the known more than to invent the new; so that the scope of that known and its compass of the new becomes a major question.

It is clear that already by 1800 Lagrange's mechanics was not containing the scope of concerns of the subject as they were then perceived (and thus its grip on the whole was still less firm than that of his Taylor-series approach on the calculus, appraised in § 3.6). While Laplace's *Mécanique céleste* was heavily larded with Lagrangism—more than its author explicitly acknowledged—it also needed geometrical thought and differential modelling to deliver the results required. And when it comes to planetary physics, with the meridianal project, cartography, and such topics, the restrictions of variational formulations had to be cast aside. The various important contributions of Legendre exemplify this situation very clearly.

These remarks carry over from the research level to the teaching of mechanics, at the *Ecole Polytechnique* and elsewhere. Authors such as Monge, de Prony and Francoeur quite explicitly moved towards geometrical and differential approaches, and even de Prony's algebra was trigonometric rather than variational.

All the tendencies away from Lagrangism will become more evident in the next three chapters, which cover the development of mechanics in the early 19th century and also the special concerns of engineers. At the same time, his tradition will still make some notable contributions, especially in § 6.3. But overall the centre of gravity (as it were) will continue to move closer to Laplace than to anyone else. In particular, in Chapter 7 we shall find him setting up an empire for molecular mechanics at just the time that Bonaparte was self-becoming Emperor Napoléon. First, however, Chapter 6 will keep track of current work in corporeal, planetary and celestial mechanics.

Chapter 6

Progress in mechanics, 1800-1815

6.1 Prefaces

6.1.1 Arago: scientific entrepreneur

In § 2.6.2 I indicated that we would see a new type of *savant* appear in French science: the professionalised careerist, using the opportunities offered by the new environment of institutional appointment and public office. Biot and Poisson have been the first of this genre; now we meet the prince—Dominique François Jean Arago.

Arago was born in 1786 into a modest land-owning family at Estagel, near the eastern side of the France-Spain border. In a nice touch of fate, the family moved to nearby Perpignan in 1795, around the time when Méchain must have been passing through, on his execution of part of the Dunkerque-Barcelona meridianal survey (§ 5.5.4). Arago entered the *Ecole Polytechnique* with distinction in 1803, and left before the end of his second year to become, on the recommendation of Laplace, *secrétaire* of the *Bureau des Longitudes* in February 1805. He was appointed *adjoint* there in 1807 (one year after Biot's appointment, and one year before Poisson's). In 1810 he began teaching at the *Ecole Polytechnique*, taking over from Monge the course for 'analysis applied to geometry' and descriptive geometry; until 1830 he also taught astronomy and geodesy, and machine theory and simple economics. In 1809 he was elected to the astronomy *section* of the *classe* (§ 6.3.6).

Arago's scientific interests lay chiefly in optics, astronomy and various aspects of geophysics, for which he had an enthusiasm at the level of his friend Alexander von Humboldt. His early work was often done in collaboration with his fellow Laplacian physicist Biot; but they were too ambitious and egocentric to build a successful partnership, and quarrelled in ways that influenced the development of Parisian science. Indeed, although he was obviously a competent mathematician, he features chiefly in our story as an entrepreneur and manipulator. Unwilling to succeed Delambre as *secrétaire perpétuel* in 1822, he blocked Biot's candidacy (§ 13.6.1), became known as the 'great elector' in the *Académie*, and succeeded Fourier as *secrétaire* in 1830 (§ 18.2.1).

As *secrétaire perpétuel* Arago somehow found time to write long *éloges* or biographical articles on Ampère, Carnot, Fourier, Fresnel, Laplace, Malus and Poisson, as well as short pieces on many other figures and an extensive essay on the young Arago. They are certainly genre items, with plentiful helpings of anecdote and allegedly verbatim quotation; I am suspicious of their probity, and normally omit them in my biographical sketches, although they are certainly a good read. It is ironic that someone who wrote so much on others has been little studied himself.[1])

Arago died in 1853, whereupon his associate J. A. Barral took over the project of preparing a selected edition of his works. Between 1854 and 1862 all 16 volumes were published, and also a parallel German edition. Such was the extent of his reputation by mid century, although much of it might have spread because of the various public and governmental offices that he held in his later years rather than the scientific and educational activities which will be our concern in this book.

6.1.2 Plan of the chapter

The broad scheme of the chapter follows that of its predecessor; the (relatively) general results are taken first, and particular branches of mechanics afterwards. § 6.2 is dominated by Poinsot: his emphasis on the couple, and his views on the principle of virtual velocities. Binet's introduction of conjugate axes is noted, and also Poisson's first textbooks on mechanics.

§ 6.3 describes an important example of interaction between Lagrange and Poisson, which occurred in the late 1800s. Extending the Lagrange/Laplace attempted proof of the stability of the planetary system, Poisson came to study certain expressions which he and Lagrange both realised were of fundamental importance for mechanics; they became known as 'Lagrange and Poisson brackets'.

The particular branches of mechanics studied in the rest of the chapter were heavily influenced by volumes 3 and 4 of Laplace's *Mécanique céleste*, published in 1802 and 1805. § 6.4 describes his (and others') contributions to celestial and planetary mechanics, with especial attention to projectiles, the moon, meteorites, comets, navigation and map-making. In addition, in the latter sub-sections notice is taken of the growing use of probability and statistics in mechanics, especially least squares with Legendre and this and other methods and theorems in Laplace. § 6.5 is a short section on the attraction of spheroids, and related work on the use and theory of pendula.

The final section, § 6.6, surveys the period with the help of two major texts: the

---

[1]) Daumas *1987a* is a moderate biography of 1943, improved in the new posthumous edition. The late A. Birembaut had been preparing a biography. Arago's (auto)biographical writings are gathered together in *Works, 1-3*: English translations of several are contained in *1857a*. Most of his surviving *Nachlass* is in *OB* (including recent finds, now at mss. 214-218). See also the minute-books of the *BL*. He cultivated a signature in which a flourish contained his name again, written vertically and in mirror image (Daumas *1987a*, 276).

second edition (1811–1815) of Lagrange's *Mécanique analytique* (as it was now entitled) and two volumes of de Prony's *Leçons de mécanique analytique* (1810–1815) given at the *Ecole Polytechnique*. There are also two interludes: 641.1 on German theoretical astronomy, and 660.1 on some aspects of mechanics outside France.

6.2 Research and education in terrestrial mechanics

6.2.1 Poinsot and the importance of the couple, 1803

During the early 1800s, while teaching in Paris *lycées*, Poinsot wrote a textbook entitled *Elémens de statique*. It was set at the lower two levels of content described in § 2.2.4, for much of it was suitable for aspirants to the *Ecole Polytechnique*. The school was so pleased with it that when he presented a copy in September 1803, they bought a second one (*EP*, *CoI*, *3*, fol. 79). In fact, the principal novelties lay in its more elementary parts; and they gave it such popularity that twelve editions appeared between 1803 and 1877 (the centenary of his birth), the first nine prepared by Poinsot himself. While there are interesting changes of detail between the editions, the basic scheme was preserved; the principal changes were the inclusion (and removal) of various later papers on related topics which he wrote after 1803 and which we shall note in due course.[1])

Poinsot began in normal style: statics was based on equilibrium, forces could be composed and decomposed, and so on. But then he broke new ground with a discussion of the 'composition and decomposition of couples' (ch. 1, sect. 2). The couple had long been known in mechanics, but regarded only as a particular system of forces. For example, cases turned up in some of Euler's writings on hydrostatics and buckling (see Truesdell *1960a*, 340–370 *passim*); while d'Alembert noted that the couple did not admit of a resultant (*1772a*, 283–284), but he did not investigate the consequences. Lagrange did not mention the notion in his *Méchanique analitique* (1788). Poinsot raised its status in mechanics with this discussion.

Poinsot was anxious to stress analogies between the basic theory of forces and that of couples. Just as a force could be moved along its line of action without al-

---

[1]) The book is Poinsot *Statics*; on the contents of the different editions see Bailhache *1975a*, 211–214. Dühring *1873a*, 421–449 recognizes the importance of Poinsot for mechanics. Few manuscripts in Poinsot's *Nachlass* relate to the book, but see *BI*, ms. 953, fols. 28–64. Lagrange praised his theory of the couple in a letter to the *Grand-Maître* of the *UN* of 1809(?) in *BN*, ms. n.a. fr. 1305, no. 46.

tering its effect, so a couple was transferable parallel to itself in its own or a parallel plane, or rotatable about its axis. Just as two (or more) forces were combined to make a resultant, so couples were composable: co-planar couples corresponded to the case of parallel forces, general couples were composed by taking the appropriate diagonal of the parallelogram formed by their axes. Further, a system of forces $\{P_j\}$ could be replaced by a resultant force R at a given point A and a couple G, since for each P_j the forces P_j and $-P_j$ might be introduced at A, yielding a force P_j to go into R and a couple $(P_j, -P_j)$ to contribute to G. Hence the equilibrium of $\{P_j\}$ depended on G, as well as R, being zero. If G and R were not perpendicular, this resolution was not unique, since G could be replaced by equivalent couples.

All these results applied to disequilibrium; Poinsot devoted his long ch. 2 to an account of equilibrium, based on the requirement that both G and R were zero. He dealt with various necessary and sufficient conditions, usually in terms of considering components of G and R. He also related his ideas to the standard conditions (522.10) that a system of forces be in equilibrium, for in addition to requiring a zero resultant as usual, he transplanted each P_j from its location (x_j, y_j, z_j) to the origin by introducing couples there; the condition that the resultant couple had zero moment gave him equations such as

$$\sum_j (x_j Y_j - y_j X_j) = 0, \tag{621.1}$$

which were the old equations of (522.10) for zero torque (*Statics*$_1$ (1803), 102–107). He also obtained Euler's torque theorem (527.1) in the course of his analysis (pp. 126–129).

Poinsot devoted ch. 3 of his book to the centre of gravity, which he defined as the point through which the resultant of the gravitational forces on each 'molecule' of a body passed, whatever its orientation. This form of definition afforded him links with his analysis of parallel forces; he proved various relations concerning the centre of gravity, including some to be found in Lagrange *1785c*.

Mention of Euler and Lagrange naturally raises the question of influence. Poinsot gave no references in this, his first, publication; but it reads like Euler in its basic principles and tightish organisation (§ 5.2.5), filtered through Monge for some proofs and the liberal use of diagrams (§ 5.2.8) and de Prony for the use of trigonometry when touches of algebra were required (§ 5.2.7). The influence of all three was also evident in the concluding ch. 4, in which Poinsot discussed various kinds of machine.

Biot reported verbally on Poinsot's book to the *classe* in November 1803 (*PV*, *3*, 27). There is no record of his comments, but he might have written the anonymous praising review in the *Moniteur universel* (Poinsot *1803a'*). The engineer

Poullet de Lisle was also pleased, concluding his review in the *Magasin encyclopédique* with the prophecy that 'one will not tarry to distinguish [this book] from the crowd, and perhaps even let it depart from the rank where the modesty of its title places it' (*1803a*, 212).[1])

Possibly encouraged by this response, Poinsot submitted a paper on the same themes to the *classe* in May 1804 (*PV*, *3*, 99): de Prony *1806a* warmly recommended it for the *Savants étrangers* in his report. The paper was published as Poinsot *1806a* in the *Journal de l'Ecole Polytechnique*, and reprinted in his *Statique* from the third to the tenth editions (1821–1861). In the preface to the second edition he said, erroneously, that it had appeared also in the *Savants étrangers*.

In the first part of his paper Poinsot summarised the main results of the first two chapters of his book, and added another which has become well known. Laplace had shown in effect that the angular momentum of a system $\mathscr{S}$ of forces relative to the origin O of coordinates took a maximal value, L, along the normal to his invariable plane (544.2). Poinsot used his theory of couples to show that if the coordinate axes lay along and perpendicular to the direction of the resultant force R of $\mathscr{S}$, then L would remain constant; but if O moved to a point O_1 perpendicularly distant p from R, then a new couple of moment Rp and direction perpendicular to R and OO_1 would have to be introduced, to combine with the original couple G to produce a new couple G_1. Further, by the law of composition of couples G_1 would be minimal when parallel with R. Thus, if the angle between G and R were φ, then the minimal couple would be given by

$$Rp = G \sin \varphi; \quad \text{and} \quad G_1 = G \cos \varphi \tag{621.2}$$

(*1806a*, art. 2).

The forces in the couple $G \sin \varphi$ were coplanar with R; and the three would resolve into a force R_1 equal and parallel to R and distant $(G \sin \varphi)/R$ from it. Poinsot called the line of action of R_1 'the *central* axis of moments', because the other axes of maximal torque lay on a hyperboloid symmetrically about it. Since he was aware of Laplace's construction of the invariable plane, it is interesting to note that his condition for determining G_1 took a minimax form, like Laplace's error theory (§ 5.5.5). Poinsot's name for the axis has been retained, although his result (621.2) is now usually presented as an especially *simple* form of finding G_1 and R_1.

Poinsot concluded his paper by applying his ideas to dynamics. He took some

[1]) A.C.M. Poullet de Lisle (1770–1849) followed a rather unusual career; for after entering the *EP* in 1796 and the *EPC* two years later, he passed much of his career not in the *Corps PC* but in the *lycées* and especially the *UN* where he was *recteur* of three of the provincial *académies* before being promoted to *inspecteur général* (see Boncompagni *1882a*). He published little, but one item was his translation of Gauss's *Disquisitiones arithmeticae* of 1801, which came out in 1807.

of the basic conservation laws, such as momentum ('force'), treating rectilinear motions like parallel forces and the principle of areas like moments of forces, including a 'Minimum maximorum area' corresponding to the plane given by (621.2). This part of the paper was suggestive of some of his later contributions to mechanics: his criticism of Laplace's invariable plane (§ 17.4.1), and his theories of the momental ellipsoid and of rolling cones (§ 18.2.4).

6.2.2 Poinsot and Ampère on the principle of virtual velocities, 1805

Poinsot and Ampère both submitted papers on the foundations of mechanics to the *classe* early in 1805. Poinsot's, which arrived in January (*PV*, *3*, 175), gained a commission composed of Lagrange, Laplace and Lacroix, which never reported. Poinsot rewrote the paper for the *Journal de l'Ecole Polytechnique*, and sent the proofs to Lagrange, who made some (rather uninteresting) remarks in the margin; it appeared unchanged in the *Journal* as Poinsot *1806b*, immediately after the paper on couples just discussed. Later he placed it, with changes, in his *Statique*, from the sixth to the tenth editions (1834-1861).

Ampère's manuscript came in a month after Poinsot's (*PV*, *3*, 183), with an addition later; after 13 months de Prony *1806b* recommended the *Savants étrangers* in his report, but it appeared, as Ampère *1806b*, in the same *cahier* of the *Journal* as Poinsot's paper and as his own study of Taylor's series described in § 4.3.3. Bailhache has described both these mechanics papers in detail in his *1975a*; it is not known if they were written independently of each other or not.[1]

Ampère offered a new proof of the principle of virtual velocities, hoping for the generality of Laplace's proof (which he apparently found satisfactory) and also the simplicity of Carnot's and Lagrange's (compare § 5.3.2). His analysis bore some relation to his paper *1806a* on the calculus of variations in its concern with dependent and independent variables (§ 4.5.2). In fact, this was perhaps the most interesting feature of the paper, for he related the degrees of freedom, and thus of independent variables of the system, to the constraints laid upon it. He even stated the principle itself in the form (522.3) of Lagrange's equations for

---

[1]) Bailhache also reproduces in an annex the texts of an early 1797 manuscript by Poinsot on the topic (from *BI*, ms. 958, fols. 157-173); the manuscript submitted in 1805 (ms. 963, fols. 148-161: see also *ENPC*, ms. 1332); and a critical edition of *1806b*, including Lagrange's comments and Poinsot's replies (from *BI*, ms. 4738) and comparison with the revised version in *Statics*$_8$ (1842).

Manuscripts relating to Ampère *1806b* are held at *AS*, *Ampère*, chs. 113-118 *passim*; note also his later proof *1826d*.

forces $\{P\}$ under constraints $\{L = 0\}$:

$$\sum Pdp + \sum \lambda dL = 0 \tag{622.1}$$

(*1806b*, 253), and considered adding the constraints successively into the analysis (p. 248). In keeping with his developing views on the calculus (§ 4.3.2), he wished to avoid infinitesimals in his proof; but instead he used some laborious algebra to calculate components of the forces $\{P\}$ in given directions, and to replace them by other forces in terms of which the equilibrium condition could be expressed. Since the proof (explicitly) assumed the law of decomposition of forces, its merit is questionable; Ampère was perhaps lucky that de Prony, and not Lagrange or Laplace, was the member of the *classe* commission who wrote the report.

Poinsot's approach was more radical, in that he sought to *avoid* using the principle of virtual velocities when establishing equilibrium. He began with 'this axiom, that if some forces currently take up equilibrium on any variable system, the equilibrium does not cease upon supposing that the system be suddenly rendered invariable' (*1806b*, 207), and so felt that only the equilibrium of hard bodies need be considered. He then argued for the equilibrium of a sequence of special cases, with the aid of the principle of decomposition of forces (*ibid.*) until 'finally this theorem follows, which is the principal object of this paper' (p. 228):

> Whatever be the equations which apply between the coordinates of the different points of the system [of forces], each of them requires, for equilibrium, that one applies at that point, along their coordinates, any forces proportional to the prime [that is, derived] functions of these equations, relative to these respective coordinates.

This result may be expressed as follows:

Theorem 622.1 Given force (X_i, Y_i, Z_i) acting at point P with coordinates (x_i, y_i, z_i), with h points in all satisfying the K equations of condition $\{L_j = 0\}$, then at equilibrium each force was given by

$$X_i = \sum_{j=1}^{K} \lambda_j (L_j)_{x_i}, \quad Y_i = \sum_{j=1}^{K} \lambda_j (L_j)_{y_i}, \quad Z_i = \sum_{j=1}^{K} \lambda_j (L_j)_{z_i}, \qquad 1 \leq i \leq h. \tag{622.2}$$

Eliminating the K multipliers $\{\lambda_j\}$ from (622.2) yielded the $(3h - K)$ equations expressing equilibrium; $(3h - K) \geq 6$, for a completely free system required the standard six equations (522.10) (pp. 228–229).

Poinsot's theorem is interesting, in that it treated the condition functions $\{L_i\}$

as potentials from which the required forces were defined as their gradients. His proof rested on the principle that 'each force must be perpendicular to the surface or to the curve upon which the point of application would only have the freedom to move, if all the other points became fixed', which was itself demonstrated in the first note of the paper (but from the composition of forces: pp. 234–237). The condition equations specified curves and surfaces in space on which points lay, and the principle above showed that for equilibrium the applied forces were normal to such curves and surfaces, thus leading after analysis to the gradient functions in (622.2).

In the second note Poinsot proved the principle of virtual velocities from his theorem; in fact, he regarded the principle as stating his theorem from a certain point of view (pp. 237–241). The proof rested eventually on differentiating the equations of condition with respect to time to yield

$$\sum_{i=1}^{h} ((L_j)_{x_i}\dot{x}_i + (L_j)_{y_i}\dot{y}_i + (L_j)_{z_i}\dot{z}_i) = 0, \qquad 1 \leq j \leq K, \tag{622.3}$$

multiplying each equation by λ_j, adding over j and rearranging produced

$$\sum_{i=1}^{h}\left(\sum_{j=1}^{K}\lambda_j(L_j)_{x_i}\right)\dot{x}_i + \sum_{i=1}^{h}\left(\sum_{j=1}^{K}\lambda_j(L_j)_{y_i}\right)\dot{y}_i + \sum_{i=1}^{h}\left(\sum_{j=1}^{K}\lambda_j(L_j)_{z_i}\right)\dot{z}_i = 0, \tag{622.4}$$

which reduced from (622.2) to the principle, in the form

$$\sum_{i=1}^{h} (X_i\dot{x}_i + Y_i\dot{y}_i + Z_i\dot{z}_i) = 0. \tag{622.5}$$

Poinsot saw d'Alembert's principle (§ 5.2.3) as another consequence of his theorem. If the forces were not in equilibrium, (622.2) would be replaced (with unit gravitational constant) by

$$\begin{aligned} \ddot{x}_i &= X_i + \sum_{j=1}^{K}\lambda_j(L_j)_{x_i}, \qquad \ddot{y}_i = Y_i + \sum_{j=1}^{K}\lambda_j(L_j)_{y_i}, \\ \ddot{z}_i &= Z_i + \sum_{j=1}^{K}\lambda_j(L_j)_{z_i}, \qquad 1 \leq i \leq h, \end{aligned} \tag{622.6}$$

which showed that the motion of the point was partitioned into that caused by the impressed forces and that due to the constraints expressed by the equations of condition (p. 233).

Various other writers considered the principle of virtual velocities. When Servois studied it in a paper *1811a* for a Turin Academy prize, he drew on Lagrange's proof using pulleys (§ 5.3.1). And the final volume of the *Correspondance sur l'Ecole Polytechnique* contained three pertinent papers: Navier *1815a* on interpreting the principle for finite displacements of the points, followed by a note Hachette *1815a* on its application to simple machines; and Rodrigues *1815a* on the principle of least action (523.9), where he used Lagrange multipliers to add in the *forces vives* terms and derive Lagrange's equations (524.9) for mechanics from

$$\int \delta T\, dt + \lambda(\delta T + \delta V)\, dt = 0. \qquad (622.7)$$

All these papers, especially Ampère's and Poinsot's, show that the foundations of variational mechanics were still not settled.

6.2.3 Poisson's *Cours* and *Traité* on mechanics, 1810–1811

Poisson did not have much time for Poinsot's advocacy of the couple; in the *Correspondance* he obtained the basic equations (522.10) for the equilibrium of a system of forces by a rather clumsy exercise *1805b* involving various projections, in which the notion of a couple, of which he was clearly aware (pp. 135, 138), was played down. Four years later he was asked by the *Conseil de Perfectionnement* of the *Ecole Polytechnique* to write up his lectures in mechanics. The principal product was the first edition (1811) of his *Mécanique*, published by Courcier; but it was preceded by a two-volume *Cours* (*1810a*). This earlier work has never been studied, although it is mentioned, as a kind of draft, in the preface to the later one.[1]) As neither work is very well-known (the second (1833) edition of the *Mecanique* is usually cited), their contents are summarised in Table 623.1. In 1811 he was 30 years old.

I shall comment on the general features of the *Traité*; a few specific points will be mentioned later. Poisson trimmed his natural Lagrangian inclinations to the educational needs of his students. 'As for the choice of proofs', he declared in his preface, 'I have not tied myself down either to the synthetic method, or to the analytic path; in a book intended for instruction, I have had above all to look for clarity and simplicity [...] often one will find geometrical considerations and for-

---

[1]) The print-run of Poisson's *Cours* was 200 copies (*EP*, III 3b, from which I have inferred the year of publication). Part of vol. 2 is in *AS, Ampère*, ch. 130; see also the list of Poisson's lectures for 1812 in ch. 392. The *Traité* was also bought for use in the school: Binet *1811b* gave it a glowing review in the *MU*. 5,000 copies were printed.

The first ever *PF* doctorate related to the *Traité*, in press at the time of the examination (23.3.1811: *AN*, AJ^{16} 5322, fol. 8). The candidate was P.-M. Bourdon, *polytechnicien* of the *promotion* of 1796, later *professeur* at a Paris *lycée* and an *inspecteur* for the Paris *Académie* and a well-known textbook writer. In his first thesis *1811a* he covered parts of rigid-body dynamics; the second dealt with Kepler's laws and the principles of gravitation.

mulae of Algebra mixed in the same question'. The mixture was found not only in proofs, for geometrical ideas were present also in the principles, the use of differentials, and in the *Traité* the use of many diagrams. He began the *Cours* with an instruction to the students to prepare their own diagrams!

Book 1 of the *Traité* started with moments, including a clumsy power-series of Fourier's functional equation (531.12) and the composition of forces (*1*, 13-33). Poisson placed the principle of virtual velocities at the end of the Book, where he gave a fairly general proof based on Laplace's argument using (531.19) (*1*, 240-251). Again, he began his treatment of dynamics with uniform velocity and acceleration (*1*, 261-288) and similarly delayed the general principles until the end (*2*, 267-306), although in between he followed Lagrange in offering d'Alembert's principle as the 'general principle of dynamics', which reduced dynamics to statics (*2*, 42-44). At times, his account was far too brief for the 'complete treatise of rational mechanics' promised in the preface; for example, the Poinsotian couple was again virtually ignored (*1*, 43-44), and angular momentum deprived of status among the principles of dynamics by the assumption of the equality of central forces of action and reaction (*2*, 270-271).

Poisson also restricted several of his physical examples to those expressible by (systems of) ordinary rather than partial differential equations. The students will have obtained a wide grasp of the range of such equations and their solution in closed forms (standard functions, and also some elliptic integrals). They could also have benefitted from his remarks in *1806a*, 100-107 on the physical interpretation of singular as well as complete solutions; that for example, if a body moved with velocity v in a resistant medium such that

$$\dot{v} = -a\sqrt{v}, \qquad (623.1)$$

then its velocity was given by the complete solution

$$v = (\sqrt{v_0} - \tfrac{1}{2}at)^2 \qquad (623.2)$$

until it stopped, but afterwards the singular solution $v = 0$ applied. He did not include such examples in his *Cours* or *Traité*, although they involve a kind of detail which students forget.

Poisson's own research in mechanics during the period 1800-1815, however, showed his adherence to Lagrangian methods. The next section shows how his and Lagrange's studies intertwined in a remarkable development of the late 1800s. First, however, this one is completed by Binet.

Table 623.1 Outline of the contents of Poisson's *Cours* (1810) and *Traité*$_1$ (1811).

The third column indicates the principal omissions ('O') of *Cours* material in the *Traité*, additions ('A') there, and reworkings ('R'). 2 *Cours* pages ≙ 3 *Traité* pages; the latter work is about one third longer in wordage.

	Traité	*Cours*	Comparison between *Traité* ('*T*') & *Cours* ('*C*')
1, 1-10	'Preliminary notions' [mainly force]	*1*, 1-7	
	Book 1 Statics		
11-33	Ch. 1 'On the equilibrium of a material point'	*1*, 7-20	
34-98	Ch. 2 'On the equilibrium of a solid body'	*1*, 20-51	
99-118	Ch. 3 'Theory of moments'	*1*, 51-64	
119-172	Ch. 4 'Application of preceding principles to heavy bodies' [esp. weight, and centre of gravity]	*1*, 64-81	A: *T*, 135-172 on centre of gravity
173-183	Ch. 5 'On friction'	*1*, 81-86	
184-230	Ch. 6 'On the equilibrium of a flexible body'	*1*, 87-100	Only funicular polygon, heavy chain in *C*: flexible chain and elastic lamina A in *T*, 204-230
231-260	Ch. 7 'Principle of virtual velocities'	*1*, 101-118	
	Book 2 Dynamics		
261-307	Ch. 1 'On the rectilinear motion of a material point'	*1*, 119-144	
308-369	Ch. 2 'On the curvilinear motion of a free material point' [inc. projectiles, motion of planets]	2, 1-56	O: *C*, 35-37 on accelerative forces, 45-56 on masses of planets and pendular deviation
370-445	Ch. 3 'On the motion of a material point on a given curve' [inc. simple pendulum, brachistochrone]	2, 56-123	O: *C*, 71-77 on motion of points on a curve R: *C*, 104-122 on brachistochrone
446-472	Ch. 4 'On the motion of a material point on a given surface'	2, 123-136	A: *T*, 464-472 on trajectories, optics
473-507	'Additions' [composition of forces (§ 5.3.2), pulley, lever, inclined plane, screw, wedge, etc.]	—	

	Book 3 Continuation of dynamics		
2, 1–41	Ch. 1 'On the mass of bodies'	*2*, 136	Almost all A in *T*
42–61	Ch. 2 'General principle of dynamics' [d'Alembert's principle, with examples]	*2*, 136–143	
62–120	Ch. 3 'On the motion of a solid body about a fixed axis' [inc. moments of inertia, compound pendulum]	*2*, 143–188	Some R and A *passim*
121–160	Ch. 4 'On the motion of a solid body about a fixed point' [Euler equations, etc.]	*2*, 188–215	
161–177	Ch. 5 'On the motion of free solid body' [inc. precession and nutation]	*2*, 216–225	
178–207	Ch. 6 'On the motion of solid body on a fixed plane' [inc. rocking, and heterogeneous bodies]	—	
208–266	Ch. 7 'On the impact of bodies'	*1*, 144–159, *2*, 225–238	This part largely R in *T*, and moved in position
267–306	Ch. 8 'General properties of motion of a system of bodies' [least action, areas, energy, etc.]	*2*, 238–269	
	Book 4 Hydrostatics		
307–317	Ch. 1 'General motions of fluids'	*2*, 270–277	
318–334	Ch. 2 'General equations of the equilibrium of fluids'	*2*, 277–289	
335–388	Ch. 3 'On the equilibrium of heavy fluids'	*2*, 289–305	A: *T*, 359–388 on linked vases, siphon and pump
389–428	Ch. 4 'On the equilibrium of floating bodies'	*2*, 306–324	A: *T*, 389–400 on equilibrium of simple bodies
429–442	Ch. 5 'Use of the barometer to measure vertical heights'	*2*, 324–334	
	Book 5 Hydrodynamics		
443–471	Ch. 1 'On the motion of a heavy fluid' [§ 8.5.7]	*2*, 335–346	A: *T*, 461–471 on water in a tube
472–495	Ch. 2 'General equations of the motion of fluids' [mostly incompressible]	—	
496–500	'Addition' [to p. 103 on moments of inertia]	—	

6.2.4 Binet's 'geometry and mechanics', 1811–1814

The geometrical *Denkweise* brought by Poinsot to his mechanics is manifest also in two papers by Binet of the early 1810s; indeed, Poinsot's influence was acknowledged in the second paper. The first introduced an approach to mechanics which is now known as 'plane quadratic moments', although only some of Binet's results are remembered. He submitted his paper to the *classe* in May 1811 (*PV*, *4*, 480), when he was in his mid twenties. Biot *1811a* quickly sent in a long and unusually warm report, part of which appeared in the *Moniteur universel*. The paper was published as Binet *1813a* in the same *cahier* of the *Journal de l'Ecole Polytechnique* as his paper *1813b* on determinants, which was to be submitted to the *classe* 18 months later (§ 4.6.1). Someone chose the unusual but telling phrase 'Geometry and mechanics' for the running heads of the earlier paper; possibly it was Poisson, who wrote up a lengthy summary *1811b* of the paper for the *Société Philomatique*, where Binet himself put a related note *1811a*.

Worried by the difficulties of integrating the equations of the motions of a rigid body, Binet began his search for a simpler form by generalising the concept of moment of inertia. Let the point (x, y, z) of the system of masses have mass m, and be distant r from a given plane α in a given direction k_1. Then $\sum mr^2$ was the plane moment of inertia of the system with respect to α and k_1, expressible as a (complicated) quadratic form in (x, y, z) with coefficients given by the direction ratios of k_1 and the normal to α. Assuming that α was fixed at O, and choosing O as origin of coordinates, the moment of inertia would vary with α; and the plane which gave the minimum value to the moment was called the 'conjugate plane' C_1 to the direction k_1. (A similarity with Poinsot at (621.2), in searching for a minimum, is noticeable; Binet was less convincing, in that he assumed that it would be unique.) Choosing a second direction k_2 within C_1, a second conjugate plane C_2 could be found; and it turned out that it contained k_1. Thus if the line of intersection of C_1 and C_2 were selected as k_3, its C_3 would contain k_1 and k_2 (*1813a*, 42–45).

The conception so far was neat; but the sums were heavy going—determinant theory and quadratic forms to us, but all done long-hand. However, Binet was able to simplify the expression for the minimal moments of inertia by appealing to a result from his later paper on determinants (*1813b*, 319–321) on the conversion of a quadratic form to a sum of squares; he cited here (*1813a*, 45) its appearance in the note *1811a*.

Binet then showed that if the axes of coordinates became the conjugate axes, then the products of plane moments were zero; thus the conjugate axes played a role similar to principal axes. Indeed, they became principal axes when they formed a rectangular system, a situation which led him to a cubic in the plane moment K

$$`K^3 - \omega K^2 + \omega' K - \omega'' = 0', \tag{624.1}$$

whose roots were the principal plane moments K_1, K_2 and K_3 (p. 51). Hence the principal moments of inertia, $(K_1 + K_2)$, $(K_2 + K_3)$ and $(K_3 + K_1)$, were the roots of

$$`P^3 - 2\omega P^2 + (\omega^2 + \omega') P - \omega\omega' + \omega'' = 0' \tag{624.2}$$

(p. 52; he cited Lagrange *Mechanics*$_1$ (1788), 391–402 for similar results). Further, Euler's torque theorem (527.1) had this analogue (pp. 53–54): if K_1, K_2 and K_3 had been calculated with respect to the direction k, and a line l were taken at an angle Θ to k and with direction cosines ($\cos\alpha$, $\cos\beta$, $\cos\gamma$) relative to the principal axes as coordinates axes, then the plane moment K_1 of the plane L with l as normal was given by

$$K_L \sin^2 \Theta = K_1 \sin^2 \alpha + K_2 \sin^2 \beta + K_3 \sin^2 \gamma. \tag{624.3}$$

Binet finished off with various geometrical considerations, especially relating his quadratic forms to conoids. In particular, he interpreted equations such as (624.3) in terms of an ellipsoid with semi-axes $(K_1^{-1/2}, K_2^{-1/2}, K_3^{-1/2})$ (p. 64). This approach to moments (of inertia) was to excite the interest of Poinsot and Cauchy in the 1820s, as we see in § 17.4.1.

Binet submitted the second paper in March 1814 (*PV*, *5*, 347). The influence of Poinsot carried even into the title, 'Memoir on the composition of forces and of the composition of moments', which was very similar to Poinsot's *1806a*. Poinsot himself reported three months later. 'The work of Mr. Binet adds nothing to the science of forces considered in itself', he judged, but regarded 'some interesting remarks' in it as sufficient to gain publication in the *Savants étrangers* (Poinsot *1814a*, 357). So the *Journal de l'Ecole Polytechnique* put it out, as Binet *1815a*.

Poinsot's appraisal of the paper is fair, but some features of the paper are worth recalling. As in the previous paper, Binet handled quadratic forms, but this time he drew on Poinsot's discoveries on couples, and also used Carnot's notations (526.1) from the 'geometry of position' to spice up the algebra. The resultant R of a set $\{P_j\}$ of forces was given by

$$R^2 = \sum_j P_j^2 + 2 \sum_{i,j} P_i P_j \cos P_i \hat{\ } P_j, \qquad i \neq j. \tag{624.4}$$

If the forces acted on a body with a fixed point O, and the perpendicular from O to P_j was of length r_j, then $\{P_j\}$ could be replaced for their momental effects by forces $\{P_j r_j\}$ acting at the appropriate points on the unit sphere about O. Hence

(624.4) became

$$R^2 = \sum_j P_j r_j^2 + 2 \sum_{i,j} P_i r_i P_j r_j \cos (P_i r_i \hat{\ } P_j r_j) , \tag{624.5}$$

where the angle was that between great circles on the sphere (Binet *1815a*, 321-336). Couples could be treated in a similar way; the resultant couple G of couples $\{G_j\}$ was given by

$$G^2 = \sum_j G_j^2 + 2 \sum_{i,j} G_i G_j \cos G_i \hat{\ } G_j, \qquad i \neq j. \tag{624.6}$$

He also considered various related forms, including bilinear ones for GR, as means of expressing, for example, the angle between G and R, components of G and R in given directions, Poinsot's central axis, and the principle of areas (pp. 337-348). The paper is a good example of algebra controlled by geometrical considerations; it is an appropriate touch that near the end he got back to Laplace's invariable plane.

Poinsot's and Binet's results apply largely to the principles of statics. Analogous work in dynamics is evident in a short book *1813b* on the rotation of a rigid body by Jacques Français. Drawing on both Poinsot's couples and Carnot's notations, he took moments of momentum about the point O of rotation, converted them to Euler's forms (525.14), and solved them in the case of zero torque about O.[1]) An interesting detail is the equation to define moments of inertia H:

$$H(x^2 + y^2 + z^2) = Ax^2 + By^2 + Cz^2 - 2Exy - 2Fyz - 2Gzx, \tag{624.7}$$

which we call the 'momental ellipsoid', and associate with work of Cauchy and Poinsot around 1830 (§ 17.4.2; § 18.2.4). But the idea is here (as it is very nearly in Binet's various quadratic forms), including a conversion to a sum of squares for principal axes and some considerations akin to an equimomental cone (pp. 11-13, 29). The credit, however, belongs not to Français, for he attributed the ideas to L. Dubuat, his colleague at the *Ecole d'Artillerie* in Metz (and a son of a famous 18th-century hydraulicist: see (853.6)). For once the members of the *classe* had been pre-empted—as we shall see in § 6.5.3, at least Poisson of their fraternity knew the book at this time.

---

[1]) For other papers of this time on rotating bodies, see Servois *1808a* (pressure on the point of rotation) and Rodrigues *1814a* (formulae for the rotational variables).

6.3 Variational mechanics: the Poisson and Lagrange brackets, 1808–1810

6.3.1 Volumes 3 and 4 of Laplace's *Mécanique céleste*, 1802–1805

Volume 3 of Laplace's *Mécanique céleste* appeared in 1802 (dedicated to 'Bonaparte, member of the *Institut National*') and volume 4 three years later (*not* dedicated to Emperor Napoléon); they were published by Courcier, successor to Duprat, who had brought out volumes 1 and 2. Biot reviewed them in the *Moniteur universel* (*1803b* and *1805b* respectively). In 1810 the work as a whole was awarded an *Institut* prize as the best achievement in astronomy and mechanics during the decade (see *Institut de France 1810a*, 8–11).

'Nothing now remains', promised Laplace, 'to fulfil the engagement which was entered into at the commencement of this work, but to give an historical account of the labours of mathematicians and astronomers on the system of this world' (*4* (1805), xxi); but this promise was to be twenty years in the fulfilling, as parts of volume 5, which also contained accounts of new work done in the interim (§ 12.6). Indeed, he found that the new volumes soon in need of addition, for later in the 1800s he published appendices to both of them. One of these appendices concerns the developments discussed in this section, which start from the studies of the stability of the planetary system reported in § 5.4.6 and developed further in the new volumes of the *Mécanique*.

6.3.2 Poisson's first study of the stability of the planetary system, 1808

Laplace discussed the perturbing forces of the second power of the masses of the planets in these new volumes (*3* (1802), 87–160; *4* (1805), 126–169). According to him, the effects of these forces were 'insensible' except for some aspects of the Saturn/Jupiter interaction and the motions of the satellites of Jupiter, where the relations (545.2, 4) respectively led to 'sensible' terms of this order.

Laplace's discussion inspired Poisson to study these forces in general in a paper *1809a* on the 'secular inequalities of the mean motions of the planets'. He looked especially at the equation for the mean anomaly u, given by

$$\int n\,dt + \varepsilon - \omega = u - \sin u, \qquad (632.1)$$

in which, unlike Laplace's (544.3), the mean angular motion n (and also the longitudes ε at the epoch and ω of the perihelion) were not assumed to be constant. Thus when seeking expression for the differentials of the orbital variables, to the first power in the mass m of (any) planet, he found, among other relations,

$$d\varepsilon = (1 - \sqrt{1 - E^2})\, d\omega + (2a^2/\mu)\, R_a n\, dt, \quad \text{where} \quad \mu := M + m, \tag{632.2}$$

R was the perturbing function, and E eccentricity (*1809a*, 14–22).

Poisson was particularly concerned with the quantity

$$\varrho := \int n\, dt = (3/\mu) \iint an\, d'R\, dt, \tag{632.3}$$

where '$d'R$' was restricted to differentials on the variables of the planet with mass m. 'It is with this formula that we shall be principally concerning ourselves in the rest of this Paper' (p. 14). If R, and indeed the orbital variables, contained terms of the order of m^2 in their trigonometric expansions (544.15), then Laplace's expression $(632.3)_3$ (introduced in *Mechanics*, *1* (1799), 642) would yield after double integration terms of the form $(m^2/p^2) \sin(pt + A)$, which would be 'sensible' if p was of the order of m (compare (545.3)).

Poisson's analysis comprised a sequence of exceedingly complicated calculations. They are not easily summarisable here, but they broadly followed Laplace's volume 3 and relied much on the expansion (544.15) of R in a series of trigonometric terms. He showed that ϱ took only periodic inequalities due to the second power of the masses (p. 25); that $d'\delta R$ (that is, the differential relative to m of the first variation of R) contained terms of the order of m^2 but was independent of the mean motions of any of the planets, so confirming the absence of secular inequalities in ϱ (p. 34); and that if the mean motion of planet m_1 contained no secular inequalities involving m_1^2 then it also possessed none of the type $m_1 m_2$, where m_2 was another planet of the system (p. 38).

'The result to which the preceding analysis has led us, suffices for the needs of astronomy', concluded Poisson, since the next order of inequalities would have to be the fourth power of masses, which 'will thus remain mingled [*confondues*] with the errors of the observations' (p. 39). However, 'it is always an interesting question of analysis, to know if the function $d'R$ never contains' secular inequalities in its terms of order m^2, and to any degree of eccentricities and inclinations (hitherto he had taken these terms only to the fourth power). Again, he based his reasoning on the trigonometric expansion of R (and also on the claim on p. 39 that $\delta\varrho = \varrho$). The expression for $d'\delta R$ fell into the form of a sum of terms each of the form

$$A\left(P \int Q\, dt - Q \int P\, dt\right), \tag{632.4}$$

where A was some constant and P and Q were products of trigonometric terms and differentials of R: when evaluated it yielded only periodic terms (pp. 43–45), so leaving the mean motion (and also the major axes of the planets) '*invariable, when one sets aside periodic inequalities*' (p. 47).

Continuing with a study of the secular inequalities of the mean longitude $(nt + \varepsilon)$, Poisson took Laplace's ordinary differential equations (546.2-3) for two planets m_1, and m_2, cast in Lagrange's coordinates (546.1)

$$h_j := E_j \sin \omega_j \quad \text{and} \quad l_j := E_j \cos \omega_j, \qquad j = 1, 2, \tag{632.5}$$

and showed that (632.2) could be converted to the form

$$d\varepsilon_1 = \tfrac{1}{2}\,[A_{12}(h_1^2 + l_1^2) - B_{12}(h_1 h_2 + l_1 l_2)]\; dt + (2a^2/\mu)\; R_a n\; dt + \mathrm{O}(E^2), \tag{632.6}$$

which when integrated would yield terms of the order $m_1 m_2 t^2$ but with coefficients too small to make them observable, as Laplace had shown.[1]) He also verified Laplace's relation (546.17)

$$m_1 \sqrt{a_1}\; d\varepsilon_1 + m_2 \sqrt{a_2}\; d\varepsilon_2 = 0 \tag{632.7}$$

with his own reading (632.6) of $\{d\varepsilon_j\}$ (Poisson *1809a*, 51).

Poisson concluded with some 'Remarks on the stability of the planetary system'. He found the relation

$$m_1 a_2 n_2 \sqrt{1 - E_1^2}\; \cos \alpha_1 + m_2 a_1 n_1 \sqrt{1 - E_2^2}\; \cos \alpha_2 = \text{constant} + \mathrm{O}(m_1^4), \tag{632.8}$$

where α_1 and α_2 were the eccentricities of the orbits (p. 53) and confirmed Laplace's assertion that such equations were not affected by the perturbing forces of order m_1^2 (*Mechanics, 3* (1803), 121-129; Laplace's equations furnished the coefficients of the invariable plane (544.2)). Further straightforward calculations confirmed the smallness of the constant value in (632.8) and related equations, and so confirmed stability after the manner of Laplace's consideration of $(546.17)_2$ (Poisson *1809a*, 55). 'We thus conclude, in ending this paper, that the stability of the planetary system is assured relative to the great axes, to the eccentricities and to the inclinations, even in having regard to the squares of the perturbing forces'.

Poisson had submitted his paper to the *classe* on 20 June 1808 (*PV*, *4*, 78), the day before his 27th birthday. On 16 August Biot reported at length and in great praise, almost demanding publication in the *Savants étrangers* and stating that 'The *classe* will note with interest the constant progression of the talent of this young geometer' (*1808b*, 92). In fact, as usual the paper appeared in the *Journal*

---

[1]) See Laplace's *Mechanics, 3* (1802), 295, 310. Laplace and Poisson calculated angles in the new centagesimal system (§ 5.5.4); Bowditch usually converted them into the orthodox system, which means that, for example here, his values of seconds of arc have to be multiplied by 3.086 to convert back to Laplace's.

de l'Ecole Polytechnique (with a summary Poisson *1808e* sent to the *Société Philomatique*); but more significant was the interest noted by Biot's co-reporters Lagrange and Laplace. For the day after the report they presented their own papers on the matter to the *Bureau des Longitudes*, and at the *classe* during the following week (*PV*, *4*, 93); and meanwhile Poisson had been elected an *adjoint* at the *Bureau*.[1])

6.3.3 Laplace's and Lagrange's adumbrations of Poisson's study, 1808–1809

Laplace rapidly published his essay in September 1808 as a 24-page appendix *1808b* to volume 3 of his *Mécanique céleste*, rather than (or in addition to) putting it out as a paper. It is included in all editions of the volume, and by Bowditch.

Laplace acknowledged the stimulus of Poisson's paper, and in running through the differential forms credited Poisson explicitly with (632.2). He gave a more succinct explanation than had Poisson of the way in which the new terms were handled, showing more clearly the bearing of the form (632.4) on the analysis by casting the variation of R in the form

$$\delta R - (\dot{R}/n)\,\delta \int n\,dt = \sum_{j=1}^{4} A_j\left(M_j \int N_j\,dt - N_j \int M_j\,dt\right), \tag{633.1}$$

where the $\{M_j\}$ and $\{N_j\}$ were partial differentials of R with respect to the variables of the problem, and the $\{A_j\}$ constants. Thus the product rule of differentiation gave (in Poisson's notation)

$$d'\left(\delta R - (\dot{R}/n)\,\delta \int n\,dt\right) = \sum_{j=1}^{4} A_j\left(dM_j \int N_j\,dt - dN_j \int M_j\,dt\right). \tag{633.2}$$

Appealing to the trigonometric expansion (544.15) of R, and thus of both M_j and N_j, Laplace showed that $(633.2)_2$ vanished because it took the form

$$\sum_{i=1}^{4}\left[\sum_{j(i)} B_j \sin(jt+\alpha)\,dt \int C_j \cos(jt+\beta)\,dt \right. \\ \left. - C_j \sin(jt+\beta)\,dt \int B_j \cos(jt+\alpha)\,dt\right], \quad \text{or} \quad \sum_{i=1}^{4} 0 \tag{633.3}$$

[1]) See *BL*, *3*, meetings of 10, 17, 24. 8. 1808. Poisson was surprisingly inactive in *BL* business.

(*1808b*, 713–717). Further,

$$d'\left((\dot{R}/n)\ \delta \int n\ dt\right) = (d'\dot{R}/n)\ \delta \int n\ dt + (\dot{R}/n)\ \delta n\ dt \tag{633.4}$$

$$= 3an\left(d\dot{R}/n \iint dR\ dt + \int (R/n)\ dt \int d'R\right) \tag{633.5}$$

from (632.3) (with $\mu = 1$), which reduced to a sum of trigonometric terms when the expansion of R was introduced. Hence $d'\delta R$ would obtain only periodic terms from the coordinate variables of the planet; and similar reasoning applied to the variables of other planets (pp. 717–726). He also summarised Poisson's analysis of ϱ in (632.3) with the equation

$$\mu\delta\varrho = 3an \iint dt\ d'\delta R + 3a^2 \iint \left(n \int d'R\right) dt\ d'R = 0 + 0 \tag{633.6}$$

from (633.5) and related results (pp. 726–727).

Among other topics discussed in the appendix, Laplace reworked his analysis of the variations of the orbital variables allowing for all powers of eccentricity and inclination, and not just to the fourth power as he had done earlier (*2* (1799), 634–725: some of that work was noted in § 5.4.5). Restricting himself to the two-planet case, he found results conformable with his previous treatment (*1808b*, 728–749).

In extending his analysis to all powers of eccentricity and inclination, Laplace was attaining a standard comparable to Lagrange's original treatment of a and nt in *1779b*. That examination had been carried only to the first power in masses; but Poisson's recent paper had for Lagrange 'reawakened an object which had much occupied me in the past, and of which I had afterwards totally lost sight' (*1809a*, 717). Lagrange's interest was somewhat different from Laplace's, in a way which typifies their differences in general; whereas Laplace wanted to check out some things in theoretical astronomy, Lagrange obtained beautiful formulae, newly simplified and symmetrical. 'Thus his [Laplace's] work and mine, leading to the same goal by different routes', commented Lagrange, 'can equally serve for the advancement of Analysis and of physical Astronomy' (pp. 719–720).

In the account about to be given, I use subscripts to distinguish the constants of integration. As in his *1779b*, Lagrange *1809a* started from the basic equations (544.5) of motion, which for Ox read

$$\ddot{x} + (M + m)\ x/r^3 = \Omega_x \tag{633.7}$$

(where Ω was now the perturbation function), and varied the six constants $\{c_j\}$ created by the integration of the reduced equations of (633.7) (where 0 was set on the right hand sides). In order to obtain six equations relating these constants, he set the first variations to zero as equations of condition:

$$\delta x := \sum_{j=1}^{6} x_{c_j}\, dc_j = 0\,; \quad \text{also for } y \text{ and } z, \tag{633.8}$$

and substituted the second variations into (633.7) to obtain in art. 3

$$\delta(dx/dt) = \sum_{i=1}^{6} x_{tc_i}\, dc_i = \Omega_x\, dt; \quad \text{also for } y \text{ and } z. \tag{633.9}$$

The previous manipulation of these equations had led to very complicated expressions; but now Lagrange, doubtless inspired by the form (632.4) used in Poisson's paper, noted that by introducing the notation

$$(x, c_i, c_j) := x_{tc_j}\, x_{c_i} - x_{tc_i}\, x_j, \tag{633.10}$$

he could obtain by elimination the equation

$$\Omega_x x_{c_i}\, dt = \sum_{j=1}^{6} (x, c_i, c_j)\, dc_j, \quad j \neq i; \quad \text{also for } y \text{ and } z. \tag{633.11}$$

$$\therefore\ \Omega_{c_i} = \Omega_x\, x_{c_i} + \Omega_y\, y_{c_i} + \Omega_z\, z_{c_i}, \tag{633.12}$$

$$\therefore\ \Omega_{c_i}\, dt = \sum_{j=1}^{6} (c_i, c_j)\, dc_j, \quad j \neq i \tag{633.13}$$

(art. 4), where from art. 6

$$(c_i, c_j) := (x, c_i, c_j) + (y, c_i, c_j) + (z, c_i, c_j). \tag{633.14}$$

Further, it followed from (633.13) by elimination (equivalent to matrix inversion) that

$$dc_i = \sum_{j=1}^{6} \varrho_{ij}\, \Omega_{c_j}\, dt, \tag{633.15}$$

where the $\{\varrho_{ij}\}$ 'will be given only by the coefficients' defined in (633.14) 'and will only be simple [but unknown] functions of the elements without t; which furnishes a very important and very useful Theorem in the Theory of the perturbations of the planets'. The results were also independent of the inverse-square form of the force law embodied in (633.7) (art. 7).

Lagrange gave especial attention to the variation δa of the major semi-axis because of the appearance of a in expressions such as (632.3) for n. After recalling his analysis in *1779b*, he used his new results. Given a system $\{m_k\}$ of masses, the integration of equations of the form (633.7) yielded constants $\{c_{ki}\}$ of integration. The contribution to the perturbation function Ω_k for the planet with mass m_k from the variation of its own constants could be written as

$$\sum_{i=1}^{6} \Omega c_{ki} \Delta c_i, \quad \text{or} \quad \sum_{i=1}^{6} (c_{ki}, c_{kj}) (\Delta c_{ki}\, d\Delta c_{kj} - \Delta c_{kj} d\Delta c_{ki}), \qquad i \neq j \quad (633.16)$$

from (633.13), where Δc_{ki} was the variation on c_{ki} (so written by Lagrange, and handled as a forward difference). Drawing on the trigonometric expansion of Ω, he imitated Poisson's reasoning over the form of (632.4) to show, like Laplace's (633.3), that each difference expression in $(633.16)_2$ was zero (arts. 11–13).

However, lack of symmetry with respect to $\{m_k\}$ in the basic equations (633.7) of motion precluded the same reasoning applying to the variations on Ω_k due to the other planets. So Lagrange adopted the procedure of his *1779a* of transferring the origin of coordinates from the centre of the sun M to the centre of gravity of M and the $\{m_k\}$ to obtain a symmetrical form for Ω. He sketched out the corresponding analysis, making some non-trivial mistakes in passage work (arts. 14–17).[1])

Lagrange read this paper to the *classe* a few days after its airing at the *Bureau des Longitudes* (*PV*, *4*, 93), and three weeks later presented an addition (*PV*, *4*, 102) on the variations of the orbital variables themselves (*1809a*, arts. 18–28). He recalled from *Mechanics*$_1$ (1788), 337–371 his complicated expressions for the rotation of a body, and noted the 'astonishing manner' in which the notation '(c_i, c_j)' simplified them.

These contemporary (but, of course, independently prepared) papers by Paris's two leading mathematicians naturally attracted attention. His duty as *secrétaire perpétuel* doubtless boosted by his interests in astronomy, Delambre began his account of the *classe*'s work for 1808 with a long summary of these latest achievements by 'two geometers of the first order' and 'their worthy student' (*1809a*, 1-9). And the student himself described the papers to the *Société Philomatique* (Poisson *1808b* and *1809c*)—very modestly, as he did not mention his own paper which had started off this line of thought.

6.3.4 Lagrange on variational mechanics, 1809

Lagrange was obviously aware of the potential value for mechanics of the new notation '(c_i, c_j)', especially its use of the form $(\alpha\beta - \gamma\delta)$ which had appeared in Poisson's (632.4), his own (633.16) and elsewhere. In March 1809 he presented a

[1]) Lagrange *1809a*, 741–749 (with notes by the *Oeuvres* anonymous editor). As usual, I have not used his notations in this discussion, not only because of primes but also because on pp. 738–741 he used 'a' for both the major semi-axis and for c_1, put the sun's mass $M = 1$ throughout, and wrote $\sum_k m_k$ as M.

paper *1809b* entitled 'On the general theory of the variation of arbitrary constants in all the problems of mechanics' (*PV*, *4*, 175), although it only dealt with dynamics. He started with the Lagrange equations (524.8), which I write as

$$dR_{\dot{\xi}_j} - R_{\xi_j}dt = \Omega_{\xi_j}dt, \quad \text{where} \quad R := T - V, \tag{634.1}$$

T was the *forces vives* function, V the potential work function (524.5) for the impressed forces on the masses and $(-\Omega)$ the (perturbation) function for the intercorporeal forces (*1809b*, art. 5).[1]) He followed the procedures of his *1809a* even to the extent of working with only three variables in (634.1); I write 'ξ_j' to emphasise the truth of the results for more variables.

Lagrange assumed to be known the solution to the reduced equation of (634.1), when $\Omega = 0\xi_1$; the six constants of integration $\{c_i\}$ were then specific functions of $\{\xi_j\}$ and t. Their variation would be the means to find the solution of (634.1) itself; for, as with (633.8-9), the first variations would supply three equations of condition

$$\delta\xi_j := \sum_{i=1}^{6} (\xi_j)_{c_i} dc_i = 0, \qquad 1 \le j \le 3 \tag{634.2}$$

(where 'd' varied only over the $\{c_i\}$), and when substituted into (634.1) they yielded, because of (634.2), the three equations

$$\sum_{k=1}^{3} R_{\dot{\xi}_k \dot{\xi}_j} \delta\dot{\xi}_k = \Omega_{\xi_j} dt, \qquad 1 \le j \le 3 \tag{634.3}$$

(Lagrange *1809b*, arts. 7-8).

Lagrange sought expressions for $\{c_i\}$ 'as we have so done for the variations of the planets in the cited Paper [*1808a*]' (art. 8). The passage-work was straightforward, though the moves were cleverly chosen. He found in the end the integral

$$\begin{aligned} &\sum_{i=1}^{3} R_{\dot{\xi}_i\dot{\xi}_i}(\Delta\xi_i\delta\dot{\xi}_i - \delta\xi_i\Delta\dot{\xi}_i) + \sum_{i=1}^{3} R_{\dot{\xi}_i\dot{\xi}_j}(\Delta\xi_i\delta\dot{\xi}_j - \delta\xi_i\Delta\dot{\xi}_j + \Delta\xi_j\delta\dot{\xi}_i - \delta\xi_j\Delta\dot{\xi}_i) \\ &+ \sum_{i=1}^{3} [(R_{\xi_i\dot{\xi}_j} - R_{\xi_j\dot{\xi}_i})(\Delta\xi_i\delta\xi_j - \delta\xi_i\Delta\xi_j)] = K, \qquad i \neq j, \end{aligned} \tag{634.4}$$

where K was a function only of the $\{c_i\}$ and their differentials and variations (art. 15). Since these were independent of each other, he could, for example, retain only Δc_1 and δc_2 as non-zero, so that

$$\Delta\xi_j = (\xi_j)_{c_1}\Delta c_1 \quad \text{and} \quad \delta\xi_j = (\xi_j)_{c_2}\delta c_2, \tag{634.5}$$

[1]) I follow Lagrange in using 'R' to denote $(T-V)$ and not the perturbation function; but otherwise I avoid his notations, especially his use of primes to serve both as subscripts and as t-derivation.

and so on, and thus convert (634.4) to

$$A = \Delta c_1 \delta c_2 (c_1, c_2), \tag{634.6}$$

where I let 'A' stand for the complicated expression to which $(634.4)_1$ was converted, and '(c_1, c_2)' was Lagrange's bracket notation for this expression itself (with the factor $\Delta c_1 \delta c_2$ cancelled off: art. 16). This value of (c_1, c_2) was different from that in (633.14) in his earlier paper; but it had similar properties, and when its companions were similarly defined he obtained exactly the old formula (633.13):

$$\Omega_{c_i} dt = \sum_{j=1}^{6} (c_i, c_j)\, dc_j, \quad j \neq i, \tag{634.7}$$

'which differ from them only in the value of the symbols' $\{(c_i, c_j)\}$ (art. 20). The aim of finding expressions for $\{c_i\}$ was now accomplished via (633.15) (art. 23).

In an 'Addition' to this paper Lagrange simplified his analysis considerably. He now realised that simply by multiplying the reduced equation of (634.1) by $\Delta \xi_j$, summing them over $j = 1, 2, 3$ and drawing on the identity

$$\Delta \xi_j dR_{\dot{\xi}_j} = d(R_{\dot{\xi}_j} \Delta \xi_j) - R_{\dot{\xi}_i} d\Delta \xi_j \tag{634.8}$$

he could replace the complicated (634.4) by

$$\sum_{j=1}^{3} [\Delta \xi_j \delta R_{\dot{\xi}_j} - \delta \xi_j \Delta R_{\dot{\xi}_j}] = K \tag{634.9}$$

(art. 25), so that (634.6) reduced via (634.5) to

$$(c_1, c_2) = \sum_{j=1}^{3} [(\xi_j)_{c_1} T_{\dot{\xi}_j c_2} - (\xi_j)_{c_2} T_{\dot{\xi}_j c_1}], \tag{634.10}$$

with similar expressions for the other constants (art. 26). He replaced R by T partly because V was free of the $\{\dot{\xi}_j\}$ but especially to emphasise the 'generality and simplicity' of the results (and their earlier versions) due to their dependence only on the *forces vives* function (art. 22). After some considerations of the bearing of the analysis on the function $(T + V)$ and a fresh derivation of (634.9), he got round to substituting (634.10) into (634.7) to obtain in art. 35

$$\Omega_{c_i} dt = \sum_{j=1}^{3} [(\xi_j)_{c_i} \delta T_{\dot{\xi}_j} - \delta \xi_j T_{\dot{\xi}_j c_i}]. \tag{634.11}$$

'It is there, it seems to me, that Analysis can give most simply for the variation of the arbitrary constants in Problems of Mechanics', Lagrange concluded (*1809b*, art. 35). However, in a brief supplement sent in some time later he noted that, since 'δ' excluded differentials with respect to t, the basic equations of motion (634.1) could be written as

$$\delta R_{\xi_j} = \Omega_{\xi_j} dt, \qquad 1 \leq j \leq 3, \tag{634.12}$$

so that (634.11) could be converted to

$$\Omega_{c_i} dt = \sum_{j=1}^{3} ((\xi_j)_{c_i} \delta R_{\xi_j} - \delta \xi_j R_{\xi_j c_i}). \tag{634.13}$$

6.3.5 The Lagrange and Poisson brackets, 1809–1810

> The application that I made [of the equations of motion] successively to the perturbations of the motion of a point attracted towards a fixed centre according to any law, and to those of the motion of rotation of a solid body, led me to identical formulae for these two problems[,] so different from each other; so that the analogous constants in the two questions, have the same differential expression; a singular result that one can consider as a theorem of mechanics. These two questions comprise all astronomy [...]
>
> Poisson *1827a*, 201

Lagrange was in his 73rd year when he produced *1809b.* It shows some signs of age; when younger he would have tidied up the material in the 'Addition', for example. But it shows that his remarkable ability to manipulate general formulae in variational mechanics was unimpaired. Delambre began his account of the *classe*'s work during 1809 with a glowing summary of the paper (*1810b*, 1–4); and Poisson informed the *Société Philomatique* not only of its contents (*1809d*) but also of his own new researches in this area (*1809h*), of which he submitted his main paper to the *classe* in October 1809 (*PV*, *4*, 266). Lacroix *1809b* praised that paper very highly in his December report, and the paper appeared that same month, as Poisson *1809f*, in the *Journal de l'Ecole Polytechnique.*

'The formulae of the last Paper of Mr. *Lagrange* are the inverses of ours', Poisson explained (*1809f*, 268); whereas Lagrange had formed differentials and variations with respect to the constants $\{c_i\}$, Poisson took the variables $\{\xi_j\}$ as base. He converted the reduced form of the Lagrange equations (634.1), by introducing new variables

$$\eta_j := T_{\xi_j} = R_{\xi_j}, \quad \text{to the form} \quad \dot{\eta}_j = R_{\xi_j}, \qquad 1 \leq j \leq 3. \tag{635.1}$$

For Poisson this was 'the simplest form that one can give' to the basic equations

(p. 273);[1]) they showed symmetry, in that the $\{\eta_j\}$ and the $\{\xi_j\}$ could each be treated as functions of the other (and also of the $\{\xi_j\}$, of course).

Poisson's approach was to assume as known two solutions of the basic equations $(635.1)_2$, in the form

$$a = f(t, \{\xi_j\}, \{\eta_j\}) \quad \text{and} \quad b = g(t, \{\xi_j\}, \{\eta_j\}) \tag{635.2}$$

and then manipulated, in Lagrangian style, the equations

$$da = 0 \quad \text{and} \quad db = 0 \tag{635.3}$$

in their expansions by the chain rule. Eventually he found on p. 280 an exact differential free from terms in t, whose integral yielded

$$[b,a] := \sum_{j=1}^{3} (b_{\xi_j} a_{\eta_j} - b_{\eta_j} a_{\xi_j}) = \text{constant.} \tag{635.4}$$

$[b,a]$ is the 'Poisson bracket', as it has become known, born out of his own use of the form $(\alpha\beta - \gamma\delta)$ in (632.4) and of intermediate versions, especially the bracket expressions of Lagrange's (633.14) and (634.10). Poisson noted immediately the obvious properties

$$[b,a] = -[a,b] \quad \text{and} \quad [a,a] = 0, \tag{635.5}$$

and also found *explicit* expressions for the coefficients $\{\varrho_{ij}\}$ in Lagrange's form (633.15) for the differentials of the constants $\{c_i\}$:

$$dc_i = \sum_{j=1}^{6} [c_i, c_j] \, \Omega_{c_j} dt \qquad j \neq i, \quad 1 \leq j \leq 6 \tag{635.6}$$

(pp. 288–289). He also recalled Lagrange's result (633.13), of which (635.6) was the converse:

$$\Omega_{c_i} dt = \sum_{j=1}^{6} (c_i, c_j) \, dc_j, \qquad j \neq i, \tag{635.7}$$

where Lagrange's expression (634.10) was the 'Lagrange bracket', expressible via (635.1) in the style of (635.4) as

$$(b,a) := \sum_{j=1}^{3} ((\xi_j)_a (\eta_j)_b - (\eta_j)_a (\xi_j)_b) = \text{constant.} \tag{635.8}$$

---

[1]) Poisson changed the letters used by Lagrange for the independent variables but preserved the use of primes for both subscripts and t-derivation; so I continue to employ the letters of § 6.3.4. Poisson's derivation of $(634.1)_2$ was marred by writing (the equivalent of) 'ξ_j' instead of '$\dot{\xi}_j$' in several places. On his use of brackets, see after (633.12).

Poisson also noted that (b,a) satisfied (635.5) (pp. 289–290), although he did not seek further properties.

The rest of Poisson's paper was taken up with reworking various important problems in mechanics. For a body moving under a central force with other perturbing forces (that is, (544.5) for planetary motion) he showed, for example, that the coefficients of Laplace's invariable plane (544.2) could be expressed in terms of Poisson brackets (p. 296), and he verified various results in Lagrange *1809a* and Laplace *1809e.* For the rotation of a rigid body about a fixed point, after a slow crawl to the Euler equations (525.15) and the basic solutions by elliptic integrals (pp. 314–323), he found expressions for the contents of integration to substitute into (635.6) (p. 336). He then applied this analysis to the particular case of precession and nutation of the earth (§ 5.5.7), and on p. 344 verified Laplace's formulae for the components of the angular velocity about the principal axes (*Mechanics, 2* (1799), 828–836).

This last section of the paper was an appendage to a paper Poisson *1809b* on the rotation to the earth, submitted the previous March (*PV, 4,* 181), approved in Lacroix's report *1809a* a month later, and summarised by Poisson himself to the *Société Philomatique* in *1809e.* As in his *1809a* of a few months earlier, he considered second-order terms to show that the displacement of the axis of rotation, and the consequent variation in the angular velocity of the earth, were undetectably small.[1])

The most striking innovation in Poisson's *1809f* was, of course, the algebra of the brackets, and it inspired Lagrange to a second stimulus from Poisson's work. He presented a new, short, paper on variational mechanics in February 1810 (*PV, 4,* 319); as usual, Poisson *1810b* described it to the *Société Philomatique.* Lagrange took up his expression (634.11) for $\Omega_{c_i}dt$; for him 'the major point of this formula is that the second member of the equation must become independent of time after the substitution of these values of $[\{\delta\xi_j\}]$' (*1810a,* art. 1). In the spirit of his calculus (§ 3.2.2) he assumed power-series expansions in t for the $\{\xi_j\}$:

$$\xi_j = \alpha_j + \sum_{k=1}^{\infty} \alpha_{jk} t^k \qquad 1 \leq j \leq 3\,, \tag{635.9}$$

which yielded such forms for $\{T_{\xi_j}\}$ as:

$$T_{\xi_j} = \lambda_j + \sum_{k=1}^{\infty} \lambda_{jk} t^k\,. \tag{635.10}$$

---

[1]) A feature of Poisson *1809b* is taken up at (1764.2) below. In a short paper published in the *Connaissance des temps, 1819g*, he simplified the theory of precession and nutation in his *1809b* by drawing on Laplace's treatment of surface harmonics for a spheroid covered by a fluid in equilibrium (Laplace *Mechanics, 2* (1799), 344–357: this work was mentioned after (552.11)).

Since $(634.11)_2$ was independent of t, one could put $t = 0$ in (635.9-10) straightaway, so that (634.11) became

$$\Omega_{c_i} dt = \sum_{j=1}^{3} ((\alpha_j)_{c_i} \delta\lambda_j - (\lambda_j)_{c_i} \delta\alpha_j), \qquad 1 \le i \le 6. \tag{635.11}$$

Now the $\{\alpha_j\}$ and $\{\lambda_j\}$ could serve as the six constants of integration for the basic equations. Hence (635.11) was convertible to 'the simplest form that is possible':

$$\Omega_{\alpha_j} dt = d\lambda_j \quad \text{and} \quad \Omega_{\lambda_j} dt = -d\alpha_j, \qquad 1 \le j \le 3, \tag{635.12}$$

where d-differentiation was used since t was involved in the variation of these constants (art. 2). The rest of the paper was devoted to properties (635.5-6) of the Poisson brackets.

The use of brackets in these papers, and in my discussion, needs explanation. Both Lagrange and Laplace had used round and square brackets, and related notations, in their previous work in theoretical astronomy, especially to represent the coefficients in the expansion (546.4) of the distance function;[1]) Poisson caused confusion when he adopted them in this work. For example, while Lagrange employed round brackets in *1809a*, art. 6 to notate (633.14), Poisson used square brackets in his summary of that paper (*1809c*, 272); and then he used them again in his *1809f*, 280 for his own bracket expression (635.4)! Lagrange clarified matters in his *1809b* and *1810a* by using round and square brackets in the way that I have followed; but then he added confusion of his own by reversing this practice when describing the new developments in the second edition of his *Mécanique analytique*, writing round brackets for the Poisson bracket (*1* (1811), 351-363) and square ones for the Lagrange bracket (*2* (1815), 81-101). Since then the notations have been used both ways round, and even sometimes with rather different expressions assigned to them. Given the notational nightmares that often attend Lagrangian mathematics, one is not surprised.

[1]) See, for example, Lagrange *Mechanics*$_1$ (1788), 243-248 and Laplace *Mechanics, 1* (1799), 583-588; I used 'b_s^{ij}' in (546.4) to avoid confusion with the concepts described here.

Maurice *1845a* described the theory presented in Lagrange *1809a* and *1809b* and Poisson *1809f*, and also analysed some of the terms that they had omitted. Although presented at the *AS* in August 1842, Maurice's paper had been written much earlier; around 1830, according to his letter of 27. 12. 1830 to Libri (*Biblioteca Moreniana* (Florence), *fonds Palagi*, ms. 433).

6.3.6 *Dénouement*

Thus an important chapter in the history of variational mechanics was launched; and the essential steps were taken quickly, for between June 1808 and February 1810 Poisson (in his late twenties) and Lagrange (in his early seventies) each wrote three papers, and the sixty-year-old Laplace contributed an essay (see Table 636.1).[1]) Poisson himself took the matter somewhat further in a paper *1816g* in the *Mémoires*, with two self-summaries *1816d* and *1816e* written for the *Société Philomatique*. Adopting Lagrange's first convention—the ones used in § 6.3.5—for the brackets, he incorporated constraints and impressed force functions into his general equations of motion, converted them into canonical forms of the type $\sum(\alpha\beta-\gamma\delta)$, and calculated both brackets for the variation of these arbitrary constants with time (*1816g*, 16–29). Among applications, he studied the variation of the major semi-axis a and mean angular velocity n. Mindful from (632.3) and elsewhere of the double integrals in expressions for ϱ, he considered $\ddot{\varrho}$ and showed that, both to terms of second and third powers of the masses, the form $\sum(\alpha\beta-\gamma\delta)$ turned up again. Hence, in a manner similar to (633.16), zero sums would obtain, and so no non-periodic inequalities arose (pp. 54–63).

As ever with new insights, there were some missed opportunities, too, which

Table 636.1 Chronology of papers described in this section.

Paper	Date of *AS* presentation	Poisson's summary	My account	Contents
Poisson *1809a*	20.6.1808	*1808e*	§ 6.3.2	Secular variations
Laplace *1808b*	22.8.1808	*1808b*	§ 6.3.3	Use of Poisson *1809a*
Lagrange *1809a*	22.8.1808	*1809c*	§ 6.3.3	Use of Poisson *1809a*
Lagrange *1809b*	13.3.1809	*1809d*	§ 6.3.4	Variational mechanics; Lagrange brackets
Poisson *1809b*	20.3.1809	*1809e*	§ 6.3.5	Rotation of the earth
Poisson *1809f*	16.10.1809	*1809h*	§ 6.3.5	Variational mechanics: Poisson brackets
Lagrange *1810a*	19.2.1810	*1810b*	§ 6.3.5	Variational mechanics: Poisson brackets
Poisson *1816g*	2.9.1816	*1816d*, *1816e*	§ 6.3.6	Variation of constants under time-dependence

[1]) For a clear, though non-technical, summary of this history, see Gautier *1817a*, 269–278. The account of Poisson *1809f* in Dugas *1950a* is mediocre. Later studies of the rotation of the earth are taken up in § 17.6.4.

were to be seized especially by Jacobi. For example, neither Lagrange nor Poisson explored the general properties of their brackets much beyond (635.5); the useful theorem

$$[a,[b,c]]+[b,[c,a]]+[c,[a,b]]=0 \tag{636.1}$$

was proved in Jacobi *1862a*, art. 26. More importantly, they did not fully grasp the importance of the constancy of the value of $[b,a]$. Immediately after stating (635.4), Poisson had remarked that $[b,a]$ 'will be in general a function of a and b, and of arbitrary constants contained in the other integrals of the equations of motion; sometimes it may happen that its value will contain neither the constant a, nor the constant b; in other cases it will not contain any arbitrary constant [...]' (*1809f*, 280–281); but it was Jacobi who noticed that if in fact $[b,a]$ was not an identity, then in general it gave a third solution, and by repetition of the process all others would follow (see Bertrand *1853a*: in some cases only combinations of known solutions are obtained). The bracket was 'The most profound discovery of Mr. Poisson', judged Jacobi in an obituary note *1840a* on Poisson; however, it 'has not been well understood either by Lagrange [...] or by its author himself'.

It would be ungracious to dwell on these missed opportunities when so much had been achieved. More curious, however, is the *classe*'s failure to give Poisson the appropriate recognition when the chance presented itself. In April 1807 Lalande died, leaving a vacancy in the astronomy *section* of the *classe*. The election did not take place until September 1809, and Poisson's current work in variational mechanics, together with his other achievements, surely made him the obvious successor; indeed, one may reasonably suspect that Lalande's death, and the resulting vacancy, had motivated Poisson to these researches in the first place. Nevertheless, only four votes went to him, against the 47 given to Arago (*PV*, *4*, 253), who, in his own later words, 'do not pretend that at the age of twenty-three years my scientific baggage was very considerable' (*1854a*, art. 45: he even claimed to have received Laplace's vote (art. 46), though his testimony is not reliable, considering his later vindictiveness against Laplace). Poisson, as we have seen earlier and will find again later, played the career game pretty hard; on this occasion he was outclassed by a real professional! He did earn a consolation prize, however, in 1810, when a commission (including Arago) awarded him the Lalande astronomy prize for 1810 (*PV*, *4*, 399, 405)—for his 'analytical papers'.

6.4 Celestial and planetary mechanics—with statistics

6.4.1 Some trends in French theoretical astronomy

Theoretical astronomy received many presentations in France following Laplace's volumes. Biot published his *Traité élémentaire de l'astronomie physique*, dedicated to Laplace, in 1805, with a three-volume second edition in 1810-1811 (Biot *Astronomy*); Delambre published his lectures given at the *Collège de France* as a three-volume treatise *Astronomie théorique et pratique* (*1814d*), preceding it with an abridged edition *1813a*; while Francoeur's elementary textbook (*Uranography*) saw its first two editions in 1812 and 1818. These works paid attention to work done since the appearance of Laplace's volumes as well as results in it; they took some note of certain contributions made by Germans during this period (on which see Interlude 641.1). However, the mathematics was restricted: in particular, very little calculus was used.

Interlude 641.1 Notes on German theoretical astronomy

In some contrast to the French penchant for complicated equations representing hopefully "exact" solutions (and even in the approximations attempted), German work at this time is notable for its emphasis on numerically convenient procedures, even if approximations are brought in at an early stage. The most important contribution, which typifies the attitude quite well, was Gauss's *Theoria motus* (*1809a*) on the motions of the heavenly bodies in elliptic, parabolic or hyperbolic orbits. Significantly, as an expression of his preference, Bowditch appended to his edition of Laplace an account of Gauss's initial moves, especially the computation of the true anomaly v in terms of the time t from equations (544.3) and others without too much logarithmic toil, and Gauss's method of determining the orbits of Ceres and her sister asteroids (see Bowditch *1834a*, 761-791, 873-910). On Gauss's theoretical astronomy, see Brendel *1929a*, 145-254; and on his early work in geodesy, Galle *1924a*, 15-57.

Piazzi's discovery of Ceres on the first day of 1801 (§ 6.4.5) is sometimes held against Hegel as disproving his claim that there are only seven planets in our system. In fact, Hegel ended his *Habilitationsschrift 1801a* by expressing a reserved view on numerical laws for calculating the positions of the orbits, and suggesting, with evident if implicit irony, that other such laws—even that in Plato's *Timaeus*—would do instead. (Ceres was detected with the help of the Bode-Titus law, but it had no theoretical grounding.)

However, other aspects of Hegel's essay show that he was lucky to obtain his doctorate for it (see Shea *1981a*).

A valuable contribution to cometary theory was Olbers's essay *1797a*. With an excellent show of geometrical *Denkweise*, he approximated to the comet's path by chording the observed arcs. Then, depending upon the comet's inclination, he projected the configuration onto either the ecliptic or (following Lambert) the plane perpendicular both to the ecliptic and the radius vector from the earth to the second observation. He calculated this variable before proceeding to the others (arts. 33–53). He also considered the errors caused by the assumption of uniform velocity in his chording technique and also in the inaccuracies of observation, and preferred Newton's method of interpolation to those in Lambert and in Laplace *1784a* (arts. 54–82). Delambre's lengthy account of Olbers's method is noted in § 6.4.6: Bowditch gave an account in English in his *1834a*, 791–830, but seems not to have known the anonymous annotated English translation of the original work (Olbers *1820–1823a'*), which is somewhat easier to follow.

Bessel's early contributions also showed numerical concern and an interest in comets. For example, his *1805a* on the computation of the time anomaly in a nearly parabolic orbit he found correction terms from Simpson's method, which Laplace had used in *Mechanics*, *2* (1799), 736–738). Later work in this area was to lead him to Bessel functions (see § 9.2.8). Again, in his *1810a* Bessel provided variants upon the method given in Lagrange *1783b* of analysing the secular variations of planetary orbits to study the perturbations of cometary paths. His treatment was to influence Encke's later analysis *1822a* of Olbers's method.

Among other contributions, Soldner *1801a* suggested that, if light were assumed to be a molecular emission (see § 7.4), then a ray would be bent when passing by a planetary body. His paper does not seem to have caused interest at this time—Laplace's nearest approach concerned the extinction of light in the sun's atmosphere (*Mechanics, 4* (1805), 556–564)—but it was recalled much later in connection with Einstein's prediction of solar bending, which was about twice Soldner's prediction. A geodeter by profession, Soldner discussed geodesics on a spheroid in *1805a*, and in the process suggested a method for handling spherical and spheroidal triangles that became a rival to those of Legendre discussed in § 5.5.4. In connection with the planned map of Bavaria, Soldner also introduced in *1810a* a cartographic procedure, which was not published until 1873 but became influential as the 'Soldner coordinates' (for information and references, see Galle *1924a*, 47–48).

By and large the French continued to love producing long, complicated expressions, and to classify their terms by magnitude. The perturbation function was a prime target; Burckhardt *1809a* is a typical study. However, Binet's still more comprehensive treatment *1812a*, containing expansions up to the seventh power in the versed sine of the inclination to the ecliptic, did not receive publication beyond a one-page summary *1812b* for the *Société Philomatique*, despite Poisson's approval *1812a* of his efforts for the *classe*.[1]) A companion French interest, in which Delambre was prominent, was the compilation of solar and lunar tables (see Wilson *1980a*, 268–304).

Of all these various studies, the following selection for this section has been made. Three problems on motion are covered, in increasing order of distance from the earth: projectiles, the moon, and meteorites and comets. The origins and distribution of comets is also considered. Then the increasing use of statistics, especially by Laplace, becomes prominent for the rest of the section. Legendre's proposal of the least squares criterion (in connection with cometary theory) is described in § 6.4.6, followed by some of Laplace's main theorems and his treatise on 'probability' in § 6.4.7; then de Prony and Puissant use statistical ideas in their planetary studies.

6.4.2 Laplace's treatment of projectiles, 1803

Laplace's theory of the motion of projectiles continued a line of research which was to lead to important results in French 19th-century mathematical physics. He sent in a paper 'On the motion of a body which falls from a great height' to the *Société Philomatique* in 1803 (soon after joining the society) and it was passed by Lacroix for publication in its *Bulletin* as Laplace *1803a*.[2]) Laplace himself reworked the material in the *Mécanique céleste*, *4* (1805), 573–596; this is the version followed here.

---

[1]) Binet's large chunk of wallpaper mathematics is kept in his *dossier personnel* in *AS*. He had submitted it in May 1812 (*PV*, *5*, 57), 15 months after its acceptance by the *BL* at their meeting of 27.2.1811.

One of the best histories of mathematical astronomy is still Gautier *1817a*, a book composed at that time by its Swiss author out of his two *PF* doctoral theses (see the certificates at *AN*, AJ^{16} 5322, fol. 49). Grant *1852a*, ch. 10 handles several pertinent themes in a prosodic way.

[2]) The paper is signed 'L. C.' at the end. The manuscript survives in Lacroix's *Nachlass* (*BI*, ms. 2396), and shows it to be in Laplace's hand, with the editorial notes by Lacroix as indicated in the text. This is an example, more easily resolvable than usual, of the difficulty with the *Bulletin* mentioned in § 2.2.4; determining whether a paper was written by the person named at its head or by the *Bulletin* editor whose initials appear at the end.

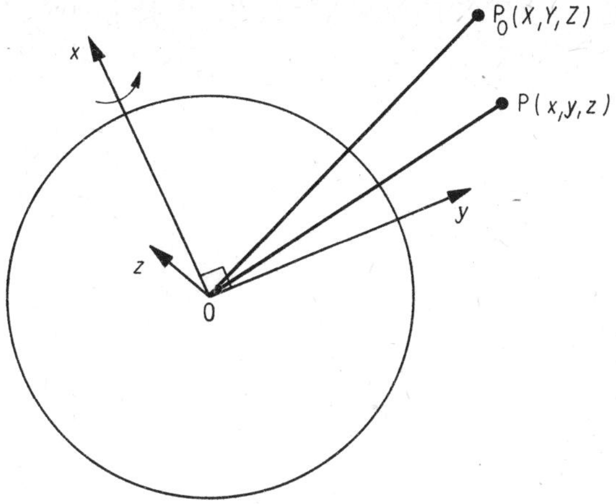

Angle	Between	lines
θ	$\mathrm{O}x$	OP_0
$\theta + \alpha u$	$\mathrm{O}x$	OP
ω	$x\mathrm{OP}_0$	$x\mathrm{O}Y$
$nt + \omega$	$x\mathrm{OP}_0$	$x\mathrm{O}y$
$nt + \omega + \alpha v$	$x\mathrm{OP}$	$x\mathrm{O}y$

Figure 642.1 Notations for Laplace's treatment of projectiles

With unclear explanations, hopefully rendered with accuracy in Figure 642.1, Laplace used both Cartesian coordinates rotating about the axis $\mathrm{O}x$ of the earth, and spherical polar coordinates r (radius vector), ϑ (τ − latitude) and ω (longitude). The body B fell from an initial point P_0 ((r, ϑ, ω) or (X, Y, Z) relative to the position of the axes at $t = 0$) to reach point P by time t, with Cartesian coordinates relative to the (changed) position of the axes given by (x, y, z) and polar coordinates by

$$(r - \alpha s, \vartheta + \alpha u, nt + \omega + \alpha v). \tag{642.1}$$

As usual, he denoted a small quantity (here small relative to the size of the earth) by α.

Taking V as the potential function for B at P, and $\varphi(\alpha s, \alpha\dot{s})$ as the function for air resistance, and setting

$$S(\alpha s, \alpha\dot{s}) := \varphi(\alpha s, \alpha\dot{s})/\alpha\dot{s}, \tag{642.2}$$

Laplace used the principle of virtual velocities to state the basic equations of motion as

$$0 = \sum_x \ddot{x}\delta x - \sum_x V_x dx + \alpha S(-\dot{s}\delta r + \dot{u}\delta\vartheta + \dot{v}\sin^2\vartheta\delta\omega) \tag{642.3}$$

(pp. 573-577). After transforming coordinates entirely into polars and borrowing from his study of stream-lines an equilibrium condition for V (*1* (1799), 213), he converted (642.3) into an equation only in δv, $\delta\vartheta$ and $\delta\omega$, so that setting each of their coefficients to zero and eliminating second-order small quantities in α and n, he obtained these ordinary differential equations:

$$\text{radius:} \quad 0 = \alpha\ddot{s} + \alpha S\dot{s} - g, \tag{642.4}$$
$$\text{latitude:} \quad 0 = \alpha\ddot{u} + \alpha S\dot{u} - gy_\vartheta, \tag{642.5}$$
$$\text{longitude:} \quad 0 = (\alpha \sin\vartheta)\ddot{v} - (2\alpha n \sin\vartheta)\dot{s} + (\alpha S \sin\vartheta)\dot{v} - (g \operatorname{cosec}\vartheta)y_\omega, \tag{642.6}$$

where g was the gravity term, to be discussed below (*4* (1805), 577-581).

Laplace handled these equations cleverly, though cryptically. The smallness of Y (and y) relative to the radius of the earth allowed him to take both g and y_ϑ as constant, so that (642.4) solved for αs in terms of the (known) S. Again, term-by-term comparison of (642.4) with (642.5) suggested the relationship

$$\alpha u = \alpha s y_\vartheta. \tag{642.7}$$

(642.6) caused more difficulty; but he noted that $\alpha v \sin\vartheta$ represented the eastward deviation of B from OP_0 and $(\alpha s \operatorname{cosec}\vartheta)y_\omega$ the deviation of the vertical at OP from OP_0, so that

$$\alpha w' := \alpha v \sin\vartheta - (\alpha s \operatorname{cosec}\vartheta)y_\omega \tag{642.8}$$

gave the deviation of B from the vertical. Further, the substitution of (642.8) into (642.6) yielded the simpler equation

$$0 = \alpha\ddot{w} + \alpha S\dot{w} - (2\alpha n \sin\vartheta)\dot{s}. \tag{642.9}$$

This mixture of w (which Laplace wrote as 'v') and s precluded him from finding solutions without knowledge of φ. He assumed the usual form

$$\varphi(\alpha s, \alpha\dot{s}) = k\alpha^2\dot{s}^2, \quad k \text{ constant; so that, from (642.2), } S = k\alpha s. \tag{642.10}$$

The equations of motion could now be solved explicitly, straightforward passage-work yielding

$$\alpha s = k^{-1}\ln\operatorname{ch}\left(\sqrt{kg}\,t\right) = \tfrac{1}{2}gt^2 - \tfrac{1}{12}kg^2t^4 + O(t^6), \tag{642.11}$$
$$\alpha w = (2n/k)t\sin\vartheta - \left(4n\sin\vartheta/k\sqrt{kg}\right)\tan^{-1}\left(\operatorname{th}\tfrac{1}{2}\sqrt{kg}\,t\right) \tag{642.12}$$
$$= \tfrac{1}{3}ngt^3\sin\vartheta\left(1 - \tfrac{1}{4}kgt^2 + O(t^4)\right), \tag{642.13}$$

where k (unhelpfully written as 'm' by Laplace) was a constant (pp. 584-589). Laplace's *1803a* essentially ended here; but in the *Mécanique* he added a section showing that if a projectile were launched vertically from the earth with velocity K, then, neglecting air resistance, it would land at a point to the west given by

$$\alpha w = -(4nK^3/3g^2)\sin\vartheta + O(n^2) \tag{642.14}$$

(*4* (1805), 591-596).

Laplace's treatment of projectiles is a characteristic example of his skill in solving differential equations sufficiently far for his purposes; as dynamics, it also shows well his ability to handle moving frames of reference. But the modern reader may be surprised by the apparent absence of the so-called 'Coriolis force'

$$n^2(r - \alpha s)\sin(\vartheta + \alpha u) \tag{642.15}$$

acting in the plane OxP and perpendicular to Ox. In fact, Laplace embodied the component of this force in his definition of the 'force of gravity' g (pp. 577-579). There was an interesting detail here, which Coriolis was to treat in general form in the 1830s, as we shall see in § 18.2.4. Indeed, Laplace had used this force also in components in his theory of tides, taken presumably from MacLaurin (§ 5.5.6).

In his textbook of 1811 on mechanics, Poisson did not present Laplace's analysis, but instead relied on a simpler argument in which the rotation of the earth was ignored but air resistance allowed for (*Mechanics*$_1$ (1811), *1*, 336-352). Erudite readers recognise it as based on Borda *1772a*. Taking a Cartesian axis system Oxy with arc length s from the origin, and adopting $(642.10)_1$ for resistance, the equations of motion became

$$\ddot{x} = -m\dot{s}\dot{x} \quad \text{and} \quad \ddot{y} = -g - m\dot{s}\dot{y}, \tag{642.16}$$

where m was essentially the constant k of $(642.10)_1$. Upon integration, $(646.16)_1$ yielded

$$\dot{x} = -ae^{-ms}\cos\alpha, \tag{642.17}$$

with a and α as initial velocity and angle of projection. When introduced into $(642.16)_2$, it was converted to

$$dp/dx = -(ge^{2ms}\sec^2\alpha)\big/a, \quad \text{where} \quad p := dy/dx, \tag{642.18}$$

in which s-integration produced

$$C - (ge^{2ms}\sec^2\alpha)\big/a^2m = p\sqrt{1+p^2} + \ln(p + \sqrt{1+p^2}), \tag{642.19}$$

with C determinable from the initial conditions. Using (642.19) to replace e^{2ms}, (642.18) gave a parametric representation for x and y in terms of integrals of p which could be integrated by various approximations.

Which methods should be used? Borda and Poisson gave some examples, including for special cases; in between them, shortly after Laplace's paper on projectiles, a long paper was written on this topic by François Français, then *professeur* at the *Ecole d'Artillerie* at La Fère. We recall from § 4.3.5 that he was a follower of Arbogast's methods, and apparently here he used them to produce solutions in very long power-series expansions. He submitted his paper to the *classe* in November 1804 (*PV, 3*, 159), whereupon Biot *1805a* sent in a praising report the following April; but the paper was not published, and on this occasion may not be missed.[1])

6.4.3 Laplace and his predecessors on the motions of the moon

> Among the numerous periodic inequalities of the same star [the moon, Laplace] distinguishes those which depend upon the solar parallax, and he makes known the values of those which have the flattening of the Earth for cause, so that, without leaving his study, the astronomer can now determine, by observation of the motion of the Moon, the shape of our planet and its distance to the Sun.
>
> Pontécoulant *World, 1* (1829), xvi

Being a rather large object relatively close by, the moon has required especially detailed study. Lunar theory, and the related three-body (sun-moon-earth) problem, were far from satisfactory at 1800; indeed, in 1840 much still needed to be done. But several contributions were made during our period, often based on the account presented in Laplace's *Mécanique céleste*.[2]) The summary here is confined to matters in which Laplace played an important role; other aspects of lunar theory will be taken up in § 17.6.8.

The first part of the theory which Laplace wrote up in detail in his *Mécanique* was its motion about its centre of gravity (*2* (1799), 933-969). Although the basic theory resembled that for the earth, summarised in § 5.5.7, in detail it was substantially different. He began with this result, discovered from observations by Cassini I, confirmed by Mayer, and analysed theoretically in Lagrange *1782a*, arts. 85-101: 'the descending node of the lunar equator always coincides with the ascending node of the lunar orbit' (p. 933), so that the planes of the moon's orbit and equator were collinear with the ecliptic (or, more precisely with the plane

[1]) The paper was lost during the 19th century and survived only in a transcription by Didion (see his *1860a*, 224-243, 247-254, for a summary account; and Charbonnier *1927a*, 192-196 for illustration and brief *résumé*).

[2]) A detailed technical history of lunar theory would be an important contribution to the history of astronomy. Waff *1976a* is an excellent study of developments between Newton and Clairaut. Gautier *1817a*, 1-128 covers much of the 18th century well; Laplace's own histories (*Mechanics, 5*, Bk. 16 (1825), ch. 1; and *Exposition*, Bk. 5) are purely prosodic. More mathematical but less historical surveys are provided in Tisserand *Mechanics, 3* (1894) and Brown and Brunn *1914a*. Most of the parts of Gautier and Tisserand relating to Euler's work are reproduced, with new commentary, in Euler *1969a'*: see also Forbes *1971a*.

through the moon's centre of gravity parallel to the ecliptic). Laplace proved this result (to within small quantities rather than the exact coincidence claimed on p. 960), and in similarly optimistic vein he claimed to show that 'the mean motions of rotation of the moon are perfectly equal to each other' (p. 939). He found an expression for the moon's libration in longitude (p. 938) and showed that 'the first principal axis of the moon', associated with the smallest moment of inertia, 'is always directed very nearly to the earth' (p. 941).

Somewhat surprisingly, Laplace's treatment of the other main aspects of lunar theory appeared later in the *Mécanique*, as Book 7 of volume 3. His results were an end-product of much effort in the 18th century, inspired by the work done around 1750 when d'Alembert, Clairaut and Euler were in close competition. Using his new method of applying Newton's laws in the respective axis directions (§ 5.2.5), Euler carried out a comprehensive study in his *1753c*, solving the equations for the three-body problem by (usually) trigonometric approximations and classifying the inequalities obtained relative to the eccentricities and inclinations of the earth's and moon's orbits. This theory was of considerable importance for 18th-century navigation theory, for Mayer used it in calculating tables for longitude at sea. In an appendix to *1753c* Euler introduced a new theory based on varying the constants of the problem – the style of mathematics which Lagrange was to place above all others. Later Euler put forward yet another theory, again employing approximation by using rectangular coordinates as independent spatial variables. The fullest account, prepared with the help of his son Jean and others, was published as Euler *1772b*; it used immediately preceding work on lunar secular acceleration and the three-body problem which had been inspired by *Académie* prizes set for 1770 and 1772. As an off-shoot of this new method, Euler introduced in *1770a* his expansion (341.2) of $f(x+y)$ in a power series in Δx, and related expansions, in order to reduce the use of approximations in determining lunar inequalities.

Clairaut's work is noteworthy for its public doubt of the veracity of Newton's inverse-square law (see Chandler *1975a*; and Waff *1976a*, ch. 5). Concerned by the apparent inability of the law to deliver the correct value for the motion of the lunar apsides, he suggested in his *1749a* adding in an inverse fourth-order term. However, he then found in his *1752a* (and elsewhere) that he could treat this motion to his own satisfaction with the law in its usual form by analysing the effect of solar attraction on the earth-moon configuration.

Like Clairaut, d'Alembert laid emphasis on obtaining time t as a function of the moon's true longitude v in the ecliptic (see, for example, his *World, 1* (1754)); and it was d'Alembert's approach which Laplace favoured (see Laplace *1798c*, 191). In the *Mécanique céleste* he took as independent spatial variables v, s (the tangent of the moon's latitude above the ecliptic) and u (the inverse of the projection of the moon's radius vector onto the ecliptic). The general equations of motion (544.5) then took d'Alembert's preferred form

$$dt = dv \Big/ \Big[hu^2\big(1 + (2/h^2)\textstyle\int Q_v/u^2dv\big)^{1/2}\Big] \tag{643.1}$$

and two lengthy equations relating u, v and s, where Q was the appropriate version of the perturbation function R of (544.6) and h a constant (*3* (1802), 368, repeated from *1* (1799), 314). After obtaining first approximate solutions (*3* (1802), 381-383), he then made various desimplifications by re-introducing some of the omitted aspects of the theory: in particular, he showed that solar action introduced second-order terms into the equation of motion largely concerned with u (p. 387). After much development via the trigonometric expansion of Q (pp. 390-469), this equation produced a solution for u (pp. 469-479) which could be substituted into (643.1) to produce after integration a formula for the moon's mean motion (pp. 510-532). After determining various constants in the analysis, he came among other results to a value for the lunar perigee 'which differs from the preceding [value, obtained from observations,] but by its four hundred and forty-fifth part' (p. 544).[1])

Laplace also studied the secular equation of the lunar perigee (and other variables). Perhaps inspired by Lagrange's interest in the inclinations and eccentricities of orbits, especially his discovery that the variation in inclination of the orbit of a planet could affect the secular variations of other planets (Lagrange *1785b*, arts. 11-19), Laplace showed that the variation in the earth's eccentricity $E(t)$ contributed to the moon's secular acceleration. As in the formula for the mean motion (though of opposite sign), the lunar perigee and its node both were expressed in the form

$$\int E^2(t) - E^2(0)dv \tag{643.2}$$

which became terms in t^2 when $E(t)$ was expanded as a Taylor series in t, the slowness of change in $E(t)$ obviating the need for terms higher than $\ddot{E}(t)$.[2])

The next problem studied by Laplace was lunar parallax, for which he used the basic equation of motion largely concerned with the variable s. He obtained a 27-term trigonometric expression for parallax (pp. 580-584) and compared its numerical values with those available in other tables, including a set recently published by the Austrian astronomer J. T. Bürg (pp. 662-665).

[1]) The extreme complications of the analysis preclude a detailed account being given here. Laplace's text is not helped by his notational shambles, concerning which note his warning about the use of 'm' (p. 378) and the use of primes to denote (for example) the sun's mass and also the earth's eccentricity (where I write '$E(t)$' and '$E(0)$' below for his 'e' and 'E').

[2]) See Laplace's *Mechanics, 3* (1802), 471-475, 542-546 (largely following his recent paper *1798c*), where he left the terms in the form (643.2). Compare his first treatment in *1788a*, arts. 7-8, where he assumed a quadratic form for $E^2(t)$.

Laplace added a refinement to his study of lunar parallax by allowing for the oblateness of the earth, which caused the moon to nutate like the earth (§5.5.7). Lagrange had introduced oblateness in his study of the moon's secular acceleration (*1776c*, arts. 10 and 20-44), but he omitted, as negligible, terms involving its inclination (art. 33). Laplace introduced them in his analysis (*Mechanics*, *3* (1802), 585-607, largely following his recent and widely published paper *1801a*), with some simplifications in the supplement to that volume (pp. 749-757). He followed his treatment of the shape of the earth (§ 5.5.5) in introducing a small quantity to denote the earth's ellipticity, and traced the effect of terms on the perturbation terms. He found inequalities in both lunar latitude and longitude that observational astronomers had noted, identifying the former as the motion caused by the earth's nutation (p. 601). He also considered the moon's (assumed) spheroidicity, but concluded that it would contribute only 'insensible' inequalities (pp. 607–616).

Laplace also studied an inequality of very long period in the moon's mean motion, of the form

sine [(2 × longitude of lunar node) + (longitude of lunar perigee)
− (3 × longitude of solar perigee)] (643.3)

(pp. 666-678; compare the form of the inequality (545.4) for Jupiter's satellites). The length of the period made its very existence problematic, and he did not commit himself to numerical values.

This was one aspect of Laplace's lunar theory to receive further attention, by Laplace and others (§ 17.6.8). Others included the absence of Lagrangian variational methods, which Poisson was, naturally, to try to supply (§ 18.2.4); and also Laplace's confident conclusion that 'The agreement of the theory with observations [...] establishes, with certainty, the constancy of the duration of a day; which is an essential element in all astronomical theories' (p. 362), which he was to look at again when he began to use Fourier's heat theory, as we shall see in § 12.4.2.

6.4.4 'Aeroliths' from the moon: Biot and Poisson on meteorites, 1803

> The comets will thus be, relative to the solar system, what the aeroliths are with relation to the Earth, to which they appear strange.
>
> Laplace *1813a*, 88

Projectiles and the moon, the two topics discussed so far, came together in a surprising way in the 1800s. A concentration of sightings of meteorites at that time re-aroused interest among scientists in various countries; in particular, Biot wrote at great length about a spectacular meteorite seen in April 1803 near l'Aigle, about 70 miles west of Paris (see especially his *1806e*).

The study of 'stones fallen from the sky' was granted the splendid name 'aerolithology' in French writings. An influential theory, attributed to Laplace in Biot *1802d* and Olbers *1803a* but first proposed (in more general terms) in Chladni *1794a* (as he noted in *1809a*, 250), asserted that the stones originated in ejections from lunar eruptions. 'Perhaps, at first sight, one will find this explanation bizarre', admitted Biot, but 'one knows that there exist volcanoes on the moon', and the stones thereby ejected would decelerate to 'the ordinary velocity of heavy bodies' when passing through the earth's atmosphere (*1802d*, 154). In reply in the *Bibliothèque britannique* to criticism there of the theory, he defended it as an hypothesis ('the word *hypothesis* signifying absolutely the same thing as *supposition*') 'considered as a contingent fact'; he did not claim the 'reality' of the hypothesis (*1803d*, 316; see also *1803e*).

Biot's exposition of this hypothesis was quoted above from a paper sent in to the *Société Philomatique* late in 1802. He continued there with a very simple mathematical analysis, in which he assumed that the meteorite travelled along the straight line joining the centres of the earth and moon, bodies which he assumed to be spherical and free from atmosphere. To find the minimal escape velocity of the meteorite, he assumed that it would come to a stop before accelerating towards earth. He formed the ordinary differential equation to express the motion, and from relevant data (taken from Lalande's *Astronomie* rather than Laplace's *Mécanique céleste*) that the meteorite would need as minimum escape velocity 7,771 *pieds* (about 2,525 metres) per second (Biot *1802d*, 156).

Quite aware that his analysis was very simple (it is a special case of the restricted three-body problem), Biot promised a more general treatment for a later number of the *Bulletin*, and concluded: 'we will give on this matter an analysis due to one of our youngest and best geometers, the c[itizen] Poisson, *professeur* at the *Ecole Polytechnique*'. But he held up Poisson's analysis for some months to see how the hypothesis fared among his readers. He sent it in to the *Société* early in 1803, in a paper *1803a* summarising Poisson's work. He judged (p. 180):

> The march of opinion in this circumstance has justified our reserve. Some people have decided, without examination, that the thing was impossible, others[,] more wise[,] have doubted with us; and it has finally happened that one has regarded as very worthy of observation, a phenomenon which at first glance one had not hesitated to reject as absurd.

He also criticised Thomas Young's recent calculation *1803a* of the escape velocity at 12,000 feet (about 3,650 metres) per second as for too large.

The 21-year-old Poisson had treated Biot's problem in a slightly different way. Taking the masses of the earth and the moon in the ratio 1: k and b and l as their respective mean radii, and setting g as terrestrial surface gravitation, a the mean distance of the earth from the moon, and v as the initial velocity of vertical projection from the moon, he found the differential equation for the location at x at time t:

$$\ddot{x} = \frac{gh^2}{(a-x)^2} - \frac{gkh^2}{x^2}, \tag{644.1}$$

with integral

$$\dot{x}^2 = v^2 + 2gh^2\left(\frac{k}{x} - \frac{k}{l} + \frac{1}{a-x} - \frac{1}{a-l}\right). \tag{644.2}$$

The smallest value for v was given by the condition that at the distance b, where equal attraction obtained between moon and earth, the projectile was stationary. This gave

$$v^2 = 2gh^2\left(\frac{k}{l} - \frac{k}{b} + \frac{1}{a-l} - \frac{1}{a-b}\right), \quad \text{where} \quad \frac{k}{b^2} = \frac{1}{(a-b)^2}, \tag{644.3}$$

so that

$$v^2 = 2gh^2\left(\frac{k}{l} - \frac{1}{a-l} + \frac{(1+\sqrt{k})^2}{a}\right). \tag{644.4}$$

Poisson used data from Laplace (*Exposition*$_2$ (1799), 23, 185–186) rather than Lalande, and obtained 2,147 metres per second as minimum escape velocity (Biot *1803a*, 181), which Biot regarded as 'a little less' than his own value but which was actually about 15 % less. Biot might have been encouraged by Olbers's calculation of 7,780 *Füsse* (about 2,330 metres) per second (Olbers *1803a*, 153), which was very close to his own.

At the end of his paper Biot promised to publish later Poisson's analysis of the ejection of meteorites at an angle to the line joining the centres of the earth and the moon; but in fact it appeared only in Poisson's full paper *1803a*, a contribution to a book-length anthology (Izarn *1803a*) on 'atmospheric lithology'. The paper was quickly translated into German. Since it is a nice example of a young man applying Laplacian theoretical astronomy of the time, an account is worth giving. Figure 644.1 would have enlightened his somewhat unclear description.

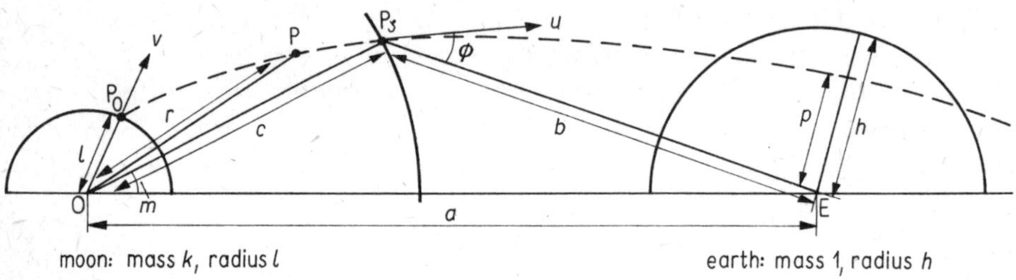

Figure 644.1 Diagram for Poisson's analysis of the path of 'aerolith' P

Setting up Cartesian axes O*xyz* at the centre O of the moon with O*x* passing through the centre E of the earth, Poisson imagined a projectile P released vertically with velocity v at point P_0 of the moon's earth-directed side, and travelling, under lunar attraction only, according to the equation

$$\ddot{r} = -gh^2k/r^2, \quad \text{with integral} \quad \dot{r}^2 = v^2 + 2gh^2k\left(\frac{1}{r} - \frac{1}{l}\right). \tag{644.5}$$

The surface $\mathscr{S}$ at which the lunar and terrestrial attractions balanced was given by

$$k\big/(x^2 + y^2 + z^2) = 1\big/((x - a)^2 + y^2 + z^2), \tag{644.6}$$

a sphere of radius $a\sqrt{k}\big/(1 - k)$ centred $ak/(1 - k)$ along Ox towards E. When P reached point $P_{\mathscr{S}}$ of $\mathscr{S}$, it was travelling with velocity u with angular separation φ from the earth's radius vector $P_{\mathscr{S}}E$ and distant c and b from moon and earth respectively. The following relationships applied there:

$$x^2 + y^2 + z^2 = c^2, \quad x = c\cos m, \quad 1/b^2 = k/c^2 \quad \text{and} \quad b\sin\varphi = a\sin m. \tag{644.7}$$

From $(644.7)_{1,2}$ c could be calculated by insertion into (644.2):

$$c^2 + \frac{2ak\cos m}{1-k}c + \frac{a^2k}{1-k} = 0, \tag{644.8}$$

while from $(644.5)_2$ u was given by

$$u^2 = v^2 + 2gh^2k\left(\frac{1}{c} - \frac{1}{l}\right). \tag{644.9}$$

Poisson's explanation of the behaviour of P after it passed through $P_{\mathscr{S}}$ was somewhat cryptic. Basically, he now discarded lunar attraction and claimed that Laplace's fundamental equations (544.10) of elliptical motion for two bodies (*Mechanics, 1* (1799), 393) gave the following expression for the perigee R of the path, distant p from E:

$$\left(\frac{1}{b} - \frac{u^2}{2gh^2}\right)p^2 = b^2\sin^2\varphi\frac{u^2}{2gh^2} \tag{644.10}$$

(*1803a*, 245–246). However, the reasoning seems to rest not only on (544.10) applied at $P_{\mathscr{S}}$, which has to read

$$u^2 = \left(\frac{2}{b} - \frac{1}{q}\right)gh^2 \tag{644.11}$$

in the current notations (together with q for the major semi-axis of the elliptical orbit) in order to avoid dimensional chaos, but also the equation given by Laplace (in *Mechanics, 1* (1799), 353) which related the velocity and location of the projectile. When this equation is applied to $P_{\mathscr{S}}$ and R (and dimensions are again watched), it yields

$$\left(2b - \frac{b^2}{q}\right)gh^2 - b^2u^2\cos^2\varphi = \left(2p - \frac{p^2}{q}\right)gh^2 \quad (= \text{constant}). \tag{644.12}$$

Now eliminating q from (644.11–12) produces not Poisson's (644.10) but

$$\left(\frac{1}{b} - \frac{u^2}{2gh^2}\right)p^2 - p = -b^2\sin\varphi\frac{u^2}{2gh^2}. \tag{644.13}$$

Despite this slip, Poisson's strategy was basically correct. In order to bring (644.10) (or (644.13)) to calculation, $(644.7)_{3,4}$ expressed b and $\sin\varphi$ in terms of c and $\sin m$ respectively; c and u were calculable from (644.8–9); and k, g and h were given quantities. This left the calculation of p as the unknown of the analysis, and the possibility that P arrived at the earth as a meteorite was expressed by the inequality $p < h$. In turn this condition supplied estimates of the initial velocity v of P and of its point P_0 of lunar launch; he felt P_0 to be specified quite well by the angle m, but he noted that a better approximation was given by $(m \pm nt)$, where n was the mean angular velocity of the moon and t the (calculable) time taken for P to travel from P_0 to $P_{\mathscr{S}}$.

Poisson took a couple of cases, one satisfying the inequality, the other not (pp. 249–251: conclusions cannot be sustained for (664.13) without knowledge of the value of c). He concluded that the hypothesis of lunar ejection 'is *physically possible*; and it is being convinced of this possibility, that Mr. de Laplace has urged the French physicists not to reject this phenomenon, as one had up to now, through lack of being able to assign a physical cause to it' (*1803a*, 253).

Biot also continued to support the hypothesis, defending it when the L'Aigle sighting of 1803 drew him into extensive studies of the subject. The mathematical analysis of the paths of meteorites was later given more sophisticated treatment (see von Niessl *1907a*); in the 1810s interest centred largely on their geological composition, and the geologist Bigot de Morogues published a history *1812a* of past and recent sightings, somewhat similar to the anthology (Izarn *1803a*) in which Poisson's paper had appeared.

6.4.5 Lagrange and Laplace on comets, 1812–1813

Bigot's book came out at a time when speculations on the related topic of comets were aired in the pages of the *Connaissance des temps*, and written by none other than Lagrange and Laplace. However, there the similarities end; for they differed both on the origins of comets and the mathematical features studied.

The new century had been marked by new planets. Piazzi had discovered Ceres on its first day, and when Olbers discovered Pallas in 1802 and found it to have an orbit rather similar to that of Ceres, he proposed that they were both fragments of a planet which had been orbiting between Mars and Jupiter but then had exploded (Olbers *1802a*). The quick discoveries of Juno and Vesta seemed to support this theory, and it led Lagrange into cosmological speculation, for the first time and in his last paper. It was read to the *Bureau des Longitudes* in January 1812 and published as *1812a*, in his 77th year.

Lagrange felt that comets were further fragments of this explosion. Further, he did not doubt 'that the Earth has suffered great catastrophes, and that the layers which form it like bark must have been lifted, broken and displaced by the action of an interior fire or by other elastic fluids contained in the globe'. Hence meteorites were possibly 'thrown principally by volcanoes situated in the polar region, and that produce the *aurora borealis* at the same time' (*1812a*, 382: compare Laplace in § 5.4.2 on catastrophes).

Lagrange's main purpose in this paper was to analyse forces, such as those released by an explosion, which could affect the orbit of a planet. He presented the basic equations without proof; presumably he obtained them from making the appropriate additions to terms in the *forces vives* equations (compare Tisserand *1890a*, 453–456). He found that if a planet were moving in a circle of radius r

around the sun and suddenly changed to an ellipse with semi-axes a and b and inclination i to the first orbit, then its old and new velocities v_0 and v_1 were related by

$$(v_1/v_0)^2 = 3 - 2\sqrt{b/r}\cos i - r/a. \tag{645.1}$$

When $a = \infty$, the orbit became a parabola, which Lagrange took to be the orbit of a comet. Its path would be direct or retrograde according as $i \lessgtr \tau$, and from (645.1) the maximum values for v_1/v_0 would be respectively $\sqrt{3}$ (when $i = \tau$) and $\sqrt{5}$ (when $i = \pi$). He went on to obtain a formula for the change of orbit from an ellipse to another ellipse, and discussed the possibilities of the orbits being parabolic, hyperbolic or elliptic if the planet broke into two nearly equal parts (*1812a*, 383-385).

Lagrange concluded by recalling Laplace's nebular hypothesis (§ 5.4.2) on the origins of planets and comets in the coalescence of matter released by a comet's collision with the sun. Regarding this as a process within 'the sun's atmosphere', he regarded it as valid only for objects travelling in the same direction in nearly circular orbits—that is, planets—and concluded (p. 389) by claiming that

> If one adjoins the hypothesis of the explosion of the planets by the action of caloric that the passage from the aeriform state to the solid state will have concentrated in its interior, one will have a complete hypothesis on the origin of all the planetary system, more in accordance with nature and with the laws of Mechanics than all others which have been proposed up to now.

Laplace could hardly let go unanswered this use of his nebular hypothesis. In February 1813, two months before the death of Lagrange, he read a short paper *1813a* on comets at the *Bureau des Longitudes.* He supported the view that had recently published by William Herschel (*1811a*, 306), that comets were formed by the condensation of matter in the distribution of matter in the universe. The elements of consistency between Herschel's general theory of condensing nebulosity and his own nebular hypothesis doubtless attracted him; and the question of the distribution of matter in the universe, which it raised, excited his developing interest in 'the application of the calculus of probabilities to natural philosophy' (to quote the title of a slightly later paper Laplace *1815a*) and led him in *1813a* to analyse the distribution of cometary paths.

Laplace began his analysis by defining the '*sphere of activity*' of the sun as the space within which solar attraction dominated the comet's path. He found that the probability that its direction of motion on entering the sphere would be within angle ω of its radius vector was $(1 - \cos\omega)$. The question of whether it would be visible on earth hinged on its velocity v, and Laplace found that the probability that its perihelion would be less than some suitable value D for visibility was

$$\left(\int_{v_0}^{v_1} (1 - \cos\omega)\, dv\right) \Big/ V, \tag{645.2}$$

where V was the maximum value, and v_0 and v_1 certain limits, all these values to be determined (p. 92). His relationship between v and ω (which involved D) seems to be sound; but the arguments for the calculation of v_0 and v_1 are very obscure, and indeed were to be criticised by Gauss and others (see Fabry *1895a*, 5-34, and Bianco *1913a*). In the end he found very long odds against a comet's path being detectably hyperbolic (p.95), assuming as his criterion of detectability that the ratio of the axes of such an orbit would be less than 100. Thus he corroborated his assessment of the conicity of the cometary paths given in his *Mécanique céleste* (*4* (1805), especially pp. 361-366 and 417-421); but the fragility of his reasoning made this paper one of his less authoritative contributions to theoretical astronomy.

6.4.6 Tracking the comets: the least-squares criterion

Comets presented various special problems for theoretical astronomy. In cosmology there was the question of whether or not they belonged to our planetary system, and the related mathematical question of the geometrical character of their paths. The eccentricities E of the paths were close to 1, in contrast to the small values for E in planetary theory; and from this difference sprang the need for alternative methods of solving the equations of motion, which were also motivated by the fact that for comets the inclinations might well not be small. There were also special difficulties in computation of the paths, due to their only occasional visibility.[1])

The last problem led to a great deal of work from Newton onwards to determine the comet's path from three observations. A useful equation linked two observations, taken at time t apart, of a comet at points C_1 and C_2 in the vicinity of the sun S:

$$6t = (r_1 + r_2 + \varrho)^{3/2} \pm (r_1 + r_2 - \varrho)^{3/2}, \quad \text{where} \quad r_1 := SC_1, \quad r_2 := SC_2, \quad \varrho := C_1C_2. \tag{646.1}$$

[1]) On various methods of determining the paths of comets, see Herglotz *1906a*, arts. 17-21 (and 7-16 on planets), and *passim* in Oppenheim *1922a*. Lambert's work is outlined by Bauschinger in Lambert *1892a'*, and appraised to some extent in Volk *1980a*.

It was obtained from Kepler's areal law and the formula for the area of ΔC_1C_2S, and was proved in Euler *1743a*, art. 13, and again in Lambert *1761a*, 40 (presumably independently). Its merit lay in its simplicity (by the standards of cometary theory), especially in being explicit in t; and in its applicability to any pair of observations, leading to relations between pairs of observations.

The interlocking character of Lagrange's and Laplace's researches, mentioned in §5.4.6, is well illustrated by cometary theory. Lagrange produced three theories around 1780. One, *1780a* on determining the path from three observations, was not easily amenable to calculations; the second, *1780b*, was tractable numerically and made no assumptions about the nature of the path but required six, rather than three, observations (this paper was published then in German translation, the French original appearing only posthumously as *1819a*); and the third, which won an *Académie* prize in 1780 and was published as *1785a*, dealt extensively with the perturbations as well as the basic path and involved lengthy manipulations and polynomials of relatively high degree (septics and octics).

Laplace was influenced by this work to publish two theories in 1784, his 36th year. In *1784b*, after rejecting Newton's method of rectifying the path (especially as used in Boscovich *1774a*) for discarding second-order small quantities in time but retaining them for latitude and longitude, he used Taylor's series and Euler's expansion (341.2) of differences to yield rather unusable expressions for latitude and longitude. In *1784a*, arts. 16–17 he again used difference expressions for latitude and longitude but directed them to interpolatory ends for calculating the comet's parameters, starting with the perihelion distance and the time of its arrival there. This work in turn inspired Lagrange to a new study *1785d* for determining the path from three observations. He largely followed his own *1780a*, but made more use of Taylor's series for approximations. Like last time, his final results were pretty grim from the numerical point of view (see, for example, art. 35 together with art. 20).

Laplace began his treatment of cometary paths in the *Mécanique céleste* with a preliminary analysis which allowed them to be elliptic, parabolic or hyperbolic (*1*(1799), 382–390), followed by the computation of a parabolic path (pp. 417–474) partly using *1784b*. Later in the work he devoted Book 9 (*4* (1805), 361–437) to an approach which recalls Lagrange *1785a* and *1785d* in some ways. Laplace also assumed the path to be elliptic (of nearly unit eccentricity), so that he could use his 'second method' of solving the equations of motions of planets (§5.4.4) to calculate the perturbations of the paths. In order to calculate them numerically he used his theory of generating functions (352.1) to furnish the required parabolic quadrature formula (pp. 380–387), rather similar to the Euler-

MacLaurin summation formula (342.3):

$$\int_0^n y(x)dx = \frac{1}{2}(y(0) + y(n)) + \sum_{r=1}^{n-1} y(r) - \frac{1}{12}(\Delta y(n-1) - y(0)) - \frac{1}{24}(\Delta^2 y(n-2) + \Delta^2 y(0)) - \ldots . \quad (646.2)$$

The theory is, once again, laborious; a dash of (646.1) might not have come amiss.

Volume 4 of the *Mécanique céleste*, which contained this theory, came out around May 1805 (*PV*, *3*, 216). Perhaps knowledge of its imminent arrival stimulated the 52-year-old Legendre to write up by March his own views on determining the paths of comets; certainly the observation of two comets later in the year motivated him to prepare a supplement on them by the following January. The whole was published as a book *1806b* by Courcier.

As we saw in § 5.5.4, Legendre was adept at interpolation, and 'I have thought that what there was better to do in the problem of comets, was to start out from the immediate data of observation, and to use all means to simplify[,] as much as is possible, the formulae and the equations which serve to determine the elements of the orbit' (*1806b*, v). Taking the sun as origin of the coördinates, and placing rectangular coordinates in the ecliptic and along its normal, the key to his method was to approximate to the coordinates (X, Y) of the earth and (x, y, z) of the comet by cubics in time t, and to base the analysis on the coefficients of these terms and on related expressions. For the case of three observations taken at equal intervals he reduced the crucial calculations to four linear equations in four unknowns (pp. 5–8), and then, acting contrarily to Laplace, he took the path to be a parabola so that he could set the semi-major axis $a = \infty$ and thus have more equations than unknowns to manipulate (p. 8).

For Legendre the solution thus obtained 'would appear to have all the exactitude and the simplicity that one can desire'; there was no point in finding more exact solutions since the observations were not accurate enough (p. 27). But he did modify his analysis for the case of two unequal intervals of observations (pp. 27–33). He then applied his methods to two well-observed comets of the 18th century (pp. 33–52), and went on to study corrections to the approximate solutions thus obtained (pp. 53–71).

In an appendix Legendre explained in detail his method of fitting a curve to data, in a proposal that has obliterated memory of the principal purpose of his book; for on pp. 72–76 he described the 'method of the least squares'. Interestingly, the example he now gave was to determine the spheroidicity of the earth from data used in the determination of the meridianal arc (pp. 76–80), in which he had been involved ten years before (§ 5.5.4). Further, when he added the supplement on the comets seen in 1805, he did not seek a foundation for the method of least squares; instead, he concentrated on modifying the basic formulae for cometary paths in various ways, especially because of the presence there, newly noted, of a small quantity in the denominator.

Thus Legendre's book is an example of a work remembered for an incidental feature rather than its main purpose. His cometary theory, while ingenious, seems not to have been widely used. In his treatise on astronomy Delambre praised it, but gave equal space to Laplace's interpolation method of *1784a* mentioned above, and most space of all to Olbers's method of 1797 (see Interlude 641.1), in which the areal result (646.1) was prominent (see Delambre *1814d*, *3*, 348–399). Biot promoted Laplace's interpolation method, reprinting Laplace's text *verbatim* and adding new material due to Mathieu (see Biot *Astronomy*$_2$, *3* (1811), additions, 185–209).

However, least squares retained interest, especially when Gauss argued in its favour, with the help of an assumed normal distribution of errors, in his treatise *Theoria motus* of 1809 on the motions of planets; for while acknowledging its appearance in Legendre's book Gauss claimed priority for the method (*1809a*, art. 186). Delambre, when summarising the contents of Gauss's book for the *Connaissance des temps*, also noted Legendre's priority of publication (Delambre *1810c*, 393); but Legendre went all Parisian *savant* over his rights. When Laplace used the method in a paper (*1811a*, art. 6), Legendre at once sent in to the *classe* in September 1811 a word-for-word transcription of the relevant part of his *1806b* (no *PV*), and had it printed as a *Mémoires* paper *1814a*; and once again Delambre gave him credit, in an extended account of least squares in his report on the work of the *classe* for 1811 (Delambre *1812a*, i–xiii). Nevertheless, when Legendre later published a second supplement *1820a* to his book, containing a new method of determining the path of a comet (using (646.1) and corresponding more closely to Olbers's theory), he appended a sarcastic and rather childish note on Gauss's conduct.[1]) Legendre was not normally a participant in the Parisian fights for rights; but on this matter he definitely caught the disease.

[1]) See Legendre *1820a*, 79–80; and Stigler *1977a* for commentary. On the history of least squares in this period, see Galle *1924a*, 4–15; Plackett *1972a*; and Eisenhart *1964a* and *1983a*. I do not distinguish errors from residuals in this book.

6.4.7 Laplace's increasing use of 'probability': the central limit theorem, least squares and the *Théorie analytique* (1812)

Veritably (it was his favourite word) [...]

Laplace, as described by Libri (*1840a*, 533) and Cournot (*1913a*, 87)

In an important respect Laplace's advocacy of the least squares criteron was superior to that given by Legendre; for he gave it a grounding in probability theory whereas Legendre's argument was merely an intuitive application of minimisation. To cast this remark in proper perspective, it is necessary to review Laplace's career in probability and statistics, if only in brief.

As was noted in § 2.5.2, Laplace entwined his interests in astronomy, differential equations and probability throughout his career.[1]) One of his earliest papers, *1774a*, dealt with 'the probability of causes by events', conceived indeed very broadly: one of its results was what we call 'Bayes's theorem' on prior and posterior probabilities, seemingly produced without knowledge of Bayes's (recently) published paper. Over the years Laplace developed a number of mathematical techniques for the theory, with applications to astronomical and geodetic observations being prominent. A few of these have been noted already: the generating function (352.1) and the generalised summation and difference formulae (352.5–8); the Boscovichean minimax theory of errors, used in connection with studying the shape of the earth (§ 5.5.5); and the recent study of the distribution of comets in the universe (§ 6.4.5). But my decision announced in § 1.4.2 *not* to explore in detail the progress of probability and statistics will be obeyed; for much of his (and others') ideas arose in connection with topics such as law, games of chance and demography, and so a detailed account would take us too far afield. Suffice it to say that by the 1800s he had made substantial contributions to regression analysis, Bayesian and non-Bayesian statistics, error theory and estimation, and the philosophy of probability. Indeed, to him more than anyone do we owe *the growth of mathematical statistics out of probability* as the daughter subject which would come to equal its parent in scope and range of interest. In common with his contemporaries, however, he used the term 'probability' to cover the whole expanding range.

---

[1]) On Laplace's contributions to probability and statistics, see especially Stigler *1986a*, chs. 3 and 4; and Gillispie *1978a passim*, building in part from Stigler *1976a*. Sheynin *1976a* and *1977a* are of some value. Pearson *1978a*, ch. 14 and Todhunter *1865a*, ch. 20 can be used but with caution. Gillispie *1980b* contains two little-known essays by Laplace.

During the late 1800s Laplace produced some of his principal results, including the grounding for the least squares criterion. The major achievement, a form of what is now called 'the central limit theorem', came in a paper presented to the *classe* in April 1810 (*PV*, *4*, 341) and soon published in the *Mémoires* as *1810b* (see art. 6). He gave the theorem no name, and indeed never individuated it as a theorem as such; indeed, his statement and proof of it were excruciating even by his standards. I hope that the following is a fair rendering; at least the notations are his.

A large number n of observations were endowed with a function $\varphi(x)$ of error x, always lying within an interval of width h and divided in the proportion $i' : i$ of positive and negative values. The abscissa q' of the 'centre of gravity' of $\varphi(x)$ was defined by him as

$$\int_{-ih/(i+i')}^{i'h/(i+i')} (x - q')\,\varphi(x/h)\,dx = 0\,. \tag{647.1}$$

Then the probability that the mean of the n errors lay within $rh/\sqrt{n}$ of q' (where r was an arbitrarily assumed value) was given by

$$\sqrt{\frac{2}{\pi kk'}} \int_{nq'-hr\sqrt{n}}^{nq'+hr\sqrt{n}} \exp\left(-\Theta^2/2kk'\right) d\Theta, \quad \text{where}$$
$$k := \int_{-ih/(i+i')}^{i'h/(i+i')} \varphi(x/h)\,dx \tag{647.2}$$

and k' was a constant defined by an integral involving φ and q'. When φ was symmetrical about the zero value of error, then $q' = 0$.

Laplace was claiming in this extraordinary result that the mean of a large collection of data could be appraised accurately by mean of the normal distribution, whatever form $\varphi(x)$ took. Briefly, he formed the n-th power of the pertaining generating function and took the coefficient of the appropriate term as the probability that the sum of the errors was $(r + nq)$ (with q arbitrary). The expression was an integral: its integrand contained $\varphi(x/h)$, which he expanded in a power series about $(x - q')/h$ and thereby "lost" φ in definitions such as those for k and k'. Returning to the generating function and treating it as a trigonometric series, he used the orthogonality properties of the functions to obtain an integral expression for the probability that the sum of the errors lay between $(r - nq)$ and $(r + nq)$. (Inspiration douptless came from his encounter with Fourier's Fourier series a year earlier; see §9.3.2.) He found that the probability eventually took the form e^{-p^2} ('p' is my symbol) multiplied by a power series whose terms could be neglected as small. Integrating e^{-p^2} over the required limits finally delivered $(647.2)_1$.[1])

In an undated 'supplement' to the paper, Laplace *1810e* related this result to the least squares criterion. He considered a set of large collections $\{c_j\}$ of 'observations' of some 'element'; each collection contained n_j members, was endowed with its own distribution of errors, and took m_j as mean value. Assuming that M was the 'true result' for the element, he used the central limit theorem to express, in terms of the normal distribution, the probability that the mean error of the observations c_j was $(m_j - M)$. Now the probability that each collection in the set took its probability was given by the product over j of these constituent probabilities. As a function $y(M)$ of M, it produced a curve which 'we will call *curve of probabilities*' (today it is called the maximal likelihood function'). The 'true value' was now defined by the condition that 'the deviation from the truth, that one can fear, is a *minimum*'. This notion, now known as the 'expected error', was written by Laplace in the form

$$\int_0^l (l - z)y(z)dz + \int_l^\infty (z - l)y(z)dz, \tag{647.3}$$

where l was a parameter whose desired value M was to be found by setting the l-differential to zero: M turned out to have the property of dividing $\int_0^\infty y(z)dz$ into equal halves. He then showed that this same value arose from applying the least squares criterion to

$$\sum_j p_j(m_j - z)^2, \quad \text{where} \quad p_j := a\sqrt{n_j/\pi} \tag{647.4}$$

was 'the greatest probability of the result given by the observations' in c_j, and a was a constant.

Thus the connection between the theorem and the criterion was established, but at the price of assuming *a priori* the normal distribution. So in a paper presented to the *classe* in April 1811 (*PV*, *4*, 475) and soon published in the *Mémoires* as *1811a*, Laplace reproved the connection by using a method of assigning

[1]) The mysteries of Laplace's presentation include $(i + i')$ 'being equal to h' and yet also being the infinite numbers of parts into which the positive and negative parts of the interval of width h was divided; i and i' being infinite and different in value from each other; the use of *pairs* of variables and constants taking numerical and integral values (such as h and $(i + i')$); unclear specification of the limits of some integrals; and the use of 'r' as a dummy variable, despite its appearance in $(647.2)_1$.

In the paper about to be described, *1810e*, I have changed a few letters as well as use subscripts.

weights (as at (556.1) with Bocovich's theory of errors). Given two collections of s observations $\{p_j\}$ and $\{\varphi_j\}$, he required of the errors $\{\varepsilon_j\}$ only that they were symmetrical about zero, and he estimated z in the system of linear equations

$$\varepsilon_j = p_j z - \varphi_j, \qquad 1 \leq j \leq s \tag{647.5}$$

by multiplying each one by some integral weight m_j. Then, drawing on maximal likelihood, the central limit theorem and the other techniques, he showed that the values of $\{m_j\}$ which minimised the 'mean value of the error to fear' were proportional to their corresponding $\{p_j\}$ and, once again, that these values were also furnished by least-squares reasoning (art. 6–7). He extended the procedure to multi-variate estimation (art. 8).

Laplace's proofs are well stocked with dubious arguments and assumptions; in addition, his references to Gauss *1809a* were rather in passing, though it does not required erudite reading to think that Gauss's work had helped to stimulate the new ideas in the first place. Nevertheless, his breakthrough was very significant, and earned the approbation in the summary of the paper given by Poisson to the *Société Philomatique*: 'the *method of the least squares of errors* [had been] already employed by several geometers, but the advantage of which had not yet been well shown' (Poisson *1811c*, 363).

Around this time Laplace felt that his ideas had proceeded sufficiently far for a treatise on the subject to be produced. Accordingly, during 1812 Courcier put out an imposing quarto volume of 464 pages under the title *Théorie analytique des probabilités*: Delambre gave it duly extensive notice in his survey of the year's work of the *classe* (*1814a*, xvii–xxi). A second edition appeared in 1814, consisting of the old text (with corrections made to some pages) and an extra chapter at the end on the probability of judgements, together with a long new preface. Supplements were added in 1816, 1818, 1820 and 1825; they were mostly versions of papers published elsewhere. In addition, a third edition was published in 1820, comprising the main text of its predecessor and the first three supplements together with an enlarged version of the preface, which itself came out in five editions between 1814 and 1825 as a separate volume entitled *Essai philosophique sur les probabilités*.[1])

The *Essai* was based upon one of the lectures which Laplace gave in 1795 at the first founding of the *Ecole Normale* (see *1795a*, lecture 10). It contained an account of his deterministic philosophy of science married to his interpretation of probability as arising from ignorance; his famous remark on our knowing the

---

[1]) 1,200 copies were printed of Laplace's *Probability*$_1$ (1812), and 500 copies of *Essay*$_1$ (1814). Poisson *1814h* announced *Probability*$_2$ (1814) to the *SP*. The print-run of *Essay*$_4$ (1819) was 500 copies; that of *Probability*$_3$ (1820) 100 more (*AN*, F^{18} 57a, Courcier, nos. 212, 220).

motions of molecules as exactly as that of comets, quoted in §7.2.1 below, comes from it. He also described his principal results and theorems, and the main applications; however, they are pretty hard to understand in this form, and precedence must be given to the mathematical accounts given in the two Books of the main text of the *Théorie.*

These Books were concerned with the panoply of probability and statistical results and arguments which Laplace had accumulated to date. The first Book presented properties of generating functions, asymptotic theory and approximations by integrals, and relationships between series expansions and difference equations: a few of them will be noted in Interlude 934.1 and §9.3.4. The second Book applied these results to a variety of situations: many of them dealt with urn problems, and encorporated a treatment of the central limit theorem and least squares (ch. 4). The proof of the theorem was a considerable improvement in clarity: the essential moves were to show that the sum of two terms symmetrically placed in the binomial expansion $(p+q)^{x+x'}$ and $2l$ terms apart was approximately given for large l by

$$\sqrt{\frac{2(x+x')}{\pi x x'}}\exp\left(-\frac{(x+x')^2 l^2}{2xx'}\right). \tag{647.6}$$

He then passed from sums of such terms to integrals once again, in order to prove that 'the probability that the difference between the ratio of the number of times that the event a can happen to the total number $[n]$ of attempts and the facility of its occurrence' was given by

$$\begin{aligned}&\frac{2}{\pi}\int_0^l e^{-\alpha^2}d\alpha+\sqrt{\frac{x+x'}{2\pi x x'}}\,e^{-t^2}+O\left(\frac{1}{x+x'}\right),\quad \text{where}\\ &t:=l\sqrt{\frac{x+x'}{2xx'}}\,.\end{aligned} \tag{647.7}$$

Thus the probability in question was asymptotically close to the normal distribution as x and x' increased (*Probability*, Bk. 2, art. 16).

The applications which Laplace made of his various results included the inclination of the planetary bodies to the ecliptic and the motions of comets as forward or retrograde (ch. 2), and the variation in heights of a barometer at the same time on successive days (ch. 5). This type of case was to gain still greater interest in his later years, as we shall see in §17.6.2.

6.4.8 de Prony on the determination of latitude and longitude, 1805

The remainder of this section is concerned with geodesy and cartography, and shows the practical man's approach, in some contrast to the theoreticians given pride of place above. In addition, evidence is furnished of early application of the least squares criterion. de Prony is the first performer, due to a paper on latitudes and longitudes published in the *Connaissance des temps* as *1805a* and also separately as a pamphlet. Recently returned from a working visit to the Po valley (and elsewhere) in Italy, he described his new approach to aid the preparation of maps; as ever, it was intended to aid calculation.

After citing Puissant's recently published treatise on geodesy (§ 5.5.8), 'the study of which one will hardly know to recommend to young geographer engineers' (art. 7), de Prony turned to his principal idea. He adopted the proposal of Clairaut *1741a* that the latitude and longitude of a point Y of the spheroidal earth should be calculated with the help of a definition of the perpendicular to the meridian at Y as the elliptical section cut by the plane through Y normal to the meridian and to the surface of the earth there. He then drew on the theory of the optimal radii of curvature to a surface, as developed by Euler and Meusnier (§ 4.7.3), to seek a sphere which could fit most closely the curvature of the earth around Y. Taking φ to be the latitude of Y, r the radius of curvature of the meridian there (the largest such radius), r' as its minimal counterpart, and α to be the flattening of the earth, he cast Meusnier's formula (473.1) for the radius $\varrho(\varphi, \chi)$ of curvature of a section making angle χ with that for r' in the form

$$\varrho((r+r')-(r-r')\cos 2\chi) = 2rr'. \qquad (648.1)$$

Now r and r' could be expressed in terms of the earth's equatorial radius a as follows (arts. 9–10):

$$r = a\left(1-\tfrac{1}{2}\alpha(1+3\cos 2\varphi)\right) + O(\alpha^2) \quad \text{and} \qquad (648.2)$$
$$r' = a(1+\alpha\sin^2\varphi) + O(\alpha^2);$$

$$\therefore\ \varrho(\varphi, \chi) = a\left[1-\tfrac{1}{4}\alpha(2\cos 2\varphi-(1+\cos 2\varphi)\cos 2\chi)\right] + O(\alpha^2). \qquad (648.3)$$

In the map conceived by de Prony the central meridian m was taken to be the one through the Paris *Observatoire* O, which took latitude K. If a perpendicular were dropped onto m to meet it at N, with latitude L, then the distance x along m from N to O was given by (648.3) as

$$x := \int_K^L \varrho(\varphi, \tau)\,d\varphi = a[\pm(L-K) \mp \alpha\sin(L-K)\cos(L+K)] \qquad (648.4)$$

(to quote his form in art. 12: '$\mp$' served in lieu of specifying $|L - K|$). By broadly similar reasoning the distance y from Y to N was expressed in terms of the longitude p of Y relative to m as

$$\tan p = \tan P/\cos L, \quad \text{where} \quad P = y\,(1 - \alpha(2 - \cos 2L))\big/a + O(\alpha^2), \tag{648.5}$$

the angle projected at the centre of the projection by YN. This angle also served to calculate the latitude of Y:

$$\sin P = \cos\varphi \sin p \tag{648.6}$$

(essentially arts. 14–16, but using 'φ' to cover his angle 'λ').

Equations (648.4–6), together with some related equations, allowed the calculation of φ and p from x and y and vice versa. de Prony discussed the numerical details, expressed in the decimal system of angles and also the "new" metre (10^7 of which, by recent definition (§ 5.5.4), constituted a quarter of m). The analysis included an interesting method of finding the centre and radius of the sphere which approximated to the area mapped. Assuming the centre to lie along the centre C of the area, with latitude φ, he determined the desired radius Q (he wrote 'ϱ' again) by taking it to be the 'mean radius' of ϱ over $[0,2\pi]$ of χ from (647.1):

$$Q = \frac{1}{2\pi}\int_0^{2\pi} \varrho(\chi, \varphi)\,d\chi = \sqrt{rr'} \tag{648.7}$$

(art. 20). He showed that this value gave the smallest errors for determining the meridianal and perpendicular arcs at C (art. 22–23); and then he related (648.7) to Legendre's recent proposal of least squares regression.

From (648.2) it followed that

$$(Q =)\,\sqrt{rr'} = \tfrac{1}{2}(r + r') + O(\alpha^2), \tag{648.8}$$

and this latter value for Q 'satisfies the condition of the minimum of the sum of the squares of errors' (art. 26). To establish this interesting and topical result, he stated that if the perimeter of the region to be mapped was represented by a fine polygon of many short sides of lengths $\{A_j\}$ in the planar sections through C, with corresponding radii of curvature $\{r_j\}$ then the mean radius r was given by a centre-of-gravity argument,

$$r\sum_j A_j = \sum_j A_j r_j\,; \tag{648.9}$$

but this was the minimal value of a least-squares function, namely

$$d\left(\sum_j A_j(z-r_j)^2\right)\Big/dz=0 \quad \text{when} \quad z=r \tag{648.10}$$

(arts. 27–29: I have written 'r' for his repetition of 'ϱ').

To give this result some mathematical backing, de Prony noted that the error in angular measure of A_j caused by replacing r_j by r was

$$\left(\frac{A_j}{r}-\frac{A_j}{r_j}\right), \quad \text{or} \ \frac{A_j(r_j-r)}{rr_j}=\frac{A_j(r_j-r)}{r_j^2}+O((r_j-r)^2) \\ =\frac{A_j(r_j-r)}{r_j^2}+O(\alpha^2). \tag{648.11}$$

Thus the error incurred over the whole polygon was given by solving for z

$$d\left(\sum_j A_j\,(r_j-z)^2/R^2\right)=0, \tag{648.12}$$

where R was *any* of the $\{r_j\}$ 'in having regard to the order of quantities which constitute the differences between the quantities' $\{r_j^2\}$ (art. 32). If one wished to follow Legendre's new method, 'it will be necessary to put the equation'

$$d\left(\sum_j A_j^2(r_j-z)^4/R^4\right)=0, \tag{648.13}$$

the square of the error in (648.12), in order to obtain (648.9) (art. 32).

These procedures, together with some further conversions of equations in terms of observed or calculated variables such as $\{\chi_j\}$, gave de Prony's paper a strong attraction for cartographers and geodeters: it also marks an early adoption of Legendre's least-squares criterion. As might be expected, Puissant responded to it in his textbook on topography, on both of the grounds just mentioned (*Topography*$_2$ (1820), 135, 180–183: note the mere index listing in $_1$(1807), 330).

6.4.9 Puissant and the new map of France

Interest in the teaching of geography, geodesy and astronomy increased around 1810, partly in connection with various geodetic and cartographic projects and also for certain courses taught at the new *Université Impériale.* Our favourite textbook writer, Lacroix, came out with an *Introduction à la géographie* in 1811, as a 'second edition' to succeed an introductory article *1804c* on geography and cartography placed at the head of the translation of an English geography text; this piece itself had drawn partly on his short history *Maps*$_1$ (1802) of map-making published in the *Mémorial topographique et militaire.* The new book fell into two parts: the first on 'critical and mathematical geography', including some basic astronomy and details of map-making (*1811b*, 1-188); the second on 'physical geography', describing many principal features of the earth and its climate (pp. 189-409). Tables, an index and maps completed the volume. The text lay at the elementary level of our scheme of §2.2.4: some geometry was used, but no calculus. A good deal of historical information was included.

Curiously, Lacroix gave very few references to Puissant, the leading academic geographer of his day. Indeed, Puissant's career was at a particularly interesting stage at this time, for he was heavily involved in a major project under way at the *Dépôt Général de la Guerre*: the preparation of a new map of France, to replace the old one produced in the 18th century (see §5.5.3, where his discussion *1812a* of its Cassini projection was cited). The new plan was agreed by Napoléon in the late 1800s, but other kinds of geographical activity initiated by the Little Man delayed substantial progress until after the Restoration. Then an administrative commission was set up, with Laplace as president (of course): it virtually folded with his death in 1827, but work went ahead until mid century. Puissant took positions of leading responsibility in the organisation, and wrote or edited the principal publications: the first two volumes of the methods and results of the surveying (*Map* (1832-1840)) and a vast collection *1832a* of tables of all the trigonometrical points used. The enterprise created a measure of general interest, beyond the confines of the *Dépôt* (§17.7.2).[1])

Somewhat surprisingly, Puissant said very little in these jumbo publications about the projection used for the new map, confining himself to topographical and surveying aspects; but he had made a detailed statement already in the 1810s (which is why my account is placed here in this book). He seems to have mooted the new projection before the scheme was in full operation; for when

[1]) Files on the project are held at *AN*: see especially F[17] 3604. Berthaut has described the history of the new map from almost all points of view apart from the mathematical details (see his *Maps*, *1* (1898), 171-340 and *2* (1899), 1-248). One device used was the *reverbère*, a form of lamp using strong reflectors, following a design apparently proposed by Fresnel (see Puissant *1826a*, 63-64).

writing on the old Cassini method in 1811, he mentioned the defects caused by its distortion of areas distant from the base meridian, and claimed that 'if therefore one wanted to construct a map which accurately made a representation of areas, and which would alter distances but little, one would choose the modified projection of Flamsteed, as I have shown elsewhere' (*1811a*, 327). It is now more usually known as the 'Bonne projection', after an 18th-century French cartographer.

Puissant had already developed many of the details, for they appeared in the 1810 supplement to his treatise *Topography*$_1$ (1807) described in § 5.5.8. He gave a further account at the end of the decade, in a rather confusing presentation across the second editions of both of his treatises: basic formulae for latitudes, longitudes, and so on for a spheroid of revolution, in *Geodesy*$_2$ (1819), *1*, 242–341; then an extensive account of the chosen projection in *Topography*$_2$ (1820), 103–135, in which some notations of the other book were adopted without explanation and the 1810 version was fairly closely followed. A summary was given in a paper *1821a* read to the *Société Philomatique.*

The projection mapped the surface of the spheroid onto the planar development of the surface of the cone with apex along the terrestrial pole CO (see Figure 649.1$_1$) and MO tangent to the spheroid along the mean parallel at M with centre R. The main features of the map are illustrated in Figure 649.1$_2$ (with different lettering from its partner: the apex was beyond this point C, and the mean parallel was AM). Other parallels (such as a and b in Figure 649.1$_1$) mapped onto concentric circles (a_2p_2, ...b_2q_2, ... in Figure 649.1$_2$). The choice of mean parallel could be made at will. Puissant chose $\frac{1}{4}\pi$ as suitable for France; in the original version Flamsteed had used the equator, thus making his projection cylindrical. In the other direction a 'principal meridian', CA in Figure 649.1$_2$, was rectified: Puissant chose the one through the Paris *Observatoire.*

The basic properties of the projection may be conveyed as follows. Imagine any point on the projection (k_1, say, in Figure 649.1$_2$), with Cartesian coordinates (X, Y) and polar coordinates (R, Θ) relative to CA and differential arc-length ds. Consider the parallel through the corresponding point of the spheroid (k_1, say) with radius of curvature ϱ and angular distance dp relative to some unit circle, and differential arc length ds_1. Then the following deduction obtained:

$$\begin{aligned} &ds = Rd\Theta\ ; \quad ds_1 = \varrho dp\ ; \quad ds = ds_1\ ; \\ &\therefore\ \varrho dp = Rd\Theta\ ; \quad \therefore\ \varrho p = R\Theta \end{aligned} \tag{649.1}$$

(since both ϱ and Θ were measured from CA). Further, if the latitude of P were H, then

$$p = N\cos H \quad \text{where} \quad N := a\Big/(1 - E^2 \sin^2 H)^{1/2}, \tag{649.2}$$

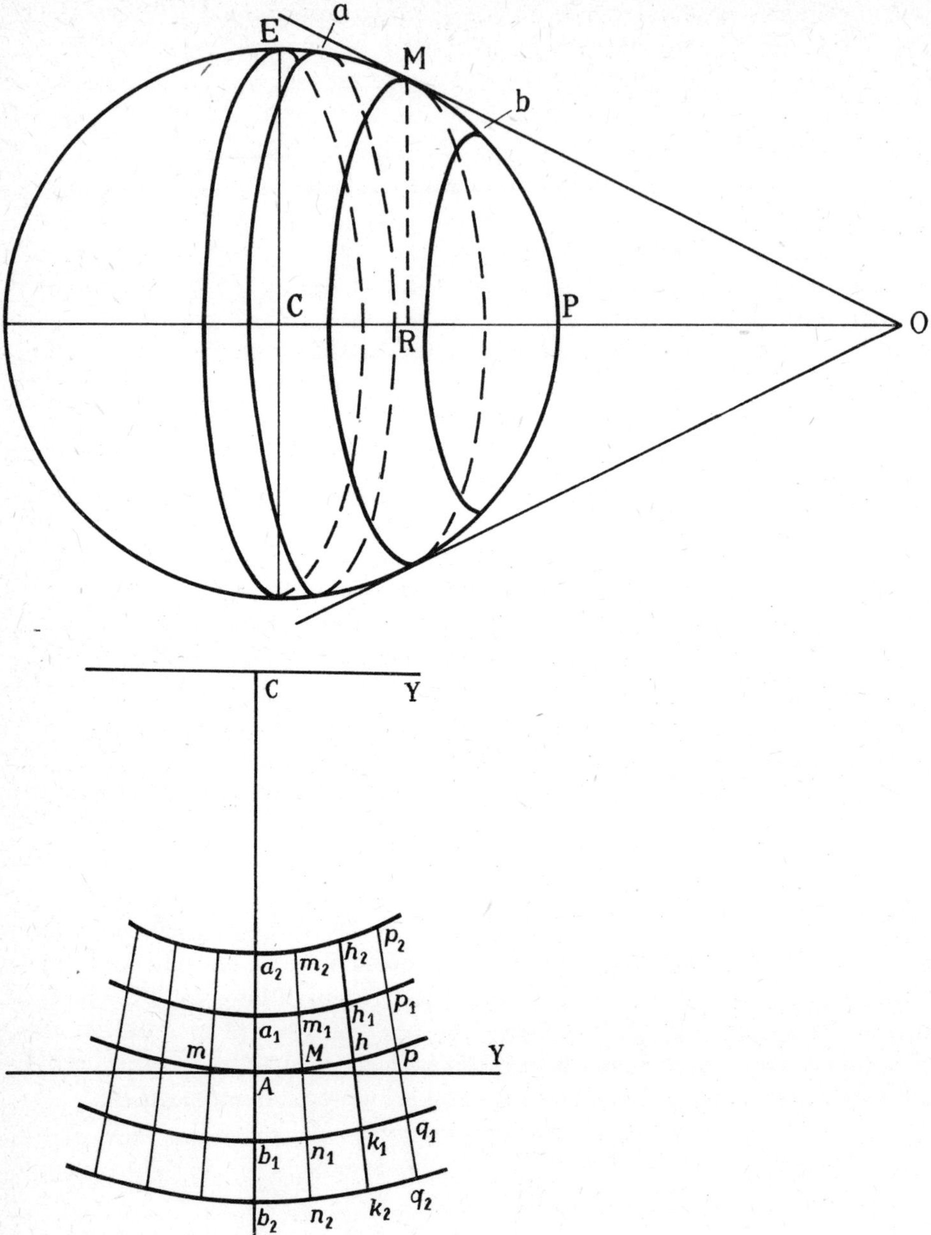

Figure 649.1 Puissant's diagrams for the modified Flamsteed projection (*Topography*$_2$ (1820), his figs. 22, 107: the latter first in $_1$, suppl. (1810), fig. 1)

with a as the length of the major semi-axis of the spheroid and E its eccentricity. Hence, from (649.1), Θ and H took the relation

$$R\Theta = \varrho N \cos H. \tag{649.3}$$

Puissant proved various other results from these equations and various results from the geometry of spheroids. For example, for a differential arc dS of a meridian (k_1k_2, say) through k_1 making an angle u with Ck_1,

$$ds = dR \sec u \quad \text{and} \quad \tan u = p \sin H - \Theta. \tag{649.4}$$

Further, if k_1 lay on the mean parallel, at latitude $h(=\frac{1}{4}\pi)$ for H, the length t of CP for R was given by

$$t = a \cot h \Big/ (1 - E^2 \sin^2 h)^{1/2} = N \cot h \tag{649.5}$$

from $(649.2)_2$ (study ΔMRO in Figure 649.1_1); thus (649.3) (with h for H) became

$$t\Theta = \varrho N \cos h = \varrho t \sin h. \tag{649.6}$$

Hence from $(649.4)_2$, $u = 0$ when $H = h$; in other words, radius Ck_1 was also the meridian in this case, so that each meridian $\{m_2n_2\}$ cut the mean parallel AM at right angles. Finally, a differential strip $d\Sigma_1$ on the map ($k_1k_2n_2n_1$, say) was nearly given by the product of arc-length s and width dR:

$$d\Sigma_1 = sdR + O(dR^2). \tag{649.7}$$

Now the corresponding strip $d\Sigma$ on the earth was the similar product of s_1 and the differential increment do of the generating ellipse of the spheroid:

$$d\Sigma = s_1 do = sdo = sdR = d\Sigma_1 + O(dR^2), \tag{649.8}$$

so that equivalence obtained for all regions where the above relations held well.

Thus this projection was reasonably conformal and equivalent, at least around the mean parallel and the principal meridian; and the simplicity of the various equations was attractive for calculation and numerical work. Less simple, however, was the determination of coordinates from latitude and longitude and *vice versa*, one of the principal requirements in map design. Although Puissant drew attention to de Prony's method using optimal radii of curvature, he himself worked with a trigonometric expansion of the latitudinal displacement Ab_1 of $P(= k_1)$ from the mean parallel AM:

$$Ab_1 = K \sum_{r=1}^{\infty} (-1)^r (q_r/r) \sin(r(h - H)) \cos(r(H + h)), \quad K \text{ constant}, \tag{649.9}$$

where the $\{q_r\}$ were defined as the coefficients in the trigonometric series

$$\left(1 + \frac{E^2}{2 - E^2} \cos z\right)^{-3/2} = \sum_{r=1}^{\infty} (-1)^r q_r \cos rz. \tag{649.10}$$

While he was able to truncate and manipulate nicely, the whole procedure was rather heavy, both for this exercise and various related ones. These included calculating the angles of intersection between meridians and parallels and their own points of intersection with the boundary of a folio of the map: in the latter case he approximated to the meridianal arcs $\{m_2 n_2\}$ by parabolae and solved the problem by using Lagrange's interpolation formula (352.14).[1])

While his aspirations were clear, and the diligence of his exegesis admirable, Puissant lacked a touch of the simplified which Legendre or de Prony might have contributed. Nevertheless, over the decades the *Dépôt Général* produced a beautiful set of maps.

6.5 Potential theory and the pendulum

The various investigations of the shape of the earth described in § 5.5 continued in the early 19th century. This section is confined to studies of the attraction of spheroids, and to the theory of the pendulum, an instrument which played an important role in the determination of the meridian and thus the spheroidicity of the earth.

6.5.1 Attractions of spheroids: Ivory's theorem and Poisson's equation

> His paper had to produce and did in fact produce a great sensation among the members of the *Académie des Sciences* [*sic*] of Paris.
>
> Saigey on Ivory's theorem (*1826a*, 93)

[1]) These details are given in the books cited just before Figure 649.1; the unclarities mentioned there are especially vexing. Sometimes Puissant did not always use the simplest versions of various formulae, either.

While still an *associé* of the *classe*, Biot presented a paper on Laplace's equation (552.2)

$$\nabla^2 V := V_{xx} + V_{yy} + V_{zz} = 0 \tag{651.1}$$

in June 1802 (*PV*, *2*, 514); it was eventually published in the *Mémoires* as *1806d*. Concerned by the difficulty of finding a general solution, he took up Lagrange's solution by power series, presented in *Mechanics*$_1$ (1788), 472-477 in connection with hydrodynamics. Three new variables (u, v, w) defined as functions of (x, y, z) would transform (651.1) into a general linear second-order equation

$$\mathscr{L}(R) = 0, \quad \text{where} \quad R(u, v, w) = V(x, y, z). \tag{651.2}$$

Its solution, following Lagrange, could be expressed as

$$R = \sum_{k=0}^{\infty} u^k \varphi_k(v, w) \Big/ k\,!, \tag{651.3}$$

where the $\{\varphi_k\}$ were some functions only of v and w, to be determined by substituting (651.3) back into (651.2) and setting the coefficient of each power of u to zero. Hence one could calculate grad R and thus the attractive forces ($-$grad V) (to use modern terms).

Now, from (651.3) $R = \varphi_0$ on the surface defined by $u = 0$, so that (651.3) could be read as defining R at any point from its values on this surface. In the particular case when $u = x$ the surface was the Oyz-plane, for which (651.1) was relatively easy to solve (Biot *1806d*, 210-215). He finished applying his method to certain ellipsoids, and in a later note *1812b* extended it to the case where the point of attraction projected into interior points of the body in the planes of the principal axes. He did not worry about the convergence of (651.3), or the functions used to change independent variables.

Biot's main procedure extended a result of Legendre *1791a* on attraction at points of an axis. 'The result of Mr. Biot', judged Legendre in a later paper (*1814b*, 158), together with his own,

> completed in a very satisfying manner the theory of the attraction of homogeneous ellipsoids, and there was little hope that this theory would be able to acquire some new degree of improvement.
>
> However, Mr. Yvory [*sic*: ...] has extended to this matter an entirely new clarity [...]

Ivory's paper, *1809a*, is an important example of foreign work stimulating the Paris mathematicians. Born in 1765 in Scotland, James Ivory was appointed to a chair in mathematics in 1804 at the Royal Military College at Great Malvern, England (see Interlude 480.1). Preoccupied until then with the management of the family flax factory, he had published very little; but this paper was the first of many on aspects of mathematical physics, including further researches on spheroids which excited a dispute with Poisson (§ 17.6.5).

Ivory dealt with the attraction of homogeneous ellipsoids, basically following Book 3 of Laplace's *Mécanique céleste* (§ 5.5.2). Many of his calculations were rather turgid manipulations of Legendre functions, but the result which Legendre was to praise brought various simplifications of the general theory of attractions. It may be stated as follows:

Theorem 651.1 Let E_1 be an ellipse with semi-axes (a_1, b_1, c_1), and $P_1(x_1, y_1, z_1)$ be a point on its surface. Let E_2 be a confocal ellipsoid of equal density, with the point $P_2(x_2, y_2, z_2)$ on its surface defined by the relation

$$x_1 : x_2 = a_1 : a_2, \quad y_1 : y_2 = b_1 : b_2, \quad z_1 : z_2 = c_1 : c_2. \tag{651.4}$$

Let the attraction of E_2 on P_1 have components (X_1, Y_1, Z_1) in the directions of the axes, and the attraction of E_1 on P_2 be (X_2, Y_2, Z_2). Then

$$X_1 : X_2 = b_1 c_1 : b_2 c_2, \quad Y_1 : Y_2 = c_1 a_1 : c_2 a_2, \quad Z_1 : Z_2 = a_1 b_1 : a_2 b_2. \tag{651.5}$$

When Ivory stated the theorem (*1809a*, 355), he did not claim it as new, for it was the conjunction of Laplace's theorems that the potentials of confocal ellipsoids at an exterior point was proportional to their masses, and that the attractive force of an ellipsoid at a point on an axis was a linear function of the length of that axis (*Mechanics*, *2* (1799), 48, 23). Poisson *1812d* noted this when summarising Ivory's paper to the *Société Philomatique*; but he gave credit to Ivory for the result, pointing out also that in fact it was true whatever law governed the attractive force. In an addendum to his paper Poisson *1812e* recorded other foreign work by reporting the receipt by the *classe* of Gauss's extract *1813c* of his great paper *1813b* on potential theory. In his own addition to this paper, Gauss acknowledged the importance of Ivory's result (which had been pointed out to him by Laplace), but criticised the paper for its use of possibly non-convergent series.[1])

Legendre was also concerned with 'Yvory''s proof, for in his paper quoted above he reworked the theorem in terms of the integral expressions for the attrac-

[1]) As the remark above on Biot exemplifies, such criticism was not normal in the 1810s (§ 3.4.2); but Gauss's concern with convergence was noted in Interlude 380.1. On Gauss *1813b*, see Cayley *1857a*, and Geppert *1933a*, 40–42; and on all papers discussed in this sub-section, see Todhunter *1873a*, *2*, 216–265. Among other responses to Ivory *1809a*, see Lagrange's manuscript in *BI*, ms. 902, fols. 65–68; and notes by Poinsot (?) in a file of Lacroix at ms. 2400.

tive forces rather than series expansions (*1814b*, 159–166: see Maurice *1833b*), and by the same means he proved a pretty result (p. 176) which in the notations of Theorem 651.1 stated that if P_1 was within E_1, then

$$\frac{X_1}{x_1} + \frac{Y_1}{y_1} + \frac{Z_1}{z_1} = 4\pi. \qquad (651.6)$$

Several of the results described above involved interior points of spheroids. Gauss used the distinction between interior and exterior points in a result which I write in vectors as

$$\int_S \frac{d\mathbf{s}\cdot \mathbf{r}}{r^3} = \begin{cases} 0 \\ -4\pi \end{cases} \text{if the point of attraction was} \begin{cases} \text{outside} \\ \text{inside} \end{cases} \text{S} \qquad (651.7)$$

(*1813b*, art. 7). As was mentioned above, Poisson reported on the extract of this paper, and seeing (651.7) there also (Gauss *1813c*, 282) may have stimulated him to think again about Laplace's equation (651.1). 'That equation in fact holds when the attracted point is situated outside the spheroid that one considers', Poisson told the *Société Philomatique*, 'or again when this being a hollow solid, the attracted point is situated in the empty interior space' (*1813i*, 389). But he then showed that the equation was invalid for points within the material, since there $\int dm/r$ became singular. To correct it he carried out a procedure which has since become standard teaching material: place a little sphere around the interior point, find the components of attraction there in each coordinate direction, and convert (651.2) to 'Poisson's equation':

$$\nabla^2 V = -4\pi\varrho, \quad \text{where } \varrho \text{ was density (constant or variable)} \qquad (651.8)$$

(pp. 389–391). He went on to solve (651.8) for simple cases, including a homogeneous ellipsoid, for which he easily obtained Legendre's (651.6) (pp. 391–392).

Some surprises attend this excellent contribution. Firstly, Poisson did not deal with surface points, although he must have realised that neither (651.1) nor (651.8) was valid there, but that $-2\pi\varrho$ is required. (The same can be said about Legendre's (651.6), which he seemed to think applied to surface points; and to Gauss's (651.7), which omitted them.) Secondly, Poisson published his modification only this once at the time, in contrast to the proliferation which he and others were wont to give to less significant results. He was to deal with $-2\pi\varrho$ in 1826, as we note at (1452.1).

Another modification to Laplace's theory was effected independently in Lagrange *1809c* and Ivory *1812a*. Laplace had shown that if a spheroid was almost a sphere of radius a whose centre O was very close to the centre of gravity of the spheroid, and the sphere and spheroid touched at a point P distant r from O, then the potential V at P satisfied the differential equation

$$\tfrac{1}{2}V + aV_r + \tfrac{2}{3}\pi a^2 = 0 \tag{651.9}$$

(*Mechanics*, *2* (1799), 85–89). However, Lagrange and Ivory showed that Laplace had overlooked possibly infinite values in an integrand which would prevent him setting an integral to zero, and also that (651.9) was true only for certain kinds of spheroid. In a sequel paper, Ivory *1812b* pointed out that Laplace's assumption that the surface of the spheroid had to be expressible as a solid harmonic was a further limitation; and he reworked the analysis for this type of spheroid in a manner which did not make that assumption.[1])

Many of these new results were assembled on one of the two doctoral theses submitted in June 1815 to the Paris *Faculté des Sciences* by the 21-year-old Olinde Rodrigues. He was one of the first Jewish mathematicians to emerge in the 19th century, and his thesis one of the early doctorates awarded in mathematics by the *Faculté*. Published as a 25-page paper *1816a* in the *Correspondance sur l'Ecole Polytechnique*, it shows that the *Université*'s standards for doctorates was fairly modest, as was mentioned in §2.4.3.

In the first part of his thesis Rodrigues ran over the general formulae for attraction, including many of the recent discoveries—Poisson's equation, Ivory's theorem, Gauss's method of finding the attraction to an ellipsoid to an external point (see Cayley *1858a*), and so on. Like a good trainee *savant*, Rodrigues was sparing in giving sources, citing only Ivory's theorem as such and referring to Poisson's account of it (Rodrigues *1816a*, 361–374).

The second part of the thesis, dealing with properties of the Legendre functions (as we call them), was more noteworthy. For example, Rodrigues pointed out that the convergence of the series expansions from the generating function was 'indispensable for the certainty of the results' (p. 375). His main concern in the part was to solve

$$((1-\mu^2)y'_n)' + (m(m+1) - (n^2/(1-\mu^2)))y_n = 0, \quad n \text{ integral} \tag{651.10}$$

(which we call 'the associated Legendre equation', and usually switch the roles of m and n). He put

$$x_n := y_n(1-\mu^2)^{-n/2} \tag{651.11}$$

[1]) Later, McCullagh *1833a* gave a simple geometrical proof of Laplace's (651.9). On all these developments, see Todhunter *1873a*, *2*, 257–266.

into (651.10) and differentiated it $(m-n)$ times to obtain the total differential equation

$$((1-\mu^2)^{m+1}x_n^{(m-n+1)})' = 0, \quad m \geq n \geq 0, \tag{651.12}$$

which integrated at once to yield

$$x_n^{(m-n)} = A + B\int (1-\mu^2)^{m+1} d\mu. \tag{651.13}$$

To find x_n he used (651.12) with the first differentiation executed, as an induction formula relating successive derivatives of x_n, starting from x_n itself when $m = n$. This brought him a relation between $x_n^{(m-n)}$ and x_n which, when substituted into (651.11), gave the complete solution

$$x_n = \left[(1-\mu^2)^m\left(C + E\int (1-\mu^2)^{-m-2} d\mu\right)\right]^{(m-n)} \Big/ (1-\mu^2)^n \tag{651.14}$$

to (651.10), where C and E were the constants of integration. He stressed that this expression would reduce to a polynomial in μ. He also found forms when $n < 0$ and $n > m$ (pp. 374–379).

Although, as we saw in §5.5.2, a considerable amount was known about Legendre functions, Rodrigues made a valuable contribution with (651.14). Apart from constants, and taking the substitution (651.11) into account, (651.14) contained for y_n both the first- and second-kind associated Legendre functions $P_m^n(\mu)$ and $Q_m^n(\mu)$, the former in a version of 'Rodrigues's formula'

$$P_m^n(\mu) = \frac{(\mu^2-1)^{n/2}}{2^m m!}((\mu^2-1)^m)^{(m+n)}, \quad m \geq n. \tag{651.15}$$

He also found from (651.10, 15) essentially the orthogonality properties of the $\{P_m^n(\mu)\}$ over $(-1, 1)$ of μ (pp. 380–382).

Rodrigues noted other orthogonality properties when he inserted y_n into his solution T of Laplace's equation (651.1). Since he was using spherical polar coordinates, T took the form

$$T = \sum_{n=1}^{\infty} r^n Y_n, \quad \text{where} \quad Y_n := y_n(\mu)(A_n \cos n\omega + B_n \sin n\omega). \tag{651.16}$$

The orthogonality of the trigonometric functions over $[0, 2\pi]$ led him to establish the orthogonality of the surface harmonics over the unit sphere (pp. 379–382):

$$\int_0^1 \int_0^{2\pi} Y_{m_1} Y_{m_2} d\mu \, d\omega = 0 , \qquad m_1 \neq m_2 .^1) \tag{651.17}$$

Rodrigues concluded by finding series expansions for the attraction to nearly apherical spheroids (pp. 382–385). He claimed on p. 376 the independence of his investigations from similar results in Ivory *1812b* and Legendre *Integral, 2* (1815), 263–275; and he made his own contributions to the theory with (651.14) and its ramifications. (One might have expected him to use (651.14) to investigate the zeros of the functions, which he could have easily done; and perhaps to have related (651.10) to Gauss's recent paper *1813a* on the hypergeometric function.) His paper is noteworthy also for its origins: while his Jewishness prevented him from studying at the *Ecole Normale*, his thesis was a significant herald of Paris mathematics produced *outside* the engineering schools.

6.5.2 Arago's and Biot's use of the Borda pendulum

As was mentioned in §5.5.4, Legendre *1806a* extended to spheroidal triangles his results on spherical triangles. He gave as one motivation the need to check if spheroidicity would introduce significant correction terms to the previous formulae, in view of the current extension of the Dunkerque-Barcelona meridian, with which his earlier work had been involved. Arago and Biot, together with others such as Mathieu, extended measurements southward between 1806 and 1808 to the Balearic Islands off the Spanish coast. In 1817 and the early 1820s, in a British-French collaboration, one or both of them travelled through Scotland and the Shetland Islands to take measurements at more northerly latitudes of the meridian.

As in the preceding project of Méchain and Delambre, the participants had various adventures (for example, it was Arago's turn to be imprisoned). Biot described their work with his usual loquacity, especially in two addresses *1810a* and *1818b* given at public meetings of the *Institut* and republished together as Biot *1818c* in the 'Histoire' department of the *Mémoires*. He and Arago published the details of their project as a joint book *1821a*, although, for reasons discussed in §7.4.7, the degree of collaboration between the two by then was low. Biot appears to have been the principal author.

1) Rodrigues's results, with its consequent expansion theorems, are hard to disentangle from a proliferation of notations: $T^{(m)}$, Z_m (an infinite series on p. 375 but finite on p. 379), Y_m and R_m (which was defined on p. 379 essentially as $P_m(\mu)$ and was used thereafter rather than (an equivalent of) $P_m^n(\mu)$). The certificates for his theses are held at *AN*, AJ[16] 5322, fol. 44.

The new project covered much the same range of interests as its predecessors: geodesy, the shape of earth, variation in gravitational attraction. But there were various differences. For example, Laplace had used his minimax method to calculate the ellipsoidicity of the earth (§5.5.5); Biot was inclined to use the new method of least squares (*1821c*, 73). More importantly, he and Arago laid more emphasis on pendulum measurements as the basic source of data. Accordingly, therefore, I confine myself to their use of the pendulum.[1])

The type of pendulum used went back to the earlier project, for the *Commission des Poids et Mesures* had asked Borda to design a refined version of the simple pendulum. The results of his work were described in a posthumous paper *1810a* written with Cassini IV and published in Delambre's report *Metrics* in 1810. Without saying so, Borda seemed to have studied Boscovich's essay *1785a* on pendulums, for he followed many of the same methods. He suspended a dense platinum globe from a long wire, and timed its oscillations against those of a pendulum clock placed immediately behind it. There were various refinements to measure accurately the length of the wire and the angle of oscillation, to allow for factors such as barometric pressure (§7.2.4) and the extension of the wire due to a rise in temperature, and to reduce to a minimum factors such as fixing the knife-edge suspension, and so on.

Borda's pendulum was, strictly speaking, a two-(sexagesimal-)seconds pendulum, as it used lengths of wire of around four metres. This was too long for Arago and Biot to carry around, and so they used a seconds-length version of about one metre. But they took pains to attain great accuracy, placing both pendulum and clock in a glass case, for example, to eliminate wind effects and obtain steady temperatures. While one respects the integrity of their endeavour, questions of precision come to mind in this decimal-point science. For example, they appeared to assume that the clocks, based on a design by L. Breguet, would be accurate enough for the purpose; and they did not consider sympathy between the observed pendulum and that in the clock.

---

[1]) Biot's manuscripts relating to the pendulum experiments are kept at *OB*, mss. E3, 14–56 (see Bigourdan *Bureau*, pt. 2 (1929), C24–C31). Files of relevant documents, including very interesting letters from Delambre to the *Ministre de l'Intérieur*, are held at *AN*, F^{17} 3212. Biot's article *1823a* on the pendulum for the *Encyclopaedia britannica* is a useful historical source: the English translation was done by Young, and relevant letters by Biot, Young and others are held in the British Library (London), Add. ms. 34613. See also Biot's letters of 22–25. 12. 1853 to Airy on the collaboration, written soon after Arago's death (Airy Papers, Royal Greenwich Observatory (Herstmonceux) Archives, ms. 375, fols. 84–86). Biot *1806g* extended Huygens's theorem on axes of synchronous oscillations. On the history of the pendulum see Wolf *1889b*, together with his annotated bibliography *1889c* (which lacks, however, two papers on pendula, Pouillet *1814a* and Deflers *1815a*, written by *normaliens* but published in the *Corr. EP*).

Borda's posthumous paper, about to be described, was given to the *BL* after his death by de Prony on 18. 8. 1804 (see *BL*, *3*, meeting of 30 *thermidor an* 12).

The theory of Borda's pendulum started off from the (well-known) haversine expression for the time T of oscillation of a simple pendulum of length l with angular displacement α either side of the vertical:

$$T = \pi\sqrt{\frac{l}{g}}f(\alpha), \quad \text{where}$$

$$f(\alpha) := \int_0^T \frac{d\Theta}{(1-\sin^2\Theta\sin^2\alpha)^{1/2}} = \sum_{r=0}^{\infty}\left(\frac{2rC_r\sin^r\frac{1}{2}\alpha}{4^r}\right)^2 \tag{652.1}$$

(see, for example, Laplace *Mechanics, 1* (1799), 58). For small oscillations α Biot felt safe to take the first two terms of this expansion and obtain

$$T = \pi\sqrt{l/g}\,(1+\tfrac{1}{16}\sin^2\alpha). \tag{652.2}$$

The problem was to convert the experimental data into usable form.

One of the principal difficulties was that one could not expect to hit exactly on the correct length of pendulum to give beats isochronous with the clock. Following Boscovich, Borda (and then Arago and Biot) used 'the method of coincidences', where the pendulum, fixed at about the right length, was set in motion with about the same angle of oscillation as that of the clock. A telescope was set up about eight metres away to observe the moments of 'coincidence', when the two pendula passed the vertical at the same moment. If J seconds passed before the next coincidence occurred, then the pendulum would have executed $(J \pm 2)$ oscillations. Conversely, the number of oscillations would be determined by the seconds recorded by the clock time.

For accurate results this length of time should be considerable; but then one could not assume that the angle α of oscillation was constant, for use in (652.2). Borda suggested on experimental evidence that the angle decreased in geometric progression from its initial value α_0 to its final value α_N during the oscillations between moments of coincidence. Thus at the nth oscillation

$$\alpha_n = \alpha_0/K^n, \quad \text{with } K \text{ a constant, given by} \quad \ln K = (N-1)\ln(\alpha_0/\alpha_N). \tag{652.3}$$

Hence, from (652.2), the total correction ϱ to be made to the N amplitudes, leaving out constants, was given by

$$\varrho := \sum_{n=0}^{N-1}\frac{1}{16}\sin^2\alpha_n = \sum_{n=0}^{N-1}\frac{1}{16}\sin^2(\alpha_0/K^n) = \frac{1}{16}\sin^2\alpha_0\sum_{n=0}^{N-1}K^{-2n}, \tag{652.4}$$

allowing for the smallness of α_0. Summing this geometric progression and using

$(652.3)_1$ when $n = N$ gave

$$\varrho = (K^2\alpha_0^2 - \alpha_N^2)\big/16(K^2 - 1)\,. \tag{652.5}$$

For small oscillations $\alpha_N \triangleq \alpha_0$, so that, from $(652.3)_2$, $\ln K \triangleq 0$. Hence

$$K^2 = \exp\ln(K^2) = 1 + 2\ln K + O(\ln^2 K)\,, \tag{652.6}$$

and (652.5) became

$$\varrho = \frac{(1 + 2\ln K)\,\alpha_0^2 - \alpha_N^2}{32\ln K} = \frac{\alpha_0^2}{16} + \frac{(N-1)(\alpha_0^2 - \alpha_N^2)}{32\ln(\alpha_N/\alpha_0)}\,. \tag{652.7}$$

Now the first term here was due to the first oscillation, for it was provided by (652.2); hence the second term covered the other oscillations, and so could absorb the first with little error if N were written for $(N - 1)$ there. So finally, the correction term could be written in two forms, either from (652.7) or the corresponding version in sines if felt more accurate or more suitable for calculation by logarithms:

$$\varrho = \frac{N(\alpha_0^2 - \alpha_N^2)}{32\ln(\alpha_0/\alpha_N)} \quad \text{or} \quad \varrho = \frac{N\sin(\alpha_0 + \alpha_N)\sin(\alpha_0 - \alpha_N)}{32\ln(\sin\alpha_0/\sin\alpha_N)}\,.^{1)} \tag{652.8}$$

If α_0 and α_N were close together, then the denominator of $(652.8)_2$ could be expanded in a power series in $((\sin\alpha_0/\sin\alpha_N) - 1)$ to yield

$$\varrho = \frac{1}{16}N\sin^2\left(\frac{\alpha_0 + \alpha_N}{2}\right) + O\left((\alpha_0 - \alpha_N)^2\right), \tag{652.9}$$

a result derivable directly from the assumption that all oscillations took the mean displacement $\frac{1}{2}(\alpha_0 + \alpha_N)$. In either case, during a period of J seconds with coincidence occurring after N oscillations, the number of equivalent oscillations would be

$$H := J\left(1 \pm \frac{2}{N}\right)(1 + \varrho)\left(1 + \frac{h}{J}\right), \tag{652.10}$$

where h was the number of oscillations to correct from mean time to the mean solar day; this factor was introduced for measurement against the new centesimal second, defined as 10^{-5} of that day.

---

1) Borda's expression for ϱ, with the numerator of $(652.8)_2$ and the denominator of $(652.8)_1$, was given without proof in his and Cassini's *1810a*, 30. Biot stated $(652.8)_2$ without proof in his and Arago's *1821a*, 455, and with a vague outline argument, including some errors, in his *1823a*, 122–123; a proof had appeared in his *Astronomy*$_2$, *3* (1811), Additions, 169–172. My argument is similar to the more elegant proof in Mathieu *1823a*, 283–286.

Arago and Biot also made various modifications to the measured length l of the pendulum. They allowed for the expansion of the wire and the globe due to temperature, the altitude at which the measurements were taken, the resultant variation in gravitational force, and barometric pressure as a factor in air resistance. These desimplifications converted l to a new value (l_1, say), so that, if J were the number of oscillations executed in a mean solar day, then from the formula for the simple pendulum ((652.2) with $\alpha = 0$) the required length (*in vacuo* at sea level with 76 centimetres' pressure) was given, via (652.10), by $l_1 H^2/J^2$.

6.5.3 Poisson and Laplace on the niceties of pendulum theory

Yet still other aspects of pendulum theory needed analysis; and Poisson took up the challenge, in a paper *1808c* published in the *Journal de l'Ecole Polytechnique* (but apparently not submitted to the *classe*). Assuming that air resistance was proportional to the square of velocity, he obtained the equation of motion

$$\ddot{s} = -g dx/ds + C\dot{s}^2, \quad \text{with } C \text{ constant}, \tag{653.1}$$

for a pendulum of unit mass located distance x vertically below the point of suspension, and at arc length s from the vertical at time t (p. 149). He easily found from putting

$$x = a(1 - \cos(s/a)) \quad \text{that} \quad s = k\cos\left(t\sqrt{g/a}\right) + O(s^2), \tag{653.2}$$

where a was the length of the pendulum. He then tried as the next approximation

$$s = k\cos\left(t\sqrt{g/a}\right) + s_1, \tag{653.3}$$

and showed from the solution that the time of half oscillation was

$$\sqrt{a/g}\,(\tau \pm \tfrac{1}{3}Ck), \tag{653.4}$$

where the positive/negative sign applied to the downward/upward oscillation. Thus he proved mathematically, apparently for the first time, '*that the time of the whole oscillation is the same as if the motion had taken place in a vacuum*' (p. 152). He also showed that the amplitudes $\{k_n\}$ of the half-oscillations obeyed the relation

$$k_{n+1} = k_n - \tfrac{4}{3}Ck_n^2 + O(k_n^3) \tag{653.5}$$

(pp. 152–153). This result did not relate well mathematically to Borda's experimental law (652.3), which was then still unpublished.

Poisson also examined the extensibility of the wire, expressed as a force $\varphi(r)$ acting along its variable length r (pp. 154–158). The equations of motion along and perpendicular to the arc were

$$\ddot{r} = g\cos\Theta - \varphi(r) \quad \text{and} \quad \dot{\overline{r^2\dot{\Theta}}} + gr\sin\Theta = 0 \tag{653.6}$$

at the moment t of angular displacement Θ from the vertical. 'The form of the function φr is unknown', he confessed, but he seems to have drawn on Hooke's law of extension to offer the form

$$\varphi(r) := g\left(1 - \frac{u}{a-b}\right) + O(u^2), \tag{653.7}$$

where a and b were the lengths of the loaded and unloaded wire and u the extension function. Replacing $\varphi(r)$ in (653.6) led to two non-linear ordinary differential equations in Θ and u, which he solved by a process similar to (653.1) for (653.3) to find that time of oscillation was still that of a simple pendulum with the length a increased by a small amount

$$O(h^2), \quad \text{where} \quad h := \sqrt{a/g}\,\dot{\Theta}\Big|_{t=0}. \tag{653.8}$$

Poisson considered again the extensibility question in an addendum *1809g*. Alerted by the Italian Paoli *1809a* to the fact that his approximations to the differential equations in Θ and u had omitted non-negligible terms and that h in (653.8) was not easy to determine in practice, he reworked Paoli's analysis to obtain (653.8) again, though with a different coefficient for h^2, and h now determined in terms of the amplitude of the oscillation.

In his *Mécanique* of 1811 Poisson did not subject the students to such niceties, concentrating on the basic theory and the effect of air resistance (*1*, 393–418). But he returned to the pendulum theory in the mid 1810s, by a rather circuitous route. Français's book on the rotation of a rigid body, which was described in § 6.2.2, ended with considerations of the case where two moments of inertia were equal (*1813b*, 45–49). Poisson gave a verbal report on this book to the *classe* in February 1813 (*PV*, *5*, 152, 168); and perhaps it stimulated him to a paper *1813e* of that year in which he took up Euler's equations (525.14) from his own book (*Mechanics*$_1$ (1811), *2*, 121–160) and solved them for the rotation of a solid of revolution about a point on its principal axis. He then used a form of this analysis in a sequel paper *1816c*, which was concerned with the effect on the motion of the pendulum of the torsion of the suspending wire. He found that its length had to be shortened by a term $O(s^4/a^3)$, where s was the radius of the suspended globe (*1816c*, 337–338). He felt this correction to be negligible in Borda's pendulum, where the use of a dense material for the globe allowed (s/a) to be quite small; he took 0.05 in a numerical example on p. 338.

Concern with the Borda pendulum continued in connection with a paper presented to the *Académie* in October 1816 by the 67-year-old Laplace (*PV*, *6*, 108). He considered several aspects of the instrument, especially the effect on the motion of the wire on the knife-edge about which the pendulum was pivoted. Taking this edge to be a half-cylinder of radius *a*, he showed after a complicated analysis that the effect was equivalent to shortening the length of the pendulum by $2a$. As a was of the order of 10^{-2} millimetres, he felt that this correction was 'sensible' for the shorter pendula (Laplace *1817a*, art. 4). He published the non-technical preface to the paper as a contribution *1816a* to the *Société Philomatique*; and Poisson adjoined a simpler analysis *1816f*, somewhat similar to the one in his own *1816c*, in which he found the same corrective term. Laplace wrote an addendum *1817b* to his main paper, showing that Huygens's theorem on the reciprocity of the centres of suspension and oscillation for synchronous oscillations was still valid with this correction; so Poisson published an addition *1817c* to his own note, showing the same thing!

6.5.4 de Prony and the reversible pendulum

Mention of Huygens's theorem and different points of suspension naturally calls to mind the so-called 'reversible' pendulum. This instrument consisted of a metal bar with two fixed knife-edge pivots and three weights, of which the lighter two were adjustable along the bar in order to produce a time of oscillation of the desired order. Introduced in Kater *1818a*, it was described by Biot as an alternative instrument to Borda's (see Biot *1821c*, 75–77; and *1823a*, 126–128); but it excited much less enthusiasm with de Prony, for reasons which go back to 1792.

When the *Commission des Poids et Mesures* had asked for a design of an accurate pendulum, they received not only Borda's plans but also a scheme *1792a* from de Prony, then 37 years old. He envisaged a body oscillating about three horizontal axes coplanar with its centre of gravity. An *Académie* commission, including Laplace, Legendre and Delambre, found the idea intriguing but rather impractical, and de Prony set it aside. However, he revived it in another form in 1798 following a suggestion by Breguet, and presented a manuscript *1800c* to the *classe* in October 1800 (*PV*, *2*, 237). But the instrument was not built, probably because of the success achieved with Borda's pendulum; de Prony published only a brief note *1801d* with the *Société Philomatique*, the two longer manuscripts remaining unpublished. Thus when he heard of new English use of this device, he claimed priority in a paper *1817a*; but he cited only the account in his textbook *Mechanics, 2* (1815), 339–342. In a pamphlet *1819c* he knew of Kater's work, and objected to the view (Kater *1818a*, 35), that the third axis was unnecessary.

The number of axes used was one difference between the two instruments; and de Prony had no moveable weights. However, their ideas were basically the same. In de Prony's instrument, if the radius of gyration of the body were k, the distances from a given point O to the centre of gravity and the pivots $\{p_j\}$ were x and $\{a_j\}$ respectively, and if the pendulum executed n_j small oscillations about P in a given time T, then the three equations for the time of oscillation

$$gT^2 = n_j^2\pi^2\left(x - a_j + \frac{k^2}{x - a_j}\right), \qquad 1 \le j \le 3 \tag{654.1}$$

could be solved for the unknowns x, g and k (though in principle with more than one solution). Then with x and k known, they could be used for determination of the seconds pendulum as required by the *Commission.* If the more accurate version $(652.1)_1$ of the time of oszillation were needed, then (654.1) could be modified to

$$gT^2 = n_j^2\pi^2\left(x - a_j + \frac{k^2}{x - a_j}\right)f^2(\alpha_j), \tag{654.2}$$

where $\{\alpha_j\}$ were the angles of half-oscillation (see *Mechanics*, *2* (1815), 330–340).

Although de Prony abandoned his pendulum, he used the ideas in two methods for regulating clocks. One was described in *1814a* for the *Bureau des Longitudes* and again in his *Mechanics*, *2* (1815), 424–434; the other featured in the paper *1817a* cited above. The first method involved moving a cursor along the axis of the clock's pendulum; the second method was a slight generalisation in which parts of the body of the instrument were moved in order to vary its moments of inertia and thus its time of oscillation. In a manner which we recognise from § 3.5.3 as characteristic, he expressed the mathematics of his second method in the form of differences (he probably preferred it to the first method for this reason), and in both papers he calculated tables indicating the magnitude of the changes required to effect given diurnal variations.

Another feature of these methods is worth noting. Since they both involved moving parts of the pendulum, they were very similar in form to the pendulum which Kater was about to announce. One can understand de Prony's pique at the recognition accorded to a foreigner of an idea which had been ignored years before by his own colleagues when he had put it forward himself.

6.6 Concluding comments

This brief final section follows the structure of § 4.8. After noting two comprehensive works published in or around 1815, some brief comments of my own are offered on the work covered in the chapter. Interlude 660.1 provides a little information on mechanics outside France.

Interlude 660.1 Mechanics outside France

The history of mechanics in the early 19th century is covered even more inadequately than is the history of the calculus, and so I cannot match even the modest aims of Interlude 480.1 here. My highly tentative impression is that, while in experimental work foreigners could at least match the French, except for Gauss and Bessel they added only footnotes to the mathematical treatments issuing from Paris. A particularly interesting study would be the balance taken in different countries between the Eulerian and Lagrangian traditions in mechanics, and the attention, if any, given outside Russia to the posthumously published papers by Euler. On some Italian work, see my *1986c.*

German ability in theoretical astronomy was noted in Interlude 641.1. Germans also made translations of most of Laplace's books: his early *1784a* (translation published 1800), the first edition of the *Exposition* (1797: the later editions seem not to have been translated); and volumes 1 and 2 of the *Mécanique céleste* (1800–1802: prepared by J. C. Burckhardt, a naturalised Frenchman and member of the *classe*). A German translation of Lagrange's *Méchanique analitique* came out in 1797 (and another, surprisingly, in 1887); no translation of the second edition seems to have been prepared. Of other textbooks discussed in this chapter, the first to receive a translation was Poisson's *Mechanics*$_1$ (1811), in 1825–1826.

Before Bowditch (1829–1839) translations of French works into English were largely confined to less ambitious versions of Laplace, especially digests of the *Mécanique céleste.* More ambitious, but in the end incomplete, was a translation of that work, with notes, by H. Harte in Dublin. Only the first volume of Laplace was done, in two volumes (1822, 1827). Later he also published translation/editions of Laplace's *Exposition*$_5$ (1824) in 1830, and of Poisson *Mechanics*$_2$ (1833) in 1842 (see my *1987b*).

The Irish involvement is significant, for a reform of mathematics there began around 1813 under the leadership of H. Lloyd, with mechanics given a prominent place (see my *1988a*). Their attention to Fresnel's optics later on is noted in §15.6.5.

6.6.1 Impressions of two surveys

The authors of the two works are Lagrange and de Prony, and they reflect in both level and content the differences between the studious theoretician and the engineer educator. Both works have been cited on occasion already, and they will be cited later.

In 1811, his 76th year, Lagrange published volume I of the second edition of his *Mécanique analytique* (as he now spelt it), but died in 1813 before volume 2 was completed. The manuscript was passed to Garnier and de Prony, who were assisted by Binet and Lacroix in its preparation for publication, which occurred in 1815. According to Virey and Potel *1813a*, 17, Lagrange had intended to prepare a third volume on astronomy; but apparently no usable materials were found. Many features and details were discussed in later decades by various writers, and in the 1850s Bertrand prepared a new version of the edition, containing 16 of these commentaries together with more editorial footnotes of his own. This version was reprinted 30 years later in Lagrange's *Œuvres*.

The basic design of the treatise was unchanged: a Part for statics followed by one on dynamics. Part 1 was similar in many details, the eight sections of 1788 (see Table 521.1) appearing again in the same order. Part 2 showed more changes in the eight sections that were completed; some of them were considerable expansions of their predecessors, both in detail and in structure. The eighth section, an extended account of the motion of mutually interacting bodies, suffered from several errors in passage-work which Lagrange did not live to correct. He died while in the middle of rewriting, as the ninth section, the passage in the old edition on the rotation of bodies. It was completed from the previous text by his editors, who also reprinted the closing two sections on hydrodynamics (§10.3.1).

The similarity with the first edition extended also to its basic principles and strategy. Mechanics remained a branch of algebra; there were still no diagrams. Although Lagrange cited several of the works discussed in this and the last chapters, rarely did his remarks surpass the *en passant*; indeed, some topics, such as Poinsot's couples, did not even gain that much attention. While several parts of the treatise were expanded, the new material usually followed on from the bases laid in 1788. Probably the most substantial new additions were the accounts of the Lagrange and Poisson brackets; they were noted in § 6.3.5.

de Prony produced several published versions of his mechanics lectures around the late 1800s, when he was in his mid fifties: two 'summaries' *1809d* and *1811a*; and a *Cours* (*1809c*), like Poisson (Table 623.1). The most important of these works was a two-volume textbook on mechanics of 1810 and 1815, almost the same years as for Lagrange's book. Entitled *Leçons de mécanique analytique données à l'Ecole Polytechnique* (but, amusingly, published by his own *Ecole des Ponts et Chaussées*), it was intended as a reworking and completion of his *Mécanique philosophique* of a decade earlier (§ 5.2.7). His scheme was different in various ways. He now divided the subject matter into statics and dynamics (like Lagrange); no longer devoted his right hand pages to columns of notations, definitions, theorems and problems, but followed the normal style; now included some diagrams; and allowed himself quite a number of digressions, so that, at about 850 quarto pages, the work was nearly double the size of its predecessor. As with Lagrange, there were references to several of the works discussed in this chapter, including accounts of his pendulum designs (as was noted in § 6.5.4) and (unlike Lagrange) a treatment of Poinsot's couples (*Mechanics*, *1* (1810), 96-107). An interesting passage ran through the story of the vibrating string problem and its solution, followed by theories of musical tuning (*2* (1815), 472-510). (Music was a passion for him, and late in his life he published a short book *1832a* on the mathematics of scales.) But otherwise the citations were again *en passant*, and the general approach largely followed that of his previous textbooks. While his approach in itself was somewhat different from Lagrange's—he made little use of variational methods, for example (§ 5.2.7)—his preservation of a chosen approach marks an important similarity in policy with Lagrange.

6.6.2 A modern impression

These books by Lagrange and de Prony suggested that the main lines of mechanics were laid down, and that the main tasks were to continue the exegeses within settled frameworks. To a considerable extent the survey provided in this chapter justifies this view: take, for example, the fresh views on the principle of virtual velocities; projectiles, meteorites and comets; or potential theory and the pendulum. Even the brackets theory of Lagrange and Poisson (§ 6.3), while an achievement of major importance, was developed as consequences of variational methods in mechanics which Lagrange himself had already made canonical.

However, certain other ideas were rather more radical in character. Poinsot's introduction of the couple, for example, changed the conception of statics and showed that a special exception to the composition of forces actually constituted a major lacuna. Again, and more wide-ranging in its consequences, was the growth of concern with probability theory and especially statistics in mechanics. Laplace was the principal instigator, although Legendre's proposal of least squares was also significant. Celestial and planetary mechanics provided almost all the scientific stimuli for these developments, a feature which guided the design of §6.4.

Thus we find an interesting tension between conservatism and innovation in mechanics in the early years of the new century. Lagrange and de Prony largely represented the former view, and Laplace the latter; and the next chapter will underline his novelties, for there we describe his new programme of molecular mechanics, which inspired several young *polytechniciens* to follow his lead. de Prony nicely captured both strands in a piece of advice which he gave to the students in one of his sets of lecture notes (*1809d*, lectures 31–33, 9):

> Those of the Students who intend to lead a career in teaching and the sciences, must devote their nightly labours to the study of the *Mécanique analytique* and the *Mécanique céleste.*

Chapter 7

Laplace's mathematicised molecular physics, 1804–1816

7.1 Prefaces

7.1.1 Physics in the early 1800s

The programme of research initiated by Laplace, which is the topic for this chapter, went beyond mechanics in its scope of reference, and took in several parts of physics. Thus it is meet to preface the account with an indication of the character of physics at that time.

At the beginning of the 19th century physics did not enjoy high prestige at either teaching or research levels. For example, it was not prominent at the *Ecole Polytechnique* (as will be described in a few paragraphs' time), the *écoles d'application* or in the *Université* (although Biot and Lefèvre-Gineau held chairs in it at the *Collège de France*); it had failed to attain the prestige of neighbouring disciplines such as mathematics, mechanics, engineering and chemistry. But textbooks were appearing; and here I shall summarise the contents of three of them, in order to show the topics covered by the subject and the general relationship between physics and mechanics.

In 1803 the physicist Haüy published the first edition of his *Traité élémentaire de physique*; a revised edition (on which I shall concentrate) appeared three years later, and a final one in 1821, shortly before his death. A substantial two-volume survey, it was an extension and revision of his lectures on physics delivered to the short-lived *Ecole Normale* (§ 2.4.2) in 1795, which had been published as his *1800a*. In a concession to the reader unusual in a French author, the new book contained a subject index.

Haüy announced proudly in his preface (*Physics*$_2$ (1806), *1*, xxviii–xxix):

> This Treatise had been designated by His Majesty the Emperor and King, to the number of those [works] which had to be of use to public instruction. No motive was more proper to excite our zeal and our efforts, than such an honorable testimony of confidence on the part of a hero, who needs only to call other men to the execution of his designs, to lift them above themselves.

Here too was a revision, for in the corresponding passage in the first edition he had brayed of the 'First Consul'.

In his textbook Haüy dealt in turn with 'the most general properties of bodies' (extension, mobility, impenetrability, divisibility: 15 pages); 'attraction' (gravity, and molecular attraction: 65 pages); 'caloric' (principles, specific heat, laws of 'elastic fluids', thermometer, gas, vapours and combustion: 117 pages); water (as liquid, ice and vapour: 77 pages); air (weight, atmosphere, transmitter of sound:

89 pages); electricity (produced by friction and heat, in air; and 'galvanic electricity': 143 pages); magnetism (action, communication, in the earth and in ironworks: 77 pages); and optics (nature, reflection and refraction, decomposition, natural vision and its enhancement by instruments: 268 pages). Most of the time he described phenomena, experiments and equipment, though there were passages on current theorising. The account included virtually no mathematics, although numerical values of physical constants and parameters were provided fairly often. This was normal for the time, reflecting the distinction between "mathematical" and "physical" sciences described in § 2.2.1 in connection with the organisation of the first *classe* of the *Institut*; indeed, some books on physics went through more or less the above material and then on into chemistry (as the physics of affinities) and perhaps even some of the life sciences. On very recent work, Haüy's book is notable in its praise of Biot's various efforts, some of which have been noted earlier and others which arise in this chapter.

In 1806 another physics textbook appeared: the first edition of the translation of the *Lehrbuch der mechanischen Naturlehre* (1805) by the Berlin physicist E.G. Fischer. It was prepared by Biot's wife, with Biot's own added footnotes and passages of text. The book, nearly 500 pages in length, was divided into eight sections: 'On bodies in general' (7 chapters); 'solid bodies' (9, including constitution and motion); heat (4, including production and propagation); 'liquids' (6, with some hydraulics); 'aeriform bodies' (5, incorporating hygrometry and barometer theory); electricity (5, mostly electrostatics but also some 'galvanism'); magnetism (2, especially magnetic dip); and light (7, mainly vision, reflection and refraction, and some instruments). Biot's notes included some simple mathematical treatments (of which the original text was largely denuded) and a few modifications and criticisms; but he also updated the text in places, and for new editions (published in 1813, 1819 and 1830) he continued to revise and expand these sections. They included discussion of the asserted constitution and modes of action of various phenomena; Fischer largely eschewed such questions, with the interesting exception of electrostatics, where he surveyed the theories to be described in § 7.5.1.

Biot began his own preface to the book (a letter to the original dedicatee Berthollet, who had suggested that the translation be done) by noting Berthollet's and Laplace's expressed regrets over the poor state of physics in France. Now none of the books available in the 1800s was explicitly directed to the needs of the *Ecole Polytechnique*, where the teaching of the subject seems to have been poor: the courses outlined in Barruel *1795a* and Hassenfratz *1798a* do not inspire confidence,[1]) although Barruel's book *1796a* of synoptic tables of branches

[1]) See Hassenfratz's notes held at *EP*, III 3b; and those made for student Clermont-Tonnerre at *Wellcome*, ms. 1665.

of physics was somewhat better. So around 1805 the *Conseil d'Instruction*, anxious to improve the situation, asked Hachette to write up his lectures on heat and electricity. The principal follower of Monge, he was largely concerned with descriptive geometry and its applications, and so he comes into his own in our story first in Chapter 8; but, as we shall see in § 7.5.5, his interest in electricity and magnetism was quite considerable. To fulfil his charge, he first produced a short book *1805a* which summarised his lectures 'on caloric and on electricity' given at an *Ecole des Sciences et des Lettres* (where, according to his preface, Poisson taught mathematics). Four years later he incorporated this material into another book *1809a* entitled *Programmes d'un cours de physique* and published by the house of Bernard. He was then in his 41st year.

Hachette's choice of topics was somewhat different from Haüy's, and included a few recent developments. The twelve lectures covered, in turn, general notions on bodies (molecular attraction, extension: 14 pages); crystallography (12 pages); 'mobility and gravity' (22 pages); capillary action (*à la* Laplace (§ 7.2.2-3): 5 pages); heat (19 pages); hygrometry and meteorology (only a one-page summary); light (propagation, reflection, refraction, dispersion, rainbow: 59 pages); electricity (electrical balance, condensor, Voltaic pile, thunder: 78 pages); and magnetism (only 10 pages). There then followed 16 pages of descriptions of certain instruments, including the Fortin barometer (§ 7.2.5) and 12 tables of physical constants. He included some simple mathematical treatments here and there, and for physical theories often relied, as was his want (§ 8.2.1), on Monge.

The relationship between physics and mechanics was recognised, although not clearly or definitively settled. Mechanics was sometimes taken to be phenomenological: laws of forces were studied, for example, without concern for their nature. (Contrast Euler in § 5.2.3, however.) The overlap was most noticeable when the philosophy of matter was involved, as in physical optics (§ 7.4).

This relationship much concerned Laplace and his followers at that time; for example, Biot's translation of Fischer was called *Physique mécanique*.[1]) Let us turn now to Laplace's programme.

---

[1]) I have found particularly useful the surveys given *passim* in Crosland *1967a*, Fox *1971a*, Frankel *1977a*, Silliman *1975a*, and a wide variety of occasional remarks and references elsewhere. A nice project awaits someone, who will have to note the changing social atmosphere, and the chances for professionalisation. Most history of physics ignores the latter; most social studies leave out the physics. An interesting project would be to study volumes 2-4 (1816-1822) of the *Dictionnaire de physique* of the *Encyclopédie méthodique*, for its (limited) awareness of contemporary developments—and identification of the authors!

7.1.2 Plan of the chapter

The story begins naturally enough with Laplace's initial studies, which grew out of volume 4 of the *Mécanique céleste*. § 7.2 starts with his supplements on capillary theory; in addition, some concurrent work on barometers and the velocity of sound played a role. Biot and Poisson were soon involved with this latter case; and Poisson made a substantial mathematical effort later on, in an analysis of the vibration of elastic surfaces in which Sophie Germain eventually won a *classe* prize in 1815—but little acclaim from anybody (§ 7.3).

§ 7.4 is devoted to the single most substantial area of success for this programme: physical optics. Arago and Biot followed the Master's call, but the major contributions came from Malus. Most of the work was physical and experimental, but there was also a measure of mathematical exploration.

The final two sections take up electricity and magnetism. Hachette's work is noted in § 7.5.5, but the main actors are again, in turn, Biot and Poisson: the latter has most of § 7.6 to himself, for his papers of the early 1810s on the mathematicisation of electrostatics. The concluding § 7.7 sums up the aspirations of Laplace and his followers at that time, when their programme was at the height of its success.

7.2 The first steps

7.2.1 Laplace's "mechanical" models of nature

> [...] I have sought to establish that the phenomena of nature can be reduced in the last analysis to actions *ad distans* from molecule to molecule, and that the consideration of these actions must serve as the basis of the mathematical theory of these phenomena.
>
> Laplace *1810a*, 235

The context of this remark was optics, but Laplace's growing view was that this theory could apply to "all" physical phenomena. More precisely, physical theories should be formulated in terms of short-range forces acting (attractively, repulsively, or both) between neighbouring 'molecules' of the pertaining matter. There is a long history of speculation concerning the 'imponderable qualities' with which all sciences should allegedly deal; Laplace's theory was related to philosophies of 'affinity' and of the intimate structure of matter which were held by several chemists of his day. Indeed, his programme was an important part of the work of the *Société d'Arcueil* which he and the chemist Berthollet directed at their neighbouring mansions in Arcueil south of Paris from the mid 1800s (§ 2.5.2).[1])

Laplace's programme manifested his desire to unify the sciences, or at least a lot of them. He may have been encouraged philosophically to attempt such a wide-ranging programme from his confidence in the uniformity of nature linked to his view of probability (§ 6.4.7). 'The regularity that astronomy shows us in the motion of comets, obtains, without doubt, in all phenomena', he wrote in a famous passage in a philosophical paper on probability which was written a few years after the programme was launched but which may well have been in his mind earlier, 'the curve described by the lightest atom, is regulated in a manner as certain as the planetary orbits; there is only difference between them, in that which in lies in our ignorance' (*1810c*, 100: reprinted in the early pages of *Essay*).

Laplace seems to have formed his programme around 1804, when he was in his mid fifties. The date is significant; for, while he launched a scientific enterprise, he may have sublimated a desire for power and influence at around the time when it was being flaunted. For if Bonaparte could make himself Emperor Napoléon in December 1804 (§ 2.3.2), why could not his former *Ministre de l'Intérieur* now become the Napoléon of science?

Some main features of Laplace's programme emerged in Book 10 of volume 4 of the *Mécanique céleste*, which may well have been written around 1804. Three of its chapters are most pertinent: on astronomical refraction, barometric pressure and 'the successive transmission of gravity'. This last topic is of interest, for in his study Laplace assumed the existence of an aetheric fluid. In fact we shall find that the programme was quite well endowed with fluids of one kind or another, in addition to the molecules and their inter-action. Their general role and status will be considered in the closing § 7.7.

Laplace did not claim his molecular philosophy of matter to be new. But his programme was novel in execution, partly for the range of topics to be covered, and partly for the role of mathematics. This took its root in his assumption, commonly adopted in his day, that the shapes of the molecules were not known, so that the central forces which were assumed to act between them could not be

[1]) The expression 'Laplacian physics', and the best account of the physical side of its activities, are to be found in Fox *1974a*. (However, the quotation at the head of this sub-section lacks the crucial word 'mathematical' in the version on his p. 100.) See also his *1971a passim* and Frankel *1977a*; my *1987a* summarises the mathematical side of the story, including its later reception. On the *Société d'Arcueil*, see Crosland *1967a*.

taken to be inverse square: they were presumed only to decrease rapidly as distance increased. Nevertheless, the range of action of the forces attached to a molecule encompassed many of its neighbours, so that cumulative effects had to be expressed, by means of the appropriate integrals.

The trans-empirical cast of Laplace's principles raises some awkward questions. For example, was the physical action mediated by the 'molecules', or by the fluid in the insensible but non-zero distances between them? Were these distances to be identified with differentials in the spirit of structure-similarity (§ 1.3.1)? How were different phenomena mediated simultaneously—by separate co-present mechanisms or by an "all-purpose" process? There was also the mathematical difficulty of the unknown force law, which he adopted as a consequence of the position over the shapes of the molecules. This view had been urged by chemists for some time; and he may also have been influenced by Clairaut's entertaining in the mid 18th century the possibility that Newton's inverse square law be modified in order to explain aspects of the motion of the moon (§ 6.4.3). He certainly knew of Clairaut's work on capillary theory when he began to analyse those phenomena in his first major mathematical exercise in Laplacian physics.

7.2.2 Laplace's theory of capillary action, 1806–1807

Laplace's studies of atmospheric refraction (§ 7.4.1) and barometer theory (§ 7.2.4) in Book 10 involved him in pressure and density (*4* (1805), 473–512, 565–572), and soon afterwards he turned to the related topic of capillary action. He wrote a string of papers and essays, which one would have expected to see in the *Mémoires*; and indeed one extract, *1806a*, begins with such a promise. However, in the end most of the papers appeared in the *Journal de physique* and the *Bibliothèque britannique*. The essays came out as two pamphlet addenda to Book 10: *1806c* of 65 pages, and its 'supplement' *1807a*, of 80 pages. I shall refer to them as his 'first' and 'second' supplements.[1]) His decision not to wait for *classe* publication surely indicates the importance which he attached to his work; and in the first supplement he even included 14 diagrams, the only ones in the entire *Mécanique*.

Biot supported Laplace's new theory very strongly. For example, in a review of

[1]) Laplace's supplements are usually now bound in volume 4 of the *Mécanique* and are so reprinted in the editions of his works, including Bowditch's (who mistranslated Laplace's name for the topic, 'capillary action', as 'capillary attraction'). On the history of capillary theory through Laplace, see Bikerman *1975a* and *1978a*; and on these and later developments, Bowditch's longer notes, Challis *1835a*, Maxwell *1876a* and Minkowski *1906a*. See also the second footnote in § 7.4.5.

the first supplement for the *Moniteur universel*, he concluded that Laplace's 'genius for invention shines out at every step, illuminated and guided by a profound knowledge of physical and chemical laws' (Biot *1806b*, 710). He also gave an extended prosodic account of the results contained in this supplement in his edition of the translation of Fischer's textbook on physics described in §7.1.1 (Fischer/Biot *Physics*$_1$ (1806), 136–146). In addition, Haüy, who carried out some experiments in capillarity for Laplace's benefit, noted Laplace's theory in his own description of the phenomena in his textbook on physics (*Physics*$_2$ (1806), *1*, 217–247).

'Clairaut is the first, and indeed the only person, who has hitherto reduced to a rigorous calculation, the phenomena of capillary tubes', claimed Laplace in his first supplement (*1806c*, 687), citing Clairaut's treatise *Earth*$_1$ (1743) on the shape of the earth. Possibly it was this praise which encouraged Courcier to publish in 1808 a new edition of the book, edited by Poisson.[1] As praise it was rather generous, especially by Laplace's standards, since the theory identified the fluid with the material of the enclosing vessel for some purposes. However, he criticised Clairaut's theory for its uncertainty over the laws of attraction, which his new view of short-range forces would clarify.

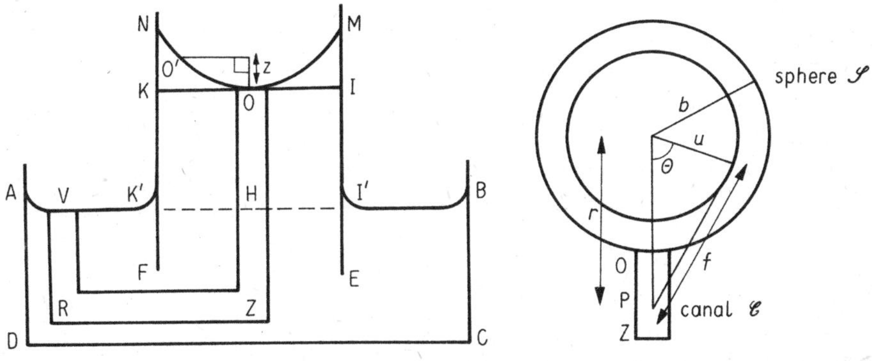

Figure 722.1 Diagrams for Laplace's capillary theory (after Laplace's fig. 1 and Bowditch's fig. 113 respectively)
Well-wetting entailed upward menisci. The tube walls NF and ME were assumed to have finite thickness.

Laplace's own theory began by considering a well-wetted capillary tube NFEM exhibiting a concave surface NOM (see Figure 722.1$_1$). For him the fluid, normally supposed in his discussions to be homogeneous and incompressible, rose up the tube in order to compensate for the lack of water in the space MON. Im-

[1]) See Clairaut *Earth*$_2$ (1808). The (few) editorial additions are anonymous, but Poisson listed the edition in his own bibliography (*1851a*, 22).

agining an infinitesimally narrow canal $\mathscr{C}$ (OHZRV in the Figure), he claimed that 'the action of the water in the tube, upon this canal, will be less on account of the concavity of its surface, than the action of the water in the vessel [ABCD] upon the same canal' (p. 690). In order to analyse this effect mathematically he went from a simple case to the general result. He calculated the cumulative 'action' of the short-range forces between the fluid in $\mathscr{C}$ and a sphere $\mathscr{S}$ of radius b lying above $\mathscr{C}$ and touching it normally at O (see Figure 722.1_2). Drawing on his earlier calculation of the attraction to a shell from an external point (*1* (1799), 288-299) he found that the action $A(u)$ of the stratum of $\mathscr{S}$ with radius u and thickness du on an arbitrary point P of $\mathscr{C}$ was given by

$$A(u)du, \quad \text{where} \quad A(u) := -2\pi\varrho u^2 \int_0^\pi \sin\Theta \int_f^\infty \varphi(q)dq\, d\Theta$$

$$= -(2\pi\varrho u/r) \int_{r+u}^{r-u} f \int_f^\infty \varphi(q)dq\, df, \tag{722.1}$$

$\varphi(q)$ was the unknown inter-molecular attraction function, and ϱ the density of the fluid (which he regrettably set equal to 1). Hence the action A of $\mathscr{S}$ on the part OP of $\mathscr{C}$ could be expressed as

$$A = \int_0^b (A(b) - A(u))\, du \tag{722.2}$$

(*1806c*, 665-669).

Laplace now simplified this expression by drawing on the property that $\varphi(q)$ was 'insensible' when q was 'sensible'. The only non-sensible value of q could be $(b-u)$, so that the four terms furnished to (722.2) by $(722.1)_2$ reduced to one, allowing (722.2) to be converted to

$$A = 2\pi\varrho \int_0^b \psi(z)\,(b-z)/b\, dz, \quad \text{where}$$

$$z := (b-u) \quad \text{and} \quad \psi(z) := \int_z^\infty p \int_q^\infty \varphi(q)dq\, dp. \tag{722.3}$$

$$\therefore A = K - (H/b), \quad \text{where}$$

$$K := 2\pi\varrho \int_0^b \psi(z)\, dz \quad \text{and} \quad H := 2\pi\varrho \int_0^b z\psi(z)\, dz. \tag{722.4}$$

$(722.4)_1$ gave a form for capillary pressure due to a minute sphere of radius b

touching the meniscus at O. H/b was much smaller than K, since in its integrand in $(722.4)_3$ either z or $\varphi(z)$ would be 'insensible': it represented the action of the meniscus itself, because it vanished when $b=\infty$ (p. 700).

Having dealt with the sphere, the general case followed from the geometry of surfaces. By similar reasoning Laplace found that the action of an ellipsoid of revolution with semi-axes b_1 and b_2 upon the canal $\mathscr{C}$ to which it was concave/convex was given respectively by

$$K \mp \tfrac{1}{2} H(b_1^{-1} + b_2^{-1}). \tag{722.5}$$

He then considered the state of equilibrium at an arbitrary point O′ of the meniscus (Figure 722.1_1), with coordinates (x, y, z) relative to a Cartesian coordinate system taking the lowest point O as origin. The difference in attraction between the two points balanced the difference of their heights; thus from (722.5) for concavity, if the principal radii of curvature were R_1 and R_2 at O′ and b_1 and b_2 at O, then

$$K - \tfrac{1}{2} H(b_1^{-1} + b_2^{-1}) + gz = K - \tfrac{1}{2} H(R^{-1} + R^{-2}). \tag{722.6}$$

Further, from the quadratic equation (473.2) for the curvature of a surface, of which the roots were R_1 and R_2, (722.6) became

$$(1+z_y^2)z_{xx} - 2z_x z_y z_{xy} + (1+z_x^2)z_{yy} = (1+z_x^2+z_y^2)^{3/2}\,(b_1^{-1} + b_2^{-1} + 2\varrho gz/H). \tag{722.7}$$

If the capillary tube, and thus the meniscus surface, were surfaces of revolution, then (722.7) could be reduced to

$$z'' + z'(1+z'^2)/v = (2b^{-1} + 2\varrho gz/H)\,(1+z'^2)^{3/2}, \tag{722.8}$$

$$\text{where}\quad v = (x^2+y^2)^{1/2}, \quad z' := dz/dv \quad \text{and} \quad b := b_1 = b_2 \tag{722.9}$$

(pp. 707–718).

Most of Laplace's further developments of capillary theory was based on (722.7–8) or on the expressions (722.6) for pressure. (722.9) itself was solved by his method of successive approximations (§ 3.3.5); putting $g=0$ in (722.8) gave as first solution the sphere

$$z = b - (b^2 - v^2)^{1/2}, \tag{722.10}$$

so that the term omitted for the first integration became

$$T := (2\alpha b \textstyle\int zv\,dv)/v = \alpha b\bigl(bv^2 + \tfrac{2}{3}(b^2-v^2)^{3/2} - \tfrac{2}{3}b^3\bigr)\big/v,\quad \text{where}\quad \alpha := \varrho g/H. \tag{722.11}$$

Hence the next approximation to the first integral of (722.9) was

$$dz/bd\beta = (1 - \alpha b^2)\sin^2\beta + \tfrac{2}{3}\alpha b^2(\sin 2\beta - \tan\tfrac{1}{2}\beta), \qquad (722.12)$$

where the new independent variable β was defined via (722.11) by

$$b\sin\beta = v + T. \qquad (722.13)$$

The final integration yielded

$$z/b = (1 - \alpha b^2)(1 - \cos\beta) + \tfrac{1}{3}\alpha b^2(1 - \cos 2\beta) + \tfrac{4}{3}\alpha b^2 \ln\cos\tfrac{1}{2}\beta, \qquad (722.14)$$

from which the height of the meniscus could be expressed in terms of its angle of contact with the capillary wall by putting $z = l$, the radius of the tube (pp. 718-725: I have put 'v' and 'β' for Laplace's overused 'u' and 'ϑ' respectively).

Among the more physically visible consequences, Laplace proved, essentially from $(722.4)_1$, the well-known experimental result that the magnitude of capillary elevation was inversely proportional to l (p. 727). He also considered the equilibriate shape of fluid trapped between two vertical planes close together (pp. 738-750). Assuming them to lie in the yz-plane, then x was dropped,

$$z_x = z_{xx} = z_{xy} = 0, \quad b_1 = \infty, \quad z' := z_y \quad \text{and} \quad b_2 := b, \qquad (722.15)$$

so that (722.7) became

$$z'' = (2\alpha z + b^{-1})(1 + z'^2)^{3/2}, \quad \text{with } z' = 0 \text{ when } z = 0: \qquad (722.16)$$

the first integration of this equation yielded

$$(1 + z'^2)^{-1/2} = 1 - b^{-1}z - \alpha z^2. \qquad (722.17)$$

As usual, Laplace avoided using Legendre's elliptic integrals to solve (722.17); instead he obtained some properties in terms of the angle of contact by taking z' to be known and treating (722.17) as a quadratic in z. In the process he came close to verifying a discovery of Taylor *1712a*, but he neither cited that work nor completed the analysis: that if two planes inclined at a small angle and with a vertical line of junction contained a fluid, then its equilibriate curve was close to that of a hyperbola. However, he went on to analyse the shape of a drop of fluid held between two such planes (pp. 756-771), quoted Hauksbee's account *1712a* of the shape, and compared his own theory with Hauksbee's data (pp. 793-799). In the second supplement he generalised the discussion to the case where the planes were made of different materials (thus possibly causing, for example, concave contact with one plane and convex contact with the other (*1807a*, 909-947).

Among other cases studied by Laplace was the shape of a large mercury drop on a horizontal plane. In (722.8) he felt that $(z')^3$ could be ignored, and thus it reduced to

$$vz'' + z' - 2\alpha vz - 2v/b = 0 \quad (\text{where} \quad z' := dz/du), \tag{722.18}$$

which is what we call 'Bessel's equation', to within a sign in z. Without providing a derivation, he gave the (to us, first) solution

$$z = \frac{1}{\alpha b\pi}\int_0^\pi \exp\left(u\sqrt{2\alpha}\cos\varphi\right)d\varphi - \frac{1}{\alpha b} \tag{722.19}$$

(pp. 971–978). Now this result was essentially Parseval's at (434.4) in his paper on his series, a work which Laplace had read in manuscript and had praised to the *classe* in 1799 … . We shall see that Fourier also found the result around this time, as (928.22), probably independently of the other two.

7.2.3 Features and extensions of the theory: Petit's doctorate, 1811

Various questions arise concerning Laplace's theory. For example, how was the differential equation (722.7) obtained? He began the second supplement with a fresh derivation of (722.7) (*1807a*, 806–834), based on the principle that a homogeneous fluid body was in equilibrium if any infinitesimally narrow canal within it, and with extremities on its surface, was also in equilibrium. He had not used this principle earlier in the treatise (it is employed in Lagrange *Mechanics*$_1$ (1788), 130–135); it yielded a laborious argument here, with less reliance on Laplacian physics but appeal to a truncated Taylor-series expansion of z in terms of x and y.

Some other physical features of Laplace's theory form a group of problems. For example, his reliance on short-range forces prevented him from using an empirical concept such as surface tension; he did not even mention Monge's paper *1789a* on surface tension, perhaps because it begins on p. 506 with a clear statement of *exactly* the type of short-range forces which grounded Laplacian physics. (However, Monge argued in empirical terms in his paper, in line with his *Denkweise* noted in § 3.3.5.) Again, Laplace assumed the density of the fluid to be constant throughout, including on its surface layers; we shall see Poisson question this assumption in § 18.2.5, in his own theory of capillary action. Laplace also regarded φ only as an *attractive* function; 'we have not noticed, in this theory, […] the repulsive force of heat. The consideration of these forces is unnecessary', he claimed (*1807a*, 1015). However, a dozen years later, more sensitive to the ubiquity of heat effects in physics, he was to modify this claim himself (§ 12.3.1).

Various questions concerning the meniscus can be posed. For example, while it was clear that the meniscus occurred because of the greater attraction of the molecules of the fluid close to those of the capillary walls than for the parts of the fluid that were centrally placed, the connection between the elevation/depression of the fluid and the concavity/convexity of the meniscus could have been explicated. Again in his first supplement he took a horizontal meniscus, and argued that at a contact point Q an outward 'intensity of action' ϱ, linear with density, was exercised by the 'particles' of the tube above Q, another force ϱ' by the 'particles' below Q, and an inward force ϱ' by the particles of the fluid itself. Thus if

$$2\varrho = \varrho', \tag{723.1}$$

these forces would cancel out (*1806c*, 776–780). In the second supplement he took account of the action in the surface fluid particles at Q acting at the angle of contact ω to the tube's surface. Reading Figure 722.1_1 now as showing OH as the capillary tube $\mathcal{T}_1$, with Q as a point "at" O on its wall, Q was subject to: a downward component $-\varrho'\cos\omega$ of action due to the surface particles; another downward action ϱ' due to the particles of fluid outside the second tube $\mathcal{T}_2$ given by HZRV; and two upward actions of value ϱ, one due to the action of the tube on $\mathcal{T}_1$ and the other on $\mathcal{T}_2$ (and thus acting on $\mathcal{T}_1$ at H). This (contrived) account gave the equilibrium condition

$$\varrho' + \varrho' \cos\omega = 2\varrho, \quad \text{so that} \quad \varrho' \cos^2 \tfrac{1}{2}\omega = \varrho \tag{723.2}$$

(*1807a*, 836–839). However, he gave no thought to the modification due to (723.1) because of the horizontal action $\varrho' \sin\omega$.

True to his experimental conscience, Laplace compared his theory with observational evidence, including some done specially for him by his *classe* colleague Haüy for the first supplement (*1806c*, 792–793) and by his *Société d'Arcueil* staff member Gay-Lussac for the second (*1807a*, 961–971, 993–1005). The results were mixed, with large variance in discrepancy; Laplace voted against the experimenters. 'The effects of the capillary action being reduced to a mathematical theory', he modestly noted, 'there is only wanting, in this interesting branch of physical science, a series of very accurate experiments' (p. 1015).

The hint to *Arcueil* members was unmistakable; but it was taken up in a surprising way. Alexis Petit became an important physicist after graduating from the *Ecole Polytechnique* in 1809, largely on the experimental side; he is best remembered today for the Dulong-Petit law of atomic heats, introduced in 1819 (§ 12.2.1). However, in 1811, his 21st year, he wrote a *Faculté des Sciences* thesis on the mathematics of capillarity (recorded at *AN*, AJ^{16} 5322, fol. 16r): it was published as *1813a* in the *Journal* of the *Ecole Polytechnique*, where he was currently a *répétiteur*.

Petit added no fundamentally new results to Laplace's theory; the most interesting aspects of his thesis are perhaps its attempts to systematise Laplacian physics and its attendant mathematics. For example, in a preliminary section he expressed some properties of the attraction function $\varphi(f)$, which was 'insensible' for 'sensible' values of the distance f. Generalising upon Laplace's integrals $(722.3)_2$, he defined recursively

$$\varphi_0(f) := \varphi(f), \quad \varphi_{n+1}(K) := \int_K^\infty \varphi_n(f)df, \qquad n \geq 0 \tag{723.3}$$

for some particular K. It followed that, like φ, $\varphi_n(f) = 0$ when $f >> 0$, and thus that in the integration

$$\int_0^K f^p \varphi_n(f)df = -\sum_{r=0}^p pP_r \varphi_{r+n+1}(K) + \varphi_{p+n+1}(K)\,; \quad \text{with } p \text{ an integer,} \tag{723.4}$$

only the last term need be retained if $K >> 0$ (pp. 3-6).

Petit used these results to derive equations of the type of (723.1) by integrating an initial quintuple integral expressing the cumulated short-range attractive forces in the capillary tube (pp. 6-13). He also found them useful in imitating Laplace's second derivation of the differential equation (722.7) (pp. 17-30), which he partly integrated by clever manipulations to obtain (722.10) for the capillary surface (pp. 17-31). Among the special cases, perhaps the most novel was his treatment of one vertical plane immersed in the fluid. (722.7) became (722.16) with b^{-1} set to zero, which integrated straightforwardly to yield an explicit form for the semi-menisci (pp. 32-34).

Just as Rodrigues's thesis of 1815 was to yield some useful additional information on Legendre functions (§ 6.5.1), so Petit's thesis of 1811 gave a corroboration, not entirely copo-cat, of the Master's great essay in Laplacian physics. 'The remarkable conformity of the results of observation and of those to which it led', the candidate concluded rather fawningly, 'assures in an incontestable manner the legitimacy of the hypothesis which serves as basis for this theory' (p. 40). He did not report any experimental data in his thesis, but he may have had in mind the used of barometers, which both he and the Master saw as an application of capillary theory.

7.2.4 Theory and tabulation for barometers

When stating in the *Mécanique céleste* the relationship between the difference r in altitude between the upper and lower stations U and L and the barometer readings h_U and h_L there, Laplace also allowed for the variation in gravity, the latitude ψ of U and L, the distance a between L and the centre of the earth, and the temperatures t_U and t_L. He stated the formula for r in the unsatisfactory form

$$r = k_1(1 + k_2 \cos 2\psi)\,(1 + k_3(t_U + t_L))\,((1 + r/a)\ln(h_L/h_U) + k_4 r) =: A + rB \tag{724.1}$$

(where the $\{k_j\}$ were known values), and only indicated verbally the conversion to the explicit form

$$r = A(1 + B) \tag{724.2}$$

that the small value of B in $(724.1)_2$ would allow (*4* (1805), 571–572). Biot and Gay-Lussac used Laplace's (724.1) to calculate the heights reached in their famous balloon ascent of 1804 from the readings of the on-board barometer (see Gay-Lussac *1804a*). Laplace reported on the flight to the *Bureau des Longitudes* on 25 August 1804, the day after its occurrence: apparently the rotation of the balloon had made difficult the taking of observations (*BL*, *3*, meeting of 7 *fructidor an* 12).

Laplace's first supplement included a determination of the correction needed when using a cistern barometer. From (722.4, 11_2) the factor $(\alpha b)^{-1}$ was the capillary effect: it had to be added on to the measured height z, since mercury gave a convex meniscus. He obtained the rather complicated formula

$$\frac{1}{\alpha b} = \frac{C}{l} - \frac{2}{l^2}\int_0^l zz\,dv, \quad \text{with } C \text{ constant (and known)} \tag{724.3}$$

and l as the radius of the tube, from taking the first integral of (722.8); the integral could be handled in terms of the angle of contact by formulae such as (722.14) (*1806c*, 802–805).

Although there was some appreciation of Laplace's formula (724.1) (see, for example, d'Aubuisson *1806a*), calculations were difficult to make from it. So Laplace *1810d* himself proposed in July 1810 to the *Bureau des Longitudes* a simpler method of correcting for capillarity by using a quadrature formula, with a table of values prepared from it by Bouvard, his handyman calculator.[1])

---

[1]) Laplace later modified his formula (724.1) in a note *1826c* read to the *BL*, for which Bouvard *1826a* produced more extended tables. For a presentation of barometer theory, and some related matters discussed in the sub-sections below to come, see Haüy *Physics*$_2$ (1806), *1*, ch. 5.

Maybe Laplace was encouraged to make his mathematics more calculable by others' work of the 1810s on barometer theory. In 1811 Biot published as a separate booklet *1811b* the section of the second edition of his *Astronomy* on barometers. He took up Laplace's suggestion in the *Mécanique céleste* that the density of the air, and thus the isothermal barometric pressure, decreased in geometric progression for equal increments in altitude (*4* (1805), 497–507). This law (on which see (741.8)) gave the relation between the barometric readings $\{H_n\}$ and the distances $\{x_n\}$

$$(H_n - H_{n+1})/(x_{n+1} - x_n) = CH_n, \quad \text{with } C \text{ constant,} \tag{724.4}$$

and for equal increments D in distance the expression

$$H_{n+1} = H_1(1 - CD)^n \tag{724.5}$$

(Biot *1811b*, 1–4: a more elaborate proof was given in Petit *1811a*, doubtless as an offshoot of his thesis). Biot considered the determination of C, both as a constant and as a sequence of constants for the layers $\{x_n, x_{n+1}\}$, and the allowance to be made for temperature and latitude. Laplace himself had used a hypothesis on the change of temperature with altitude (*Mechanics*, *4* (1805), 566); Biot's treatment was equivalent, for it brought him to Laplace's (724.1), with an indication of the conversion to (724.2) (*1811b*, 22–23). Among his tables he used the one for capillarity calculated by Bouvard in Laplace *1810d* (cited above).

Biot's interest in barometers was stimulated largely by his geodetic observations. Perhaps he chose (724.5) out of the available laws for its mathematical similarity to Borda's law (652.3) for the pendulum. Poisson did not use (724.5) in his account of barometry in his textbook on mechanics: he assumed a constant temperature and obtained an equation slightly different from (724.1) and within $O(r/a^2)$ of it (*Mechanics*$_1$ (1811), *2*, 429–442).

de Prony was also much concerned with barometer theory. In a note *1799c* for the *Société Philomatique* he drew on his paper *1796b* on the thermal expansion of gases, of which we noted the numerical aspects in § 3.5.3. He altered his formula for altitude to allow for the thermal expansion of the air trapped in the barometer's cavity, as expressed by his supposedly 'true form' (353.1)

$$z = \mu\varrho^T, \quad \mu \text{ and } \varrho \text{ constants,} \tag{724.6}$$

of the law for the volume z of air as a function of temperature T—a form which Gay-Lussac was to criticise, as we saw in § 3.5.3.

de Prony's suggestion did not catch attention; he had not allowed for several of the factors involved in Laplace's (724.1), and since the expansion constant ϱ was only slightly greater than 1, (724.6) would simplify to the linear function of T used in (724.1). But he maintained his interest, and in a paper *1813a* for the *Bur-*

eau des Longitudes he reported on various aspects of his use of barometers in measurements in France and Italy. He used tables prepared by the German geodeter J. Oltmanns using (724.1), praising them for their use of logarithms. But he also wanted to simplify the logarithmic term in (724.1) itself by using

$$q := \frac{h_L - h_U}{h_L + h_U}; \quad \text{thus} \quad \ln\left(\frac{h_L}{h_U}\right) = 2\sum_{i=0}^{\infty} \frac{q^{2i+1}}{2i+1} = 2q + O(q^3) \tag{724.7}$$

(pp. 306, 312–313: he incorporated into h_U a correction term apparently due to (724.6)). Discussing also the consequence of errors in the measurement of h_U and h_L, he used the expansion in (724.7) to produce an expression for the error Δz in z in terms of the errors Δh_U and Δh_L (pp. 309, 312–313).

The impression left by barometer theory at this time is somewhat indifferent. The basic theory, as encapsulated in Laplace's (724.1), was largely known by 1800. Later developments varied between Laplace's uncalculable (724.3), manoeuvres surrounding de Prony's (724.6–7) (in which the effect of latitude was omitted), and later some statistical arguments (one by Laplace is described in § 17.6.3). There were also disputes about barometric practise (in de Prony *1813a*, for example). And, while there was such attention to errors, the *savants* did not discuss the error of capacity in cistern barometers, caused by the rise and fall of mercury in the tube. Perhaps this was a compliment to the instrument designed by their Paris colleague Fortin in 1800, in which this error could be avoided; but if so, they might have said so.

7.2.5 Correcting the velocity of sound

One of the starkest discrepancies between theory and experiment in 18th-century mechanics was the velocity of sound. Careful experiments by an *Académie* commission in 1738 had established a value of 1,038 *pieds* (337.18 metres) per second, after correction for barometric pressure and other effects; but current theory, based on Newton's suggestion (*Principia*$_3$ (1726), Book 2, prop. 48) that

$$\text{velocity} = (\text{pressure}/\text{density})^{1/2}, \tag{725.1}$$

gave values around 20% less than this. Newton had fudged his theory to fit exactly with an experimental value (equivalent to 348.08 metres per second) of his day; various suggestions had been made during the 18th century to develop his underlying idea concerning the compression and extension of the air (see Dyment *1931a* and Finn *1964a*). For example, a 'light conjecture' of Lagrange was that the 'elastic force' of the air was proportional to the *m*th power of its density, rather than the form adopted in (725.1), and he found 1.36 for the value of m (*1762e*, art. 55); but he gave no argument to back up his theory.

A major contribution came from Laplace, as an offshoot of his interest in heat, which had developed in the 1770s in connection with some experiments on specific heats conducted with Lavoisier. In their first paper *1784a* they were agnostic as to the nature of heat; but in a later paper *1793a*, written by Lavoisier on their joint behalf shortly before his decapitation, they spoke freely of caloric, the heat-bearing substance. In notes published in Berthollet's *Essai de statique chimique* Laplace *1803b* used caloric theory to offer analyses of the compression of gases, and also proposed to Biot that sound caused the compression and expansion of the air through which it passed, in turn creating heat and cold respectively. Biot stated Laplace's theory of caloric to the *Société Philomatique* in these words: 'The atmospheric air, when one condenses it, loses a part of its latent heat which passes to the sensible heat; and by contrast when one rarifies it, it takes back a portion of the sensible heat which it converts to the state of latent heat' (Biot *1802c*, 116). Thus Laplace supported a material rather than a vibratory theory of heat, as he was to do during the heyday of Laplacian physics. The general situation over theories of heat around 1800 is summarised in § 9.2.1.

In a paper published in the *Journal de physique*, Biot *1802b* attempted to mathematicise Laplace's theory. The density of air changed during the propagation of sound from 1 to $(1 + s)$, with s small; at constant temperature so would its 'elasticity' (or pressure) ε, by the Boyle-Marriotte law. But ε would be further modified by the factor $(1 + ks)$ due to the supposed heating, to become

$$\varepsilon = gnH(1 + s)\,(1 + ks) = gnH(1 + (1 + k)s) + O(s^2), \qquad (725.2)$$

where g was the gravitational force (at sea level), H the barometric pressure, and n the density of mercury relative to air.

Biot now imitated Lagrange's analysis of the motion of a compressible fluid and its reduction to the wave equation for the plane propagation of sound (*Mechanics*$_1$ (1788), 496–503: Lagrange's treatment is described in § 10.3.1). Biot took a 'molecule' of air at (x, y, z) with variable density ϱ, moving with (low) velocity (p, q, r) under no impressed forces, for which the basic equations reduced for the x-direction to

$$\varrho_t + (\varrho p)_x + (\varrho q)_y + (\varrho r)_z = 0 \quad \text{and} \quad \varrho p_t + \varepsilon_x = O(p^2), \qquad (725.3)$$

with its companions for y and z. He took the differential of (725.2) in the form

$$d\varepsilon/\varrho = gnH(1 + k)ds/(1 + s), \quad \text{since } \varrho = 1 + s \text{ in motion}, \qquad (725.4)$$

and found the integral

$$E := \int d\varepsilon/\varrho = gnH(1 + k)\ln \varrho. \qquad (725.5)$$

Now, from $(725.3)_2$ and its companions,

$$dE = d\varepsilon/\varrho = \sum_x \varepsilon_x dx \Big/ \varrho = -\sum_x p_t dx, \tag{725.6}$$

which could be converted via the assumption of a velocity potential φ into

$$dE = -\sum_x p_t dx = -\sum_x \varphi_{xt} dx = -d\varphi_t, \quad \text{so that} \quad E = -\varphi_t. \tag{725.7}$$

He now took $(725.3)_1$ in the form

$$(\ln \varrho)_t + \sum_x p(\ln \varrho)_x + \sum_x p_x = 0 \tag{725.8}$$

and used (725.5, 7) to replace $\ln \varrho$, p, q and r and obtain the wave equation in the form

$$\varphi_{tt} = gnH(1+k)\nabla^2\varphi + O(p^2). \tag{725.9}$$

Thus the velocity v of sound was given by

$$v = \sqrt{gnH(1+k)} \tag{725.10}$$

(Biot *1802b*, 178–181, with 'ϱ' written for his 'Δ').

As with Lagrange's m, Biot had no genuine explanation in physics for his factor k; but he calculated its value by taking known data for g, n and H (atmospheric pressure) to obtain 915 *pieds* (or 297.28 metres) per second for $\sqrt{gnH}$. He related it to the *Académie*'s experimental determination of 1,038 *pieds* per second by introducing these data into (725.3):

$$1{,}038 = 915\sqrt{1+k}; \qquad \therefore\ k = 0.2869. \tag{725.11}$$

According to G. Amontons's experiments on heat generation, this value corresponding to an increase to temperature of 68.8° Réaumur (or 82.6 °C). In Biot's own later experiment *1809e*, using enclosed volumes of gas, he tried to detect this generation of heat.[1]) But theoretical understanding had to wait for further experimental work on heat in the 1810s, as we shall see in § 12.2.

Although Biot acknowledged Laplace as the source of the idea underlying (725.2), Laplace published only the two notes in Berthollet's *Statique chimique* mentioned above. Laplace claimed there to have checked his theory by calcula-

[1]) In companion studies Biot, with the help of Malus, Bouvard and others, used the current development of the Paris water supply (§ 8.5.3) to measure the velocity of sound in air and in metal as transmitted in the iron of the connected conduit pipes (see Bell *1973a*, 191–196). Biot gave the short story *1808c* to the *MU*, and the full version *1809f* to the *Société d'Arcueil* (and the *CI* in October 1808: *PV*, *4*, 132–133).

tion, although he did not present it in Book 10 of the *Mécanique céleste*, which would have been an appropriate place. Perhaps he felt unable to offer an argument better than Biot's; but it would have afforded him an opportunity to proselytise Laplacian physics, since the heat effect asserted to occur could be expressed as a short-range inter-molecular action, as he was to allow for (but then decided to ignore) in his capillary theory (§ 7.2.2). Under these circumstances it is not surprising that his chief mathematical supporter was to study sound and its propagation in the 1800s.

7.2.6 Poisson on the wave equation, 1807

Poisson submitted a long paper on sound to the *classe* in August 1807 (*PV*, *3*, 563); it was his first major mathematical essay. Biot came up with a long report *1807a* two months later, with the usual recommendation to the *Savants étrangers* which Poisson converted into an appearance in the *Journal de l'Ecole Polytechnique* as *1807b* with a summary *1807c* for the *Société Philomatique*.

Acoustics had long been a preoccupation in science. We saw in § 5.2.5 theoretical studies advanced in the creation of the wave equation, and in § 3.3.2–3 the various solutions offered it. Among further advances in the mathematical treatment, Lagrange's *1762d* is interesting for its (vaguely) Fourier-integral transforms (compare § 9.3.2), while Euler's studies of echoes included in *1767b*, arts. 39–41 the use of pulses embodying functions which took non-zero values for only one value of x (though he did not explicitly define them as functions, perhaps because they were quite outside the normal type of function used in the 18th century (§ 3.3.2)). Elsewhere, Euler's *1766d* on the vibration of membranes incorporated the contribution to Bessel-function theory noted in § 3.4.4.[1])

Beginning his paper with a review of 18th-century work, Poisson circumscribed his own analysis to fit within the constraints of Laplacian physics, and Lagrangian mathematics. 'To set the question in the most general point of view', he wrote, 'Let us consider an elastic fluid of which all the molecules are subject to the action of any forces'. It 'is the case of all forces of nature' for the impressed forces on a molecule to admit a potential q, and its (small) velocity to take similarly a velocity potential φ; thus, like Biot, he drew on Lagrange's analysis of compressible fluids and took for his basic equations $(725.3)_1$ and an inte-

[1]) The best account of mathematical acoustics in the 18th century is Truesdell *1955a*: see also his *1960a passim*. An important primary source is the correspondence between Euler and Lagrange of 1759–1760, of which the most scholarly publication appears in Euler *Works*, *(4A) 5* (1980), 419–444. There are various histories of physical acoustics, but they skate over the mathematical details.

gral drawn from the full versions of $(725.3)_2$ and its companions:

$$\varphi_t + (\varrho\varphi_x)_x + (\varrho\varphi_y)_y + (\varrho\varphi_z)_z = 0\,, \tag{726.1}$$

$$q - \int dp/\varrho = \varphi_t + \tfrac{1}{2}(\varphi_x^2 + \varphi_y^2 + \varphi_z^2) \tag{726.2}$$

(Poisson *1807b*, 329–331).

The paper was devoted to a sequence of solutions of these equations, usually in special cases, and their consequences for acoustic theory. I shall confine myself to a few examples of its Lagrangian and Laplacian traits.

Taking ϱ and the temperature to be constant, and assuming no impressed forces, Poisson combined (726.1-2) by means of a special case of an analysis given at (726.8) below, and obtained the three-dimensional (isothermal) wave equation

$$\varphi_{tt} = a^2\nabla^2\varphi + O(\varphi_x^2)\,, \quad \text{with } a^2 \text{ constant.} \tag{726.3}$$

Naturally he sought for a functional solution, and found one by transforming coordinates to the spherically polar $(r, \cos^{-1} u, \omega)$ and thus (726.1) into the equation

$$(rR)_{tt} = a^2(rR)_{rr}\,, \quad \text{where} \quad R := \int_0^{2\pi}\int_{-1}^{1} \varphi du\, d\omega\,. \tag{726.4}$$

$$\therefore\ rR = f(r - at) + F(r + at)\,. \tag{726.5}$$

This form depended on knowledge of φ; it was obtained by applying the same transformation of variables to (726.3), and finding a solution in terms of surface harmonics (pp. 334–339, 343–347).

Poisson's discussion of the velocity of sound largely followed that of Biot, including the same explanation of the heat effect by short-range action between the molecules of air (p. 332). But he could draw on recent experimental work by the *Société d'Arcueil* for numerical data. In particular, he used Arago and Biot's recent work on refraction in gases (§ 7.4.1) to calculate the density of mercury relative to air and obtain from $(726.3)_2$ a revised value of 279.29 metres per second for the theoretical value of the velocity of sound at 0 °C. Now Gay-Lussac *1802a* had found from his experiments on gases and vapours (§ 3.5.3) that the volume of air increased by a constant proportion 0.00375 per °C; thus for 6 °C its velocity would be

$$\begin{aligned}&279.29 \times \sqrt{1 + 6\alpha}\,, \text{ or } 282.42 \text{ metres per second,}\\ &\text{where } \alpha := 0.00375\,.\end{aligned} \tag{726.6}$$

To match this value with the experimental datum of 337.18 metres per second at 6 °C found in 1738 by the *Académie*, the correction factor (that is, k in $(725.11)_2$)

had to be recalculated (*1807a*, 362):

$$282.42 \times \sqrt{1+k} = 337.18\,; \qquad \therefore\ k = 0.4254\,. \tag{726.7}$$

Poisson also found a formula for the velocity of sound in air by an extension of Biot's Laplacian reasoning. The passage of sound compressed the air slightly to change density from D to $D(1+\gamma)$ and 'elasticity' from gh to

$$p = gh(1+\gamma)\,(1+k\gamma) = gh(1+(k+1)\gamma) + O(\gamma^2)\,, \quad \text{with } \gamma \text{ small}, \tag{726.8}$$

as in Biot, where $ghk\gamma$ indicated the change caused by heating; and the velocity a of sound was given by Biot's (725.10) as

$$a^2 = gh(1+k)/D. \tag{726.9}$$

To find the compression ratio of air, Poisson now took it at density D, pressure gh and temperature Θ, and compressed it to density $D(1+\gamma)$ and temperature $(\Theta+\omega)$, $\omega > 0$ and assumed to be proportional to γ. At constant temperature Θ the pressure would have increased to $gh(1+\gamma)$, by the Boyle-Marriotte law; thus by Gay-Lussac's result, it was actually given by

$$p = \frac{gh(1+\gamma)\,(1+\alpha(\Theta+\omega))}{1+\alpha\Theta} = gh\left(1+\gamma+\frac{\alpha\omega}{1+\alpha\Theta}+O(\gamma^2)\right). \tag{726.10}$$

Thus by equating (726.8) with (726.10),

$$k\gamma = \frac{\alpha\omega}{1+\alpha\Theta}, \quad \text{giving for the compression ratio}$$
$$\frac{1}{m} := \frac{\gamma}{\omega} = \frac{\alpha}{k(1+\alpha\Theta)}. \tag{726.11}$$

Setting in (726.10) $\Theta = 6$ and the values of α and k from (726.6_2, 7_2) yielded $m = 115.99$ or the compression ratio $\frac{1}{116}$ (p. 363): this figure was to be widely quoted by physicists for many years. Further, (726.11) allowed (726.9) to become

$$a^2 = \frac{gh}{D}\left(1+\frac{\alpha m}{1+a\Theta}\right), \tag{726.12}$$

an expression involving known or measurable quantities (pp. 360–363).

The most interesting theoretical part of Poisson's paper was his study of the isothermal propagation of sound along the x-axis (pp. 364–370). (726.1–2) now converted, via $(726.3)_2$, to

$$a^2 \ln(\varrho/\varrho_0) + \varphi_t + \tfrac{1}{2}\varphi_x^2 = 0 \quad \text{and} \quad \varrho_t + \varrho\varphi_{xx} + \varrho_x\varphi_x = 0\,, \tag{726.13}$$

where ϱ_0 was the value of ϱ at $t = 0$; eliminating ϱ yielded on p. 364 the non-linear horror

$$\varphi_{tt} + 2\varphi_x\varphi_{xt} + \varphi_x^2\varphi_{xx} = a^2\varphi_{xx}. \tag{726.14}$$

To solve this equation he applied the method known now as a 'contact transformation' but then often called 'Legendre's transformation' (as was mentioned in § 3.3.5). In this method new dependent variables were introduced and treated as the independent variables of the problem. Poisson introduced

$$z := \varphi_t - a\varphi_x + \tfrac{1}{2}\varphi_x^2 \quad \text{and} \quad w := \varphi_t + a\varphi_x + \tfrac{1}{2}\varphi_x^2, \tag{726.15}$$

which reduced (726.14) to

$$z_t + z_x(a + \varphi_x) = 0 \quad \text{and} \quad w_t = w_x(a - \varphi_x) = 0, \tag{726.16}$$

respectively. Treating φ_x now as a function of z (or w) and of t, (726.16) integrated to yield the functional solutions

$$z = f_1\left(x - at - \int_0 \psi_1(z, t)dt\right) \quad \text{and}$$
$$w = f_2\left(x + at - \int_0 \psi_2(w, t)dt\right). \tag{726.17}$$

Substituting these solutions into (726.15) and drawing on $(726.13)_1$ yielded

$$2a\varphi_x = z - w \quad \text{and} \quad 2a^2 \ln(\varrho/\varrho_0) = -z - w, \tag{726.18}$$

a solution of which the further development depended upon the initial conditions; he concentrated on the case where propagation was initiated over an interval $[0, \alpha]$ of x. The physical significance of this work was that hitherto 'one regards the velocities $[\{\varphi_x\}]$ of the molecules of air as very small' and so worked to $O(\varphi_x^2)$; but even if they were not so small, as in propagation in a thin 'cylindrical channel', 'the velocity of sound is independent of the magnitude of that of the molecules of air as well as of the cause which produces the sound; [...] which conforms to experience' (*1807c*, 21). He was to return to these considerations in 1823, as we shall see in § 12.3.4.

7.2.7 Parseval on the wave equation, 1806

In the historical preamble to his long paper on sound Poisson omitted any mention of two recently published papers on solutions to the wave and related equations. Parseval had submitted the first to the *classe* in July 1801 (*PV*, *2*, 373: compare pp. 262–263) and the second in August 1803 (p. 692). Lacroix reported on both papers, in his *1801b* and *1804b* respectively, hoping for publication which, as with all of Parseval's papers (§ 4.3.4), occurred in the 1806 volume of the *Savants étrangers* (*1806a* and *1806c*).

Poisson's "ignorance" of these papers is especially interesting with regard to the second one, for it also contained a use of Legendre's transformation to solve (726.14). Using Euler's notations, with

$$p := \varphi_x \quad \text{and} \quad q := \varphi_t, \tag{727.1}$$

Parseval set on p. 526

$$\varphi = px + qt - \omega, \quad \text{where} \quad d\omega = xdp + tdq, \tag{727.2}$$

and used Legendre's method to show that ω satisfied the linear partial differential equation

$$\omega_{pp} - 2p\omega_{pq} + p^2\omega_{qq} = k\omega_{qq}, \quad \text{where} \quad k := gh \ \text{(of (726.8))}. \tag{727.3}$$

He now invoked an integral form of solution to (727.3) developed in Laplace *1782a*, arts. 18-20. He expanded the exponential function in its integrand as a power series in a dummy variable s, took only those terms free of s (like his treatment (434.10) of the Lagrange series), and finally came up with the functional solution for ω:

$$\omega = \psi(p + q\sqrt{k}) + \xi(p - q\sqrt{k}), \tag{727.4}$$

which was placed into (727.2) to lead to a solution for φ (Parseval *1806c*, 524-534). de Prony had described another version of Parseval's analysis in the *Mécanique philosophique* (*1800a*, 346-354), praising its author; maybe this had alerted Poisson originally.

In the rest of his paper Parseval took up the analysis of fluid flow in a thin horizontal canal given in Lagrange's *Mechanics*$_1$ (1788), 487-512. He drew on the discussion of the wave equation provided in his first paper, *1806a*. There he had obtained the functional solution to the equation by obtaining it recursively from the one-dimensional case to those of higher dimensions. Given the wave equation for three dimensions in the form

$$u_{tt} = au_{xx}, \quad v_{tt} = a(v_{xx} + v_{yy}) \quad \text{and} \quad w_{tt} = a(w_{xx} + w_{yy} + w_{zz}), \tag{727.5}$$

he substituted the functional solution

$$u = \varphi(x + \sqrt{a}\,t, y, z) + \psi(x - \sqrt{a}\,t, y, z) \tag{727.6}$$

of $(727.5)_1$ into $(727.5)_2$ to obtain the power-series solution

$$v = \sum_{r=0}^{\infty} a^r \left(a^r \int^{(2r)} u_{y^{2r}} (dt)^{2r} \right)_{a^r} \Big/ r!; \tag{727.7}$$

a similar expression for w in terms of v, using z, solved $(725.5)_3$ (*1806a*, 382–390). He now drew on the proof of "his" series (434.6), in that he noticed that v was the sum of the terms free from s (a new variable) in the expansion of $T(1+s)$, where

$$T(s) := \sum_{r=0}^{\infty} (as)^r \int^{(2r)} u_{y^{2r}}(dt)^{2r}. \tag{727.8}$$

$$\therefore \ v = \text{real part of } \left[\frac{1}{\pi}\int_0^{\pi} T(1+e^{-iv})\, dv\right], \text{ where } \quad s := e^{-iv}. \tag{727.9}$$

To find an explicit form for $T(s)$ he took the two series

$$P(s) := \sum_{r=0}^{\infty} (as)^{r/2} \int^{(r)} u_{y^r}(dt)^r$$

and

$$Q(s) := \sum_{r=0}^{\infty} (-1)^r (as)^{r/2} \int^{(r)} u_{y^r}(dt)^r, \tag{727.10}$$

and showed that they satisfied the partial differential equations

$$P_t - \sqrt{as}\, P_y = R(x, t) := u_t \quad \text{and} \quad Q_t + \sqrt{as}\, Q_y = R(x, t), \tag{727.11}$$

with known solutions (including functions of integration) respectively

$$P(s) = \int R\left(x, y + \sqrt{as}\, t\right) dt + f_1\left(x, y + \sqrt{as}\, t\right), \tag{727.12}$$

$$Q(x) = \int R\left(x, y - \sqrt{as}\, t\right) dt + f_2\left(x, y - \sqrt{as}\, t\right). \tag{727.13}$$

$T(s)$ could now be evaluated from the relation

$$2T(s) = P(s) + Q(s): \tag{727.14}$$

w was determinable from v by similar means (pp. 390–395).

Parseval's procedures well exemplify the points made in § 3.3.5 on preferred forms of solution to differential equations. They show originality, in manipulating a method of solving differential equations inductively on the number of independent variables which has become known as 'the method of descent'. His paper compares well with a rather wearisome piece, written by Biot with some help from Brisson, submitted to the *classe* a year earlier than Parseval's (*PV*, *2*, 169) but published as Biot *1803f* in the *Mémoires* after his election to full membership, on power-series solutions in general and for the wave equation in particular. (Note his reference to Parseval in the letter to Lacroix in § 20.1.) However, Parseval's paper made no impact, a state of affairs probably attributable to the outsider status of both its author and the journal in which it appeared.

7.3 Poisson versus Germain on elastic surfaces, 1811–1815

7.3.1 Sound and elasticity: the *classe* prize for 1811

When one reads the word 'elastic' in the literature of this time, it is important to bear in mind the scope of its reference: not merely elastic lines, surfaces and solids, but also the elastic aetheric fluids which (for some *savants*) existed and permitted the transmission of sound, held together the rings of Saturn, and so on and on. Similarly, an 'elastic body' might refer to the aether (for those disposed towards the existence of such things); or it could be a piece of light or fluid; or perhaps an ordinary object such as a drum-head. With a bit of luck the discussion might apply to many of these cases at once.[1]) This point is evident in the rest of this chapter and also later, especially in Chapter 15.

This section is centered on a dispute between Germain and Poisson over theorising about the vibration of an elastic surface. The tale started in 1802 when the German physicist E. F. F. Chladni described (again) in his book *1802a* on acoustics his sand-pattern technique of finding the nodal lines on an elastic membrane set into vibration by the stroke of a bow. As the book was in German, the French were at first only partially aware of his work; for example, Biot confessed his knowledge as confined to the accounts in the *Bulletin* of the *Société Philomatique* (*1803f*, 24). But when Chladni came to Paris in 1808 to demonstrate his clavicylinder, a new instrument rather like a large accordion (see his *1809a*), de Prony, who was deeply interested in music, got together with colleagues from his *classe* and composers from the *beaux-arts classe* and wrote a report *1808a*. Further, when Chladni's book came out the following year in his own French edition *1809b*, de Prony prepared a lengthy report *1809b* for the *classe* (while Biot put a long review *1810b* in the *Mercure de France*). de Prony also wrote the preamble *1809a* for a *classe* prize problem on the vibration of elastic surfaces, with entries to be received by October 1811 (*PV*, *4*, 170).

The choice of topic for this prize (transferred from an unawarded prize of 1808 in galvanism) probably reflects not only the genuine interest in the phenomena but also an opportunity for Laplace to further the cause of Laplacian physics. Laplace himself hinted as much in a paper on optics of this time, from which came the quotation at the head of § 7.2.1 (*1810a*, 288):

---

[1]) Truesdell's histories bring out well the range of 'elastic' phenomena in the 18th century. Todhunter *1886a*, completed by Pearson, has a much narrower conception of elasticity, and even then is excessively brief on the 18th century, although there are some useful accounts of work done between 1800 and 1850.

> To determine the equilibrium and the motion of an elastic lamina[,] naturally rectilinear and bent along some curve, one has supposed that, at each point, its spring is in inverse ratio to the radius of curvature. But this law is only secondary and derives from the attractive and repulsive action of the molecules, following a function of the distance [...]

Poisson would be the man expected to take the hint; but when October 1811 arrived, only one entry was in, and it was not by him. The judges found its differential equations to be incorrect, and reset the prize for October 1813 (*PV*, *4*, 562); and then the comedy began.

Firstly, Poisson at last gained election to the *classe*, in March 1812 (in his 31st year). The election was surprising, in that he joined the *section* of general physics, a subject to which he had not yet made any contributions (see Home *1983a*): he received 30 votes to the 24 accorded to the engineer Girard (*PV*, *5*, 40). Then in October 1813 he was nominated to the elasticity prize commission (pp. 250-251): it gave the one entry received an honourable mention, but set the prize again for October 1815 (p. 292). Then at last Poisson came up with his own paper on the problem; but when he read it in August 1814, Legendre objected that a paper was being read on a topic currently under prize offer. Poisson continued reading, but Legendre's point was felt worthy of examination by a special commission (pp. 385-387), which never reported .... So Poisson sent in his paper for the *Mémoires*, where it appeared as *1816h*; he also quickly published a summary *1814a* with the *Société Philomatique*. Hachette reprinted this in his *Correspondance sur l'Ecole Polytechnique*, adding in an optimistic footnote that it 'will be very useful to the young geometers who compete for the prize'.

Hachette's advice may have been genuinely intended, for students or recent graduates of the school in 1814 included Sadi Carnot, Chasles, Duhamel, Morin and (a splendid touch) Lamé and Saint-Venant; but it would have asked much even of them to produce an entry, and indeed only one paper was received, again from the same author as before. The commission reported to the *classe* at the end of 1815: it still did not accept the mode of derivation of the differential equation, but it praised the correlation of theory with Chladni's experiments. 'The author', it announced after opening the covering letter, was 'Madamoiselle Sophie Germain of Paris' (*PV*, *5*, 595)—as doubtless it had known all along.

7.3.2 Laplacian elasticity: Poisson's molecular model, 1814

The first part of Poisson's paper contained the equilibrium of a flexible and non-elastic surface. It had recently been studied in Lagrange *Mechanics*$_2$, *1* (1811),

156–162, with a simplified version in his own textbook (*Mechanics*$_1$ (1811), *1*, 212–230). But in this paper, 'I arrive at the same equation by a different method, which has the advantage of showing to which particular restriction it is subordinated: in effect it supposes each element of the surface equally stretched in all directions' (*1816h*, 170).

Impressed forces (X, Y, Z) at point m (x, y, z) of a surface of constant thickness caused forces to act over the 'infinitely small element' at m given by

$$(Xk\,dx\,dy, Yk\,dx\,dy, Zk\,dx\,dy), \quad \text{where} \quad k := (1 + z_x^2 + z_y^2)^{1/2}. \tag{732.1}$$

The short-range inter-'element'al isotropic forces T acting through the sides of an infinitesimal cuboid set parallel to the yz-plane had, from the geometry of surfaces, direction cosines

$$(1 + z_y^2)^{1/2}/k, \quad -z_x z_y/(k(1 + z_y^2)^{1/2}), \quad z_x/(k(1 + z_y^2)^{1/2}) \tag{732.2}$$

and magnitude

$$T\,dy(1 + z_y^2)^{1/2}, \tag{732.3}$$

since (presumably) each point was dy away and the force would be $T\,dy \sec \Theta$ if $\tan \Theta := z_y$. A similar analysis of the forces T' acting at m parallel to the xz-plane led to expressions like (732.2–3). Hence, resolving in each direction, the equilibrium equations took on p. 179, after some tidying, the form

$$Xk + (T(1 + z_y^2)/k)_x - (T' z_x z_y/k)_y = 0, \tag{732.4}$$

$$Yk - (T z_x z_y/k)_x + (T'(1 + z_x^2)/k)_y = 0, \tag{732.5}$$

$$Zk + (T z_x/k)_x + (T' z_y/k)_y = 0. \tag{732.6}$$

The unknowns T and T' could be eliminated from these equations; in particular, if $T = T'$, then the equation for flexible surfaces became

$$Z - pX - qY + T[(1 + z_y^2)\, z_{xx} - 2z_x z_y z_{xy} + (1 + z_x^2)\, z_{yy}]/k^2 = 0 \tag{732.7}$$

(p. 179), which reduced to Laplace's (722.7) for the capillary surface when the impressed forces were suitably allowed for (p. 185). Again, the wave equation (727.5)$_2$ for two dimensions could be obtained by inserting the relevant details into (732.7) (p. 191).

The second part of the paper dealt with elastic surfaces S' in equilibrium. 'Elastic forces' between particles m and m' distant h apart had to be admitted: the 'repulsive force' was $f(h)$, where f was not known. The cumulative forces on m from all points m' within its sphere of activity had as component

$$\iint [(x - x')\, \varepsilon^2 f(h)/h]\, \omega \tag{732.8}$$

in the x-direction (with similar expressions for y and z), where ε was the constant thickness of the surface and the double integral covered all 'elements of the surface' ω of the points $\{m'\}$. These expressions led Poisson to add in to (732.7) the term

$$U := \varepsilon^2 \iint [(z - z') - z_x(x - x') - z_y(y - y')] \, (f(h)/h) \, \omega \tag{732.9}$$

(pp. 192, 191–193, in a misnumbering of the pages).

Poisson subjected (732.9), and the modified (732.7), to clever transformations. Following a tradition prior to Euler's formulation of Newton's equations (§ 5.2.5), he set up coordinates at m along the normal and in the tangent plane, converted some expressions into polar coordinates, evaluated integrals of the type of Petit's (723.4) (though without using Petit's work) and obtained an expression for T (pp. 193–208). Then he took the Taylor expansion of z, and other quantities, in terms of the variables of the coordinate system just described, and ended up on p. 215 with 'the principal concern of this paper':

$$\begin{aligned} Z - z_x X - z_y Y + (\varepsilon^2 a^2 \pi - 2\Pi)\, kH + \tfrac{1}{2}\, \varepsilon^2 b^2 k\pi\, GH \\ = \frac{1}{8}\, \varepsilon^2 b^2 \pi \bigg(\frac{(1 + z_y^2)}{2k}\, H_{xx} - \frac{z_x z_y}{k}\, H_{xy} \\ + \frac{(1 + z_x^2)}{2k}\, H_{yy} - z_x\, HH_x - z_y\, HH_y + kH^3 \bigg), \end{aligned} \tag{732.10}$$

where $\quad G := (z_{xx} z_{yy} - z_{xy}^2)/4k^4 \quad$ (p. 211), $\qquad$ (732.11)

$$H := [(1 + z_y^2)\, z_{xx} - 2 z_x z_y z_{xy} + (1 + z_x^2)\, z_{yy}]/2k^3 \quad \text{(p. 211)}, \tag{732.12}$$

$$a^2 := \int_0 \alpha^2 f(\alpha)\, d\alpha, \quad b^2 := \int_0 \alpha^4 f(\alpha)\, d\alpha \tag{732.13}$$

(repulsion integrals: p. 203)

and $\quad \Pi := \int (X dx + Y dy + Z dz)\quad$ (assumed exact differential: p. 194). (732.14)

In a non-technical account for the *Journal des savants* of Germain's and Poisson's *pas de deux*, Biot claimed to have found the equation a long time before and to have presented it in public lectures (*1817b*, 146). He cannot have had in mind (732.10) but the equation for a plane surface, which Poisson obtained from (732.10) by placing the surface in the xy-plane and setting

$$z_x = z_y = 0; \quad \therefore G = 0 \quad \text{and} \quad H = \tfrac{1}{2}\,(z_{xx} + z_{yy}), \tag{732.15}$$

$$X = Y = 0, \quad Z = -\varepsilon z_{tt}; \quad \therefore d\Pi = -\tfrac{1}{2}\, \varepsilon z_t^2 \triangleq 0, \tag{732.16}$$

and $\quad \tfrac{1}{2}\, \varepsilon^2 a^2 \pi + \tfrac{1}{16}\, \varepsilon^2 b^2 \pi H^2 - \Pi = V, \qquad$ (732.17) where V was the tension at the boundary of the surface (*1816h*, 216–220: on p. 220 he argued from (732.17) that V was constant, but he did not discuss how the boundary was held). Then (732.10) reduced on p. 220 to an equation which I express via the Laplacian operator as

$$\varepsilon z_{tt} - V\nabla^2 z + \tfrac{1}{32}\varepsilon^2 b^2 \pi \nabla^2\nabla^2 z = 0. \qquad (732.18)$$

'I will conclude this paper by making known an interesting property of the surface [S] in equilibrium', stated Poisson. From (473.2), if ϱ and ϱ' were the principal radii of curvature, then

$$1/\varrho + 1/\varrho' = H \qquad (732.19)$$

(as in (722.6) for capillary action). The property in question was that

$$\delta \iint_S kH^2 \, dx \, dy = 0, \qquad (732.20)$$

and thus that the integral took an extremum (actually a minimum) in the equilibriate state (pp. 221–225). He showed that this property held for some other integrands involving ϱ and ϱ', thus generalising upon the property of minimal potential energy stored in the surface, 'which was first conceived by D. Bernoulli, and which Euler then verified at the end of his treatise on *isoperimeters*' (p. 225, having in mind Euler *1744a*, ch. 6 for the verification, and the opening to its first addition for the acknowledgement to Bernoulli). However, Poisson sought no solutions to the differential equations that he had found.

7.3.3 Non-Laplacian elasticity: Germain's geometrical model

> [...] it is asserted that she [Germain] could never form the conception of space, except by the means of time; this was her own mode of expressing, to the writer of the notice, a state of mind by which he accounts for another fact, that she had very little aptitude for pure geometry, and a great attachment to the theory of numbers.
>
> de Morgan *1842b*, 175 (drawing on some unpublished communication)

The Germain affair has been studied in some detail. Although much more can be done,[1]) I shall restrict myself to the principal features, especially as contrasted with Poisson.

Germain's first entry was submitted in September 1811, when she was 35 years old (and five years older than Poisson). She took her inspiration from Euler's paper *1782a* on the vibration of elastic bands and rods, in which he had formed an integro-differential equation like (525.1) and differentiated it to obtain the equation of motion, which I write as

$$\sigma y_{tt} + T y_{ss} + B y_{s^4} = 0, \tag{733.1}$$

where T was (constant) tension, σ the density and B what we would interpret as the bending modulus (art. 4). Little in the paper was new, but Germain was probably attracted to the detailed discussion of the simple modes of vibration for various configurations of the rod, determined via the roots in ω of equations such as

$$\operatorname{sh}\omega \sin\lambda\omega \sin(\lambda-1)\,\omega = \sin\omega \operatorname{sh}\lambda\omega \operatorname{sh}(\lambda-1)\omega, \quad \lambda \text{ a parameter} \tag{733.2}$$

(art. 48). Euler analysed this equation in detail only for $\lambda = \frac{1}{2}$, but Germain attempted a treatment for all rational values of λ. Not surprisingly, given the state of affairs in the theory of equations (§ 4.6), she found difficulties, and sought the help of Legendre in correspondence.

Germain also tried to imitate Euler's derivation of (733.1) to obtain the equation of motion of the elastic surface. She found an integro-differential equation, which in effect equated the balance of moments with the means of resisting deformation, itself expressed via (732.19). Differentiation produced an equation which reduced for small vertical displacements z to

$$z_{tt} + K(z_{x^4yy} + z_{xxy^4}) = 0, \quad \text{with } K \text{ constant.} \tag{733.3}$$

[1]) The most detailed biographical study is Bucciarelli and Dworsky *1980a*: while excellent on the social and personal aspects, it is light on technical details, especially concerning Germain's second entry. Todhunter's account is largely restricted to a (good) critical study of the third entry (*1886a*, 147-160), in the published form Germain *1821a*. Kirchhoff *1850a* discussed her models; so does Dahan *1987a*.

Germain's principal manuscript collections are these:

AS: manuscripts of all three entries;

BI, ms. 2381, drafts and versions of some entries;

BN, ms. fr. 9118 and n.a. 4073, various letters; 9114-9117, main collection of manuscripts, including notes and drafts; n.a. fr. 5166 and 9544, volumes containing a few notes and draft letters;

ENPC: de Prony's notes on her *1821a* inserted in his own copy; ms. 715 (but currently lost), manuscript of her posthumous *1880a*;

Darmstaedter: letter from Fourier, 1821;

Biblioteca Riccardiana (Florence), *Fonds Palagi*, ms. 432: nine letters to Libri (see my *1984a*);

British Library (London), manuscript of a note on the vibration of plates, submitted to the *AS* on 1. 9. 1823 (*PV*, *7*, 525) and published in *1880a*, 60-64, kept in their copy of *1821a*.

Some of the letters cited above, and also others, are printed in both editions of her *Works* (which, however, does not contain her scientific writings).

Lagrange, who sat on the commission, found this equation unbelievable. He doubted that mean curvature expressed all the properties of a deforming elastic surface, and, even granting this premise, he obtained from it not her (733.3) but

$$z_{tt} + \nabla^2\nabla^2 z = 0.^{1)} \tag{733.4}$$

Poisson's own equation (732.18) was to include a boundary force term.

In her second entry, of 1813, Germain aimed at Lagrange's (733.4), and by Lagrange's variational methods. Her method was rather grim; but the discussion of special modes of vibration, *à la* Euler, of (733.2) for special values of λ, and so on, led to some comparisons with Chladni's experimental results which the commission felt good enough to deserve an 'honourable mention'.

The third, prize-winning, entry of 1815 was revised afterwards (with some help from Fourier) and published by Courcier as the book Germain *1821a*, and it is this version which is described here. She now spoke of the principal radii of curvature R and R' when the surface was in a 'natural' condition, and r and r' when under constraint. She argued vaguely that the 'action of the forces of elasticity' was proportional to the change in mean curvature, which I write as

$$H - H_0, \text{ where } H := 1/r + 1/r' \text{ (as in (732.19)) and } H_0 := 1/R + 1/R' \tag{733.5}$$

(pp. 1-8, 12). "Thus" their moment over the elements dm of the surface was given by

$$-\iint_S N^2(H - H_0)\,\delta H\,dm, \tag{733.6}$$

where N^2 was an elastic parameter, proportional to the fourth power of the thickness of the surface S. In addition, dm apparently extended itself in its upper 'part' and compressed in its lower 'part', and so exercised a force

$$\tfrac{1}{2}\iint_S N^2(H^2 - H_0^2)\,\delta dm \tag{733.7}$$

to restore itself (pp. 12-16). Hence the general equation of motion was

$$\iint_S N^2(H^2 - H_0^2)\,\delta dm - \iint_S N^2(H - H_0)\,\delta H\,dm = 2\iint_S (r - R)_{tt}\,\delta(r - R)\,dm. \tag{733.8}$$

She did not actually state this equation: I have put it together from pp. 18 and 27, taking in also a change in reference for 'r' on p. 24.

---

1) Lagrange's note was published first by Navier in his *1826f*, 149-150 (presumably from the version in Lagrange's *Nachlass* at *BI*, ms. 912, fol. 175) and later in Germain *Works*$_1$ (1879), 337 and $_2$(1892), 301 (presumably from the version in her *Nachlass* at *BN*, ms. fr. 9118, fol. 34).

Germain then took the case of the circular cylinder. After putting the relevant details into (733.8) and making some mistakes in the variational calculus, she obtained an equation which reduced to Lagrange's (733.4) for a plane surface (when the radius of the cylinder was infinite) (pp. 19–27). She also found an equation for the elastic annulus which differed from that in Euler *1766e*, art. 10 in the sign of a term; so she devoted some space to Euler's "mistake" instead of noticing her misuse of the variational calculus which had led to the discrepancy (pp. 28–30).

Germain's method of solving these equations followed Euler's, of inserting boundary conditions and seeking the simple nodes. Unfortunately, quite apart from the particularity of her solutions, there are such errors in passage-work (those on pp. 36–37 are just careless, like the over-use of '*A*' and '*S*' there) as to render worthless several of her analyses of the equations corresponding to Euler's (733.2), and the comparisons between her theoretical predictions of the nodes and Chladni's and her own experimental results. Since the prize was awarded to her for the corresponding parts of her third entry, one wonders if the commission was guilty of careless reading or excess chivalry—or both.

7.3.4 Contrasts

Socially, Poisson and Germain were far apart: he a *classe* pace-setter, working within Laplacian physics but nevertheless agreeing for a prize to be awarded to a mere woman who used old-fashioned empirical conceptions and made mistakes. Their personal contacts were slight: a few letters were exchanged, but when Germain sent Poisson a list of questions with space left on the sheets for reply, Poisson returned the sheets unannotated.[1)]

The intellectual distance was no less great, and shows well the difference between ways in which the same class of phenomena can be studied. Poisson's trans-empirical Laplacian world of short-range repulsions is hard to interpret as elasticity. Firstly, how many properties of elasticity are they supposed to encompass? Germain thought, falsely but understandably given Poisson's presentation, that they only covered extension (Germain *1821a*, 8). Secondly, effects due to the thickness of the "thin" surface were left out (as Poisson was to admit for his own work in *1828h*, 360); indeed, the degree of ε in the equations (ε^2 in Poisson's (732.10) but ε^4 in Germain's (733.6)) would need discussion, as we shall find

[1)] See *BN*, ms. fr. 9114, fols. 228–234. Her draft letter in ms. n.a. fr. 9544, fols. 21–22 might also be addressed to Poisson; an exchange of letters in 1816 is published in Germain *Works*$_1$ (1879), 344–348 and $_2$(1892), 307–310. 500 copies of her book *1821a* were printed (*AN*, F^{18} 87, no. 80). See also her pamphlet *1826a*.

Navier thinking in § 15.2.5 and § 15.5.4. Thirdly, did the mathematical structures of Poisson's integral definitions (732.10) describe the nature of some elastic action (§ 1.3.1), and would they be invariant under coordinate transformations? Fourthly, was a molecule assumed to be cuboidal in shape, with edges dx, dy and dz? Finally, the nature of the impressed forces (X, Y, Z) was not discussed by either party; for example, whether or not they included couples, and so (for example) whether or not torsion was included in these "general" theories of elastic surfaces.

In various ways Poisson's model brought him close to the notional mode § 132.3 of mathematical physics. Biot assessed Poisson's derivation of the differential equation (732.18) as 'much more daring and more general' than his own, but added, with a confidence that is difficult to emulate, that it was 'also more certain' (*1817b*, 146).

One of the surprises of the story is that Germain, rather than Poisson, used (or tried to use) Lagrange's variational methods to obtain the basic equations; for largely she worked within Euler's geometrical tradition, where variations were used more sparingly. For her the principal radii of curvature were central to her analysis, whereas Poisson brought them in at the end of his paper (for a variational purpose). For her Chladni's experiments were a principal aim, even to be repeated; Poisson mentioned Chladni only at the beginning of his paper.

But Germain's vague reasoning brought her also to the notional mode of mathematical physics. So did some of her solution methods, especially a (mercifully unpublished) display of all simple modes written out in large sheets suitable for conversion into wallpaper; even more grotesque expressions than those in Plana's paper *1815a* on elasticity, which Poisson correctly appraised as 'such a complicated equation, that it seems impossible to make any use of it'.[1]) She also drew on Poisson's result (732.20) on the minimal stored energy of the surface, and on some related reults, to argue that the 'elastic force' could be proportional to $(1/r - 1/r')$ instead of $(1/r + 1/r')$ (*1821a*, 8-9); but since r' is not infinite, the change surely makes some difference to the nature of the hypothesis of which 'I am going to try to prove the legitimacy' (p. 1). She also made no use of Binet's paper *1815b*, submitted to the *classe* in August 1814 (*PV*, *5*, 390) and praised by de Prony in a lengthy report *1814b*. Binet presented various difference formulae related to curved surfaces and discussed the basic elastic properties, ending with a criticism of Lagrange's analysis of the flexible surface (mentioned in § 7.3.2). It was Poisson *1816i* who took up Binet's ideas, in a short paper on the torsion of

[1]) Poisson *1816h*, 170; Plana's paper appeared in the *JEP*, presumably with Poisson's blessing. Wallpaper manufacturers should consult Germain's designs at *BN*, ms. fr. 9116, fols. 204-211. On Plana, see § 17.6.6, footnote.

an elastic line of which part was fixed—an aspect of elasticity omitted from his own earlier study *1816h* of the elastic surface.

A lot of work need to be done in elasticity theory. We shall see some of the next round of achievements in § 13.4 with Fresnel, and especially in Chapter 15 with a battle royal between Navier, Cauchy and Poisson—but not Germain.

7.4 On Laplacian optics

> One sees that these facts are extremely favourable to the system of Newton on the emission of light, and by contrast seems to destroy the system of waves proposed by Huygens, and sustained since by Euler and by other celebrated *savants*.
>
> Biot *1806a*

At this point in our story I pick up the science of optics; but I shall be dealing only with those aspects which lay within the growing realm of Laplacian physics. In fact, this was the dominant interest in French optics in the early 19th century, and a massive literature is involved; thus I must be strict in selection, as ever.[1]) The starting point, once again, is Book 10 of Laplace's *Mécanique céleste*.

7.4.1 Laplacians on the refraction of light in gases, 1804–1806

The first topic discussed by Laplace in Book 10 was atmospheric refraction, which would change the observed position of the heavenly bodies. He assumed

---

[1]) Lalande provided a useful, though surprisingly non-mathematical, survey of 18th-century optics in Montucla *History*, *3* (1802), 427–605. A clear indication of the understanding of geometrical optics around is given by the new edition *1802a* of Lacaille's elementary textbook of 1756 on optics, which was produced by some unnamed *polytechniciens*. Its mathematical limit was elementary calculus, as in Poisson's calculation *1802c* of the incident angle to the horizon to see a given colour of the rainbow. This result is often attributed to Airy *1838a*.

Early-19th-century optics is covered in detail in reports written soon afterwards: Arago *1825b*, Brewster *1830a*, J. Herschel *1830a* (which was rapidly translated into French), Lloyd *1835a* and Young *1824a*. However, they sometimes restate the Laplacians' achievement in terms of Fresnel's later wave theory, or mix the two together. They are not always clear in indicating the types and cuts of crystal lamina to which a particular theory or class of phenomena applies, the kinds of incident light used, or the configurations employed: these difficulties arise especially in writings on polarisation. The best writings on Laplacian optics are the more recent texts cited in the next sub-section on Malus.

that light was composed of tiny 'particles' attracted by some unknown force $\varphi(r)$ when at distance r from the centre of the earth (assumed to be spherical, of radius a), thus pulling it off its rectilinear course. In order to find the differential equation of the path, he first took the case of a molecule P approaching the plane surface $\mathcal{S}$ of a semi-infinite transparent solid $\mathcal{T}$ of density ϱ. Writing '$\Pi(s)$' for the 'action' on ϱ of a stratum of $\mathcal{T}$ parallel to $\mathcal{S}$ and distant s from P, he produced at once the differential equations

$$\ddot{x} = 0 \quad \text{and} \quad \ddot{s} = -\varrho \int_s^\infty \Pi(u)\, du, \tag{741.1}$$

where x was a coordinate variable parallel to $\mathcal{S}$ (*4* (1805), 438–449). Integration, and the introduction of the velocity of P without $\mathcal{T}$ and also within it (when $(741.1)_2$ took the + sign), produced Snell's law in the form

$$\frac{\text{sine (incident angle)}}{\text{sine (refracted angle)}} = \left(1 + \frac{4K\varrho}{n^2}\right)^{1/2}, \text{ where } \quad K := \int_0^\infty \int_s^\infty \Pi(u)\, du\, ds \tag{741.2}$$

and n was the incident velocity of the molecule (p. 453). From (741.2) he derived various inequalities concerning whether light would be refracted or reflected (pp. 457–464).

Having checked the validity of some basic properties of light under his interpretation, Laplace now sought the differential equation of the path of a ray. He took the atmosphere to be composed of layers of varying density $\varrho(r)$ which bent the ray by their successive 'action'. Since this action was (assumed to be) of very short range, 'we shall suppose that [it] is represented by'

$$\Pi(s) := Qe^{-Ns}, \quad Q \text{ and } N \text{ constants, } N \text{ large (he wrote '}i\text{').} \tag{741.3}$$

Hence the force acting on the ray at distance r from the centre of the earth,

$$\varphi(r) := \int_0^s (\varrho(r-u) - \varrho(r+u))\, \Pi(u)\, du, \tag{741.4}$$

reduced, after expanding $\varrho(r)$ in a Taylor series, to

$$\varphi(r) = -2K\varrho'(r) + O(s^3) \tag{741.5}$$

(pp. 449, 465–467). This led him on p. 469 to a vile first-order ordinary differential equation between ϱ and the angle Θ between the tangent to the path of the ray and the zenith of the observer, whom it would eventually visit (p. 469).

This analysis clearly heralds Laplacian physics: the phenomenon treated in terms of molecules with short-range forces between them; integrals such as $(741.1_2, 4_2)$, which Petit was to systematise in (723.3); and so on. At this stage

Laplace was more concerned with knowing the function $\varrho(r)$ in order to integrate his differential equation. Accordingly he tried four 'hypotheses', equivalent to

$$\varrho(r) = \varrho(a) \quad \text{(constant density: pp. 473-477)}, \tag{741.6}$$

$$\varrho(r) = \varrho(a) \exp(-a(r-a)/rl) \text{ (constant temperature: pp. 477-497)}, \tag{741.7}$$

$$\varrho(r) = \varrho(a) + A(r^{-1/m} - a^{-1/m}) \quad \text{(pp. 497-507)}, \tag{741.8}$$

$$\varrho(r) \triangleq \varrho(a)\,(1 + B(r-a)/r) \exp(-C(r-a)/r) \quad \text{(pp. 507-521)}, \tag{741.9}$$

where a, A, B, C, l and m were known parameters. (The third hypothesis was incorporated into (724.4), in connection with barometer theory.) In contrast with his confident statements made in many other places, he was philosophically cautious about these hypotheses, showing perhaps the greatest enthusiasm over the fourth for incorporating elements of the second and third (p. 507); a less imperial author might have written that he was guessing. When he proceeded to analyse astronomical refraction at high altitudes, he was glad to obtain results 'independent of any hypothesis on the constitution of the atmosphere' (pp. 523-524); while for refraction at low altitutes he adopted the second hypothesis (741.7), to $O(r^2)$ (pp. 542-550).

In order to increase knowledge of $\varrho(r)$ Laplace thought that 'it will be useful to make a greater number of experiments' on the refractive powers of gases (pp. 471-472). So Biot and Arago filled a prism (previously used by Borda in similar work) with various gases, placed it in Laplace's senatorial chamber at the *Palais du Luxembourg*, and looked through it at the lightning conductor on top of the *Observatoire* about 1,500 metres away (squares 8A-8E of Figure 263.1). They carried out careful experiments, using both a Borda circle (§ 5.5.4) and a Fortin barometer (§ 6.6.3). For theory they followed Laplacian optics, describing the *forces vives* of the molecules, allowing for barometric pressure by Laplace's formula (724.1) and calculating the refractive indices from (741.2) (with $\varrho(r)$ instead of the constant ϱ there). The indices for compound gases were calculated from those for their constituents by the centre-of-gravity formula (see Frankel *1977a*, 59-64).

Biot and Arago published their work in a paper *1806a* in the *Mémoires*. The results obtained, and the properties examined, coroborated received knowledge in this area, and thus added greatly to the promise of Laplacian physics; Biot expressed his confidence in a summary *1806a* published in the *Moniteur universel*, in words quoted at the head of this section. Their work also manifested some personal aspects; for Biot presented it to the *classe* under his own name (*PV*, *3*, 328, 338), with only an acknowledgement to Arago, but his 20-year-old collaborator objected (see Arago *1859a*, 702-703). This was the first of a series of disputes between these ambitious young careerists; a crucial one is described in § 7.4.7 below.

7.4.2 Malus's first researches in optics, 1807

Enter now the chief Laplacian optician. Etienne Louis Malus was born in Paris to the Treasurer of France in 1775. After studying briefly at the *Ecole Génie* at Mézières in 1793, he entered the foundation year of the *Ecole Polytechnique* and then participated as a soldier in the Egyptian campaign (§ 2.6.2). Upon his return to France in 1801 he was posted to various provincial garrisons; but he managed to spend much time in Paris from 1807, and moved there permanently two years later. He was elected to the elder Montgolfier's place in the *classe* in August 1810 (*PV*, *4*, 373) and provisionally appointed *Directeur des études* at the *Ecole Polytechnique* the following year, having already served for six years as graduation examiner in descriptive geometry. But within a few months of his definitive appointment, in February 1812, he died. This was an especially sad loss; it would have been fascinating to mark his reaction to Fresenel's wave theory when it began to develop a few years later! Although his research interests centered on optics, his competence was wide, as his *Ecole Polytechnique* appointment testifies; his manuscripts show especial interest in mathematical questions.[1])

Malus had studied optics while in Egypt, especially with his *polytechnicien* classmate Lancret, whose expertise in differential geometry was noted in § 4.7.4. In April 1807 Malus submitted his first paper to the *classe* (*PV*, *3*, 516); dealing with various properties of reflection and refraction, it showed, like Lancret's work, considerable ability to handle the geometry of surfaces *à la* Monge. Six months later Lacroix gave an acute report *1807b* recommending the paper for the *Savants étrangers*; it finally appeared there as Malus *1810b*, having already been published in a somewhat different version as *1808a* in the *Journal de l'Ecole Polytechnique*.

Although concerned with geometrical optics, in which no ontology of light need be entertained, Malus seemed to be sympathetic to the Laplacian view. Indeed, his best-known result—that rays reflected or refracted once from a surface were normal to another surface (both papers, art. 4), but *not* after further reflections (*1810b*, art. 12) or refractions (*1808a*, art. 27; compare *1810b*, art. 18)—had

---

[1]) Malus's small *Nachlass* is held at *BI*, mss. 1692-1694; there are significant materials also in his *dossier personnel* at *AS*. Arago was a close friend, and his biography *1855a* contains information unavailable elsewhere. A notebook on problems in calculus and mechanics was once possessed by J.A. Barral, the editor of the works of Arago, from whom he presumably obtained it. In 1896 the Paris publishing house Hermann presented it to the Swedish mathematician G. Mittag-Leffler, and it is now kept at the *Institut Mittag-Leffler* (Stockholm).

Welcome attention has been paid to Malus in recent years. See Chappert *1977a* (which contains a nearly complete list of publications and manuscripts), and and Frankel *1974a* and *1976a*. On double refraction, which is soon to be discussed, see Buchwald *1980b*, 332-357.

to be corrected later, for the result was shown by Dupin *1822e* (and by others) to be true also of further wave surfaces (see Chappert *1977a*, ch. 5). Had Malus himself thought in terms of wave surfaces, then he would probably have recognised the extension of his theorem for himself.

Thereafter Malus worked directly in Laplacian optics. His next paper was a criticism and reworking, from the Laplacian standpoint, of the recent researches on the refractive indexes of opaque bodies reported by the English physicist W. H. Wollaston in a paper *1802a*, of which a summary had been translated into French in 1803. Malus submitted his paper in November 1807 (*PV*, *3*, 619), gained Gay-Lussac's approval *1808a* for it two months later, and published it in the *Journal de l'Ecole Polytechnique* as *1809c* (and later, unchanged, in the *Savants étrangers*). He drew on Laplace's view that light reflected off an opaque body without entering it, and also his interpretation of Snell's law (741.2) (*Mechanics*, *4* (1805), 454-459); namely, that if l and l' were the refractive indices of two media, and Θ the angle of incidence, then in the formula

$$l^2 \cos^2 \Theta = \gamma(l^2 - l'^2), \quad \text{where} \quad 0 \leq \gamma \leq 1, \tag{742.1}$$

'reflection will start when $\gamma = 1$' for transparent bodies, while the corresponding value of γ for opaque ones was ½ (Malus *1809c*, 221-222). Describing also his own experiments, he regarded his work using opaque and transparent wax as 'a positive proof of the hypothesis on which the analysis is founded' (p. 223).

7.4.3 Malus on 'polarisation' and his prize problem "victory", 1808-1809

The crucial period in Malus's career occurred at the end of 1808. Probably under the influence of his and Wollaston's work, and Häuy's earlier study *1791a* of double refraction in Iceland spar, the *classe* had announced at the end of 1807 a prize problem on double refraction, with entries to be received by the beginning of 1810 (*PV*, *3*, 632-633). This gave Malus, for whom it was doubtless intended, plenty of time; and on 12 December 1808 (*PV*, *4*, 144) he reported to the *classe* the discovery that has most preserved his name in science. As his friend Arago later recalled (*1825b*, 858-859),

> [...] as he happened to be decomposing, by means of a rhomboid of carbonate of lime [Iceland spar], towards the end of the year 1808, the light of the setting sun, reflected from the glass of the windows of the [Palais du] Luxembourg, he first observed the difference of the two images in different positions of the rhomboid [...].

Biot recounted this story in *1820c*, 112.

Good fortune of residence had attended Malus's discovery. He had carried out the experiment at his home at the northern end and the western side of the *rue d'Enfer* (square 9D of Figure 263.1), where he was suitably placed to receive the reflected evening sun from the windows of the *Palais du Luxembourg* (square 8E). The light will have come in to him *over a relatively long period of time*, from window to window. Indeed, it would be fun if the crucial rays were reflected from *Laplace's* window on the south side of the *Palais*, from where Arago and Biot had observed the lightning conductor of the *Observatoire* in 1806 (§ 7.4.1)!

Malus had discovered polarisation by reflection (or, more precisely, partial polarisation, according to Arago's further recollection). Its significance lay not only in the novelty of the phenomenon but also in its simplicity. Hitherto polarisation had been known only of doubly refracted rays, and thus appeared to be connected with double refraction; but now it had to be regarded as a more fundamental property of light. For Malus it formed a crucial support of Laplacian optics; he even regarded it as a refutation of (Huygens's) wave theory, which apparently would not allow that 'the phenomena of reflection are different for some given angle of incidence' (*1809e*, 260). Indeed, the word 'polarisation' was introduced by Malus under the influence of his Laplacian view. He began using the word around 1811, but his explanation of it appeared only in a posthumous paper (*1814a*, 106–107):

> These observations lead us to conclude that in these circumstances the light acquires properties independent of its direction relative to the surface which reflects it, but uniquely relative to the sides of the vertical ray, and which are the same for the south and north sides, and different for the east and west sides. In giving the names of poles to these sides, I will call polarisation the modification which gives properties to the light relative to these poles. Up to now I have delayed admitting this term into the description of physical phenomena which are in question; I have not dared introduce it in the papers where I have published my latest experiments; but the varied effects [*variétés*] which this new phenomenon offers, and the difficulty of describing them, forces me to admit this new expression, which simply signifies the modification that the light has suffered in acquiring new properties which are not relative to the direction of the ray, but only to the sides considered at right angles and in a plane perpendicular to its direction.

This passage is a charming example of the nervousness with which a scientist introduces a new technical term. In fact the word *'Polarisation'*, and especially *'Po-*

larität', had already appeared in the writings of German thinkers such as Goethe and Schelling, as a general philosophical category to individuate opposites (such as attractive and repulsive forces, for example); but Malus's statement has the flavour of personal discovery, and indeed he is not likely to have come across such foreign literature. If he was set off by an analogy, it is much more likely to have been with the (magnetic) poles of the earth, or the oscillations of a mounted magnetic needle.

When the *classe* was told of Malus's discovery on 12 December 1808, it was quick to respond. Most unusually, Laplace wrote the report on Malus's paper, and even did so in a week, in a recommendation *1808a* which appeared at once in the *Moniteur universel.* The *Société Philomatique* printed the paper itself the next month as Malus *1809a*, with a longer account *1809d* appearing in the *Société d'Arcueil* journal.

Malus had another important success at this time; but this one also brought him experience of the dark side of Paris scientific life. It was a proof of Huygens's construction of the extraordinary ray in double refraction, using the least action principle (the details are given at Figure 745.1). Laplace mentioned this achievement in his report *1808a* just mentioned, although it was irrelevant to the subject; but a month later he presented to the *classe* his own version of the proof, playing down Malus's work (*PV*, *4*, 159). This was quickly printed in the Arcueil journal as Laplace *1809d*, with two different summaries appearing elsewhere (*1809a* for the *Société Philomatique*, and *1809b* in the *Journal de physique*); then his main paper came out again in the *Mémoires* (with the date of presentation printed as '1808' instead of '1809'). Poisson praised 'Laplace's' proof in his textbook on mechanics (*Mechanics*$_1$ (1811), *1*, 466–468).

In this way the 23-year-old Malus was upstaged by his 59-year-old Master; for his own derivation was published only after Laplace's, in his prize paper on double refraction. This work was submitted in September 1809 and was awarded the prize at the end of the year (*PV*, *4*, 292), rather ahead of the deadline of January 1810 that had been set; it was published as *1810c* in a (rather obscure) book form, and in the *Savants étrangers* in the following year (with the same printing). Thus the prize must have seemed to be a consolation: according to Arago, 'despite [Malus's] well-known reserve, he expressed himself, on all occasions, with a bitterness of which one would not have believed him capable' (*1855a*, 154).

7.4.4 Malus's empirical intensity laws

As the prize had been set for double refraction, Malus discussed his new discovery rather briefly in his prize paper (*1810c*, arts. 48, 50–53); but he described at length his theoretical and experimental investigations of double refraction in various materials. The paper is rich in interesting details, some not very well known; I confine myself to a brief survey of his principal concerns.

Wherever possible, Malus worked with rhombohedrons of crystal, cut along its natural faces; the 'axis' $\mathscr{A}$ of the crystal, $\mathscr{C}_1$, was defined as the diagonal joining the faces enclosing three obtuse angles, which were measured with the aid of a Borda circle (arts. 2–5). An incident ray i striking a natural face at the normal would be refracted doubly, as usual; but only the extraordinary ray e would show angular displacement, for the ordinary ray o continued in the direction of i. The plane $\mathscr{P}$ containing o and e (and thus also i) was called a 'plane of the principal section'; it was parallel to $\mathscr{A}$ (art. 5).

Malus gave much attention to the case where i fell at the normal on $\mathscr{C}_1$ to produce o and e; these in turn were doubly refracted when passed through a second crystal $\mathscr{C}_2$ of Iceland spar, set with its faces parallel to those of $\mathscr{C}_1$, producing two pairs of rays which he denoted by 'oo', 'oe', 'eo' and 'ee'. If ε were the angle between the principal sections of $\mathscr{C}_1$ and $\mathscr{C}_2$, the law for the intensities of these rays, which he announced in art. 43 as empirical laws, may be written as

$$F_{\text{oo}} = F_{\text{ee}} = K\cos^2\varepsilon \quad \text{and} \quad F_{\text{eo}} = F_{\text{oe}} = K\sin^2\varepsilon, \quad K \text{ constant.} \qquad (744.1)$$

If i, while still lying in $\mathscr{P}$, entered $\mathscr{C}_1$ at angle Θ to the normal, then o and e continued to remain in $\mathscr{P}$; otherwise there was angular displacement from $\mathscr{P}$. The formulae for the intensities of the two pairs of rays were more complicated than (744.1), being dependent not only on Θ but also on the angle between the incident plane and the principal sections, but they showed properties similar to (744.1). Writing in Carnotian fashion (526.1),

$$(l_1, l_2) \wedge (l_3, l_4) := \text{angle between planes containing lines } (l_1, l_2) \text{ and } (l_3, l_4), \qquad (744.2)$$

I express the results as follows. If $\mathscr{A}_1$ and $\mathscr{A}_2$ were the axes of $\mathscr{C}_1$ and $\mathscr{C}_2$ respectively, and angles α and β were defined by

$$\alpha := (\mathscr{A}_1, \text{o}) \wedge (\mathscr{A}_2, \text{oe}) \quad \text{and} \quad \beta := (\mathscr{A}_1, \text{e}) \wedge (\mathscr{A}_2, \text{eo}), \qquad (744.3)$$

then the intensities were given in art. 44 for the refracted rays by

$$\begin{aligned} F_{\text{oo}} &= K_1 F_{\text{o}} \cos^2\alpha, \quad F_{\text{oe}} = K_2 F_{\text{o}} \sin^2\alpha, \\ F_{\text{eo}} &= K_3 F_{\text{e}} \cos^2\beta \quad \text{and} \quad F_{\text{oe}} = K_4 F_{\text{e}} \sin^2\beta, \end{aligned} \qquad (744.4)$$

where the $\{K_j\}$ were constants to be determined (art. 44). Similarly, denoting the rays reflected within $\mathscr{A}_1$ by appropriate capital letters and defining

$$\gamma := (\mathscr{A}_1, \text{o}) \wedge (\mathscr{A}_1, \text{oE}) \quad \text{and} \quad \delta := (\mathscr{A}_1, \text{e}) \wedge (\mathscr{A}_1, \text{eo}), \qquad (744.5)$$

his laws for the intensities of reflected rays (art. 46) may be stated:

$$F_{\mathrm{oO}} = K_1 F_0 \cos^2 \gamma, \quad F_{\mathrm{oE}} = K_2 F_0 \sin^2 \gamma,$$
$$F_{\mathrm{eE}} = K_3 F_{\mathrm{e}} \cos^2 \delta \quad \text{and} \quad F_{\mathrm{eO}} = K_4 F_{\mathrm{e}} \sin^2 \delta. \tag{744.6}$$

For Malus the 'intensity' of a ray was to be understood as the 'quantity of molecules' displaying the pertaining property. Each 'molecule' possessed a rectangular system (a, b, c) of principal axes, and could rotate in the bc-plane, of which the normal a lay in the direction of propagation. Then if the angle of rotation was ψ, the 'quantity of molecules' which turned so that either a or b was perpendicular or parallel to the 'repulsive forces' acting on them was proportional to $\sin^2 \psi$ or $\cos^2 \psi$ respectively. From this 'unique law' all the intensity laws, and the phenomena of total and partial reflection and double refraction, could be deduced (as he claimed in *1809b*, 344).

Malus also discussed exceptional or singular cases; for example, when only single refraction occurred. (He assumed that crystals had one such direction; it was left to Brewster *1818a* to discover, by more careful experimentation, that some crystals possessed more than one direction, and to Fresnel to call them 'optical axes' (§13.4.3) and so avoid the Laplacians' confusing habit of calling them 'axes' even though they were not the same as $\mathscr{A}$.) Some of the properties were, of course, known long before; Malus's work was distinguished partly by its systematic presentation, partly by its attention to the reflectedly polarised rays covered by (744.6), and especially for the mathematical grounding furnished to the science by Laplacian physics. It is this last feature to which we now turn.

7.4.5 Laplacian double refraction, 1809

One of the merits seen in Laplacian physics was its capacity to draw on the principles and methods of mechanics. In this spirit Malus began his study of double refraction (*1810c*, art. 18) by adopting the principle (523.9) of least action. For a 'molecule' travelling with velocity u when distant s from its point of impact O with the face of the crystal $\mathscr{C}$, then across a path encompassing refraction,

$$0 = \delta \int u^2 dt = \delta \int u ds = \int (u \delta ds + ds \delta u). \tag{745.1}$$

He now set up a Cartesian coordinate system with the xy-plane parallel to the face of $\mathscr{C}$ and the xz-plane parallel to its axis $\mathscr{A}$, and also (in effect) spherical polars s, Θ and ω (the angular displacement from the x-axis of the projection of i on the face). He did not assume the path of o to be rectilinear but expressed $(745.1)_4$ in terms of differential geometry. Straightforward manipulation of the trigonometry converted (745.1) to an equation of the form

$$0 = P\delta x + Q\delta y + R\delta z, \tag{745.2}$$

in which, as usual (§ 5.2.2), setting the coefficients of the variations to zero gave the basic equations of motion after refraction. Writing 'u_0' and 'u_i' for the velocities of o and i at O, and applying subscripts similarly to the coordinate variables of each ray, I state the first integral of the equations furnished from (745.2) by P and Q as follows:

$$\begin{aligned} u_0 x_s = u_0 \sin \Theta_0 \cos \omega_0 &= u_i \sin \Theta_i \cos \omega_i \\ &+ (u_0)_{\omega_i} \sin \omega_0 \operatorname{cosec} \Theta_0 - (u_0)_{\Theta_i} \cos \omega_0 \cos \Theta_0, \end{aligned} \tag{745.3}$$

$$\begin{aligned} u_0 y_s = u_0 \sin \Theta_0 \sin \omega_0 &= u_i \sin \Theta_i \cos \omega_i \\ &- (u_0)_{\omega_i} \cos \omega_0 \operatorname{cosec} \Theta_0 - (u_0)_{\Theta_i} \sin \omega_0 \cos \Theta_0. \end{aligned} \tag{745.4}$$

Malus actually marked his variables for incidence with primes, and used the unadorned letters as the values at refraction.

'One knows that in ordinary refraction the velocity of the refracted light is constant', Malus continued, 'whatever be the direction of the ray' (art. 19). Thus $(u_0)_{\Theta_i}$ and $(u_0)_{\omega_i}$ could be set to zero in (745.3–4), which produced, after manipulation, Snell's law and the perpendicularity of the plane (i, o) to the face of the crystal; that is,

$$u_0 \sin \Theta_0 = u_i \sin \Theta_i \quad \text{and} \quad \tan \omega_0 = \tan \omega_i. \tag{745.5}$$

Malus based his analysis of e (art. 21) upon Huygens's construction, which is illustrated in Figure 745.1. HDE was an ellipse traced on the surface of the face and centred at C, and HME the semi-ellipsoid of revolution created by the rotation of HDE about CE. The incident plane NCR cut the ellipse along BK, where

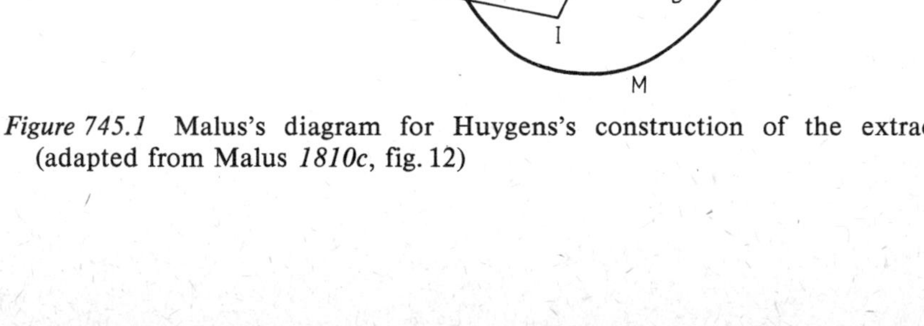

Figure 745.1 Malus's diagram for Huygens's construction of the extraordinary ray (adapted from Malus *1810c*, fig. 12)

CK was of length $u_0 \times 1$ (for some unit interval of time); in this plane OC was constructed as perpendicular to i, and O determined along it by the perpendicular dropped from K. KT was the line in the plane of the ellipse perpendicular to CK. Then if the plane through KT and tangent to the ellipsoid touched it at I, CI was the path of the extraordinary ray e. In addition, the ordinary ray o was given by the direction CZ (not shown in Figure 745.1), where Z was the point of contact of the tangent plane from K to a sphere centered at C, which itself would touch the ellipsoid at a point specifying the optical axis of the crystal.

'In order to verify this construction by the results of experiment it is necessary to translate it into an analytical expression to which one can directly apply numbers', continued Malus (art. 21). His long proof (arts. 21–25) drew heavily on the equation of motion given from (745.2) with $R = 0$. His reasoning relied much on the definitions in art. 21 of the major/minor axes a and b as the inverses of the extraordinary/ordinary refractive indexes (art. 22), so that, from $(745.5)_1$,

$$b := \mu_o^{-1} = u_i \big/ u_o \quad \text{and} \quad a := \mu_e^{-1} = u_i \big/ u_e. \tag{745.6}$$

Malus's verification of Huygens's law is ingenious; but it hardly corroborates the principles of Laplacian physics, since it is *founded*, via (745.6), on Huygens's construction. Indeed, it is hard to see how the ellipsoid is to be interpreted in Laplacian physics anyway. For Huygens, a wave theorist of light, the structure-similarity (§ 1.3.1) is clear; the ellipsoid is a wave-front. But the construction merely tells Malus that e travels along OE, just as his theorem on caustic surfaces (§ 7.4.2) says only that the reflected and refracted molecules have reached certain positions at the same time. The English polymath Young, who advocated a type of wave theory of light at this time,[1]) criticised molecularist theories of light in ways such as this, for example in an article *1809a* to which the Laplacians did not reply. In view of French readiness to note foreign work when it suited them—for example, Ivory (§ 6.5.1) or Wollaston (§ 7.4.2)—their silence here may be construed as deliberate.

[1]) On Young's theory see G. Cantor *1978a*. It is not the same as Fresnel's; for example, it was far less mathematical. An important common feature, however, was the principle of interference, which 'first occurred to the translator of this article in the room and at the table on which he is now writing', as he charmingly put it in an annotation to his translation of Arago's account (*1825b*, 848) of Fresnel's later (and independent) adoption of the principle (§ 13.2.2). Among many other concerns of Young was capillary theory (§ 7.2.2–3), which Laplace came eventually to acknowledge in *1806b* (unlike his ignorance of Leslie *1802a*). In general Young's scientific work wanted for mathematical refinement, so that in those respects even he was outflanked by the *savants*.

Another important British figure in optics was the Scottish scientist Brewster. In addition to finding that some crystals were not uniaxial (§ 7.4.4), he discovered polarisation effects independently of Arago (see Brewster *1818a*, 203) and suggested, as an experimental law, that the refractive index was the tangent of the angle of polarisation (*1815a*, 127). In 1819 he invented the kaleidoscope, as a consequence of studying polarisation by many reflections. See also Table 746.2.

Young's article was a review of Laplace's paper *1809d* on double refraction, in which Malus's ideas were given their first publication. After describing Huygens's law ('the phenomena [...] are inexplicable in his hypothesis': p. 268) and briefly mentioning Malus's derivation and his experiments (pp. 268–269, 274), Laplace "thought of" using the principle of least action to derive the laws of double refraction. The first part of the analysis is quite easy to read after Malus's; even some of the letters are the same.

Laplace's treatment of e was different from Malus's, for he tried to derive Huygens's law rather than verify it. Since the behaviour of e was symmetrical about its (assumedly unique) optical axis, u_e could be an even function of the angle V between e and this axis. Presumably bearing in mind the form of the intensity laws, Laplace chose as 'the simplest case' for this function

$$v_e = \beta^2 + \alpha^2 \cos^2 V, \quad \text{with} \quad \alpha^2 \text{ and } \beta^2 \text{ constants,} \tag{745.7}$$

and from his equivalents to Malus's basic equations of motion (to be read now for e rather than o, of course) found formulae to determine Θ_e and ω_e (pp. 279–282). In his 'it is easy to see' mood (§ 5.4.3), he dispensed with proving the final expressions 'to which Mr. Malus is led in an elegant manner, and which the geometers will easily [*sic*] take from the construction of Huygens' (pp. 281–282). His analysis matched up with Malus's by setting in (745.7)

$$\beta^2 = 1/a^2 \quad \text{and} \quad \alpha^2 = 1/b^2 - 1/a^2, \tag{745.8}$$

which yielded from (745.6) with $v_i = 1$

$$v_e^2 = 1/a^2 - (1/b^2 - 1/a^2) \cos^2 V = v_o^2 - (1/b^2 - 1/a^2) \sin^2 V. \tag{745.9}$$

The congruence of Huygens's construction with Laplace's and Malus's analyses is due basically to the fact that Huygens drew on Fermat's principle of least time while the other two used the principle of least action, a generalisation of Fermat's principle (as Laplace noted on p. 275). Ampère was to make clearer the relationship, and also to find a somewhat more general result, in a paper presented to the *classe* in March 1815 (*PV, 5*, 471) and published in the *Mémoires* as *1816a*; for he used the principle of least action to furnish a geometrical construction equivalent to Huygens's for determining the direction of e, when the ellipsoid was not one of revolution.

In his *1809d* Laplace did not give short-range forces a major role in the analysis. So in his 'Note' *1810a* to the paper (the essay containing the credo of Laplacian physics quoted at the head of § 7.2.1), he provided a new analysis in which he began by stating Snell's law in a form akin to the conservation of *forces vives.* if a molecule L of light travelled with velocity v in a medium, where it would be acted upon by a molecule dm of the material distant f from L with (attractive or repulsive) force $\varphi(f)$, then

$$v^2 = a^2 + 2\iint dm\varphi(f)df, \tag{745.10}$$

where a was the velocity in a vacuum and one integral ranged over all molecules dm (whose presence in (745.10) causes a dimensional chaos typical of Laplace). Since, as usual, $\varphi(f)$ was 'insensible' when f was 'sensible', dm would affect L only during their period of close proximity. Without argument, he asserted that the acceleration applied to L as it approached dm would be cancelled out by its deceleration on departure, so that v would be constant throughout the medium, and independent of its previous passage (if any) through other media. He claimed that this argument applied to ordinary and extraordinary rays alike (*1810a*, 286–288).

Laplace presented no motivation for his result, but he might have been encouraged by the experiments on the constancy of the velocity of light in the universe, carried out at his request by Arago in 1806.[1]) He may also have been influenced by Malus; his 'Note' is undated, but as it was published only with the 1810 *Mémoires* printing of *1809d* and not in the 1809 Arceuil journal, it may have been written after Malus submitted his prize paper in September 1809. Now Malus's paper contains in art. 49 an analysis of refraction modelled on Laplace's treatment described at (741.1), in which a molecule of light was subject to attractive and repulsive 'total actions' by the crystals, from which its refractive velocities could be calculated, along lines which Laplace was to imitate in (745.10).[2]) However, the extent to which double refraction was ground in Laplacian physics remained modest; for neither Laplace nor Malus tried to derive from its principles Malus's intensity laws stated in (744.1-6).

Despite such limitations, Laplace's hopes for his scientific programme were greatly strengthened by its success in optics. His *1810a* was to be his most ambi-

---

[1]) The history of these experiments is curious. Arago submitted a manuscript *1806a* to the *CI* in September 1806 (*PV*, *3*, 417), which Delambre took three years to approve in his report *1809c* (see also Delambre *1810b*, 47). Despite this recommendation, the manuscript was never published. In December 1810 Arago, now a member of the *CI*, presented a new paper describing fresh experiments (*PV*, *4*, 399); but this paper did not appear until *1853a* and was omitted from his *Oeuvres*. As in his experiments with Biot on refraction in gases, he used a prism to measure the refraction of solar or stellar light (and thus calculate the velocity from Laplace's (741.2), if desired). He acknowledged help from Duchayla (§ 5.3.2).

[2]) Malus repeated this analysis in an interesting paper on 'the influence of the forms of molecules of light in various optical phenomena', read in 1809 to a scientific society in Strasbourg (where he was then stationed) and published in its journal as *1811a* (but largely ignored by historians of Malus). The same volume contains an essay *1811a* on double refraction by the Alsatian Kramp, which was also an entry for the prize and received an honorable mention (see *PV*, *4*, 292: Kramp is not named). Kramp confined himself to Iceland spar and (largely) to the crystallographic aspects of the phenomena.

tious statement on it; the quotation at the head of § 7.2.1, and the remark on elasticity in § 7.3.1, came from it, and we shall meet it again in § 9.3.1 in connection with heat diffusion. He even claimed that 'The consideration of the mutual action of the molecules of matter, even furnishes a direct proof of the principle of virtual velocities' (p. 295). Briefly, a system of masses effected their mutual attraction by means of successive actions of the points comprising the lines, curves and surfaces along which they were contrained to move (rather like the 'successive transmission of gravity' noted in § 7.2.1). If the system were free of forces due to friction or elasticity, these actions would cancel out (essentially for being equal and opposite), thus leaving the 'action' of the impressed forces, which would be zero when the system was in equilibrium (pp. 295–298). The proof is rather similar to the one presented as (531.20) in the *Mécanique céleste*, and thus about as (un)convincing; but it reflects well his state of optimism in 1810.

7.4.6 Depolarisation and its consequences, 1811–1816

A massive amount of research in optics was carried out by some of the Laplacians in the 1810s. Biot was especially active, with some enormous papers in the *Mémoires* and a stream of summaries sent to the *Société Philomatique* and elsewhere. This sub-section is confined to some of the consequences of two discoveries made by Arago in 1811 (when he was 25 years old) and published in a paper *1812a* in the *Mémoires*.

While studying Newton's rings, Arago let an incident ray i of plane-polarised natural light strike at the normal a thin polarising lamina of mica and then traverse a rhombohedron of Iceland spar set with faces parallel to the lamina. When the principal sections of the lamina and the spar were parallel or perpendicular, the light would remain white through its course; but if either were rotated in the plane of its faces, two rays o and e would appear in the spar.

Arago described the rays o and e as 'depolarised' (*1812a*, 10) because the presence of the mica allowed passage of the light through the Iceland spar in certain of its configurations which otherwise would block passage. The phenomena that he and Biot were to discover, and to theorise about, were divided into types, with especial interest in cases where i was already polarised. Such a version of the case described above fell under a theory which Biot called 'mobile polarisation'. However, if the polarising lamina was made of rock crystal (or quartz), for example, a different ensemble of effects was observed which Biot attempted to account for (especially in *1819a*) in his theory of 'rotatory polarisation'. For both theories the depolarisation was partial: that is, some properties of natural light were *not* restored.

Table 746.1 summarises Arago's principal discoveries and Biot's explanations for these cases, using mica and quartz, as examples of the polarisers. The term 'chromatic' or 'coloured' polarisation usually referred to any cases where o and e were coloured, especially when complementarily so in quartz.

Table 746.1 Brief outline of Arago's experimental discoveries and Biot's theories on the behaviour in Iceland spar of o and e, produced by passing plane-polarized light through a thin polarizing lamina (for which mica and quartz have been chosen as exemplifying materials).

Topic	Effect for mica lamina	Effect for quartz lamina
Basic phenomenon	o and e have complementary colours	Exhibits coloured o and e
Rotating spar in its own plane	Changes intensity of colours	Changes colour: equivalent to changing thickness of polariser
Rotating polariser in its own plane	Changes intensity of colours	Has no effect
Using a thicker polariser	Changes colcurs: still complementary	Changes angle of rotation proportionately
Name of Biot's theory	'Mobile' (or 'moveable') polarisation	'Rotatory' (or 'rotary') polarisation
Principal idea	Molecules of e (only) oscillate uniformly in lamina perpendicular to direction of propagation	Molecules rotate axes of polarisation through a particular angle, either to right or left
Colours distinguished by	Different *rates* of uniform oscillation	Different angles, and also velocities, of rotation ('rotatory dispersion')
Principal writings by Biot	*1814b*, 60–106; *1816c*, *4*, 388–498	*1814b*, 241–263; *1816c*, *4*, 499–538; *1819a* (§ 13.3.2)

Biot varied his theories in their various presentations: Table 746.1 does not carry the details. His most developed theory of mobile polarisation was expressed in a vast and disorganised paper *1814b* in the *Mémoires* which was presented in five parts between November 1812 and May 1813, straddling his 39th birthday. Impressed by Newtons' theory of 'fits' to explain the refraction of molecules of light, he took up Malus's view that the molecule could rotate in the bc-plane while propagating along its normal a (§ 7.4.4) and proposed that it oscillated in the spar between its initial orientation, given by the angular separation between the plane of polarisation of and the axis of the spar, and an orientation given by $(2p - \alpha)$, where p was the angle between $\mathcal{P}$ and the axis of the mica. On this (frail) theoretical reasoning he then gave these expressions for the intensities F_o

and F_e of o and e:

$$\left.\begin{matrix}F_o\\F_e\end{matrix}\right\} = O\,\frac{\cos^2}{\sin^2}\,\alpha + E\,\frac{\cos^2}{\sin^2}\,(2p-\alpha)\,, \tag{746.1}$$

where O and E were the intensities of o and e produced when i was doubly refracted in the mica (*1814b*, 105: I have put 'p' for his 'i'). He manipulated these equations to cover certain cases (pp. 125–137) and to evaluate O and E (pp. 160–167), and managed to obtain some good predictions (see, for example, pp. 168–179), a success which led him to claim that his intensity laws 'do not offer an empirical evaluation, but a theory founded upon the facts' (p. 369).

Biot's paper, and indeed all his writings on optics of this period, were important for the physical aspects of the subject; for he revived Newton's ideas and extended the range of phenomena to which they could be applied, and especially gave prominence to the periodic properties of light. However, his theories were unsatisfactory in certain ways. Some of his principles were only empirical guesses. He put forward different rules for the action of polarisation for crystals of thickness less than and greater than a certain thickness, and so he had to make a sharp distinction between polarisation in 'thin' and 'thick' crystals which of course those defining properties cannot sustain. Again, Laplacian opticians usually maintained an ontology of distinguishing molecules corresponding to different colours, as if the colour spectrum were split into discrete levels; maybe Biot had in mind Malus's idea of 'quantity of molecules' (noted after (744.6)) to indicate the quality of a particular colour.

Further, while working within Laplacian physics, Biot could not fully mathematicise the theories in the way laid down by the Master. For example, while he used the mathematics of short-range forces to analyse the motion of the particles of light (pp. 73–88), he did not so derive his intensity laws, but argued with Malus's law (744.1) (pp. 103–105, 122–125). As he rather lamely put it when summarising his paper (in the third person) to the *Société Philomatique* (*1813a*, 211),

> Mr. Biot tried at first to represent this phenomenon by the same formulae that Malus had already given for the intensities of the beams transmitted by [Iceland spar]. He saw that this law did not apply there any more. He sought the modifications that had to be made, and, by multiplying the observations in all directions [*sic*], he found the two formulae [(746.1)], which represent all the phenomena.

In contrast to his optimism over mobile polarisation, Biot was much more circumspect over his theory of rotatory polarisation in his vast paper *1814b*. 'I only give [it], if you like, as a mathematical conception', he confessed, although he claimed at once that it 'gives perfectly an account of all the phenomena' (p. 256). However, 'I do not claim any more to attribute to it the same certainty as to the theory of oscillations' expressed in his theory of mobile polarisation: 'I even only give it, if you like, as a simple hypothesis' (p. 258). He was not to develop it in more detail until 1818 when, as we shall see in § 13.3.1, it had to contend with Fresnel's wave theory of light.

Another proposal of Biot made at this time was more enduring, although even then his Laplacian terms were to be short-lived. In a widely published note *1815a* summarising a paper *1816d* in the *Mémoires*, he wrote:

> [...] new experiments have made me discover that, in a large number [of crystals], the extraordinary ray is attracted towards the axis instead of being repulsed. So that, in relation to this property, the crystals must be divided into two classes, the one that I call *of attractive double refraction*, the other *of repulsive double refraction*. Iceland spar is part of this latter class; rock crystal is contained in the other. Besides it has seemed that the force, be it attractive, be it repulsive, always emanates from the axis of the crystal and always follows the same laws; so that the formulae of Mr. Laplace always apply.

Laplacian optics was still preserved, including Laplace's derivation of Huygens's law from his proposal (746.7) for the velocity v_e of e in terms of its angular displacement from the optical axis of the crystal; and Biot made somewhat clearer the question of whether forces on the molecules of light were repulsive or attractive. But his 'names involve a hypothesis which has no existence in nature and must altogether be abandoned', in the later judgement of Brewster (*1830a*, 572). Brewster was a molecular theorist in the 1810s, but he came to prefer the neutral terms 'negative' and 'positive' to distinguish the crystals. These names—of which Brewster's have endured—and their relationship to Huygens's ellipsoid, are indicated in Table 746.2.

Table 746.2 Terms and properties for the classification of doubly refracting crystals.

Type of ellipsoid	Refractive indices (from (745.6))	Terms for the crystal	
		Biot's	Brewster's
oblate: $a > b$	$\mu_0 < \mu_e$	repulsive	negative
prolate: $a < b$	$\mu_0 > \mu_e$	attractive	positive

7.4.7 An important row: Arago versus Biot, June 1813

The social consequences of Arago's and Biot's researches concern Arago's initial paper *1812a* reporting the discoveries, and Biot's first paper *1812d* on mobile polarisation (of which *1814b*, just discussed, was the successor). Arago presented his paper in August 1811 (*PV*, *4*, 514); by the end of the month he had a notice *1811a* in the *Moniteur universel*, and between October and December a three-part summary *1811b* was in print with the *Société Philomatique*. Biot's paper arrived in the following June (*PV*, *5*, 59). However, he had kept the *classe* informed of his progress, and in a footnote attached to the title of *1812d* he mentioned that he had been away from Paris when Arago had presented his paper and knew only the published extracts. He then claimed that at the *Bureau des Longitudes* he had invited Arago to establish the chronology of his discoveries and theories by means of sheets lodged with the *classe* at the *Institut*, 'and Mr. Arago himself seemed to acceed to this; but as he has since failed to complete this formality, I believe that to have to recall the extracts cited above as containing all the results which were publically known as belonging to Mr. Arago, at the time when I read my paper, 15 February 1813'. Biot's new date for his paper is interesting, for there is no record of him presenting it to the *classe* that Monday (*PV*, *5*, 166-167)!

Arago *1812a* and Biot *1812d* were published consecutively in the *Mémoires*. Arago was furious when he saw Biot's footnote, and in June 1813 they sent in a joint note from Arcueil asking for a recall of all copies of the *Mémoires* so that the first page could be replaced. The affair came up a few days later at the *classe* (*PV*, *5*, 224-225), where it was witnessed by Duchayla, whom we met in §5.3.2 proving the composition of forces in 1804; he had also helped Arago two years later in his experiments on the velocity of light (§7.4.5, footnote). Duchayla described the *contretemps* in a letter to his friend Poinsot, who happened to be out of town at the time, acting as admissions examiner for the *Université Impériale*:

> I will tell you that your friend and colleague Biot shits himself more and more [*B..t se crotte de plus en plus*]. This is how he has just crowned his scientific piracies. At the time of the justified complaint of Arago, on the matter of the experiments relative to polarisation, Mr. de la P[lace] induced him [Biot] to place a preamble containing a historical summary at the head of his paper, to appease Arago. He read this preamble, Arago[,] if he was not satisfied, had well resolved to remain calm. Well, I declare that on receiving the volume of [*Mémoires* of] the *Institut*, Arago saw no more the preamble, but in its place a note at the bottom of the page, in which he was indicted.

After relating the chronology of events described above, he continued:

> You judge easily the indignation of Arago. He brings the affair to the *Bureau* [*des Longitudes*], where the facts contained in the note relative to the *Bureau* were unanimously declared false. He gave a summary of B..t's piracies before these gentlemen, and his friend[,] his companion of glory Soponis [Poisson], immovable up to then, has at last been forced to abandon him in saying[:] I did not know that at all.[1])

The page of Biot's paper was replaced (in nearly all copies of the *Mémoires*, anyway), but the relationship between Arago and Biot had been permanently damaged.[2]) Personal rows, uncertainly grounded theories; this is an appropriately foreboding note on which to end our survey of the heyday of Laplacian optics. The curtain rises again on the subject in Chapter 13.

7.5 On Newtonian electricity and magnetism

The spectrum of topics discussed in this chapter now broadens still further to take in electricity and magnetism. In addition, a certain change in emphasis occurs, in that every topic described so far was noticeably, even substantially, informed by mathematical analyses. By contrast, electricity and magnetism were only just beginning to receive such treatment. Laplacian physics was to provide stimuli, but the mathematicisation stayed at a more preliminary level, based on laws of inverse-square central action. The state and status of physics at this time was described in § 7.1.1: within that context the general position on electricity and magnetism is summarised in the next sub-section, and then the influence es-

[1]) This letter from Duchayla (and another one) survive in Poinsot's *Nachlass* at *BI*, ms. 953, fols. 162–163; extracts appear below as § 20.4, from my *1981b*, 676–678. Biot's footnote is transcribed on p. 676, and also in my *1981a*, 103. Arago published some manuscripts pertinent to his side of the dispute (on which note *PV*, *5*, 41, 43 for April 1812) in his Works, *10*, 75–84. See also *BL*, *4*, meeting of 16.4.1813, where this 'phenomenon absolutely foreign to the new paper of Mr. Biot' rather found in his favour.

Poinsot's preservation of Duchayla's letters is weird; they are kept in a dossier of letters sent to him at Besançon in connection with his duties as admissions examiner for the *UN*. Why did he keep this dossier in his personal papers instead of handing it to the *UN* administration, as he seems to have done with all the other such files that he accumulated over the years of fulfilling these duties?

[2]) The *AS pochette* for the meeting of 21. 6. 1813 contains a transcription by Delambre of a disavowal by Biot of ill-intent against Arago. Someone has crossed it through, and it does not appear in *PV*, *5*, 224–225.

pecially of Coulomb, and the work of Biot, are recorded. The next section is given over to Poisson's contributions to mathematical electrostatics.

7.5.1 'Electricity' and magnetism in the late 18th century

The first detail to fix is that up to 1800 the word 'electricity' referred almost always to what Ampère was to call 'electrostatics' (§ 14.2.2): the detection and use of electrical charge in the universe and its harness into the manufacture of electrical instruments and machines. During the 18th century the subject developed enormously, from a scrappy collection of casual observations and tentative remarks to an impressive range of experimental routines and theoretical ideas. The Leyden experiment on electrical discharge, the Leyden jar, lightning conductors, galvanic piles and electrophores were well established components of the electrician's cabinet. (However, awkward ontological questions arose; for it was not clear if the fluids produced by lightning, frogs' legs, batteries, and so on, were just similar or in fact identical.) Progress in the area of magnetism was perhaps less spectacular, but there was continuing interest in the behaviour of magnets in various situations, in terrestrial magnetism, and in related topics such as the design of compasses for navigation.

By and large electricity and magnetism were regarded as separate disciplines, with at most only general points of similarity between them. An exception was the St. Petersburg scientist F. Aepinus, who had published an important book *1759a* on both areas, in which he offered as an alternative to theories of 'effluvia', fluids and the like then prevelant, an explanation in the style of Newton's *Principia* postulating that both electricity and magnetism were the result of certain kinds of action at a distance. His theory was a pretty symmetry, for he supposed that repulsion acted between the corpuscles of ordinary matter and the particles of the electric or magnetic fluid, and attraction occurred between corpuscles and particles (arts. 26–31). In sentiments anticipatory of Laplace (§ 7.2.1) he did not claim to know the law relating these forces to the distances of separation of the particles (arts. 30, 112), although he assumed specific forms in a few given cases (arts. 299, 334). These forms enabled him to *quantify* the pertaining analysis; for an important novelty of the book was the role assigned to mathematics. It consisted largely of finding expressions for 'forces' exerted by one body upon another in terms of the quantities of fluid in each one, with linear relations between force, mass and fluid taken to apply.[1])

[1]) Aepinus's book has recently appeared in an English translation *1979a* prefaced by a splendid preface (Home *1979a*) on Aepinus and on 18th-century electricity and magnetism. Among the various histories of electricity, Heilbron's study *1979a* of electricity in the 17th and 18th centuries is not likely to be surpassed in this one; but see also Bellone *1980a*, chs. 5 and 6. On Coulomb's contribution, see Gillmor *1971a*, ch. 6.

As far as electricity was concerned, Aepinus was extending the ideas of Franklin of the previous decade; in particular, he also assumed the existence of only one electric fluid as the transmittive medium of electric forces. However, his adoption of a Newtonian, force-oriented theory replaced the electric atmospheres which Franklin imagined to surround physical bodies; attraction and repulsion played the roles assigned by Franklin respectively to the excess and deficiency of the atmosphere. But soon after the appearance of Aepinus's book there developed a preference among some electricians to speak of two electric fluids, or two electricities, each active in its own right but cancelling out when the carrying bodies were brought together (§ 7.5.5).

7.5.2 The importance of Coulomb and Volta

Enter now Coulomb, who serves as a father figure for this and the next sections of this chapter. From the mid 1770s he began a series of researches in electricity and magnetism which was maintained until the close of his research career 25 years later. He began by winning an *Académie* prize of 1777, his 42nd year, on the design of compass needles, with a paper which was published in the *Savants étrangers* as *1780a.* As is to be described in § 8.4.1, this essay included a study of the effects of friction caused by mounting the needle in pivots or chapes, a led to his introduction of the torsion balance. Then during the 1780s, when a member of the *Académie*, he presented a string of seven important papers on electricity and magnetism, and continued with a few related papers until the early 1800s.

Coulomb's attempts at theorising are distinguished by their Aepinean desire to banish the mysterious effluvia and like substances, and to work with laws of force for electricity and magnetism. However, in contrast to Aepinus's assoziation of electricity and magnetism, he regarded them as separate types of phenomena; further, unlike Aepinus's interest in magnetism, he worked more in electricity, and also tended to prefer a two-fluid theory of electrical action. As an experimenter, he went far beyond Aepinus and indeed many of his predecessors; for example, one purpose to which he applied the torsion balance (§ 8.4.1) was to try to determine the law relating force and distance. The equipment that he designed belonged to a type called 'the electric balance'. By suspending a pith ball from a wire of which the torsional properties were known, fixing another one to a vertical rod so that they were at the same level and charging them both electrostatically, he found from three experiments that the angular displacement of the first ball was more or less inversely proportional to the square of its initial angular separation from the other ones, thus suggesting a respectably Newtonian inverse-square law for repulsive electric forces (Coulomb *1788a*, 572). In the second paper he extended this result to attractive forces by allowing a suspended paper disc to oscillate in the neighbourhood of an oppositely charged metal globe and invoking the theory of its simple harmonic motion (*1788b*, 578-587); and also to magnetic forces by suspending horizontally a long magnetic needle and studying the relationship between the angles of torsional displacement and of equilibrium relative to the (torsion-free) magnetic meridian (pp. 600-611). His later papers included a study of the distribution of charge on conducting bodies; we shall see Poisson give them a mathematical setting in §7.6.

Coulomb's studies to date were publicised in a non-mathematical presentation of Aepinus's theory of electricity and magnetism, and its later developments, by Haüy in a book (*1787a*, 39-46, 176-189): most of the first 38 pages of the part on electricity appeared as a paper *1787b* in the *Observations de physique* (later, the *Journal*). When he wrote his textbook on physics in the early 1800s (§7.1.1) Haüy gave electricity 143 pages in the second edition (*Physics*$_2$ (1806), *1*, 363-445, *2*, 1-58). He dealt in turn with 'electricity produced by friction or by communication' (including the Leyden experiment), 'natural electricity' (including lightning conductors), 'electricity produced by heat' and 'galvanic electricity'. This last topic involved very important recent developments; for in 1800 Volta announced his invention of the pile, a battery which could produce galvanic electricity to an extent not previously known, thus opening up new possibilities for the development of the subject and its applications. As Haüy recalled (*2*, 2, although predating the event by a year and anticipating the self-crowning by three), when Volta came to Paris in the autumn of 1801, among other reasons to demonstrate his new instrument, Bonaparte, *membre de la classe* 'and his Majesty the Emperor and King, after the reading of a Paper where Volta showed his discovery, proposed to award a gold medal to this physicist' (compare *PV*, *2*, 425). The event took place a month later at a meeting in December, at which Biot read a report published later in the 'Histoire' department of the *Mémoires* as *1804e* (*PV*, *2*, 440). Prizes for galvanism were proposed regularly thereafter, sometimes with a large sum of money attached; however, as we saw in §7.3.1, they were not always awarded.

7.5.3 Biot on aspects of electrostatics, 1800-1801

Biot once again, the every-ready battery for the problems of the day: fresh from difference and mixed equations (§4.4) and soon to tackle the velocity of sound (§7.2.5), the passage of meteors (§6.4.4) and the diffusion of heat (§9.2.3), in August 1800 he sent in to the *classe*, as an *associé*, a report of his experiments 'on the motion of the galvanic fluid' (*PV*, *2*, 382), publishing part quickly as *1800b* with the *Société Philomatique* and rehearsing the full story *1801b* in the *Journal de physique.* The interest of these studies, mostly made with small piles, lies partly in its disclosure of Laplace's interest in electricity, which is now little known. This concern goes back to Laplace's and Lavoisier's finding *1784b* that electricity was emitted with change of state between liquid and gas; Laplace had also written the *Académie* report *1787b* on Haüy's book *1787a* on Aepinus's theory (where it is printed after the 'preface'). Now, according to Biot *1800b*, Laplace's own recent experiments at the *Ecole de Médécine* led him to think that the larger were the plates of the pile, the greater was the rate of emission of galvanic fluid at a slower velocity. Biot sought to explore Laplace's hypothesis, and found some confirmation for it, along with other properties of the fluid. In a way not unusual for its time, he distinguished this fluid from other known sources of 'electricity'.

This study was entirely experimental; soon afterwards Biot exhibited his mathematical debt to Laplace in a short paper *1801a* placed with the *Société Philomatique*, in which assumed the inverse square law for electrostatics, and applied potential theory to analyse the distribution of the electric fluid inside an ellipsoid E of revolution. He argued that repulsion would thrust all interior molecules to the boundary, to form an 'infinitely thin layer' with interior surface I forming an ellipsoid confocal with E when equilibrium was achieved. Taking E in the form

$$x^2 + m(y^2 + z^2) = K^2 \tag{753.1}$$

and the forces at any point $H\,(a, b, c)$ of I to be (P, Q, R), then equilibrium was sustained if

$$Pda + Qdb + Rdc = 0\,. \tag{753.2}$$

For the proof of the confocality of E and I he invoked Laplace's treatment in the *Mécanique céleste* of potentials on spheroids (*2* (1799), 51-62) to find formulae for (P, Q, R) as the difference between the actions on H of E and of I which reduced, on the assumption of infinite thinness, to the cases

$$P = 4\pi\varrho\omega a\,, \quad Q = 4\pi\varrho\omega mb \quad \text{and} \quad R = 4\pi\varrho\omega mc\,, \tag{753.3}$$

where ϱ was the assumedly constant density of the fluid, and ω a small quantity associated with its thinness. Now (753.3) satisfied (753.2), since the equation involved was a differential of (753.1) at the point H:

$$ada + m(bdb + cdc) = 0\,. \tag{753.4}$$

Biot's analysis is corrigible enough. One problem was that the variation in density of the fluid meant that its inner and outer surface of equilibrium were not confocal ellipsoids, so that the equilibrium condition was difficult to interpret. He was aware of problems here, but did not resolve them; and he seemed to be untroubled by the constancy of ω. His friend Poisson was to take up these questions in greater detail a decade later, as we shall see in §7.6.2.

7.5.4 Biot and Humboldt on terrestrial magnetism, 1805

In his account of terrestrial magnetism, Haüy (*Physics*$_2$ (1806), *2*, 104-106) reported on the recent studies by—of course—Biot. These included the invariance of magnetic force with altitude, found during his balloon ascent of 1804 with Gay-Lussac (§7.2.4) and confirmed in Gay-Lussac's own ascent (*1804a*, 75-78). Another collaboration of note occured soon afterwards with Alexander von Humboldt, at 35 years old five year Biot's senior, just back from an expedition of the Americas with the French botanist A. Bonpland. Humboldt had measured magnetic dip: Biot, adjoining those results to his own measurements in the French Alps (perhaps made during his trips as admissions examiner for the *Ecole Polytechnique*), sent in to the *classe* in December 1805 an important joint paper on the variation of terrestrial magnetism with latitude (*PV*, *3*, 168). A short summary (Humboldt and Biot *1804a*) soon appeared with the *Société Philomatique*, quickly followed by the full story *1804b* in the *Journal de physique*, including a chart of isodynamic zones. Apparently Laplace came in again, suggesting the following calculation of the angle of dip. Coulomb had shown that magnetic forces obeyed the inverse square law, and so had justified his comparison of the oscillations of a magnet with the motion of a simple pendulum, and from the usual formula $(652.1)_1$ for pendulum oscillation it followed that the magnetic force F, which was analogous to gravitational attraction, was proportional to the square of the rate of oscillation of the needle (Coulomb *1788b*, 584). Thus if the angle of dip were I, and M oscillations were observed in a given time in the magnetic meridian and P in the perpendicular plane, then

$$\sin I = (F \sin I)/F = P^2/M^2 \tag{754.1}$$

(Humboldt and Biot *1804b*, 431-432).

The most significant calculation in the paper was that of the angle of dip of a magnet. Figure 754.1 shows a section of the circular earth of radius r, through the centre C and by the magnetic axis Cx; the 'austral' (south) and 'boreal' (north) poles were located at A and B respectively. A 'molecule of austral fluid' M, located at magnetic latitude v (or $\tau - u$) relative to the magnetic equator Cy, executed a force represented as (X, Y) in the Cartesian frame Cxy, which was the

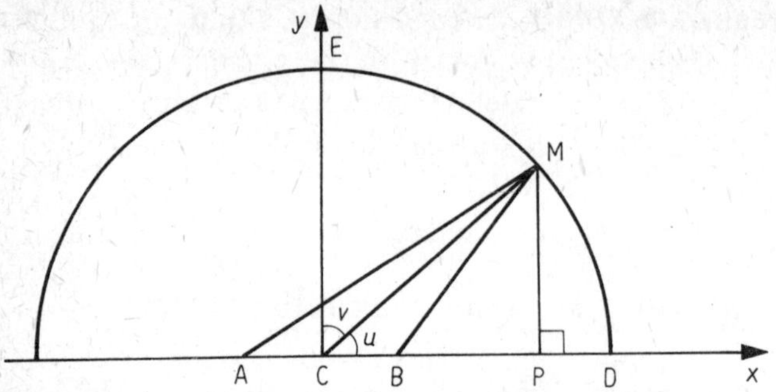

Figure 754.1 Calculation of magnetic dip (Humboldt and Biot *1804b*, 440)

resultant of attraction of M to B and repulsion of M from A; thus it was given, via the inverse square law, by

$$X = \frac{F(x-a)}{D^3} - \frac{F(x+a)}{D'^3}, \quad Y = \frac{Fy}{D^3} - \frac{Fy}{D'^3}, \tag{754.2}$$

where F was the magnetic force at unit distance. In addition,

$$\left.\begin{matrix} D^2 \\ D'^2 \end{matrix}\right\} = r^2 \pm 2ax + a^2; \quad \text{and a constant } K \text{ was defined as} \quad K := a/r. \tag{754.3}$$

Then by straightforward manipulation and trigonometry, the angle of inclination β of M to Cx was given by

$$\tan\beta = \frac{Y}{X} = \frac{\sin u}{\cos u - (K(D'^3 + D^3)/(D'^3 - D^3))}. \tag{754.4}$$

Experimental studies suggested that $K \triangleq 0$; but then the K-term in (754.4), with D and D' replaced by (754.3), took the form 0/0. However, l'Hôpital's rule came the rescue, and furnished the value $\frac{1}{3}\sec u$. Hence (754.4) became

$$\tan\beta = \frac{\sin u}{\cos u - \frac{1}{3}\sec u} = \frac{\sin 2u}{\cos 2u + \frac{1}{3}} \tag{754.5}$$

(pp. 440–446). Thus the angle of dip I, defined in terms of the latitude v (that is, $\angle$ECM) of M by

$$I = \tau - (\beta - u) = \pi - \beta - v, \tag{754.6}$$

was given from (754.5) by

$$\tan\beta = -\tan(I + v) = \frac{\sin 2u}{\cos 2u + \frac{1}{3}}. \quad \therefore \tan(I + v) = \frac{\sin 2v}{\cos 2v - \frac{1}{3}} \tag{754.7}$$

(Humboldt and Biot *1804a*, 242).

$(754.7)_2$ is not in its simplest form. Bowditch *1818a* showed that it reduced to

$$\tan I = 2\tan v, \qquad (754.8)$$

a decade after Krafft *1809a* had told the St. Petersburg Academy that he guessed that $\tan I$ and $\tan v$ were in some proportion α, used Biot's and Humboldt's data to find 2.12 as a mean value for α, and decided to take $\alpha = 2$ anyway! Mathematics can be useful sometimes.

7.5.5 Hachette's interest in electrostatics in the 1800s

Then in his early thirties, Hachette also took up a strong interest in electrostatics, in the early 1800s. He worked with C. B. Desormes, a former student of his in the founding year of the *Ecole Polytechnique* and by profession largely a chemist (we shall meet Desormes again in § 12.2.3 in connection with some later studies of heat). Their writings dealt mainly with developing a type of electric battery called a 'doubler', in which discs were brought in and out of contact (by some means or another) with an electrometer or similar instrument, in a cycle of moves which roughly doubled the charge on some or all of the discs after each cycle. Their two main papers appeared in the *Annales de chimie: 1802a* described, in a lecture of December 1802 to the *Société Philomatique*, a version produced for use at the *Ecole Polytechnique*; and *1804a*, on their special design, was submitted to the *classe* in November 1803 (*PV*, *3*, 25) on their behalf by Lacroix (of all people), who praised it in a report *1804a* of the following January.[1]) In addition, Hachette *1805b* described in his *Correspondance sur l'Ecole Polytechnique* experiments in which they floated an electric pile on a small raft in calm water and found that it did *not* take up any particular position, unlike the orientation of a lightly magnetised steel bar towards the terrestrial poles in the same circumstances.

As was recorded in § 7.1.1, Hachette expanded his short book *1805a* into his 'Programmes' on parts of physics, for use at the *Ecole Polytechnique.* The chapter on electricity was the longest in the book (*1809a*, 133-210), and there were some pictures on making lightning conductors (pp. 228-232); in his survey of the instruments he mentioned their new doubler (pp. 164-166). As regards theories of electricity, he stated both the one- and two-fluid theories (pp. 133-140) and kept

[1]) The *SP* published the bulk of Lacroix's report: Hachette quoted a passage in his *1832a*, 96, and the manuscript is kept at *BI*, ms. 2400. Hachette's role in buying equipment for the *EP* is revealed in the file VII 2c2, box 1.

a notable balance (as it were) between them, using each in different places; and presumably it was his insistence that the two-fluid theory was removed from the published syllabus of the physics course for the academic year 1808–1809 (*EP, Col, 4*, meeting of 15 September 1808). His view was rather unusual for the French, who then were usually two-fluidists. For example, Aepinus had an argument that if there was only one fluid and two charged bodies were in equilibrium, then the fluid-fluid repulsion and the two fluid-body attractions had to be balanced by a body-body *repulsion* (*1759a*, arts. 24–31); Haüy saw this conclusion as a defect in the one-fluid theory (*Physics*$_2$ (1806), *1*, 397–399).

As with many figures of the time, Hachette was less interested in magnetism. He gave it only ten pages in his *Programmes* (*1809a*, 211–220), although he added in the design of a marine compass (pp. 233–234). Haüy was rather more expansive, with 77 pages in his textbook, dealing with terrestrial magnetism and the magnetic effects found in mines as well as the basic principles of attraction and repulsion (*Physics*$_2$ (1806), *2*, 58–134). As regards the character of magnetism, he gave his vote on pp. 58–60 once again to theories oriented around intermolecular forces, and in particular to Coulomb's recent conversion of Aepinus's view (*1759a*, arts. 3, 84) that the magnetic fluid was found throughout a body into the theory that each molecule of the body was itself a small magnet (Coulomb's seventh paper, *1793a*, art. 30) and that there were two magnetic fluids (*1801a*, art. 2).

7.6 Poisson's inauguration of mathematical electrostatics, 1812–1813

7.6.1 The motivations to Poisson's studies

It is clear that the mathematical aspects of these studies, when present at all, were modest in scope and purpose. The first major exercise was produced in the early 1810s by Poisson, as one might expect. The circumstance are quite curious, and seem to relate to his continuing non-membership of the *classe*. We noted in § 7.3.1 his pantomime with Germain over the prize problem of December 1812 for elastic surfaces; at the same public meeting another problem was set, on 'the manner in which electricity is distributed at the surface of electric bodies' (*PV, 4*, 560–563). Doubtless it also was directed towards Poisson's mathematical skills; but the outcome was unexpected.

In anticipation of his performance over the other prize, Poisson submitted a paper on electricity to the *classe* on 9 March 1812 (*PV*, *5*, 31); Lagrange, Haüy and Biot were asked to report. By the next meeting Biot had prepared it, but it was overtaken by other events; for another subject discussed at that meeting was the election of a successor to Malus. As was described in § 7.3.1, Poisson was elected a week later (p. 40), and so could now *present* this paper to the *classe*; and for good measure he sent in a supplement the following August (p. 79). A second paper arrived in September 1813 (p. 242).

Having now obtained the *entrée* desired for so long, Poisson did all the correct Parisian things with these two papers. He published the opening pages of each one as summaries, *1812f* and *1813h* respectively, which appeared in the *Journal de physique* and the *Correspondance sur l'Ecole Polytechnique:* the *Société Philomatique* took freshly written summaries *1812c* and *1813k*. For once even the full papers appeared quickly in the *Mémoires*, as *1812b* and *1814g* (at 92 and 112 pages respectively); Delambre gave the first one generous attention in his notice of the work of the *classe* for 1812 (*1814a*, v-xi), and also covered the second one in his report on the activities of the following year (*1814b*, 262-266).

Although Poisson's first paper was presented in two parts (the date of the first part being misstated as '9 May' on the title page), the published version seems to be an integrated text. In its opening section, which also served as its summary paper, he claimed—in a rather parochial way—that two-fluid theories were generally accepted (he used the 18th-century terms 'vitreous' and 'resinous' to name them); but in his ensuing mathematics he did not attempt to reflect the structure (§ 1.3.1) and indeed often spoke of 'the electric fluid' in general terms, and of the 'free fluid' to refer to the excess of one fluid over the other one at a given point. Assuming inverse-square laws for electrostatics, his purpose was to use them in an analysis of the pertinent potential theory (to use the modern term, which he did not employ). His principal inspiration for both theory and data was Coulomb's fifth paper, *1789a*, where Coulomb had used mathematics only as far as calculating the ratios of density of the electric fluid in terms of the inverse-square electrical 'action'.

After presenting the general principles of this theory in the first paper, Poisson concentrated on two special cases of distribution of the fluid: in a spheroid, especially one which was nearly spherical; and on two spheres in the presence of each other, including making contact, but also at a great distance apart. My summary account follows the same order, with the theory of the spheroid treated in the next sub-section and the two-spheres problem occupying § 7.6.4. The second paper continued work on this latter question in various ways, as we shall see in § 7.6.5. Both papers are rich in details, concerning the use of potential theory, the solutions to equations, and the comparisons with Coulomb's data; indeed, they constitute one of Poisson's finest achievements, and deserve a fuller study than is possible here.[1])

7.6.2 The first paper, 1812: the potential theory of the 'electrical field'

For his first problem Poisson set up spherical polar coordinates (r, ϑ, ω) at the centre C of the spheroid $\mathscr{S}$, and invoked Laplace's presentation of potential theory from the *Mécanique céleste* (§ 5.5.2) to state the potential V at a point $O(x, \vartheta, \omega)$ of the surface point $O'(r', \Theta', \omega')$ of $\mathscr{S}$:

$$V = \int_0^{2\pi} \int_{-1}^{+1} Xy'r'^2 d\mu' d\omega', \tag{762.1}$$

where $\mu := \cos\Theta,\ \mu' := \cos\Theta',$

$$X = \left[r'^2 - 2r'x\left(\mu\mu' + \sqrt{1-\mu^2}\,\sqrt{1-\mu'^2}\,\cos(\omega - \omega')\right) + x^2\right]^{-1/2}, \tag{762.2}$$

and y' was the thickness of the electric fluid at O' (*1812b*, 14–17). y' was his principal target for study; it was his way of imagining the presence of the fluid, although he readily admitted as 'the same thing, the intensity of electricity' (*1814g*, 165). In order to allow for its capacity both to attract and repel O' (or equivalently, to allow for the presence of both fluids), he let y' take positive and negative values (*1812b*, 18). While very small, $y' \neq 0$, and so he stressed the need to distinguish the exterior and interior surfaces of the fluid, proposing that the forces acting at O' from each be found by expanding X in convergent power series (552.4) of x/r' and r'/x respectively and evaluating V_x, V_μ and V_ω at $r' = x$ in each case (p. 20).

Poisson applied his theory to a nearly spherical spheroid, defined on p. 23 by

$$r = a(1 + \alpha t(\mu, \omega)) \quad \text{and} \quad y = b(1 + \alpha z(\mu, \omega)), \quad \alpha \text{ small, } a, b \text{ constants,} \tag{762.3}$$

and t and z unknown functions, and found fault in Biot's analysis (§ 7.5.3). For on assuming an expansion of t in spherical harmonics $\{T_j\}$ and using in (762.1),

[1]) I am grateful to John Clark (Townsville, Australia) for sending me drafts of his study of Poisson's electrostatics and its background. The historical literature on the papers consists of rather brief summaries.

he found the following expression, to $O(\alpha^2)$, for V_x:

$$-V_x = (4\pi b/(1+2\alpha t))\left(1+\alpha\sum_{j=1}^{\infty}(j+1)\,T_j\right) \tag{762.4}$$

$$= (4\pi b/(1+2\alpha t))\,(1+\alpha(2t+z)) = 4\pi b(1+\alpha z) + O(\alpha^2). \tag{762.5}$$

In plain words, at a point of the surface of a spheroid the radial 'electric' repulsion was proportional to the thickness of the fluid there. Since y was proportional to r, by (762.3), it followed that repulsion at a pole and the equator of an ellipsoid were in the ratio of the corresponding axes. 'This last result is precisely contrary to that which is announced in the old *Bulletin* of the *Société Philomatique*, n° 51, page 23', he added, without mentioning Biot's name (p. 27); and he pointed to the need to respect ω as truly variable, which the author in the *Bulletin* had not done (§ 7.5.3).

'[W]hat one calls a *level surface*, consists in the function which we have designated by V', Poisson pointed out, using the standard term (§ 5.5.2); as the surface of the spheroid was such a surface, the resultant force at any of its points lay along the normal there *(1812b*, 29). Among other properties was one 'which has been communicated to us by Mr. Laplace': 'a purely synthetic demonstration' (p. 5), and pretty general, that the repulsive electric force was proportional to its thickness. In my Figure 762.1 an infinitely thin layer $\mathscr{L}$ of fluid was cut by a line Aa normal at A to its exterior surface BAE to strike the interior surface at a. The

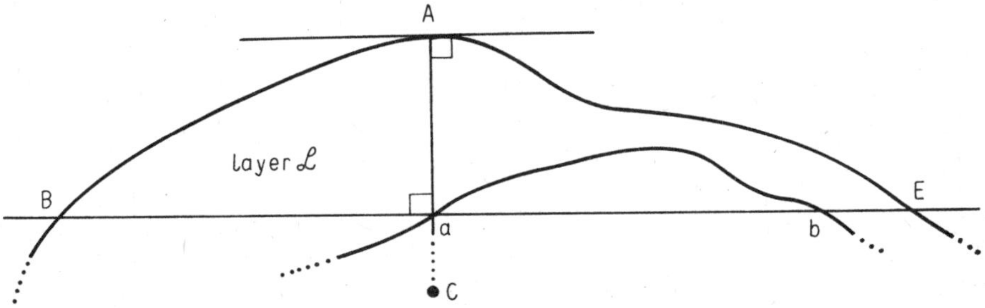

Figure 762.1 Action of a layer of electric fluid

'actions' of $\mathscr{L}$ at A and a along Aa were R and R' respectively; and when the plane BabE normal to Aa cut $\mathscr{L}$ into a small and a large segment (the small one shown as aBAEb, the large one comprising the rest), then their actions at a along Aa were given by s and S respectively. Now taking all actions as attractions, 'to fix the ideas', a was pulled into $\mathscr{L}$ by the *difference* between the large and the small segments, while A was similarly drawn by their *sum*. Thus

$$R' = S - s \quad \text{and} \quad R = S + s + O(y^2), \quad \text{so that} \quad R - R' = 2s. \tag{762.6}$$

To find s, take any point C along Aa as shown, and set up spheres there of radii Ca and CA and thus create a spherical shell $\mathscr{S}$ of thickness y. From $(762.6)_3$, its action s' at a for this *special* case, was given, via

$$R = 4\pi y \quad \text{and} \quad R' = 0, \quad \text{by} \quad 4\pi y = 2s'. \tag{762.7}$$

(Maybe $(762.7)_1$ led Poisson to see the need for 'Poisson's equation' (651.8) for interior points to a surface a year later.) Further, since s' was independent of the distance AC, it equalled that of the original layer $\mathscr{L}$. Thus

$$s = s'; \quad \text{and so from } (762.6_2, 7_2), \quad R - R' = 4\pi y. \tag{762.8}$$

Further, the action T in any direction in the tangent plane at A was constant along Aa 'on observing that in this direction the action of the little segment can be supposed null'; and in particular, when the fluid was in equilibrium,

$$T = 0; \quad \text{and also} \quad R' = 0, \quad \text{so that from } (762.8)_2, \quad R = 4\pi y, \tag{762.9}$$

the promised proportionality between R and y (pp. 31-34).

This proof is interesting in various ways. It smacks of Laplace's 'it is easy to see' mode, especially at the supposed generality of Figure 762.1 (which Poisson did not provide, but which seems to reflect the thinking), the discard term in $(762.6)_2$, the equality $(762.8)_1$ of s and s', and the constancy of T. It attracted the attention of the 25-year-old Cauchy, who soon came up with a superior proof—and, typically of him, of a more general result.

7.6.3 Cauchy's reproof of Laplace's theorem on potentials, 1815

Cauchy published his short paper *1815d* with the *Société Philomatique*; at its head he stated that it had been submitted to the *classe* in March 1815, although no record of this has survived. After noting Poisson's 'synthetic demonstration', he showed 'how one can deduce it from the general formulae of attraction'. Essentially the proof hinged on considering a quartet of neighbouring points: R(X, Y, Z) on the surface ('A' in Figure 762.1); O(x, y, z), also on the surface; and M(x_1, y_1, z_1) and N(x_2, y_2, z_2) along the normal Ra, respectively inside and outside the fluid. He used the Euler notations 'p' and 'q' for z_x and z_y; they took the values P and Q at R, so that the tangent plane at R was given by

$$z - Z = P(x - X) + Q(y - Y). \tag{763.1}$$

$$\therefore x_1 - X + P(z_1 - Z) = 0 \quad \text{and} \quad y_1 - Y + Q(z_1 - Z) = 0. \tag{763.2}$$

He then calculated the component A of attraction of the fluid at M along a direction set at *arbitrary* angle Θ to Ra. For convenience he took the z-axis in this direction, and found (eq. (5))

$$A = \iint_{\mathscr{S}} e(z - z_1)\,(1 + p^2 + q^2)^{1/2}/\mathrm{OM}^3 dx\, dy, \tag{763.3}$$

where $\mathscr{S}$ was the surface and e the 'attractive force' at O; the same form applied for the attraction B at exterior point N, with ON replacing OM and z_2 for z_1.

To evaluate A, Cauchy took 'a very small quantity' α and set

$$Z - z =: \alpha, \quad \text{so that} \quad Y - y_1 = -Q\alpha \quad \text{and} \quad x - x_1 = -P\alpha \tag{763.4}$$

from (763.2). For O he put

$$x - X = \alpha x', \quad y - Y = \alpha y' \quad \text{and} \quad z - Z = \alpha z', \tag{763.5}$$

where x', y' and z' were not necessarily small. Then 'very nearly' (that is, taking O to lie on the tangent plane at R),

$$z' = Px' + Qy'; \quad \text{and also} \quad (1 + p^2 + q^2)^{1/2} = \sec\Theta \quad \text{and} \quad e = E \tag{763.6}$$

(the value of e at R). Hence the dominant part A' of A, given by the portion of $\mathscr{S}$ close to R, was given by eq. (10):

$$A' = -E\sec\Theta \int_{-\infty}^{\infty}\int_{-\infty}^{\infty} (Px' + Qy' + 1)/\varrho^3 dx'dy', \tag{763.7}$$

$$\text{where} \quad \varrho := ((x' - P)^2 + (y' - Q)^2 + (Px' + Qy' + 1)^2)^{1/2}. \tag{763.8}$$

Obviously the oddness of $(Px + Qy')$ reduced to zero its contribution to A': as for the rest, a change of variable produced

$$A' = -E\cos\Theta \int_{-\infty}^{\infty}\int_{-\infty}^{\infty} x'[\sec^2\Theta + ((1 + P)^2 + 2PQt + (1 + Q)^2t^2)\,x'^2]^{-3/2} dx'dt, \tag{763.9}$$

$$\text{where} \quad t := y'/x'. \therefore A' = \mp 2\pi E\cos\Theta \tag{763.10}$$

after evaluation. The choice of sign was determined by $Z > z_1$ or $Z < z_1$ respectively; and for each case the opposite sign obtained for the attraction B' at the exterior point N. Hence

$$A' - B' = \pm 4\pi E\cos\Theta. \tag{763.11}$$

He ended by claiming that (763.11) applied also to $(A - B)$: no demonstration was supplied, but presumably he appealed to the rapid increase in the size of ϱ^3 when O moved further away from R.

7.6.4 Poisson's first paper: mathematics and data for two spheres

Poisson's study of a spheroid had to remain a theoretical exercise, as no data were available; but for the problem of distribution of electric fluid on two spheres, which occupied the rest of his first paper, he had Coulomb's data (§ 7.5.2) on which to call. Arranging his spherical polar coordinates so that the polar line joined the centres of the spheres, the distribution of the fluid was then cylindrically symmetrical, so that V was independent of ω. Hence the potential theory simplified to two assumed expansions. Firstly, the thickness y of the fluid was given by

$$y = \sum_{r=0}^{\infty} A_j P_j(\mu), \qquad (764.1)$$

where the $\{P_j\}$ were generated as in (552.5):

$$(1 - 2\mu x + x^2)^{-1/2} = \sum_{r=0}^{\infty} P_j(\mu)\, x^j, \text{ with } ((1-\mu^2)\, P'_n)' + n(n+1)\, P_n = 0\,. \qquad (764.2)$$

Secondly, in terms of the function

$$\varphi(u, x) := \sum_{j=0}^{\infty} A_j P_j(\mu)\, x^j/(2j+1)\,, \qquad (764.3)$$

the potential V at $O(x, \mu, \omega)$ of the sphere $\mathscr{S}$, with centre C and radius a, was expressed as

$$\begin{aligned} V &= 4\pi a \varphi(\mu, x/a) \quad \text{if} \quad x \le a, \quad \text{and} \\ V &= (4\pi x^2/a)\, \varphi(\mu, a/x) \text{ if } \quad x \ge a \end{aligned} \qquad (764.4)$$

(*1812b*, 36–39). The two expansions (764.2_1–3) were connected by the relation

$$\begin{aligned} y(\mu) &= (2x\varphi_x(\mu, x) + \varphi(\mu, x))|_{x=1}\,; \\ \therefore\ y(1) &= (2xf'(x) + f(x))|_{x=1}, \quad \text{where} \quad f(x) := \varphi(1, x) \end{aligned} \qquad (764.5)$$

(pp. 45,39). Finally in these preliminaries, the excess E on $\mathscr{S}$ of one type of electric fluid over the other type was given, by means of (762.1) and (764.1), as

$$E := \int_0^{2\pi} \int_{-1}^{+1} y a^2 d\mu\, d\omega = 4\pi a^2 A_0 = 4\pi a^2 \varphi(1, x)|_{x=0} = 4\pi a^2 f(0) \qquad (764.6)$$

when (764.3, 5_3) were employed; thus A_0 was the (mean) thickness of the fluid over $\mathcal{S}$ (pp. 47–48). All these equations and definitions applied also to the second sphere $\mathcal{S}_1$, with centre C_1 and radius b: its parameters were, in turn (x_1, μ_1, ω_1), $\Phi(\mu_1, x)$, $\{B_j Q_j(\mu_1)\}$, V_1, Z and $F(x_1)$.

An important principle for Poisson stated that, in equilibrium 'if we only want that the action of two fluid layers be null for all points contained within the interior of one and the other sphere', constant potentials $4\pi h$ within $\mathcal{S}$ and $4\pi g$ within $\mathcal{S}_1$ had to be assumed. (764.4) then yielded the equations

$$a\varphi(\mu, x/a) + (b^2/x_1)\,\Phi(\mu_1, b/x_1) = h, \tag{764.7}$$

$$(a^2/x)\,\varphi(\mu, a/x) + b\Phi(\mu_1, x_1/b) = g \tag{764.8}$$

(p. 42); and they played a role in an important task for Poisson, namely to check that Laplace's result $(762.8)_2$ was verified for these two spheres. When O was outside both $\mathcal{S}$ and $\mathcal{S}_1$, its potential U was given by $(764.4)_2$ as

$$\begin{aligned} &U = V + V_1, \quad \text{where} \\ &V = (4\pi a^2/x)\,\varphi(\mu, a/x) \quad \text{and } V_1 = (4\pi b^2/x_1)\,\Phi(\mu_1, b/x_1) \end{aligned} \tag{764.9}$$

(p. 42). Now the force acting at O could be regarded as composed of a radial component R along OC and a component T perpendicular to R and lying in OCC_1, the senses of the directions depending on the types of fluid involved. From potential theory, these components were given by

$$R = -U_x = -V_x - (V_1)_{x_1}(x_1)_x - (V_1)_{\mu_1}(\mu_1)_x, \tag{764.10}$$

$$\begin{aligned} T &= -U_\vartheta/x = \sqrt{1-\mu^2}\,U_\mu/x \\ &= (\sqrt{1-\mu^2}/x)\,[V_\mu + (V_1)_{x_1}(x_1)_\mu + (V_1)_{\mu_1}(\mu_1)_\mu] \end{aligned} \tag{764.11}$$

(p. 48), where the relations between x, μ, x_1 and μ_1 were provided by trigonometry:

$$CC_1 =: c = x\mu + x_1\mu_1 \quad \text{and} \quad x\sqrt{1-\mu^2} = x_1\sqrt{1-\mu_1^2} \tag{764.12}$$

(p. 41). To effect the verification, he used $(764.9)_3$ to express (764.7) in the form

$$4\pi a\varphi(\mu, \chi/\alpha) + V_1 = 4\pi h, \tag{764.13}$$

and differentiated it partially with respect to x and μ. Setting $x = a$ in the resulting equations to eliminate $(V_1)_{x_1}$ and $(V_1)_{\mu_1}$ yielded from (764.10–11)

$$\begin{aligned} &R = -V_x + 4\pi a\varphi_x(\mu, x/a)|_{x=a} \quad \text{and} \\ &T = (\sqrt{1-\mu^2}/x)\,[V_\mu - 4\pi a\varphi_x(\mu, x/a)]|_{x=a}. \end{aligned} \tag{764.14}$$

To relate these results to the U of $(764.9)_2$, he noted that when $x = a$, $\varphi_\mu(\mu, a/x) = \varphi_\mu(\mu, x/a)$ and $\varphi_x(\mu, a/x) = -\varphi_x(\mu, x/a)$, (764.15)

so that

$$\begin{aligned} V_\mu &= 4\pi a \varphi_\mu(\mu, x/a)|_{x=a} \quad \text{and} \\ V_x &= -4\pi\varphi(\mu, 1) - 4\pi a \varphi_x(\mu, x/a)|_{x=a}. \end{aligned} \tag{764.16}$$

$$\therefore T = 0 \quad \text{and} \quad R = 8\pi a \varphi_x(\mu, x/a) + 4\pi\varphi(\mu, 1) = 4\pi y \tag{764.17}$$

from $(764.5)_1$. Hence the force was radial; and since interior forces were zero (by the assumption underlying (764.7)), Laplace's result $(762.8)_2$ on the difference in potentials was validated (pp. 48–52).

Poisson's first attempt to produce a solution to the equations dealt with the equilibrium potentials at points lying along CC_1. In this case $(764.12)_1$ became

$$c = x + x_1, \quad \text{with} \quad -a \le x \le a \quad \text{and} \quad -b \le x_1 \le b, \tag{764.18}$$

and (764.7–8) reduced, with the help of the definition of f in $(764.5)_3$, to

$$\begin{aligned} af\left(\frac{x}{a}\right) + \frac{b^2}{c-x} F\left(\frac{b}{c-x}\right) &= h \quad \text{and} \\ \frac{a^2}{c-x_1} f\left(\frac{x}{c-x_1}\right) + bF\left(\frac{b}{x_1}\right) &= g \end{aligned} \tag{764.19}$$

(pp. 41–43). To solve for f and F he took the special case of spheres in contact, so that, using $(764.18)_1$,

$$a = 1 \text{ (for simplicity) and } c = 1 + b. \tag{764.20}$$

The solution of (764.19) was a brilliant piece of work. Eliminating F from the two equations, and using (764.20), produced for f

$$f(x) - \frac{b}{b + (1+b)(1-x)} f\left(\frac{1+b-x}{b+(1+b)(1-x)}\right) = h - \frac{gb}{1+b-x}. \tag{764.21}$$

This functional equation far surpassed any studied hitherto for difficulty (compare § 4.3); Poisson was silent on the details of his means of solution. He solved the reduced equation (with $(764.21)_2$ set to zero), I imagine, by noticing that it could be expressed as

$$\begin{aligned} &(1-x) f(x) = (1-k) f(k), \quad \text{where} \\ &k := (1+b-x)/(b+(1+b)(1-x)), \end{aligned} \tag{764.22}$$

and so took the solution

$$f(x) = P(x)/(1-x), \quad \text{where } P \text{ satisfied} \quad P(x) = P(k). \tag{764.23}$$

Now he added in the first term, h, of $(764.21)_2$ and tried as solution

$$f(x) = \frac{h'}{1-x} \int_0^1 \frac{t^{\frac{m}{1-x}+n} - 1}{1-t}\, dt, \tag{764.24}$$

where h', m and n were constants to be determined. Whence came this form? It belongs to the world of elliptic and related integrals—indeed, it is similar to a beta function—but I know of no result of the time which would have suggested (764.24) as a possibility. Nevertheless, it worked (modulo some unasked questions about multi-valuedness, and singularity at $t = 1$): substitution into (764.21) (with the last term dropped) produced

$$n = 1, \quad m = b/(1+b) \quad \text{and} \quad h' = bh/(1+b). \tag{764.25}$$

He found a similar form to solve (764.21) for the second term only. Hence 'for the complete integral of the equation' (764.21),

$$f(x) = \frac{P(x)}{1-x} + bw\left(h \int_0^1 \frac{t^{bw-1} - 1}{1-t}\, dt - g \int_0^1 \frac{t^{bw-m} - 1}{1-t}\, dt\right) \tag{764.26}$$

(pp. 52–54), where I have introduced, as an auxiliary variable,

$$w := ((1+b)(1-x))^{-1}. \tag{764.27}$$

Poisson then sought various consequences for electrostatics. He showed that putting $x = 1$ and $x = b$ in (764.7–8) gave $g = h$ (as was to be expected), and that the first term of the solution should be suppressed for taking an infinite value at $x = 1$ (pp. 54–55). (764.26) then reduced to

$$f(x) = bhw \int_0^1 \frac{t^{bxw}\left(t^{-\frac{1}{1+b}} - 1\right)}{1-t}\, dt; \tag{764.28}$$

a similar form obtained for F by substituting (764.27) into $(764.18)_1$:

$$F(x) = \frac{hw}{b} \int_0^1 \frac{t^{xw}\left(t^{-\frac{b}{1+b}} - 1\right)}{1-t}\, dt \tag{764.29}$$

(pp. 54–55). To find the thickness y of fluid he expanded the integrand of (764.26) in a power series of $(1-t)$, obtained expressions for $f(x)$ and $f'(x)$, and placed them in $(764.5)_2$ to show that $y = 0$ when $x = 0$. Thus, 'whatever be the ratio of the radii of the spheres which touch, there is never any electricity at contact' (p. 57). As he already noted in his introduction (p. 8), this finding was in line with Coulomb's report in the fifth paper (*1789a*, arts. 14–32 *passim*).

Next, Poisson considered the consequences of separating $\mathscr{S}$ and $\mathscr{S}_1$ after this contact. From (764.6$_5$, 26), the thickness A of fluid on $\mathscr{S}$ at equilibrium was given by

$$A = f(0) = \frac{bh}{1+b}\int_0^1 \frac{t^{-\frac{b}{1+b}} - 1}{1-t}\,dt; \tag{764.30}$$

a similar expression obtained from (764.29) for the thickness B on $\mathscr{S}_1$:

$$B = F(0) = \frac{h}{b(1+b)}\int_0^1 \frac{t^{-\frac{b}{1+b}} - 1}{1-t}\,dt. \tag{764.31}$$

Now Coulomb had found data for the value of B/A (*1789a*, art. 11); and to compare them with his theory Poisson drew on one of Euler's theorems on gamma functions to simplify the ratio of (764.30–31) a bit and to expand $(1-t)$ in (764.29) in a power series and thus integrate (764.30) term by term for the rational values of b ($\frac{1}{2}$, $\frac{1}{4}$ and $\frac{1}{8}$) involved. The three values for B/A which he found differed from Coulomb's data by +6.5%, +0.8% and −9.1% respectively (pp. 57–60). Upon executing a similar calculation with $x = -1$, he obtained an expression for the ratio of thicknesses of fluid at points diametrically opposite the point of contact of the spheres and found similar orders of closeness to Coulomb's data (p. 66).

In order to consider points of the diameters of $\mathscr{S}$ and $\mathscr{S}_1$, Poisson expanded $(1-t)$ in the integrals of (764.28) for $f(x)$ in a power series in t and obtained a rather complicated series expansion of $f(x)$ in powers of x. This could be converted into a series for $\varphi(\mu, x)$, thanks to a relation furnished from $(764.5)_3$; and then an expression was obtained for y by appeal to $(764.5)_1$. The results (pp. 72–75) were rather horrendous; but in the cases where $b = 1$, 2 and 4 they simplified down to known series so that numerical values could be calculated for values of μ such as -1, 0, $\frac{1}{2}$ and $\frac{\sqrt{3}}{2}$ (that is, $\Theta = 200°$, 100°, $66\frac{1}{3}°$ and $33\frac{1}{3}°$, to cite from p. 76 the decimal division of the circle (§ 5.5.4) which, unusually for the 1810s, he persisted in using in both of his papers). Since h was still unknown, these values could only be taken in proportion with each other, but were still comparable with Coulomb's data; the same orders of discrepancy were disclosed (p. 80) as were noted in the previous paragraph.

Poisson's final exercise in the first paper took the case where the spheres were at a great distance apart. g and h both applied, and so the original equations (764.7–8) were invoked. After some ingenious Taylor-series expansions in the small increment b/c, and other routines, to obtain expression for $f(x)$ and $F(x)$ up to $O((b/c)^3)$, he drew on (764.30–31) for the mean thicknesses A and B of fluid on $\mathscr{S}$ and $\mathscr{S}_1$. This gave him expressions for $\varphi(\mu, x)$ and thus, via $(764.5)_2$, for thickness y in terms of A, B, a, b, c and μ: a similar formula obtained for thickness z on $\mathscr{S}_1$ in terms of these constants and μ_1. However, he was unable to draw many conclusions from these results (pp. 84–92), and nothing to relate to Coulomb's earlier studies. Nevertheless, it gave him a line to pursue in his second paper.

7.6.5 Poisson's second paper, 1813: important new integrals

Doubtless Poisson's first paper would have won the *classe* prize, had it been submitted as such; it certainly deserved it. The second paper, of September 1813, is of rather less interest for electrostatics, but it contains some quality mathematics. Recalling that he had solved the functional equations (764.7–8) for f and F only when the spheres were in contact or at a great distance apart, he now sought general solutions. Writing (764.21) for f in the form

$$f\left(\frac{x}{a}\right) - \frac{b}{k - cx} f\left(\frac{c - ax}{k - cx}\right) = \frac{h}{a} - \frac{gb}{ac - a^2 x}, \quad -1 \leq x \leq 1,$$

where

$$k := \frac{c^2 - b^2}{a} \tag{765.1}$$

and 'ax' was written for 'x', and bearing in mind his solution (764.23) to the reduced equation, he tried for this reduced equation the form

$$f(x) = \varphi(x)/(1 + mx), \tag{765.2}$$

where m was to be chosen so that x disappeared from the coefficients of the unknown function φ. This condition produced the quadratic for m

$$cm^2 + (k + a)m + c = 0 \tag{765.3}$$

with roots m_1 and m_2, and solution for φ:

$$\varphi(x) = ((1 + m_1 x)/(1 + m_2 x))^n \tag{765.4}$$

for some n. Substituting (765.4) into (765.1) via (765.2) led to the equation

$$b^{2n-1} = (k + \dot{m}_1 c)^{2n-1}, \quad \text{with unique solution} \quad n = \tfrac{1}{2}. \tag{765.5}$$

Thus the solution to $(765.1)_1$ was

$$f(x) = P(x)\,((1 + m_1 x)/(1 + m_2 x))^{-1/2}, \tag{765.6}$$

where P was an arbitrary periodic function, as required by the theory of difference equations (§ 3.5.1) (*1814g*, 174–177).[1])

As in the first paper at (764.24), Poisson now took on the first term on the right hand side of $(765.1)_1$. We recall that there he had produced from nowhere that integral solution; this time he proceeded in more analytical fashion, and one typical of him, in proposing a power-series solution with functional coefficients to be determined:

$$f(x) = (h/a) \sum_{j=0}^{\infty} \psi_j(x)\, b^j, \quad \text{with} \quad \psi_0(x) = 1. \tag{765.7}$$

When introduced into $(765.1)_1$ (without its last term), (765.7) converted to the difference equation

$$(k - cx)\,\psi_{n+1}(x) = \psi_n((c - ax)/(k - cx)) \tag{765.8}$$

for which, 'on considering with attention this condition', he proposed the solution

$$\psi_n(x) = (A_n + B_n x)^{-1}, \quad A_n,\ B_n \text{ constants}. \tag{765.9}$$

Substituting in (765.9) produced a pair of second-order linear difference equations with constant coefficients for $\{A_n\}$ and $\{B_n\}$; the usual method (351.2) gave their solutions, so that (765.9) became

$$\psi_n(x) = (\alpha_1 - \alpha_2)/[(\alpha_1 + a - cx)\,\alpha_1^n - (\alpha_2 + a - cx)\,\alpha_2^n], \tag{765.10}$$

where α_1 and α_2 were roots of the quadratic in α

$$\alpha^2 + (a - k)\,\alpha + b^2 = 0. \tag{765.11}$$

Again like last time, he went through the same procedure for the other term on the right hand side of $(765.1)_1$ and so found as a 'complete and general' solution of (765.1) the sum of (765.6) (with (765.3)), (765.7) (with (765.10)), and the com-

[1]) I have written 'Φ', 'm_1' and 'm_2' for Poisson's unhappy choices 'f'', 'm' and 'm''. While I have spoken of (765.1), and related equations, as functional equations, they have properties similar to those of difference equations.

panion to (765.7):

$$f(x) = \frac{P(x)}{(c-(k+a)x+cx^2)^{1/2}}$$

$$+\frac{h(\alpha_1-\alpha_2)}{a}\sum_{n=0}^{\infty}\left(\frac{b^n}{(\alpha_1+a-cx)\alpha_1^n-(\alpha_2+a-cx)\alpha_2^n}\right) \quad (765.12)$$

$$-\frac{g(\alpha_1-\alpha_2)}{a}\sum_{n=0}^{\infty}\left(\frac{b^{n+1}}{(c-(\alpha_2+a)x)\alpha_1^{n+1}-(c-(\alpha_1+a)x)\alpha_2^{n+1}}\right)$$

(pp. 177-182, with 'α_1' and 'α_2' for his 'α' and 'α'').

Poisson considered the immediate consequences of (765.12): the discarding of the first term in the case (764.20) of contact between $\mathcal{S}$ and $\mathcal{S}_1$, the companion series solution for $F(x_1)$ in this case, the passage from $f(x)$ to $\varphi(\mu, x)$ via (764.5$_3$,2) and the thicknesses y and z of the fluid on $\mathcal{S}$ and $\mathcal{S}_1$ from (765.5) and its companion for Φ (pp. 182-190). He then considered the convergence of the power series in (765.12), finding that each one, both for $f(x)$ and $F(x_1)$, could be written as a power series in

$$\beta := \alpha_1/\alpha_2, \quad \text{where} \quad \alpha_1 < \alpha_2 \quad \text{by specification.} \quad (765.13)$$

'Now it is visible that these series will be convergent all the times that β will be less than unity' (p. 193); however, he did not mention cases where the denominator of a term could be zero.

Poisson now applied his results to two specific cases, of which one was given by $a = 1$, $b = \frac{11}{4}$ and $c = \frac{19}{4}$. He truncated each series to its first four terms and expressed y as a linear combination of the mean thicknesses A and B of fluid (764.30-31) deposited on $\mathcal{S}$ and $\mathcal{S}_1$ respectively:

$$y = 1.2248\,A - 1.5567\,B. \quad (765.14)$$

He chose this example because Coulomb had found that after charging $\mathcal{S}$ and $\mathcal{S}_1$ positively and separating them to this distance, $\mathcal{S}$ was lightly negative (*1789a*, art. 24). Thus, to complete his calculation, he had to determine B/A—which brought him back to his integral expressions in (764.30-31) (pp. 208-210). Fortunately, the integrals in this case could be expanded in rapidly convergent series of known integrals; for example, since

$$x^{1/15} = \exp(\tfrac{1}{15}\ln x) = \sum_{j=0}^{\infty}(\tfrac{1}{15}\ln x)^j/j!, \quad (765.15)$$

then for B,

$$\int_0^1 \frac{x^{1/15} - x^3}{1 - x^4} dx = \int_0^1 \frac{1 - x^3}{1 - x^4} dx + \sum_{j=1}^{\infty} \int_0^1 \frac{(\ln x)^j}{1 - x^4} dx/(j!\, 15^j)\,, \qquad (765.16)$$

where each component integral was either a known result (in Euler) or could be found by approximation. Truncating to these series to the third term produced

$$B = 0.8107\, A, \text{ so that, in (765.14), } y = -0.0372\, A\,, \qquad (765.17)$$

the slight negative charge which Coulomb had found (p. 212).

Series solutions in general, integral solutions in the case of contact; how did they relate? In pondering this question Poisson came to one of his most enduring contributions to mathematics. In a 'Digression on the definite integrals and on the summation of certain series', he noted Euler's factorisation of exponential and trigonometric functions into infinite products of quadratic forms and realised that by applying the processes of analysis to these results he could produce new ones, including summations of series and evaluation of definite integrals pertinent to his current studies. Specifically, starting from Euler's theorem in the *Introductio* (*1748a*, *1*, art. 159)

$$\frac{\operatorname{ch} p - \cos\Theta}{1 - \cos\Theta} = \left(1 + \frac{p^2}{\Theta^2}\right) \prod_{j=1}^{\infty} \left[\left(1 + \frac{p^2}{(2j\pi - \Theta)^2}\right)\left(1 + \frac{p^2}{(2j\pi + \Theta)^2}\right)\right], \qquad (765.18)$$

Poisson took logarithms, differentiated with respect to p, and used the integral

$$\int_0^{\infty} e^{-qt} \sin pt\, dt = p/(q^2 + p^2) \qquad (765.19)$$

to replace each term in (765.17) by an integral. This gave him

$$\frac{\operatorname{sh} p}{\operatorname{ch} p - \cos\Theta} = \frac{2p}{\Theta^2 + p^2} + 4\int_0^{\infty} \frac{\operatorname{ch}\Theta t \sin pt}{e^{2\pi t} - 1} dt, \qquad 0 \le \Theta \le \pi, \qquad (765.20)$$

together with various related integrals (pp. 212–216). These results were of importance from several points of view.

Firstly, there was the forms of the integrals themselves, which took a more intimate relation when, for example, a cousin to (765.19) was multiplied through by e^{-mp}, $m > 0$ and integrated over $[0, \infty]$ of p to produce, with the help of

(765.18),

$$\int_0^\infty \frac{\operatorname{sh} p\, e^{-mp}}{\operatorname{ch} p + \cos \Theta} dp = 2 \int_0^\infty \frac{t \operatorname{ch} \Theta t}{(m^2 + t^2) \operatorname{sh} \pi t} dt \tag{765.21}$$

(p. 218). We shall pursue the consequences of these forms in § 11.4.1, when Poisson used them as the 'Poisson integral' in studying the convergence of Fourier and other series.

Secondly, some special results emerged, of interest in their own right as well as of use in electrostatics. This one stood out: putting $\Theta = 0$ in (765.20) yielded

$$\frac{e^p + 1}{e^p - 1} - \frac{2}{p} = 4 \int_0^\infty \frac{\sin pt}{e^{2\pi t} - 1} dt, \tag{765.22}$$

and multiplying through once more by e^{-mp} and p-integrating over $[0, \pi]$ produced, after some passage work,

$$2C = 2 \int_0^1 \frac{x^{m-1} - 1}{1 - x} dx - \frac{1}{m} + 2 \ln m = 4 \int_0^\infty \frac{t dt}{(e^{2\pi t} - 1)(m^2 + t^2)}, \tag{765.23}$$

where $x := e^{-p}$ and $C := \int_0^1 \frac{dx}{1 - x} - \int_0^1 \frac{dx}{\ln(1/x)}$, (765.24)

that is, Euler's constant 0.57716 (for which, incidentally, Poisson cited two sources in Euler on p. 222).

Thirdly, these various evaluations of integrals could be used to sum certain series pertinent to electrostatics. In particular, in (765.21) Poisson put

$$p = 2n\delta + \ln k, \quad \text{where} \quad \delta := \ln \beta, \quad k > 0, \quad \beta > 0, \tag{765.25}$$

multiplied through by $\gamma^n (\gamma < 1)$, and summed over n to produce

$$\sum_{n=0}^{\infty} \frac{\gamma^n}{1 - k\beta^{2n}} = \frac{1}{2} \sum_{n=0}^{\infty} \gamma^n - \sum_{n=0}^{\infty} \frac{\gamma^n}{2n\delta + \ln k}$$

$$- 2 \int_0^\infty \frac{\sum_{n=0}^{\infty} \gamma^n \sin (2n\delta + \ln k) t}{e^{2\pi t} - 1} dt; \tag{765.26}$$

this came down, after a lot of passage work and some ancillary results, to

$$\sum_{n=0}^{\infty} \frac{\gamma^n}{1-k\beta^{2n}} = \frac{1}{2(1-\gamma)} - \int_0^{\infty} \frac{k^t}{1-\gamma e^{2\delta t}}\,dt$$

$$- 2\int_0^{\infty} \frac{\sin(t\ln k) - \gamma\sin(\ln k - 2\delta)t}{(e^{2\pi t}-1)(1-2\gamma\cos 2\delta t+\gamma^2)}\,dt \qquad (765.27)$$

(pp. 226–228). Further, the sum in $(765.27)_1$ was expressible as a double sum and so summable in reverse order:

$$\sum_{n=0}^{\infty} \frac{\gamma^n}{1-k\beta^{2n}} = \sum_{n=0}^{\infty}\sum_{m=0}^{\infty} \gamma^n k^m \beta^{2mn} = \sum_{m=0}^{\infty} \frac{k^m}{1-\gamma\beta^{2m}}\,. \qquad (765.28)$$

This result allowed γ and k to be permuted in integrals in which they both appeared (p. 229).

After this display of virtuoso manipulation and blithe rigour, Poisson could fulfil his aim of converting into integrals his series solutions for $f(x)$, $F(x_1)$, y, z and the rest, because the series could be expressed within the form of $(765.27)_1$ and so draw on the various integrals. In particular, the series expansion of $f(x)$ mentioned after (765.12) was stated as the arithmetical difference of two such series, and so was converted to a quartet of integrals and a pair of terms by grace of (765.27) (actually with γ and k permuted: pp. 230–233). From then on all the previous calculations could be reworked: in turn, of thickness of fluid A, of $F(x_1)$, β, y and $\varphi(\mu, x)$ (pp. 233–242). Of particular interest was the case of contact between the two spheres, where he twice invoked (765.23) with different expressions assigned to m, eliminated the C-term between them, and thereby converted an expression found for $f(x)$ from the use of (765.28) into the integral solution (764.26) which had appeared from nowhere in the first paper (p. 234–237).

The rest of the paper was concerned with analysing the case where $\mathcal{S}$ and $\mathcal{S}_1$ were very close together, so that the two roots α_1 and α_2 of (765.11) were nearly equal and $\beta \triangleq 1$ in (765.13), thus allowing various approximations and truncations to be employed (pp. 241–274: Bowditch *1818b* pointed out an error of passage-work on Poisson's p. 245, and noted its consequences). His discussion was oriented more to questions of mathematics than to electrostatics, but on p. 252 he interpreted sparking in terms of large values for y and z being caused by small values in a denominator, and later on he calculated some numerical proportions for $y{:}A$ in the case where $a = 1$ and $b = 3$ (pp. 265–268). He did not compare his results with any experimental works, however.

I have not rehearsed any of the derivations in these last two paragraphs, for while usually straightforward as routines, their symbolic complications preclude easy summary of their content. Throughout, though, they show particularly well Poisson's great skill as a manipulator of mathematical expressions.

7.6.6 The isolated achievements of Poisson

These two papers were pioneering milestones, extending mathematics to a new area of physics. Note, however, that they were not the *first* such major extension: priority belongs to Fourier's work in heat diffusion, already several years old in creation by 1812 but still unpublished—and coolly received in Paris, as we see in Chapter 9. I have placed Fourier's work there because it helped to motivate some substantial changes in thinking in mathematical physics in ways which this achievement of Poisson failed to match. Why was he unsuccessful? Two different kinds of reason suggest themselves.

Firstly, while Poisson extended the mathematicisation of electrostatics to bring it into the realm of potential theory, he did not adorn it with the mathematical trappings of Laplacian physics: the multiple integrals of cumulated inter-molecular action which we have seen in §7.2-4. Further, apart from the 'level surfaces' of constant potential, the analysis seems strangely devoid of the "feel" of electricity. It is not necessary for mathematics to speak in favour of one or another constitution of the electric fluid(s)—we shall see Fourier in similarly agnostic mood over the nature of heat in §9.2.1—but the solutions, series or integral, do not reflect any of their properties (§1.3.1). Again, while Poisson had made a notable contribution to the concerns of Coulomb's fifth paper, he had not addressed major questions in electrical theory in general, such as the performance of electricity in a machine or in nature. Finally here, the theory was formidably difficult even for mathematicians to follow, and required much even of Poisson himself to produce quantitative predictions; and it lacked general results of the power that Green was to introduce with his potential function (Interlude 1754.1).

Secondly and maybe more importantly, Poisson's work failed to make a major impact because Paris showed only a patchy interest in electricity and magnetism, especially in its mathematical aspects. This seems a surprising statement to make, yet it is exemplified by these two papers, *which did not excite any immediate major response in Paris.* While every other area of mechanics and mathematical physics was pursued there with a vigour matched only by skill and persistence, the worlds of electric and magnetic fluids only aroused interest from time to time among the *savants,* Poisson *included.* This will will be borne out again in Chapter 14, when the story of electromagnetism and electrodynamics is rehearsed, and where we shall take up this branch of mathematical physics once again.

7.7 Concluding comment: the aspirations of the Laplacians

Laplacian physics was receiving much publicity around the mid 1810s. The syllabi of the *Ecole Polytechnique* show its presence in the teaching there (see Fox *1974a*, 107-108); and it was also prominent in Biot's long textbook on experimental physics (Biot *Physics*) based on his teaching in the Paris *Faculté des Sciences*, which appeared in three editions between 1817 and 1824.

Although long, the textbook was called a 'Précis' because it contained the non-mathematical parts of a four-volume *Traité* on physics which Biot had published in 1816, his 43rd year (Biot *1816c*). Over 2,400 octavo pages in length, the *Traité* must have taken him a month or two to write. He divided it into seven Books, beginning with a survey of instruments and then dealing in turn with acoustics, electricity, magnetism, light and heat. Intended as a full-scale work in the sense described in §2.2.4, it reflected his own preferences in physics; for the Books on light were disproportionately long at nearly 1,000 pages, with a vast discussion of the properties of mobile polarisation, and some attention to rotatory (now called 'successive') polarisation.[1]) This concentration on light emphasised still further the Laplacian character of the whole work. Its mathematical level was rather less than Poisson's textbook of 1811 on mechanics (§6.2.4); a typical example of his style is barometer theory, where he used a slightly simplified version of Laplace's basic formula (724.1) (Biot *1816c*, *1*, 101).

Perhaps the second volume gives the clearest impression, for it contains the Books on acoustics (208 pages) and 'electricity' (341 pages), in which Biot's own participation had not been as substantial as in optics. His account of the propagation of sound included a run through his and Poisson's development of the theory (§7.2.5), including Gay-Lussac's 0.00375 and Poisson's $\frac{1}{116}$ (*2*, 13-31), and a quite extensive survey of Chladni's experiments, including the nodal lines on elastic surfaces, although he passed over the competition between Germain and Poisson (§7.3). Otherwise, he discussed in detail musical temperament and the theory of wind instruments.

Biot's discussion of 'electricity' began with a typical (§7.5.5) Parisian advocacy of a two-fluid theory of electrostatics (p. 217), followed by an account of Coulomb's work on the laws of repulsion and attraction (pp. 226-243) and the laws of distribution of electricity in equilibrium (pp. 263-315). He praised Poisson's two

---

[1]) For these topics see respectively Biot *1816c*, *3*, 146-512; and *4*, 1-252; and *4*, 253-599. Sparse in references and acknowledgements, in the best French tradition, the treatise (and some of Biot's contemporary writings on optics) provoked a clenched-teeth response in Brewster *1830a*, 490-496.

papers in several places between pp. 291 and 312, but understandably he quoted only the simplest of Poisson's expressions for the thickness of the fluid(s). Thereafter he was largely concerned with a survey of electrical devices and machines, including on pp. 361-370 a detailed statement of the accumulation of charge in a condensor (mathematical in its use of geometric series); however, he left out the doubler (§7.5.5). He also discussed natural phenomena, such as thunder (pp. 442-459).

Reflecting the rather smaller interest of the time noted in §7.5.1, Biot's account of magnetism (*3*, 1-145) was shorter than that of electricity. He was careful only to press analogies with electricity, and would point out differences such as the presence of interior 'action' in a magnet even in equilibrium (p. 14). He gave an account of terrestrial magnetism (pp. 15-39), and of Coulomb's experiments with the torsion balance (pp. 29-36) and establishment of an inverse square law (pp. 63-69).[1]) His final chapter dealt with terrestrial magnetism, in which his collaboration with Humboldt featured prominently; he quoted on p. 132 his formula $(754.5)_2$ for magnetic dip, although he did not provide a proof.

During the 1810s Laplacian physics came to reach an impressive level of development in research, with consequences for its diffusion. The Master himself played a role here; for in the 1813 edition of his *Exposition du système du monde* he included a lengthy chapter on 'molecular attraction' (*Exposition*$_4$ (1813), 315-357). Clearly, for him the programme was sufficiently well developed to sustain a popular presentation before the public.

As we now know, the programme reached its peak at this time, and thereafter began to suffer difficulties and competition, which will be a major feature of several chapters below. But even without this hindsight, certain weaknesses in the facade can be detected. I shall refer to two questions: the role of fluids, and the use of mathematics.

As was mentioned in §7.2.1, and as we have seen since, Laplace's programme made use of fluids to supplement the activity of the molecules. Indeed, in parts of his *Mécanique céleste* written admittedly before the programme was launched, he concluded his lunar theory by considering the perturbations on the earth and moon which could be caused by an aether surrounding the sun (*3* (1802), 678-698), and assumed that there would be 'an infinitely thin stratum of fluid, spread upon the surface of these rings' to keep them in equilibrium (*2* (1799), 492). Again in Book 10, in his discussion of 'the successive transmission of gravity', he stated that 'We shall suppose that the sun is surrounded by a fluid' (*4* (1805), 609), working in the tradition of his time of assuming an aether and seek-

---

[1]) In connection with Coulomb's work on both electricity and magnetism, Biot cited manuscripts by Coulomb here and there (for example, *3*, 106). Unfortunately, they seem to be now lost.

ing its properties. 'If light consists in the vibrations of an elastic fluid', for example, 'the proceding analysis will give the effect of its resistance upon the motion of the planets and comets' (p. 629). He then went on to calculate its possible velocity of transmission as 'at least a hundred millions of times greater than that of light', so that 'mathematicians may suppose, as they have heretofore done, that the velocity of the gravitating fluid is infinite' (p. 645). Some of this work consistuted an equivalent to calculating the gravitational constant, in that some fundamental action of gravity was being appraised (compare *Exposition*$_1$ (1796), *2*, 19-24; or $_2$(1799), 185-187).

As the programme developed, the Laplacians continued to deploy fluids, especially magnetic and electrical ones for Biot (§ 7.5.3-4) and Poisson (§ 7.6): we shall also find in § 9.3.1 that caloric played a role in their studies of heat. But the question posed in § 7.2.1 arises again: were these fluids the site for inter-molecular action, or only an occasional component of the programme?

When the programme was under way, Biot expressed his cautious opinion on the use of fluids in electricity, magnetism and heat. In a general philosophical essay (*1809a*, 113-114), he wrote:

> In order to explain [these phenomena], physicists have imagined certain elastic fluids endowed with attractive and repulsive properties, and capable of penetrating all bodies or only some of them [...] By means of these suppositions one can, to a certain point, represent the majority of phenomena [...]; but there still remain many of them which lend themselves with difficulty to these explanations, and others which escape them entirely.
>
> Consequently the true [*véritables*] physicists admit the consideration of these fluids solely as a convenient hypothesis, to which they are very careful not to attach ideas of reality, and which they are ready to modify or to abandon entirely as soon as the facts show themselves to the contrary.

Let us consider now the use of (pure and applied) mathematics in Laplace's programme. The question related to the status of fluids, for we saw that Biot's magnetism and Poisson's electrostatics involved only a partial mathematicisation of the phenomena. They both used some properties of decomposition of forces, and Poisson also drew on a little potential theory; but there was no sight of the cumulative action integrals.

Speaking more generally, one strength of Laplacian physics was its ability to draw on principles and methods of mechanics, as was stressed in § 7.4.5; and both there and elsewhere we have seen effective deployment of many of the main ideas of that subject such as decomposition laws, conservation of *forces vives*, and the principle of least action. However, it was not yet proven that "all" areas of physics could be so expressed; or, looked at from the other way, whether the reduction of physics to these "mechanical" assumptions and procedures could be successfully achieved.

So we leave the programme now, and await its future from Chapter 9 onwards. First, however, let us catch up with the engineer *savants* during the Revolutionary period.

Chapter 8

Aspects of engineering mathematics, 1800–1816

8.1 Prefaces

8.1.1 Engineering in French science: de Prony's *Nouvelle architecture hydraulique*

As was made clear in Chapter 2 and elsewhere, the new system of engineering schools was introduced principally to supply engineers for public works and military needs. This chapter and Chapter 16 are the principal places in this book for discussions of French engineering, but its concerns permeate several other sections. We shall see in § 19.3.1, as a major feature of French physical science, that about half the *savants* with whom we are concerned worked largely on problems motivated by engineering, while the other half concentrated on the more theoretical branches of mathematical physics and gave engineering limited attention, or even none at all.

The emphases laid in this book relate most closely to the theoretical branches, so that the treatment of the engineering components is very selective; in fact, the topics chosen are those which have maximal bearing on the calculus and mechanics described elsewhere, and/or on the philosophical issues raised in § 1.3-4. As a contribution to the history of French engineering in general, therefore, it is very partial. But the prominent position accorded to engineering in French science at this time should emerge. From the modern point of view, it is civil and mechanical engineering. Most of it arose in connection with projects in public works; but the place of military engineering will be considered in the next subsection, and a growing concern in the late 1820s with industrial and commercial engineering is noted in § 16.6-7.

de Prony advocated the prominence of engineering with especial vigour, even before the founding of the *Ecole Polytechnique*. In particular, between about 1787 and 1793 he attempted to edit a survey of civil engineering and its profession in a volume entitled 'Science of the roads and bridges engineer, dykes and embankments, canals and maritime ports'. It was to be published as a supplement to the *Encyclopédie méthodique*, and in plan followed the *encyclopédiste* style of a 'Preliminary discourse' followed by a series of dictionary entries; but, although lists of entries were prepared and some even written, only the discourse appeared, as a pamphlet de Prony *1788a*, with most of it reprinted in a 'tableau' *1789a* added to the last volume of mathematics in the *Encyclopédie*. He saw the survey as covering six aspects: history; 'administration and public works; relevant sciences (including descriptive geometry, interpolation and hydrodynamics)'; a 'descriptive'

Table 811.1 Outline of the contents of de Prony's *Nouvelle architecture hydraulique* (1790–1796)

Section	Topic	Page
	Preliminary notions (velocity, momentum, 'force', etc.)	
	Part 1 (1790): Mechanics 'for Artists in general'	1
1	Statics (composition of forces; equilibrium; virtual velocities; centres of gravity; levers, pulleys, wedges; equilibrium of arches)	49
2	Dynamics (motion of a body under forces; translation and rotation; percussion; motion in machines)	169
3	Hydrostatics (equilibrium of (in)compressible fluids; applications to air bubbles, dams, earth pressure; hydrometer; 'elastic fluids, and principally air'; (water-)pumps, pistons, valves)	245
4	Hydrodynamics (fluids in vessels; fluid flow out of narrow and wider orifices; pressure in tubes; impact and resistance of fluids)	328
5	Machines and motors (factors involved in equilibrium friction with inclined plane, ropes, screws (including experiments); 'On forces' of men and horses; vapour, and its use in the motion of machines)	427
	End of text	575
	72 pages of tables, separately paginated	
	Part 2 (1796): Steam engines	
	Preamble' (apparatus and experimental techniques)	1
	Description of two engines 'at double effect'	35
	Other engines not of this type (including the one at Gros-Caillou) [compare § 16.5.1]	73
	Advantages of the 'double effet' type; their construction	106
	Rectilinear motions produced by circular motions	123
	Expansion of gases [essentially *1796b*: § 3.5.3]	152
	End of text	196

account of the use of theory in engineering practice; the 'engineering arts'; and a bibliography.[1])

This project failed, doubtless because of the social and political difficulties of the time; but in 1790, de Prony's 36th year, he achieved success on his own, when Firmin Didot published the first volume of his first major treatise. Entitled 'New hydraulic architecture', in fact it covered all the major concerns of civil engineering of that time with the exception of canals and rivers (which are discussed in § 8.5.4). Then six years later the second volume was published; shorter

[1]) For relevant correspondence and materials, see *ENPC*, mss. 1011–1016 and 1057–1063 *passim* (especially as some files are lost), 1366, 2630(6) and the file at 2000 *carton* 89. On the treatise about to be described, see ms. 1012 for two letters of 1788 from Didot to de Prony.

in text but even more lavishly endowed with diagrams, it was devoted entirely to steam engines. It lacked the 'Preliminary discourse' which had been promised in the first volume; however, it did contain Garnier's comments *1796a* on that volume. The whole work is cited as de Prony *Hydraulics*; a German translation appeared in 1798 and 1801.

Table 811.1 indicates its principal contents, in a manner which imitates that used in Table 521.1 for Lagrange's *Méchanique analitique* of 1788. While there is a natural (and conscious) overlap with Lagrange's concerns in several subjects, the overall emphases fell, of course, on engineering artefacts, including reports of experiments and tests. The book is also rather different from his later textbooks, which were described in § 5.2.7 and § 6.6.1; although intended for educational purposes (and praised as such in an anonymous review de Prony *1790a'* in the *Moniteur universel*), it is really a full-scale treatise in the sense described in § 2.2.4, and would have been too demanding (and probably also too expensive) for normal teaching use. But it is a fine survey of engineering mechanics at the end of the 18th century, with attention paid to the mathematics when necessary, extensive footnotes on various topics, and useful surveys of the recent, and even more distant, history of his subject.[1])

8.1.2 Some remarks on mathematics applied to military purposes

> But of this I can assure you, that there is not a movement of any body of Men however small whether of Horseback or on foot, nor an operation or March of any description nor any service in the field that is not formed upon some mathematical Principle, and in the performance of which the knowledge and practical application of the Mathematicks will be found not only useful but necessary. The application of the Mathematicks to Gunnery, Fortification, Tacticks[,] the survey and knowledge of formal Castrematation etc. cannot be acquired without study.
>
> The Duke of Wellington to his son Duoro, 1826
> (Wellington *1952a*, 44-45)

It was mentioned in § 2.3.5 that some of the specialist engineering schools were military; thus it is meet to add here a further group of preparatory remarks, on the place of military mathematics in this book. At the research level, various of

[1]) During the 1810s de Prony hoped to publish a third volume, on machines for lifting water (see Firmin Didot's catalogue for 1817 in *BN*, call-mark Q^{10}); but it never materialised.

the topics described have obvious military uses (for example, Laplace's treatment of projectiles in § 6.4.2), and no special point will be made about them. In addition, the motivations to a piece of research from a military need will be recorded where appropriate, in the usual way.

From the point of view of education, the situation is somewhat more complicated. The enthusiasm of the *Ecole du Génie et d'Artillerie* at Metz to criticise the teaching of the *Ecole Polytechnique* for excessive emphasis on abstract theory was mentioned in § 2.4.4, and a particular example was cited in § 4.3.2. The difference of roles for mathematics between the *Ecole Polytechnique* and the military schools is well exemplified, although not explicitly discussed, in a *Cours de mathématiques* for use in such schools prepared by four *professeurs* at the *Ecole Militaire* at St. Cyr and published in 1809 (a year after the school moved there from Fontainebleau). I cite this book as Puissant *1809a*, after one of its authors; his distinguished role in the propagation of cartography and topography was indicated in § 5.5.8 and § 6.4.9. In their book, which followed their St. Cyr teaching, they presented arithmetic, algebra up to the resolution of cubics, basic geometry, descriptive geometry and trigonometry with especial reference to surveying, and simple statics, dynamics and hydrostatics. The level is at most that of the first year at the *Ecole Polytechnique* and at times that of its *aspirants*; for example, no calculus was used at all. Thus it followed the tradition for such books from the the 18th century; in particular, the *Cours de mathématiques* of Bézout mentioned in § 2.4.2 as a best-seller (Bézout *Course*), was originally prepared in the 1760s for military schools. However, that work contained some calculus, in the parts on mechanics.

The perception at the *Ecole Polytechnique* of military needs is exemplified by a report prepared by Lacroix for the meeting of the *Conseil d'Instruction* on 19 *pluviôse an* 13 (8 February 1805), concerning the courses in topography and fortifications. Recalling the origins of the school and its success, he followed the Laplacian line, that general common principles should be taught there (including construction, attack, and the defence of positions), with the special theories presented at the *écoles d'application*. For example, according to Lacroix 'the theoretical part' of designs was 'the most important in the course on fortification, or rather the course of military art; and it is its generality which renders it useful to the greatest number of students and which distinguishes it also from the teaching of the *Ecole spéciale du Génie et d'Artillerie*' at Metz.

This line of thought sounds irreproachable, and for the particular case cited might also be satisfactory; but the general imbalance between l'*Ecole de l'X* (§ 2.3.5) and the real world of armies still remains rather vast![1]) Nevertheless, the

[1]) Lacroix's report is conserved at *EP*, III 3f: it is reproduced in § 20.2.

training of the military in France was widely recognised as important; for example, in the letter quoted at the head of this sub-section Wellington told his son that of foreign languages 'the most important of all for an officer to learn is the French language'. Indeed, as a youth Wellington had studied at the military college at Angers during the 1780s—an enrolment which the French doubtless came to regret!

8.1.3 Plan of the chapter

§ 8.2.1 provides brief biographies on two *savants* who gain prominence now: Hachette and Girard. § 8.2.2 indicates some of the organisation and publications especially devoted to engineering.

The technical account begins in § 8.3 with a short review of Monge's descriptive geometry. Although it did not fulfil his aspirations, it furnished useful techniques for many branches of engineering; two of them are described in § 8.3.3–4, and others arise from time to time elsewhere in this book. § 8.4 surveys some aspects of engineering materials and structures, with especial emphasis on earthworks, while § 8.5 tackles the subject of water-flow, including its measurement and the design of canals. Girard and de Prony are prominent in these two sections; and in § 8.5.7 Hachette returns, when his researches on the flow of water out of orifices were appraised by the theoreticians Poisson and Cauchy.

A significant omission from this chapter is the general theory of machines and steam engines. This topic aroused especial interest in the 1820s, and so is held over until Chapter 16. Further, Interlude 860.1 is regrettably perfunctory on engineering outside France.

8.2 Some principal figures and organisations

8.2.1 Two further engineering savants: Hachette and Girard

In § 2.5 we were introduced to de Prony, Coulomb and Monge, whose activities were strongly directed to engineering. Here we meet a couple of their friends.

Monge's most devoted follower was Jean Nicolas Pierre Hachette, who was born in 1769 to a bookseller at Mézières, a town in the Ardennes near the Belgian border, now known as 'Charleville-Mézières'. It contained a famous military school (of which the Metz school just mentioned was the successor) and he taught there in the late 1780s after studying at the *Université* at Rheims. In 1792 he taught at a *collège* at Collioure, on the Mediterranean coast; then his close association with Monge began, not just as mathematical colleagues but also as ardent supporters of the Revolution. They taught together at the *Ecole Polytechnique*, and issued textbooks and joint papers on descriptive geometry and its applications. We saw Hachette's interest in physics, especially 'electricity', in § 7.5.5: in 1806 he launched a course in machine theory (§ 16.3.5). He also edited the *Correspondance sur l'Ecole Polytechnique*, as was recorded in § 2.2.3.

The cessation of the *Correspondance* was caused by Hachette's retirement in 1816, when his Revolutionary ardour began to cause him difficulties. He retired (or perhaps, as he was only 47, was retired) from the *Ecole Polytechnique* in the Restoration reorganisation of 1816; the new regime "forgot" to give him his pension. (However, he kept the adjoint chair in the Paris *Faculté* that be had gained in 1810.) He was elected to the *Académie* in November 1823, but the King refused to confirm the election (*PV*, *7*, 583, 604); so five years later he declined suggestions to stand for a vacancy in the *section* for rural economy. Eventually he was elected to the *section* for mechanics in October 1831 (*PV*, *9*, 706, 714); but within three years this dangerous figure was dead. For his students they 'had in him not only a teacher, but a friend, a father'.[1])

Hachette, like Monge, largely made his career in education. We turn now to a contemporary whose activities, like Coulomb's, were mostly concerned with public works projects.

Pierre-Simon Girard was born in 1765 in Caen, Normandy, one of many children of a clock-maker. He studied at the *Ecole des Ponts et Chaussées* before the Revolution, and became a member of the *Corps*. His interests lay largely in hydraulics, especially canal engineering and related topics; he helped de Prony in the early 1790s with the dictionary described in § 8.1.1, and studied the Nile river as chief engineer on the Egyptian campaign of the late 1790s (§ 2.6.2). On his return to Paris in the early 1800s his main preoccupation until his death in 1836 was the design and execution of the Ourcq canal to improve the water communications and supply in Paris (§ 8.5.4). He also worked on various other projects; in particular, he designed in the 1820s, as a continuation of the Ourcq scheme, a canal near Soissons to link the Seine, Aisne and Ourcq rivers. He was also in charge of gas lighting in Paris in the 1810s (compare Biot *1817a*). His other main

---

[1]) The words of the obituary Dégérando *1834a*. No significant study of Hachette's career exists; his self-'Notice' *1831c* is the most substantial account, written for his *AS* election. Interesting documents on his pension, and pensions in general, are to be found in *EP*, VI 1b2, and *Smith*; his *PF* career is filed at *AN*, F^{17} 20922; correspondence on his non-election in 1823 is kept at *ENPC*, ms. 889; and his letter declining to stand in 1828 is held in his *dossier personnel* at *AS*. A miscellaneous file is kept at *Darmstaedter*.

interest was in the strength of materials on which he published an important book—his first—in 1798 (§ 8.4.3).

Girard's work was quite well appreciated by his colleagues: he was a respectable second to the elections of Malus in 1810 and Poisson in 1812 to the *section* for general physics of the *classe* (*PV*, *4*, 373; *5*, 40). His election to that *section* in June 1815 (*PV*, *5*, 519) was very appropriate, as he succeeded the hydrographer P. Levèque.[1])

8.2.2 Organisations and publications for engineering

Engineering papers appeared in the various journals mentioned already: the *Mémoires*, the *Journal de l'Ecole Polytechnique*, and so on. There were also publications connected with public works, and projects run by, or for, organisations such as the *Corps des Ponts et Chaussées* (§ 2.4.1); for example, the *Receuil polytechnique des ponts et chaussées*, which appeared between 1804 and 1807 in regular numbers, provided news on many engineering projects. The military side was also important, through the *Ministère de la Guerre*. We noted in § 5.5.1 the ministry's *Dépôt Général de la Guerre*, which covered geodesy and related fields, publishing a *Mémorial* as its journal from 1802: the ministry also put out from 1803 an occasional journal entitled *Mémorial de l'Officier du Génie*, run by an army organisation called the *Comité des Fortifications*.

In § 2.2.2 it was mentioned that the *Académie de la Marine* was suppressed in 1793. Dupin *1815c* proposed in a pamphlet that it be re-established, but he was not successful; however, the year before the marine engineers had been formally incorporated into a *Corps des Ingénieurs Hydrographiques de la Marine* (see Rollet de l'Isle *1951a*, 1-11), and in 1819 the *Annales maritimes et coloniales* was launched (to receive publicity later in Dupin *1832a*).

All of these bodies and publications were normally devoted to public works and military engineering, but several organisations dealt with commercial and industrial engineering: the *Conservatoire des Arts et Métiers*, founded in 1799 as both a depository for artefacts and a (rough) equivalent to a modern technical college; the journal *Annales des arts et manufactures*, launched by the publisher Blaise in 1800, which put out several volumes a year and was so successful that it was reprinted twice by 1818; and so on. However, more detailed consideration of

[1]) Richet *1933a* provides a breezy summary of the life of his great-grandfather: little attention has been given to Girard's work. No *Nachlass* survives, although manuscripts, and correspondence with de Prony, are held in *ENPC* (some are cited below) and a large file of professional and career documents is preserved at *AN*, F^{14} 2235^{1}.

these organisations is deferred until Chapter 16, where they are shown to have played quite a significant role in the 1820s. For now, we return to 1800, and the heyday of descriptive geometry.

8.3 Perspectives on descriptive geometry

In § 3.3, § 4.7.3 and elsewhere we have taken note of calculus applied to the geometry of surfaces; here we deal with its brother subject, descriptive geometry. Its history has been well treated:[1]) I provide here only a general indication of its content and place in French science, followed by two interesting applications of the early 19th century. Its historical background is lengthy, with connections with art and architecture; but I start with the military world of Mézières.

8.3.1 Teaching descriptive geometry at the *Ecole Polytechnique*

Even before entering the *Ecole* at Mézières, the young Monge had become deeply interested in draughtsmanship, and while studying there he started simplifying methods of engineering drawing. When he taught at the school he began to develop some basic ideas of descriptive geometry in connection with stone-cutting, carpentry and related fields. As military mathematics, it may have been classified by the authorities for many years; thus while he used its methods in some of his research papers of the 1780s, and presumably also in his book *1794a* on the design and manufacture of cannonry, he was able to publish his methods only when he gave a course at the short-lived *Ecole Normale* in 1795. Those lectures appeared in the *Séances des écoles normales* that year as the first edition of his *Géométrie descriptive*, the major textbook on the topic for its time.

When the *Ecole Polytechnique* was launched, Monge's stress on the need for practical courses was adopted, and so his beloved descriptive geometry featured largely in the syllabi. But from 1800 onwards Laplace's desire to have more theoretical syllabi, described in § 2.3.4, began to prevail. Thus while in 1801 descrip-

[1]) The histories available discuss the subject from various points of view. See especially Booker *1963a* (on various types of engineering drawing: chs. 9-10 on Monge's principles); Loria *1921a* (general history: chs. 4-5 on Monge and his followers); Paul *1980a* (on Monge and descriptive geometry at the *EP*); and Taton *1951a* (chs. 2-4 on descriptive geometry and its connections with the calculus). Daston *1986a* outlines the general context of French geometries. The range of theories called 'descriptive geometry' is well conveyed in Rowe and McFarland *1953a*.

tive geometry had taken 40 % of the first-year class contact time against 26 % for analysis and mechanics, within five years the proportions were reversed (26 % to 43 %) and remained roughly of this order (Paul *1980a*, 124). Although the subject was used in other courses, the decrease in emphasis is fairly represented by these figures. Indeed, as the years went by descriptive geometry came to be taught only during the first year of the course, although 'analysis applied to geometry' still featured in both years.

At first the *Professeur* was Monge, but when his health began to fail he was succeeded in 1810 by Arago, who taught the course until 1830. Hachette was the principal assistant to both of them until his "retirement" in 1816; presumably he did not succeed Monge, because since 1806 he had been conducting a new course in the theory of machines (§ 16.3.5).

8.3.2 Descriptive geometry: practice or theory?

> [...] descriptive Geometry will carry in the most complicated operations the evidence which is its characteristic, and in its turn, Analysis will carry into Geometry the generality which is peculiar to it.
>
> Monge *Geometry*$_3$ (1811), art. 10

In Monge's conception of descriptive geometry, space was, in effect, divided into four quadrants by means of two perpendicular planes. A three-dimensional object O was projected onto these planes, and adjacent half-planes were then rotated about their line of intersection so as to form one plane containing the real (rather than apparent, or perspectival) projections of O. The theory as such described the various means of obtaining these projections, and the relationship between them; and also the converse problem, of (re)forming O (or at least, discovering its basic properties) from given projections. When, as was often the case, O possessed curved surfaces, the projections contained curved lines, so that the geometry of curves and surfaces, with its attendant calculus, had to be developed as a tandem theory. For many practical situations, approximations by straight lines or well-studied curves (such as conic sections) had to be made.

In principle O could be any object; and in practice a wide variety of situations was covered by the theory. Monge's plan of 24 lectures for his initial course at the *Ecole Polytechnique* in 1794–1795 (which is transcribed from the manuscript in Taton *1951a*, 94–95) contained a number of applications of the theory to stone-cutting and the construction of arches, wood-cutting and carpentry, shadows, perpective, topography and simple machines (involving conversions between rectilinear and circular motion). In 1812 Hachette gave a more extended summary of these applications in his first 'Supplement' to the third edition of Monge's textbook, of which he was the editor (Hachette *1812b*, iii-vii: the print-run of Monge's edition, incidentally, was 2,000 copies).

However, when one looks at the book itself, a markedly different impression of descriptive geometry comes over. That edition, for example, contains 162 pages of Monge's main text and 118 of Hachette's supplement; yet the only genuine applications given were Hachette's summary just described and an account by Monge of a few problems in topography (*Geometry*$_3$ (1811), arts. 95–102). *Otherwise this textbook on engineering mathematics was pure geometry* (known as 'stereotomy'). Typical constructions and questions are: 'Being given the projections of a line and the traces of a plane, construct the angle that the line and the plane form between them' (art. 21); 'Through a given point, set a plane tangent at the same time to the surfaces of two given spheres' (art. 41); and 'Construct the intersection of two conic surfaces of any bases' (art. 76). While the reader can—sometimes—readily see physical interpretations, nothing was said about them in the text. Indeed, the applications mentioned above were rather pathetic: one of them required the engineer out in the field to determine his altitude from those of points marked on a map and within his range of visibility by imagining conoids towering above him and yielding suitable curves of intersection. This procedure seems inferior to standard triangulation, especially when it is raining.

Thus this practical subject seems rather theoretical in practice: even notional, perhaps, in the sense of § 132.3. Anyone using Monge's book would have to work out the practicalities for himself. (The fourth edition, published in 1820 and edited by his niece's husband Brisson, was rather more useful, as it contained lecture notes by Monge on shadows and perspectives.) The lectures may also have been of limited utility: Monge's published outline *1795a* gives that impression, as do students' lecture notes that I have seen.[1]) But Lacroix's textbook *Curves*, published in seven editions between 1795 and 1840, is similar in character to Monge's in concentrating on purely geometrical properties.

This situation seems to have arisen because of considerations such as those stated by Monge at the head of this sub-section (remarks which go back to 1794, according to Taton *1951a*, 81). He wanted his subject to be a pukka mathematical discipline, and so presented it with these emphases; but instead it fell short both of that level and also that of practical engineering mathematics. For in this latter respect the geometry hinders more than it helps, as the topography example above shows; it must also have made the lectures rather boring. Even the ba-

[1]) See the notes taken by Lomet (§ 5.5.8) of the course taught by Monge at the *EPC* in 1794–1795 (*ENPC*, ms. 150), which seems to have been still less practical than the course given at that time at the *EP* (and discussed in the text). See also the notes of Hachette and of student Atthalin of courses given by Monge at the *EP* during the early 1800s (*EP*, III 3b).

sic division of space into quadrants led to somewhat cumbersome procedures, and in several countries it gave way to the so-called 'direct method', which is based on the direct construction of projections (plan, elevation, profile) without intermediate conceptions or procedures. But in addition, research in descriptive geometry did not lead to new intellectual sources of the depth seen in other aspects of mathematical physics. The main advances either took place in the applications rather than in the conceptions of the subject itself; or they lay in the twin subject of the geometry of surfaces. The correction to Malus's theorem on normal surfaces, noted in passing in § 7.4.2, is one example; in the rest of this section two others are briefly described.

8.3.3 Brisson on the geography of canal design, 1802

We met Brisson first in § 4.3.5 speculating on algebraic solutions to differential equations, and then in § 7.2.7 helping his brother-in-law Biot with the wave equation. But his editorship of the fourth edition of Monge's textbook, noted a few paragraphs ago, is much more typical of his interests, for in profession he was an important member of the *Corps des Ponts et Chaussées*, and in the 1820s taught at the *Ecole*.

With his colleague P.L. Dupuis-Torcy, Brisson submitted a paper on the geography of canal design to the *classe* at two meetings in June and July 1802 (*PV*, *2*, 515, 520). It was recommended for the *Savants étrangers* by Lacroix in October in his report *1802a*; however, while surely in time for the 1806 volume of that journal, it did not appear there. Only part of it was published, in the *Journal de l'Ecole Polytechnique* as Dupuis-Torcy and Brisson *1808a*, immediately after Brisson's article on differential operators; by then his co-author was dead.

The paper was an attempt to apply descriptive geometry to the planing of canals. The author's principal concern was with *'canals at the point of division'*, which connected two (or more) rivers across a chain of hills. They began by simplifying terrain into lines marking river beds and hill ranges and serving as edges of polyhedra, whose slopes, planar if possible, represented banks, plains, and the slopes of hills. They also defined 'orders' of rivers and hills in a recursive manner: 'a chain or a basin of any order, is entirely composed of a sequence of chains or of basins of the immediately lower order' (p. 264). While the definition makes poor graph theory (for example, if a tributary T to a river R has its own tributary t, then all other tributaries to R have to be of second order even if, unlike T, they have no ts), its basic strategy is clear.

After setting up this simplification of reality (pp. 262–269), the authors now projected it 'horizontally' onto sea and other levels, and 'vertically' onto appropriate great circles. They marked on these projections the turning points (which corresponded to absolute or relative optima, or saddle points), and they located the lines of greatest slope. They then argued that these points and lines, and curves of highest-order river and range, would normally suffice to locate the principal features of the terrain for the design of canals (pp. 269–281). For example, the 'point of division' of a canal, whither the water was brought so as to divide along its two parts, should be located at a 'relative maximum or minimum' of the 'greatest sinuosities' of the terrain, and its course determined relative to the principal slopes of the pertaining terrain. These criteria (which left out the use of canal tunnels, a possibility to be appraised relative to soil geology rather than physical geography) should minimise, or at least reduce, the lengths of the canal and the number of required locks. In addition, in concentrating on the main rivers of the region, they sought to produce a canal designed to run close to major towns and villages, and so increase its economic potential (pp. 281–288).

While lacking in quantitative details, the proposal in this paper forms an intriguing extension to the applicability of descriptive geometry. During his professional career Brisson designed canals in various parts of France, doubtless with these considerations in mind, and they surely influenced the report on the interior navigation of France written by the *Directeur* of his *Corps* (Becquey *1821a*).

Around that time Brisson himself wrote a detailed account of the total navigation system of France, which was submitted to the *Académie* in September 1827 (*PV*, *8*, 594). At the end of the year Dupin sent in a long report (but only dealing with canals) which was published as a short book *1828a*. Brisson died in October 1828; in the following year his essay appeared posthumously as *1829a*, under the editorship of A. Duleau (an excellently named colleague at the *Ecole des Ponts et Chaussées*: see § 15.2.4), with the earlier paper included as an appendix.

8.3.4 Dupin on the stability of floating bodies, 1814

Dupin himself was deeply concerned with naval engineering in his early career. Later he turned to educational and social aspects of engineering, and to such effect that he is a major figure in Chapter 16. His biography is postponed until § 16.3.4, therefore; but it is meet to describe here one example of these early interests.

Like Brisson, Dupin was devoted to Monge and descriptive geometry; as was mentioned in § 8.2.1, he published the first biography of Monge, in 1819. After graduating from the *Ecole Polytechnique* as top student in 1803, he went straight into the *Corps Maritime*, where he worked on the design and restoration of ports at various places abroad (§ 8.4.6). But he also researched intensively on descriptive geometry and the geometry of surfaces. On his return to France in 1812, he published a volume *1813a* called 'Developments of geometry [...] to follow the descriptive geometry and the analytical geometry of Mr. Monge'. Dupin was then 29 years old.

One application which Dupin made of his results was to the stability of vessels. He submitted a paper on this topic to the *classe* in January 1814 (*PV*, *5*, 300); Carnot reported at length in August, in a most interesting appraisal which was published in Gergonne's *Annales* as Carnot *1814a*.[1]) Dupin himself published the report in two of his later books; one of these contained also his paper, which is cited as *1822b*. He had taken care already to publish a summary paper *1815b* in the *Correspondance sur l'Ecole Polytechnique*, and Delambre had given him attention in a *classe* notice (*1818a*, liv-lv).

The stability of floating bodies was usually presented in textbooks as part of hydrostatics. The 'plane of floatation' determined the shape of the displaced water; stability depended upon the relationship between the metacentre and the centre of gravity of the body; and so on. Dupin possibly learnt his material as a student from de Prony's lectures and the *Mécanique philosophique* (*1800a*, 234-289); Poisson gave a similar account in his textbook (*Mechanics*$_1$ (1811), *2*, 389-428). Further, Bossut reprinted some of his studies of the 1760s on the trimming of ships in his book *1812a* of miscellanea.

The mathematics can become quite tricky; even the floatation of an object with triangular cross-section leads to a quartic to determine the positions of equilibrium (on pp. 394-395 of his textbook Poisson nicely used Descartes's rule of signs (§ 4.6.3) to determine their number). Dupin saw the potential of the geometry of surfaces to produce more general theory, especially avoiding the usual assumption that the body was symmetrical in shape. While still a student at the *Ecole Polytechnique* he had produced some useful theorems on second-order curves, and in a paper *1807a* published in the *Journal* (in the volume containing also Brisson's papers, as it happens) he had presented various results which served as the geometrical basis for his study of stability of floating bodies.

When a body B, assumed to be convex in shape, floats, a 'plane of floatation' p could be traced on its surface, and the 'centre of the hull' γ determined. If the distribution of material within B were changed, so could its position of floatation, and thus the planes p and the centres γ. Dupin explored the relationship between p and γ, and their consequences for stability, using elements of descriptive geometry and especially his theorems of *1807a*. Here, by way of an example, is one of his basic considerations (*1822b*, 18-22).

---

[1]) Poinsot was the co-reporter, and his draft report survives at *BI*, ms. 954, fols. 192-199v. A letter to Dupin from Carnot of 8.8.1813 on floating bodies is held in Carnot's *dossier personnel* at *AS*, and one from Dupin to Lacroix of 9.7.1814 at *BI*, ms. 2396. Habicht *1974a* and *1978a* are excellent surveys of Euler's ship theory in general; our period has been largely ignored.

Figure 834.1 shows two 'centres of the hull' γ_1 and γ_2 for positions of floatation given by planes a_1c_1 and a_2c_2. Assuming that the total weight of B was unchanged, the volumes below a_1c_1 and a_2c_2 were equal. Thus the volume a_1Fa_2 equalled c_1Fc_2, and so their moments about $\gamma_1\gamma_2$ were equal also. This meant that $\gamma_1\gamma_2$ was parallel to the line joining their centres of gravity m and n. When a_1c_1 and a_2c_2 were infinitesimally separated, $\gamma_1\gamma_2$ became the tangent plane to the curve of 'centres of the hull', parallel to the corresponding plane of floatation. Dupin then proceeded to various theorems of stability; for example, that it would be stable/unstable if the distance between γ and the centre of gravity of B was locally minimal/maximal (pp. 48–50). He drew heavily on the theorems on orthogonal surfaces proved in his book *1813a*.

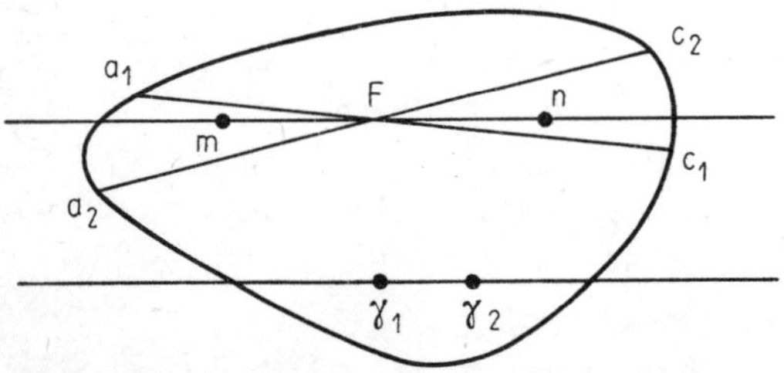

Figure 834.1 Dupin's diagram for a floating body (after his *1822b*, fig. 5)

That book, as its title (quoted earlier) indicates, was a development of Monge's descriptive geometry. The book *1822a* containing the paper on floating bodies was a continuation of *1813a*, and it also contained two papers (*1822c* and *1822d*) on the planning of roads. In those papers Dupin made use of optima (in effect, aspects of geodesics), as in his paper on floating bodies—and also, indeed as in Dupuis-Torcy and Brisson's analysis of canal design, which, perhaps surprisingly, he did not cite. He also considered other aspects of road design in these papers.

Both Dupin's and Brisson's work brings some excitement to descriptive geometry and its related fields. Had such studies received more publicity in the teaching of the subject, it might have received more acclaim. As it was, however, the main emphases were laid on the stuff described in § 8.3.2. Booker has admirably summed up the consequences as follows (*1979a*, 113):

> [...] if one is the academic type, one may [...] wonder how this can be developed for its own sake, just to see what turns up; if one is, on the other hand, a practical man, one may well wonder whether this doctrine, beautiful as it is, has any practical use. Who wants to find the centre of a sphere which will pass through four points in space?

8.4 On the engineering science of materials

> Moreover, the Sciences are monuments devoted to the public good; each citizen owes to them a tribute proportional to his talents. While the great men, carried to the summit of the edifice, draw & put up the higher floors, the ordinary artists scattered in the lower floors, or hidden in the obscurity of the foundations, must only seek to improve what cleverer hands have created.
>
> Coulomb *1776a*, introduction

The contributions of Coulomb to engineering now come into their own, for his studies of the 1770s and 1780s dominate this and the next section. Most of the pertinent papers were reprinted in a posthumous volume *1821a* entitled *Théorie des machines simples*. Bachelier was the publisher: he had been selling offprints for years, and the evident durability of the work doubtless encouraged him to produce this version. 1,000 copies were run off (*AN*, F^{18} 87, no. 12).

8.4.1 Coulomb's studies of friction and torsion

In the whole realm of physical phenomena, one of the most difficult to analyse in the 18th century was friction. There seemed to be many parameters involved in its effect, and its was hard both to find theoretical relationships between them and to devise experiments where only a few would be examined at once. Even its status within Newtonian mechanics was not clear; if it involved 'forces', then they had to be proportional to some distance r, but it was not clear what the 'r' might refer to, or if -2 was the correct index for it.

The history of friction is an important example of engineering influencing science, for the study was heavily motivated by the growing use of machines in society. And it was within this tradition that Coulomb advanced the subject in the 1770s with systematic and careful studies together with modest, but useful and effective, mathematical analyses.[1)]

Coulomb's first published paper, *1776a*, which provided the quotation above, covered several problems in engineering statics, mainly concerning friction and

---

[1)] Gillmor provides a useful guide to the history of friction and torsion, and references to other histories, in his *1971a*, chs. 4–5. Heyman *1972a* discusses some of Coulomb's applications of friction, but his translation there of Coulomb *1776a* is too free. Bell *1973a*, 168–179 provides a succinct summary of Coulomb's contributions.

rupture: it is the seminal paper for this section. In particular, it led him to extensive studies of 'simple machines', in which behaviour was governed by dynamical friction. He wrote up a massive paper which won an *Académie* prize in 1781 (his 46th year); like its predecessor, it appeared in the *Savants étrangers*, as *1785a* (§ 16.3.4). This paper was the source for most of the extensive section 'On machines and motors' in de Prony's *Hydraulique* (*1* (1790), 427-574), and it was to give its name to Bachelier's posthumous book.

Coulomb started out from Amontons's law that 'the resistance due to friction is proportional to the pressure' (the compressive force between the contacting bodies), although he knew that it was not fully general (*1776a*, art. 4). In his *1785a* he studied static and dynamic sliding friction for various rough and smooth surfaces, and then examined the stiffness of ropes and the role of friction in the performance of axles. He considered various aspects of the surfaces in contact (area, type(s) of material, roughness, graininess, velocity of motion, use of lubricants, and so on) as well as other factors, such as temperature and air pressure.

Although Coulomb drew various conclusions from Amontons's law about statical and dynamical friction (*1785a*, 95-100), he could not derive from it any equations, which instead were induced from experimental results. A good example is his study in art. 35 of the force F required to move pieces of oak lubricated by tallow as a function of the time T over which they are first allowed to remain in static contact. His experiment led him to think that

$$F = A + mT^{\mu}, \quad \text{where } A, m \text{ and } \mu \text{ were constants,} \tag{841.1}$$

was 'the simplest expression that one could imagine' for the relationship; but since the experimental data suggestes that $\mu \triangleq 0.2$, the falsehood then followed that F and T would go to infinity together. So he suggested instead the form

$$F = (A + mT^{\mu})/(C + T^{\mu}), \quad \text{with } C \text{ also a constant.} \tag{841.2}$$

An important context in which Coulomb studied friction was the action of pivots and chapes, which he used in his mounting of needles in his studies of magnetism. His first paper on magnetism (§ 7.5.2) considered the action (*1780a*, ch. 4), and a few years later he presented to the *Académie* his experiments 'on the force of torsion and on the elasticity of metal wires' in a paper *1787a* which is notable for his introduction of the general torsion balance (and indeed for the experimental subtleties evidenced in general).

Coulomb suspended an object, smooth and cylindrical to minimise air resistance, from a wire made of the material to be tested, and fixed a pointer on its base to measure its angular deflection when set in azimuthal oscillation. The mathematical basis of his theory lay in the assumption that torque, or 'momentum of the force of torsion', was proportional to its angle of deflection $(A - S)$ from its initial value A. If a particle p of the cylinder W was at distance r from the axis of suspension and moved with velocity v, then, 'by the principles of Dynamics' (namely, angular momentum),

$$n(A - S)dt = \left(\int_W pr^2\right)dv, \tag{841.3}$$

where n was the torsional constant of the material of the wire. If the cylinder were of radius a, and its circumference moved with velocity u, then

$$v/r = u/a; \quad \text{and also} \quad u/a = dS/dt \tag{841.4}$$

so that (841.3) became, after one integration,

$$n(2AS - S^2) = (u^2/a^2) \int_W pr^2. \tag{841.5}$$

Hence, using (841.5) in $(841.4)_2$, he found

$$dt = \frac{a}{u}\, dS = \frac{dS}{\sqrt{2AS - S^2}} \sqrt{\int_W pr^2/u}. \tag{841.6}$$

Now
$$\int \frac{dS}{\sqrt{2AS - S^2}} = \sin^{-1}\left(\frac{A}{S} - 1\right), \tag{841.7}$$

so that the time T of oscillation of W between its extreme positions was given by

$$T = 2\int_0^A (dt/dS)dS = \pi \sqrt{\int_W pr^2/n}. \tag{841.8}$$

The integral expressed the moment of inertia of W with respect to its axis. As this quantity was $\frac{1}{2}Ma^2$ for the cylinder, where M was its mass, (841.8) reduced finally to the form

$$T = \pi\sqrt{Ma^2/2n}, \tag{841.9}$$

from which n could be calculated (*1787a*, arts. 5-7).

In a noteworthy desimplification of his analysis Coulomb allowed for non-linear torsion by modifying his expression for torque to

$$n(A - S) + \sum_r \mu_r(A - S)^{m_r}, \tag{841.10}$$

where the $\{\mu_r\}$ and $\{m_r\}$ were to be determined, both in number and magnitude (art. 8). (Was this expression the inspiration for de Prony's 'veritable form'

(353.1) for gas expansion?) Coulomb used (841.10) mainly in an interesting study of frictional resistance imparted to W when it rotated in fluids (art. 18). His experimental work was largely concerned with the verisimilitude of (841.10) for various materials and the possible involvement of other factors (such as the length of the wire). The most interesting of his general conclusions are perhaps those in art. 28, where he commented on the difficulty of achieving close repeatability in experiments, and concluded that air resistance 'only enters into our experiments, for quantities that one can neglect'.

Coulomb returned to these studies in later papers. His further thoughts on fluid resistance are described in § 8.5.2; here we take up a paper which he sent in to the *Académie* in July 1789 and which was eventually published as *1797a*. Entitled 'On the friction of the point of pivots', it took up his analysis of friction for rotating magnetic needles (*1780a*, arts. 72–73) and examind in general the measurement of resistance to rotatory motion. He rotated an open brass hoop H, loaded with weights at each end in order to give the system a low centre of gravity, about a lens of some smooth material mounted under its apex, and counted the number of revolutions to cessation. He had had second thoughts on air resistance, for he executed some of his experiments in *vacuo* inside a bell jar (*1797a*, art. 3).

Theoretically Coulomb proceeded much as in (841.3), angular momentum now indicating that

$$Fdt = -du \int_{\mathrm{H}} pr^2/a, \tag{841.11}$$

where F was the 'momentum of resistance' (or frictional torque) of the resisting force (Coulomb wrote 'A' again). Integration yielded

$$2Fx = (u_0^2 - u^2) \int_{\mathrm{H}} pr^2/a, \quad \text{where} \quad \dot{x} := u. \tag{841.12}$$

If $x = X$ when $u = 0$, then it followed that

$$u_0^2/2X = F\Big/\left(\int_H pr^2/a\right) = \text{constant} \tag{841.13}$$

(art. 4). He then used his experimental data on checking the constancy of $u_0^2/2X$ (or, rather, a proportion to it, since in effect he determined the initial *angular* velocity by noting the time taken to execute the first revolution: art. 6). He varied the surface pressure P by increasing the weights on the ends of the hoop, and found in arts. 21–22 that the reducing number of revolutions suggested the law

$$F^3 \propto P^4. \tag{841.14}$$

Coulomb's studies of torsion have some similarity at (841.8) with pendulum theory, as he noted in *1787a*, art. 6. Conversely, his study of friction and pivots was a valuable background to the niceties of pendulum theory that Laplace and others were to introduce in the 1810s (§ 6.5.3). Above all, his introduction in (841.3) of the torsional constant n for a material was a major conceptual innovation.

8.4.2 Coulomb on the rupture of beams

In his paper *1776a* on engineering statics, Coulomb studied the 'cohesion' of materials, the 'rupture' of bodies, and the 'resistance' of masonry pillars (arts. 5-9). While Euler and others had written regularly on the bending of flexible and elastic bodies, the strength of materials itself had not been much studied for some time.[1]) Coulomb's contributions were noteworthy for presenting results (not necessarily new ones) of some generality, and for reviving interest in the topic.

Coulomb's basic propositions for the equilibrium of a collection of coplanar forces acting on a plane figure resting on a line AB stated that the resultant would be perpendicular to AB and equal and opposite to the reaction force exerted by AB (arts. 1-3). Thus, for a rectangular horizontal beam held at one end carrying a weight φ at L (see Figure 842.1), the 'tensions' PQ acting on the upper

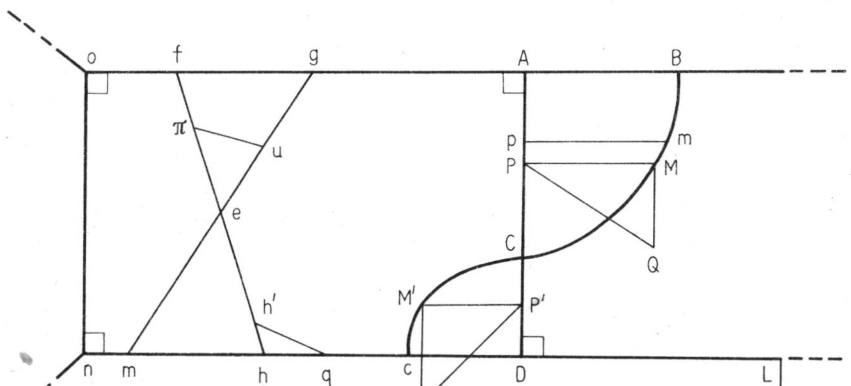

Figure 842.1 Coulomb's diagram for the bending of a horizontal beam (after *1776a*, fig. 6, incorporating corrections to the lettering)

[1]) Useful guides to the history of strength of materials up to Coulomb are provided in the chs. 3 of Gillmor *1971a*, Heyman *1972a* and Timoshenko *1953a*. Related parts of the history of elasticity are covered in Truesdell *1960a*, parts 1, 2 and 4 *passim* (pp. 396-401 on Coulomb *1776a*).

part of the beam at an arbitrary vertical section AD would balance the 'compressions' P′Q′ acting in the lower part (art. 7). The horizontal and vertical components of these forces determined triangles PMQ and P′M′Q′ and thus the curve BmMM′c, which intersected AD at C, the point common also to the neutral line (to use a modern term: Coulomb had none). Taking moments about C for the beam produced the equilibrium equation

$$\text{`}\int \text{Pp.MP.CP} = \varphi \times \text{LD'}. \tag{842.1}$$

Equation (842.1) applied to Coulomb's basic propositions 'whatever be the relation between the extension of the elements of a solid & their cohesion', he added, stressing its generality. He took two materials, acknowledging Bossut's and Viallet's study *1764a* of the construction of dykes as his source for the distinction between the two (still art. 7).

For 'perfectly elastic wood' the neutral line would lie down the centre of the beam, so that when the small 'element' ofhn deformed into ogmn (assumed also to be rectilinear, in some contrast to the earlier generality), the triangles efg and emh were congruent. Thus, setting 'T' for the tensile and compressive forces acting within this small region (Coulomb wrote 'δ'), and neglecting shear, the moments of triangles efg and emh would balance the moment of φ at L, so that

$$\tfrac{1}{6}\, T(fh)^2 = \varphi \times \text{nL}. \tag{842.2}$$

For a material like stone, 'composed of stiff fibres, or which are susceptible neither to compression, nor to stretching', the neutral line lay along the bottom section nL. Assuming that fracture occurred around h as pivot, he found that

$$\tfrac{1}{2}\, T(fh)^2 = \varphi \times \text{nL} \tag{842.3}$$

to be the replacement for (842.2). However, his experimental results refuted it, and so he rejected his hypothesis concerning the nature of stone, and in rather vague terms suggested that just prior to fracture the pivot moved to some point h′ on fh with fracture occuring along a line h′q (art. 7).

To understand the process of fracture more fully Coulomb next analysed the rupture of a stone column of square section (arts. 8–9). I follow his second treatment, which included internal friction. In Figure 842.2 the column was about to crack along CM due to the critical load P, which was just equal to the resistance S to shear of the material (Coulomb wrote 'δ' again) and the retarding force due to friction. Balancing forces along the side RQ of the force triangle FRQ, he found

$$P \sin x = Sa(a \sec x) + (P \cos x)/n, \tag{842.4}$$

where $x := \angle \text{DCM}$ and n^{-1} was the coefficient of friction. (This is Coulomb's ver-

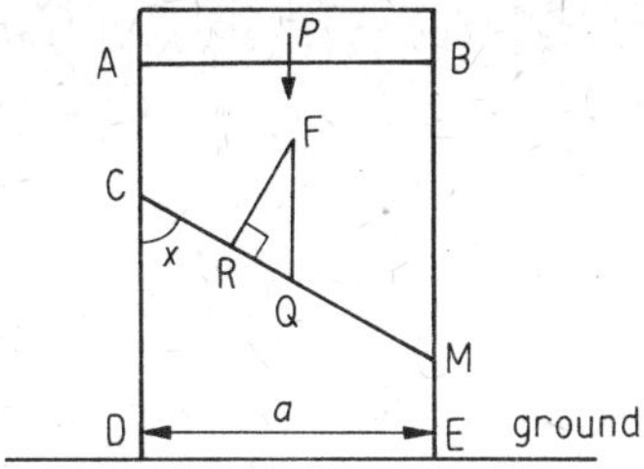

Figure 842.2 Coulomb's diagram for the rupture of a stone column (after *1776a*, fig. 5)

sion of the modern 'Coulomb's equation' for the strength of a material.) Hence

$$P = Sa^2 \sec x/(\sin x - n^{-1} \cos x). \tag{842.5}$$

Coulomb's paper carried the phrase 'rules of *maximis* and *minimis*' in its title, and such an application was made now; for by differentiating (842.5) with respect to x gave for the maximal critical load this equation for x:

$$\tan x = n/(\sqrt{n^2+1} - 1). \tag{842.6}$$

He then compared this result with Musschenbroek's experimental results (cited from *1736a*), and felt satisfied with a discrepancy of 8 % in the value of P for brick (art. 9). Later readers, such as Navier (in Bélidor *1813a*, 170–186) were to be less tolerant.

Mention of Musschenbroek naturally brings to mind the 'Euler-Musschenbroek buckling formula', which in turn recalls Euler's work on this problem. In fact, several of Euler's papers on buckling had not been published, or even written, when Coulomb prepared his paper; but it is doubtful whether Coulomb would have used them anyway, for Euler (and also Lagrange in papers of the 1770s) studied buckling within their tradition of the mathematics of elasticity, and so launched themselves into heavy sums involving elliptic integrals and Bessel functions in series expansions.[1]) The difference between the theoretical analyst whose calculations are hard (though by no means impossible) to implement and the practical engineer whose theoretical assumptions are hard to accept (and impossible in Coulomb's art. 8, when 'one disregards, to simplify, the weight of the column' in analysis of rupture ignoring internal friction), is well exemplified here: several of the philosophical issues of § 1.3.2 are present. There is even a difference of problem, or at least a difference of emphasis within the problem-area;

[1]) These studies are described in Truesdell *1960a*, 345–370. This tradition in elasticity theory was noted briefly in § 7.3.3, and the extent of knowledge of Bessel-function theory was pointed out in § 3.4.4.

for Euler wanted to see how the column would *bend*, but Coulomb was primarily concerned with how it would *break*.

8.4.3 Girard on the strength of materials, 1798

Girard now enters our story, on account of his research, begun in 1787, on the strength of materials. His book *1798a*, published during his 34th year, was the most extended account of the subject to date. A German translation appeared in 1803, and some of the results were used by J.M. Sganzin in his lectures of that time at the *Ecole Polytechnique* on 'stereotomy applied to roads and bridges' (*EP*, III 3b). In a discussion of methods of 'calculating the force of pieces of wood', Navier *1813b* discussed pertinent parts of Girard's book and their historical background (see § 15.2.2 for the context of his interest).

Girard's book began with a lengthy historical survey encompassing both theoretical and experimental studies, with Coulomb's *1776a* singled out for especial praise (*1798a*, xxxvi-xxxix). There then followed four sections: general formulae for the resistance of solids, and their application to 'hypotheses' on the extensibility of fibres (pp. 1-77); solids of 'equal' (or homogeneous) resistance, loaded and unloaded (pp. 78-120); experiments on the resistance and elasticity of oak and deal (pp. 121-188 and nine tables), and the flexure of 'perfectly' and 'imperfectly' elastic bodies (pp. 189-220 and two tables).

Much of the theory presented in the first two sections ran through results (especially Euler's) on the bending of solids, adopting Galileo's hypothesis that the fibres of the material were inextensible, and also the Marriotte-Leibniz hypothesis that the fibres were both extensible and compressible and that, by Hooke's law, their resistance to extension was proportional to the extension. Since Girard wished to allow for non-small deflections, he had to resort to some cumbersome analyses (for example, the power-series solution on pp. 73-77).

Unfortunately one assumption from the past which Girard accepted (p. 7) was James Bernoulli's claim (*1706a*, Lemma 4) that the moment required to bend a beam a given amount was independent of the position of the neutral line. It is somewhat strange that this "result", which intuitively seems counter to the behaviour one would expect in elastic materials, was normally accepted throughout the 18th century down to Girard's time. In Girard's case it vitiated against the efficacy of his careful experimental studies on wood, reported in the third section of his book. For example, it led him to think on p. 66 that 'experiment has confirmed' Euler's original view that the 'absolute elasticity' (or bending modulus) of a beam was proportional to a cubic product of its cross-sectional dimensions (*1744a*, 1st addition, art. 40) and dismiss his later theory involving quartic powers, which was proposed on the grounds that the effect was internal rather than superficial. Euler offered this alternative first in *1759b*, arts. 5,43 (Girard cited a later appearance): the issue related to the study of buckling noted in the last subsection.

Such issues may have been masked for Girard by the more fundamental problem that real wood is not a "perfect" material anyway. For example, he reported that 'The duration of the action of the load with which the solid is charged influences the expression of its elasticity' (p. 164), a variation which he attributed to atmospheric changes (p. 166). There were differences in performance between different beams of the same material, the effect of fatigue, and so on. Thus in his final section he looked more closely at two aspects 'Of the circumstances of inflection'. Firstly, he tried to analyse the effect on a perfectly elastic cantilever beam of a weight mounted over a pulley and heaving up on its free end over a period of time. He wrote down a partial differential equation to represent the motion; its solution depended on the expressions adopted for loading and flexure, and he found results for special cases (pp. 189-203). He mentioned that one result agreed with earlier work (as reported in de Prony's *Hydraulics, 1* (1790), 219-223), but otherwise his theory was rather speculative. He concluded the section with an attempt to analyse the effect of settling ('*tassement*') on the flexure of a beam by applying the theory of 'aeriform fluids', of which the compressibility was interpreted as describing the internal behaviour of the material of the beam (pp. 203-220).

Girard gave a copy of his book to the *classe* on publication, and de Prony wrote a report *1798d* in March 1798, with Coulomb as co-reporter. The report was very praising, so presumably they passed over (or missed) Girard's views on the neutral line and properties of bending. Both reporters were still active in this and related areas: we now consider de Prony's current studies.

8.4.4 Coulomb and de Prony on the stability of arches

The last article of Coulomb's *1776a* was devoted to 'the equilibrium of arches, in having regard to friction and to cohesion' (art. 18). He concentrated on two sources of rupture which had pointed out by his predecessors; sliding at a join between sections of the arch, and rotation or tilting at a join.[1]) For the section GmMa of the arch in Figure 844.1, its weight φ and the horizontal thrust ('*pression*') A on aG at f would have to balance to prevent sliding. Setting n^{-1} for the

[1]) For valuable summaries of Coulomb's contributions and their prehistory, see Gillmor *1971a*, 109-115; and Heyman *1972a*, ch. 6.

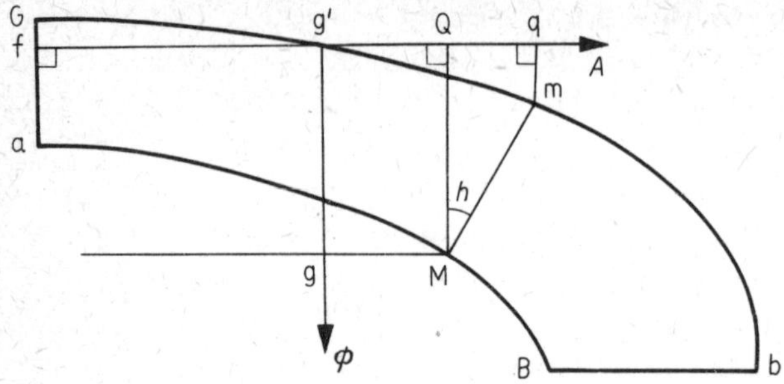

Figure 844.1 Coulomb's diagram for the stability of arches (after *1776a*, fig. 14)

coefficient of friction and C for the 'cohesion' of the masonry, he resolved forces along the join mM (of length z) to find this relation between A and φ when mM was on the point of sliding down:

$$\varphi \cos h = A \sin h + n^{-1}(\varphi \sin h + A \cos h) + Cz, \tag{844.1}$$

where $h := \angle$QMm. At the point when mM would slide up,

$$A \sin h = \varphi \cos h + n^{-1}(\varphi \sin h + A \cos h) + Cz. \tag{844.2}$$

By the theory of maxima and minima, differentiating (844.1–2) with respect to h would produce the value of h required to yield a maximal value for A.

To prevent rupture by rotation, Coulomb claimed (presumably from his 'preliminary propositions' reported in § 8.4.2) that the resultant force acting on section GmMa would have to pass between M and m. However, since he took account of the tensile strength of the mortar along Om, this condition did not have to apply. If rotation about M were to be avoided, then by taking moments about M he obtained the inequality

$$B \times \mathrm{MQ} \geq (\varphi \times \mathrm{gM}) - C'z^2, \tag{844.3}$$

where B was the resultant thrust from Ga assumed to act at f, and C' was the 'cohesion' of the masonry (he wrote 'δ'). Similarly, taking moments about m to avoid rotation there, he found

$$B \times \mathrm{mq} \leq (\varphi \times \mathrm{g'a}) + C'z^2. \tag{844.4}$$

He did not optimise the limiting case equalities in (844.3–4) with respect to h, since he did not know the position of f; instead, he discussed the limits within which B could lie.

By contrast with Coulomb, de Prony preferred to follow the approach advocated by La Hire and pursued in Bossut *1778a* (a paper written before Coulomb's but published later), in which the arch was regarded as composed of a sequence of wedges. de Prony presented a version of this theory in his *Hydraulique* (*1* (1790), 152–168), citing Bossut's paper as 'the best that we know on the statics of arches in general' (p. 168). Doubtless he was attracted by the possibility of casting the theory in his beloved difference equations (§ 3.5.3), an opportunity which he siezed in his *Mécanique philosophique* (*1800a*, 136–139). He also used this theory when he wrote his report *1801e* on experiments carried out by himself and others over the previous four years on the stability of the cupola on the *Panthéon* in Paris. Further, he outlined, in a paper *1810a* in the *Journal de l'Ecole Polytechnique*, a new graphical method of splining (as we would now say) adjacent wedges of circular arcs so as to yield a 'continuous' (to us (§ 3.3.2), differentiable) overall curve. The method was modelled on the idea of a singular solution to a differential equation being obtained by eliminating the parameter a between the general solution $V=0$ and $V_a=0$ (§ 3.3.6). Mindful of the limitations caused by using circular arcs, he devised a simple rule of construction involving an ellipse (or a hyperbola) and a straight line. He took their equations in Cartesian coordinales x and y, eliminated x, y and dy/dx between them and their derivatives, and from the resulting relationship between the constants obtained the required differentiable second-order curve as the *envelope* of a certain sequence of contructed straight lines, after the manner of a singular solution.

de Prony was obviously fond of this work, for he reprinted both papers in 1834 in the recently founded (§ 18.2.1) *Annales des ponts et chaussées*. However, the theory of arches to which it related had the disadvantage of furnishing inaccurate estimates of the thrust line in the arch, as Coulomb had pointed out at the end of his *1776a* when criticising La Hire. While Coulomb's own theory may seem simple regarding the properties of masonry, the possible behaviours in an arch, and its mathematical method, it was not simple-minded, and it gained some appreciation among French engineers of the following generations, as we shall note in § 17.7.6.

8.4.5 Embankments and earth pressure theory

Coulomb's analysis of the stability of arches in *1776a* was preceded by his treatment in arts. 9[bis]–15 of embankments and the thrust of soil. This part of his paper has been widely praised; I note here some of its early influence upon the French.[1])

[1]) Among Coulomb studies, valuable surveys of this work are given in Gillmor *1971a*, 100–107, and Heyman *1972a*, ch. 5. A useful overview is provided by Golder *1953a*. Another valuable history, which has been missed by later historians, is Kötter *1892a*.

Coulomb saw the problem of earth pressing against a vertical retaining wall as strongly similar to that of the rupture of beams. Assuming a homogeneous soil, he set up in Figure 845.1 a force diagram like that in his Figure 842.1. In equilibrium the horizontal reaction A of the wall CE (presumed to act at F), the weight φ of the wedge of earth CBH which was about to slide, and the reaction R on the wedge due to friction and cohesion in the soil, were in balance. While aware that rupture might occur along a curve such as BG, he drew on experience to assume that rupture would occur in planes parallel to BH and thus that CBH could serve as a typical wedge, with B arbitrarily located along CE.

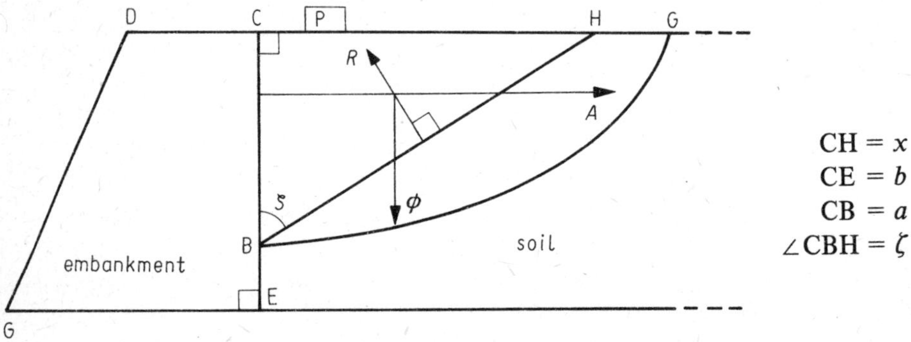

Figure 845.1 Coulomb's diagram for earth pressure (after *1776a*, fig. 7)

Proceeding as (842.3), then, Coulomb resolved forces along BH to find that for limiting values of the horizontal force A the equations

$$A(x \pm an^{-1}) = \tfrac{1}{2}\varrho ax(a \mp xn^{-1}) \mp F(a^2 + x^2), \quad \text{with} \quad \tfrac{1}{2}\varrho ax = \varphi, \tag{845.1}$$

where n^{-1} marked once again the coefficient of friction of the soil and ϱ its density, F was the force due to the cohesion of the soil (Coulomb wrote 'δ' once more), and the upper/lower sign represented the case of the wedge sliding down/-up. As before, he now drew on the theory of extrema to find the maximal value which A could attain for downward slippage by differentiating (845.1) with respect to x and obtaining, as in (842.5),

$$nx = -a + a\sqrt{1+n^2} \quad \text{and thus} \quad A_{\max} = K_1a^2 - FK_2a, \tag{845.2}$$

$$\text{where } K_1 = \varrho\left(\sqrt{1+n^2} - 1\right)^2/2n \quad \text{and} \quad K_2 = 2\left(\sqrt{1+n^2} - 1\right)/n. \tag{845.3}$$

The expression $(845.2)_1$, interestingly, was independent of cohesion. If friction were ignored, then n would be infinite, leading in $(845.2)_1$ to the equation

$$x = a \quad \text{and thus} \quad \zeta = \pi/4 \tag{845.4}$$

(arts. 9[bis]-10). This result had been known before, but as an empirical law. Furthermore, in contradiction to past lore, Coulomb did not equate the angle of slippage calculable from (845.4) with the angle of natural repose of the soil (the state which applied if no retaining wall was present).

Coulomb now turned to the strength of the embankment GECD. Implicitly assuming that the differential increase in pressure at B was a function of the pressure from C down to B, he used $(845.2)_2$ to show that the moment T of the thrust A about G was given by

$$T := \int_0^b (b-a)\, dA/da\, da = \frac{1}{3} K_1 b^3 - \frac{1}{2} \varrho K_2 b^2. \tag{845.5}$$

To prevent the embankment from turning over at G, its moment there would have to exceed T. He did not produce the simple inequality to express this result but took as an example the special case (845.4), took GECD to be a rectangle, and assuming (in rather strange conjunction) that the embankment was both 'solid and indivisible' and also as the same density as the rupturable earth, obtained a limiting ratio of about 1:10 for the proportions of width to height (art. 11). He also refined his analyses by allowing for the earth to be carrying a load additional to its own weight (art. 13), considering the frictional forces along CE (art. 14), and reworking the calculation of maximal thrust of non-cohesive soil for the curve BG rather than the line BH (art. 15).

Coulomb's approach was adopted in various of de Prony's writings. He used it in his *Hydraulique* (*1* (1790), 286-292) and summarised the principal features in a brief paper *1799d* for the *Société Philomatique*. Then he treated the theory in more detail in the *Mécanique philosophique* (*1800a*, 300-313) and especially in a short book *1802c* (which was distributed among the students at the *Ecole Polytechnique*). He showed again his penchant for trigonometric and numerical methods.

de Prony laid more emphasis on the distinction between slippage and repose, suggesting that the angle θ for the latter was given by the coefficient of friction so that (in our notation)

$$\tan\theta = n^{-1}. \tag{845.6}$$

His own version of (845.1), with upper signs chosen, read:

$$A = \tfrac{1}{2}\varrho a^2 \tan\zeta \tan(\theta-\zeta) - Fa(\tan\zeta + \tan(\theta-\zeta)). \tag{845.7}$$

Differentiating with respect to ζ gave for maximal thrust

$$2\zeta = \theta \quad \text{and thus} \quad A_{\max} = \tfrac{1}{2}\varrho a^2 t^2 - 2Fat, \quad \text{where} \quad t := \tan\tfrac{1}{2}\theta. \tag{845.8}$$

$(845.8)_2$ was his trigonometric version of Coulomb's $(845.2)_2$ (de Prony *1802c*, 5-9).[1])

de Prony then worked out a series of calculations which one could use in practice in embankment construction. If a_0 were the maximum depth to which one could dig perpendicularly down before the bank thus exposed would collapse, then $(845.8)_2$ would apply with A_{max} set at zero, thus yielding

$$C = \tfrac{1}{4}\varrho a_0 t. \qquad (845.9)$$

Substitution of this value of C into $(845.8)_2$ gave a simple formula for A_{max}; when placed in (845.7) with A set to zero it yielded a quadratic for $\tan\zeta$ in terms of a and a_0, which provided information on the height of an embankment retaining no earth (as when under construction, for example). Coulomb's (845.5) on the possibility of overturning was also employed (pp. 9-21). Like a true difference analyst, de Prony also discussed the case where the embankment was topped by soils of different densities (pp. 21-26). He continued with an account of a graphical method for determining the elevation of a retaining wall from a specially prepared diagram, on which one marked the relevant values of the parameters (the heights and slopes of the walls, and the densities of their materials and of the body of matter to be retained: pp. 31-38). He gave a rather more detailed account of the method in a 12-page pamphlet *1802b*, which was also published in the *Annales des arts et manufactures*.

This publicising of Coulomb's approach helped to bring it into general circulation. For example, Navier *1809a* wrote a long review in the *Moniteur universel* of de Prony's book and the pamphlet (the piece was presumably commissioned as a review of the treatise Mayniel *1808a*, which he criticised over earth pressure and embankments): he also gave an account in one of his notes in Bélidor *1813a*, 91-98. But the next advance in theory, and the last to be discussed here, came from an unexpected source.

We saw in § 4.3.5 that Arbogast's 'calculus of derivations' was refined by Jacques Français, his follower. Français's teaching at the military school at Metz in the 1810s included earth pressure theory, and he found Coulomb's and de Prony's work limited in scope and Mayniel's account unreliable anyway. So he gave the subject his own treatment in a distinguished paper, written in the

---

[1]) In some versions of this theory de Prony's angle ξ is presented as its right-angle complement. The lettering has been changed as follows:

Here:	A	Φ	Θ	f	a	a_0	ϱ	F
de Prony:	P	Q	τ	n	h	$h,$	π	γ

1810s and published in the *Mémorial de l'Officier du Génie* as Français *1820a*.[1]) A partial translation into German was published in Vienna in 1828.

Français wished to examine the consequences of allowing the internal wall of an embankment or rampart to slope. de Prony had continued to assume that the reactive force A was horizontal and had not allowed for the change of shape of the wedge of earth. So Français reworked the Coulomb/de Prony theory using Figure 845.2.

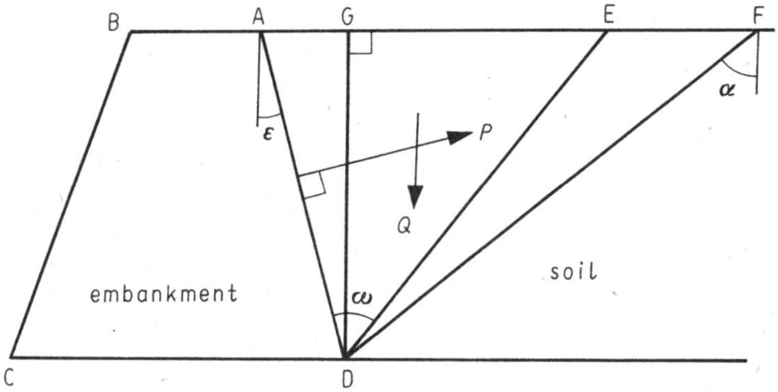

$\angle$ADG $= \varepsilon$
$\angle$GDF $= \alpha$
$\angle$EDA $= \omega$
GD $= h$

Figure 845.2 Français's diagram for earth pressure (following *1820a*, fig. 1, but cast in mirror image to ease comparison with Figure 845.1)

The resistive force P perpendicular to AD and the weight Q of the wedge ADE led to forces acting downwards along the edge DE of slip; but in equilibrium they were balanced by forces contributed by the cohesion and friction of the soil, which took coefficients c and f respectively. Thus in equilibrium, resolving along DE,

$$P \sin \omega = Q \cos(\omega - \varepsilon) - f(P \cos \omega + Q \sin(\omega - \varepsilon)) - ch \sec(\omega - \varepsilon), \qquad (845.10)$$

where h was the height GD of the embankment. Setting $\angle$GDF equal to the complement of the angle of friction (as we now normally designate it) and introducing ϱ as the density of the earth yielded the equations

$$f = \cot \alpha \quad \text{and} \quad 2Q = \varrho h^2(\tan(\omega - \varepsilon) + \tan \varepsilon) \qquad (845.11)$$

and thus converted P to the following trigonometric form:

[1]) Français sent a manuscript version of *1820a* to de Prony; it is now kept at *ENPC*, ms. 913. It is dated 18. 7. 1816, with an addition of 6. 6. 1817.

$$P = [(\tfrac{1}{2}\varrho h^2 + ch\tan\alpha)\tan(\omega - \varepsilon)\tan(\alpha - \omega) + (\tfrac{1}{2}\varrho h^2\tan\varepsilon - ch)\tan\alpha]\,(1 - \tan\varepsilon\tan(\alpha - \varepsilon))\cos\varepsilon. \quad (845.12)$$

Differentiating to find the maximum value for P produced

$$2\omega = \alpha + \varepsilon \quad (845.13)$$

as the value of ω which, when placed in (845.12), yielded

$$P_{\max} = [\tfrac{1}{2}\varrho h^2(\tan\tfrac{1}{2}(\alpha + \varepsilon) - \tan\varepsilon)^2 - 2ch\tan\tfrac{1}{2}(\alpha + \varepsilon)\,(1 - \tan\varepsilon\cot(\alpha + \varepsilon))]\cos\varepsilon. \quad (845.14)$$

This result reduced to de Prony's $(845.8)_2$ when $\alpha = 0$, as it should do (Français *1820a*, 160-164).

Français then repeated various of de Prony's calculations for his own results; determining c from $P_{\max} = 0$, taking moments about C to investigate overturn, and so on (pp. 164-172). Some of the formulae were inevitably complicated, although he tried his best to convert them, *à la* de Prony, into forms suitable for logarithmic computation. Presumably these difficulties dissuaded him from analysing curves, rather than planes, of slippage. But he went on to consider other aspects of practical embankment building; for example, the effect of settlement on the density of the soil and the preferable shape(s) of an embankment (pp. 172-184), and the possibility that the embankment would slip upon its (horizontal) base (pp. 182-188). He provided a table of values for his equivalent to de Prony's a_0 of (845.9) (pp. 188-194), and, in an addition requested by the *Comité des Fortifications*, some considerations of the foundations on which the embankment was built (pp. 195-206).

Français brought to this research the liking for algebra which we noticed in § 4.3.5. Indeed, his paper possibly extended the usefulness of algebra in this context near to its limit; later work would rely more on the calculus or upon geometrical methods, as we shall see in § 17.7.6 in a paper of the 1820s written by Lamé and Clapeyron.

8.4.6 Dupin on the non-linear flexure of timber, 1813

In the brief notes on Dupin's early career in § 8.3.4, it was mentioned that he entered the *Corps Maritime*. This professional activity led him to study the flexure of timber, in connection with ship-building. In April 1813 he submitted to the *classe* the results of his experiments carried out two years earlier at the arsenal at Corfu (*PV*, *5*, 199): three months after submission, the engineer Sané, in a report *1813a* co-signed by Carnot and de Prony, found the piece to be 'worthy of praise and encouragement', and it appeared in the *Journal de l'Ecole Polytechnique* as Dupin *1815a*, after a summary *1814a* had come out in Gergonne's *Annales*.

This research is notable in the history of experimental mechanics for its careful study of the deflection y of a simply supported wooden beam as a function of its loading m, shape and dimensions (see Bell *1973a*, 15–22). Of particular interest is Dupin's assertion that y was a *non-linear* function of m: 'I say that one should have this general equation' for the central loading of all the possible types of wood:

$$y = H + (m-1)\,\delta' + \tfrac{1}{2}(m-1)\ (m-2)\,\delta'', \tag{846.1}$$

where H was the value for y when $m = 1$ (he worked in units of four kilogrammes), and δ' and δ'' were 'two arbitrary constants dependent upon the force and the dimensions of the wood' (*1815a*, 146).

In form (846.1) was cast as a sum of differences, which must have pleased Dupin's former teacher de Prony. As a relation between y and m, it gave a parabola; but for y as a function of horizontal displacement x from the centre, he proposed from his experiments the 'comparative hyperbola'

$$a^2y^2/b^2 = x^2 + 2ax, \tag{846.2}$$

where a and b were constants (pp. 175–177), as a 'hypothesis', '*without fear of making appreciable errors in practice*' (p. 179).

Much of Dupin's experimental work was concerned with these two equations, and other equations and notions related to them. Among other results, he found experimentally that y varied inversely with the cube of the thickness of a beam of rectangular cross-section, and he drew on the view that the deflecting force at a point was proportional to the curvature there to prove this finding by means of a differential model (pp. 163–166). He also found that the neutral line ('invariable fibre') was constant in length under flexure (p. 166).

Once again the contrast between the engineer and the theoretician is evident; for we are far away from the differential equation (732.10) for an elastic surface which Poisson was then seeking. And in this case the theoreticians won in influence; for this *linear* theory captured the attention, not only with Poisson and Germain at this time but also later with Navier, Cauchy and again Poisson. This is the tale to be told in Chapter 15.

8.5 On the engineering studies of fluids

8.5.1 Theory and practice in the mathematical theories of fluids

The general theory of hydrodynamics is to be taken up in § 10.3, when we see Poisson's and Cauchy's own methods of solution, developed in the 1810s. But some comments on the state of knowledge are required here in preface to the discussion of fluid flow in this section.

Lagrange's *Méchanique analitique* of 1788 was usually taken as the authoritative source. As was indicated in Table 521.1, its last 85 pages were devoted to the general principles of hydrodynamics, together with the motion of 'incompressible' and 'compressible and elastic' fluids. He produced the general equations of motion and the condition for incompressibility; we saw Laplace use forms of these equations in § 5.5.6 in connection with tidal theory. Among other results, Lagrange modified his equations for non-uniform density, discussed the possibility that differentials were exact (and so admitted a potential) and examined some cases of first-order boundary conditions. His solutions, and other manipulations, tended as usual to employ power series, although he gave the functional solution (332.2) to the wave equation.

de Prony gave a more detailed account of fluids in his *Mécanique philosophique* (*1800a*): 110 pages on hydrostatics followed by 164 on hydrodynamics (though we recall from § 5.2.7 that his text usually appeared only in the left hand pages). His account of hydrostatics concentrated on various pressure laws, including a fair amount of the geometry of surfaces for cases such as submerged bodies. For hydrodynamics he covered more or less the same cases as Lagrange, although, as with his treatment of dynamics, he preferred to teach an algebraically expressed version of the approach using differential models rather than from variational principles. He also reported some results obtained since the publication of Lagrange's book. Francoeur's textbook on mechanics followed its usual style, noted in § 5.2.8, of broadly following de Prony's approach, but also providing diagrams;

in 70 pages he surveyed the principal properties of incompressible and heavy fluids, and water-flow through orifices (*Mechanics*$_1$ (1802), 337–410). Poisson's *Cours* and *Traité* of 1810–1811 covered roughly the same ground as Lagrange for hydrostatics (together with the theory of floating bodies, as was recorded in § 8.3.4), although it was somewhat slight on hydrodynamics. The two topics together took up about 20% of the *Traité* (see Table 623.1).

None of these works gave a fair impression of the achievements of the late 18th century in these topics; Daniel Bernoulli and Euler are especially unlucky. Lagrange's narrow conception of mechanics is one of the main causes, as was discussed in § 5.1.1. Montucla's history of mathematics provided some compensation, although the account written by Lalande was not very detailed or technical (Montucla *History*, *3* (1802), 679–719).

However, our major concern here is with the engineering aspects of fluid-flow. Hydraulics (and hydrography) were prominent topics in France in the late 18th century (see Russo *1964a* on its educational aspects), and figures such as Borda, Bossut, Chézy, Coulomb, Dubuat and de Prony devoted much effort to it. As one might expect, de Prony's *Nouvelle architecture hydraulique* included a substantial review of the results (see Table 811.1); indeed, it is still one of the best sources of information.[1]) It serves as the starting point for the topics studied in this section: fluid resistance, water-flow through orifices, water-flow in canals, and floating bodies.

8.5.2 Coulomb on the resistance of fluids, 1800

A basic component of hydraulics is the determination of laws governing the resistance of the fluid to motion, if possible in general situations or at least in special cases or for certain types of fluid. 'The geometers who were occupied with the theory of fluids, from Newton up to Messrs. Euler and d'Alembert', wrote de Prony, 'made such great efforts to submit their analysis of motion independent of any hypothesis. Unfortunately their attempts were found to be unfruitful', despite the power of the calculus (*Hydraulics, 1* (1790), 411). In particular, for resistance as a function of velocity v, various hypotheses had been proposed, the choice partly depending upon the type of fluid and body (if any) involved. In

[1]) The history of 18th-century hydraulics is not widely covered. Some of its mathematical and experimental aspects are noted *passim* in Rennie *1834a* and *1835a*, Forchheimer *1905a* and Truesdell *1954a* and *1955a*. Valuable histories were written at that time or soon afterwards: Bernard *1787a* on rivers, Gauthey *1816a* on canal-building, and his editor and grand-nephew (§ 15.2.2) Navier *1811a* on the construction of harbours along the Seine.

Book 2 of his *Principia* Newton had suggested forces proportional to v, $v^{3/2}$ and v^2, and some linear combinations thereof; his successors mostly dropped $v^{3/2}$, but some tried v^n for an n to be determined. de Prony gave an account of several of the most recent studies in his treatise (pp. 378–410).

Coulomb became attracted to the problem in the late 1790s, during his sixties; for he realised that his torsion balance, already used in other researches, could be employed here to obtain experimental data of greater accuracy than seemed otherwise possible. We noted his first studies in § 8.4.1; in a paper presented to the *classe* in May 1800 (*PV*, *2*, 172) and quickly published as *1801b*, he reported on his new 'experiments designed to determine the coherence of fluids and the laws of their resistance in very slow motions'.

Coulomb set up a torsion pendulum, suspending from a vertically mounted wire a thin vertical shaft on which were mounted the horizontal disc for rotation in a vessel of the liquid under test, and also, further up so as to be above the fluid line, a protractor from which the angular displacements could be measured. He broadly followed the torsion theory described above in § 8.4.1; if an element distant r from its centre were moving with angular velocity u at a moment when the angular displacement had decreased to S from its initial, maximal, value A, then, much as in (843.1),

$$n(A - S)\,dt = du \int_{D} \mu r^2, \tag{852.1}$$

where n was the torsional constant of the wire (arts. 7–9). From his experiments, described later in the paper, he was convinced that the force F due to fluid resistance was given by

$$F = au + bu^2, \quad \text{where } a \text{ and } b \text{ were constants.} \tag{852.2}$$

Thus (852.1) had to be modified (in art. 12) to

$$(n(A - S) - au - bu^2)\,dt = du \int_{D} \mu r^2. \tag{852.3}$$

Equation (852.3) was subject to a rather curious treatment (art. 14). Essentially, Coulomb took the integral of (852.1) as an approximate solution for u, substituted it for u in the terms involving resistance in (852.3), and put $u = 0$ on the right hand side to let S then correspond to a full oscillation (art. 12). He then claimed from experimental evidence that the decrease in A satisfied the relation

$$\Delta A = 2A - S \tag{852.4}$$

(he actually wrote 'dA'!), which simplified the solution just mentioned to the form

$$\Delta A = (m + pA)\,A, \tag{852.5}$$

where m and p were (basically) constants related to c and b in (852.2) (art. 14). Claiming again from experiments (described in arts. 42–46) that $b << a$ and so could be ignored, then p could also go, thus giving as solution to (852.5):

$$A_1 := A - \Delta A = A\,(1 - m). \tag{852.6}$$

Thus after q oscillations, the amplitude A_q was given (in art. 16) by

$$A_q = A\,(1 - m)^q. \tag{852.7}$$

The rest of Coulomb's paper was devoted to an account of ten series of experiments; the data obtained were used to calculate m and p in (852.5). He described various advantages of his procedure over other work on water-flow: for example, that the disc could be suspended at any chosen depth, its (slow) rotation minimised turbulence and other aberrant effects, the apparatus could easily be adjusted to give long or short oscillations (long ones gave the more reliable results), the temperature in the vessel could be varied, and so on (see Gillmor *1971a*, 167–174). While the theory was obviously rudimentary compared with the imposing lore of Lagrangian hydrodynamics, its potential for good experimental work impressed Coulomb's fellow engineers, who contributed to other aspects of water-flow in the first decade of the new century.

8.5.3 de Prony and the hypotheses for measuring water-flow, 1802

> The Engineer Officers must attach some importance to the research of the best method for gauging the capacity of flowing water. The inundations, the sheets, the scours, and the other manoeuvres of the water that they fashion for the protection of fortresses or of camps; the water conduits which the needs of the garrisons require, the diversions of these waters into the headquarters, the establishment of the canals which connect several places together, close off the land to detachments, and favour the operations of armies; in a word, the normal service of the Engineer Officers requires that they can measure products or consumptions of water, now by approximative and rapid procedures, now by slower, but more rigorous, methods.
>
> de Prony *1804e'*, art. 1

At their invitation, de Prony read a report in February 1802 to the *Assemblée des Ponts et Chaussées* on measuring the flow of water. It was published later in the year as a short book *1802d*, with financial support supplied by the *Ecole Polytechnique* (see *EP, CoI*, 2, fol. 177): a long extract *1802e* appeared as a paper in the *Annales des arts et manufactures*. The topic also interested the military, who published in the *Mémorial de l'Officier du Génie* in 1804 an 'Extract' of de Prony's book together with comments by one captain Ferraudy and an appraisal by Recicourt, the director of fortifications. The 'Extract' is actually a freshly written *précis*, possibly not by de Prony himself; but I cite the ensemble of pieces collectively as de Prony *1804e'*. The quotation above is its opening: the author was then in his late forties.

A principal means of measuring water-flow was by its passage through an orifice; but the calculations depended on its shape, size, position and direction. de Prony discussed the problem at some length in his *Hydraulique*, theorising on flow through horizontal and vertical orifices of 'any' shape and reporting experimental work on water-flow through orifices and tubes and overflow from filled vessels (*1* (1790), 328-376: see also his leaflet *1806c*, written for students).

In his new report de Prony emphasised two hypotheses concerning the behaviour of water. One, called the 'parallelism of slices', asserted that when water leaves a vessel by an orifice of larger than infinitesimal width, 'the molecules of the water descend from the constant level towards the orifice, by vertical directions, and with the same degree of velocity; so that each molecule, arriving in the plane of the orifice, is animated by a velocity due to the height which corresponds to it' (*1804e'*, art. 25). Daniel Bernoulli, one of the early advocates of this hypothesis, was aware of its limitations (*1738c*, 13); but Poisson was to make effective use of it (§ 8.5.7), and de Prony was always to remain an enthusiast (§ 17.4.2).

The other hypothesis concerned the '*contraction of the fluid vein*', caused when the orifice 'is at only a small distance from the upper surface of the fluid'. 'The molecules of some slice' now 'follow more or less inclined curves, and move with unequal velocities'; thus 'in many cases, they take there the shape of a *funnel* or *conoid cavity*' (de Prony *1802d*, 4-5). But he was cautious about the general efficacy of the theory: he emphasised that the formulae about to be presented assumed in their derivation a large head of water above the orifice (or upstream of the sluice) lying in placid state, for example (pp. 5-7).

The formulae themselves were standard from the 18th century or even earlier. de Prony used the parallelism hypothesis to prove in *Hydraulics*, *1* (1790), 334-338 (and state in *1802d*, 6-9) that the flow q per second through a vertical rectangular orifice of base b and height c with centre located k below the level of water in a paraboloidal vessel was given by

$$q = b(\sqrt{2k+a} - \sqrt{2k-a})/3\sqrt{g}, \tag{853.1}$$

where g was gravitational acceleration (in suitable units, the new metric system, of course, being preferred). Cavitation was too hard to analyse mathematically; he quoted Newton's contraction factor 0.71 for tiny orifices, but preferred Bossut's experimental result of about 0.62 for somewhat larger circular orifices.[1]) He furnished extensive tables to help calculate q from (853.1) (*1802d*, 34–68); his discussion of g drew on the recent determination of the meridianal arc (§ 5.5.4), Borda's pendulum (§ 6.5.2) and (by implication) his own large tables (§ 3.5.4).

A related method of measuring water-flow, more suitable for use on rivers and canals, was to excavate a basin and allow water to flow out through a sluice gate. de Prony designed a particular form of this device, recommending a length of at least 30 metres, preceded by Π-shaped channels to aid the achievement of stagnation and equipped with a Pitot tube to measure the height of the water (de Prony *1802d*, 16–19; *1804e'*, arts. 31–39). He also chose a mathematical procedure amenable to calculation. Assuming (without argument) that the drop z in level of the water could be related to time t (during outflow) by quadratic interpolation, he produced the formula

$$z = \left(\frac{t}{t_2 - t_1}\right)\left(\frac{t_2 - t}{t_1} z_1 - \frac{t_1 - t}{t_2} z_2\right) \tag{853.2}$$

in terms of the observed readings z_1 and z_2 at t_1 and t_2. Now if the sluice were of area ω, and v the mean velocity of flow through it, then volume $v\omega\, dt$ of water would flow through it in a time interval dt. This volume would also be given by Sdz, where S was the (constant) area of the basin, so that

$$v\omega\, dt = Sdz. \tag{853.3}$$

Hence the flow q of water was given, via (853.3) and the differential of (853.2), as

$$q = \omega v = S\left(\frac{dz}{dt}\right)_{t=0} = S\frac{t_2^2 z_1 - t_1^2 z_2}{(t_2 - t_1)\, t_1 t_2} \tag{853.4}$$

(*1802d*, 20–21). Drawing on his lectures on numerical methods at the *Ecole Poly-*

[1]) de Prony *Hydraulics*, *1* (1790), 368–370; *1802d*, 13–14. Strangely, he ignored Borda's study *1769a* of water-flow in which the 'Borda mouthpiece' (a tube inserted into the orifice) was popularized and its effect on contraction discussed: on this paper, and its neglect, see Truesdell *1955a*, lxiii to lxxvi.

technique (*1796a*, 256–259), he modified (853.4) for the case of n readings taken at uniform intervals of time (*1802d*, 22–25, 30–32).

Equation (853.4) has the advantage that all its quantities are (relatively) easy to measure. However, the underlying theory is not as free of hypotheses as de Prony wished to imply. For example, the evaluation of dz/dt at $t = 0$ seems in danger of over-estimating q; a better value might have been $t = \frac{1}{2}(t_1 + t_2)$, when (853.4) becomes

$$q = S(z_2 - z_1)/(t_2 - t_1). \tag{853.5}$$

Again, (853.2) appears from nowhere, and Ferraudy queried it in the *Mémorial* article, on the grounds that it made the unlikely assumption that the mean velocity was a linear function of time. Ferraudy also found relationships between the values of q incorporated in the formulae for n observations provided by different values of n, and was concerned with the shape of the orifice and the effect of cavitation on the variation of velocity with depth (de Prony *1804e'*, arts. 40–43). This last matter had been discussed by de Prony in arts. 11–12 (although not in his *1802d*): he appealed to the empirical laws (in Dubuat *1786a*, arts. 390–391) that the mean velocity v, the surface velocity U and the bed velocity V were related by

$$2v = U + u \quad \text{and} \quad \sqrt{u} = \sqrt{U} - k \tag{853.6}$$

(where the constant k depended upon the units of measurement). Overall Ferraudy felt sufficiently uneasy about de Prony's method to propose that the water be run off from the basin into a pit for direct measurement of volume (de Prony *1804e'*, art. 43).

Recicourt rounded off the *Mémorial* article by appraising seven methods of measuring water-flow. They included Ferraudy's idea (satisfactory for an approximate evaluation, but refinements rather laborious); direct calculations of ωv of (853.4) (simple, but somewhat suspect in view of the need for Dubuat's (853.6)); use of a formula such as de Prony's (853.1), together with corrections for contractions and other circumstances (needed few observations, but corrections laborious to handle); use of de Prony's (853.4) or its generalisation to n observations (simple observations, but used a theory queried by Ferraudy); and that method together with the time noted to refill the basin to the previous level as a check on the observations made.

This last option was Recicourt's proposal and his preference, but he ended his survey by warning that 'When one has so many methods to obtain the same result, it is a certain sign that none of them is to be found to be perfect'. The readers of the *Mémorial* were left to make up their own minds.

8.5.4 de Prony on water-flow in canals, 1804

de Prony returned to the analysis of water flow in 1804 in a short book entitled 'Physico-mathematical researches on the theory of flowing waters'. Despite the similarity of the title to that of the 1802 predecessor, the purpose here was rather different: a relatively general theory of water-flow was provided in order to aid engineers in the design of pipes and canals. His discussion was wide-ranging (frankly, rather rambling); perhaps he was trying to make up for the surprising omission of canals and rivers from his *Hydraulique*. It is cited as de Prony *1804f*.

The book received considerable attention at the time, including extensive admiring reviews in the *Moniteur universel* from Biot *1804d* and in the *Magasin encyclopédique* from Girard *1805a*, and a notice in Delambre's report on the year's work of the *classe* (*1804a*, 98–99); de Prony published an eight-page pamphlet *1804c* of his own. Later Hachette wrote a summary *1815c* in his *Correspondance sur l'Ecole Polytechnique* (see § 8.5.7 on the context).

de Prony's survey of water-flow tried to take account of viscosity and related properties of the water, and fluid resistance. It is not easy to summarise, especially with the wealth of notations employed—so great indeed, that two large tables of symbols were tipped in (*1804f*, 1, 31). Two of the most important results, however, well convey his methods and intentions.

We saw that de Prony had appealed in 1802 to Dubuat's formulae (853.6) relating the mean, surface and bed velocities. Perhaps in view of the reservations expressed by Ferraudy, he had thought further about these laws. The mean velocity U and surface velocity V should definitely satisfy the properties

$$V = U \quad \text{when} \quad V = 0 \quad \text{and} \quad V = \infty; \tag{854.1}$$

and so he offered, instead of Dubuat's $(853.6)_1$, the relation

$$U = V(V + a)/(V + b), \tag{854.2}$$

where a and b were constants to be determined from observations (p. 75). *How* this equation was obtained remained a mystery, but presumably he appealed to the Newtonian tag on his title page: *'In this (experimental) philosophy propositions are deduced from phenomena and are reduced to universals* [generales] *by induction'*.

In the same philosophical vein de Prony appraised Coulomb's empirical law (852.2) of fluid resistance as 'the glory of this discovery'. From this law he derived the second result which we consider. In a rectilinear closed canal of uniform section, U should be constant, and the water resistance be proportional to the tangent slope I. He used these special conditions to simplify an equation (on

p. 45) for water-flow down to

$$gRI = \alpha U + \beta U^2, \tag{854.3}$$

where g was gravitational acceleration, R the ratio of sectional area to perimeter, and α and β constants to be determined, once again, from observations (pp. 58–62). The equation could be made explicit in U by the usual formula for the roots of a quadratic equation (pp. 64–67), when it took a form similar to one given in 1775 by his predecessor Chézy (see Mouret *1921a*, 241–246).

The introduction to de Prony's book was obviously written afterwards, since it discussed the calculation of the constants of these equations and other numerical matters. The account is particularly interesting, as it constitutes perhaps the first use by another of Laplace's minimax theory of errors described in § 5.5.5.

Laplace's presentation in the *Mécanique céleste* in 1799 had been 'purely analytical', wrote de Prony, 'and as the considerations and the geometrical constructions are much more familiar than abstract analysis to a great number of engineers', he cast it in a form more appropriate for their *Denkweise* (*1804f*, xvi). Given any linear relationship

$$Z = \alpha + \beta X \tag{854.4}$$

for which the "best" values for α and β were to be found from the given data, he plotted them as points in the XZ-plane and constructed the enclosing polygon, which used some (perhaps all) of them as vertices. Then he took each side in turn, drew a parallel line through the most distant vertex of the polygon and measured the distance d_j between the two lines in the Z direction. He then determined the (assumedly unique) minimum of these $\{d_j\}$, and took as best fit line the one mid-way between the side of the polygon involved and its constructed parallel (pp. xviii–xxi). This gave him a geometrical representation of Laplace's minimax condition (555.5).

de Prony also treated Laplace's weighted version (555.6) of Boscovich's criterion of absolute deviation (pp. xxi–xxiv), although in view of Laplace's silence in the *Mécanique céleste* over Boscovich he probably did not know of Boscovich's treatment. He used a simple trigonometric interpretation of the method to show that a minimum was attained.

In order to apply these methods to the equations produced on the body of the book, de Prony converted those equations to the linear form of (854.4). (854.2) for mean velocity became

$$V/(V-U) = \alpha + \beta V, \quad \text{where} \quad \alpha := b/(b-a) \quad \text{and} \quad \beta := 1/(b-a), \tag{854.5}$$

while (854.3) for water-flow was rendered as

$$gRI/U = \alpha + \beta U. \tag{854.6}$$

The values obtained for α and β from the two methods (equal weighting being given to the data in the second method) differed noticeably for (854.5) (pp. xxvii–xxxii). Rather than query the use of linear regression, he weighted the data in the second method to obtain a third set of values; but he told the reader only that 'various means' were used (p. xxix), so there is no way of appraising the merit of these calculations. As so often with de Prony, the aspiration is admirable, but the achievement mixed.

de Prony had the opportunity to use his formulae a few years later when in 1810 he joined a commission to study the Pontine marshes (an area south-east of Rome down to Terracina). He carried out his main researches in 1811 and 1812, and discoursed on the project in January 1815 to a public meeting of the *Institut*; but his report did not appear until 1822, when Firmin Didot brought it out as a large book, with the plates issued in the following year (*1822–1823a*). The book began with his 1815 lecture, and the main text consisted of discussions of the geography and geology of the area, its development since the 1770s and its state as of 1811, and his proposals for its future; but most relevant to us was an extensive introduction, in which he cited his 1802 and 1804 publications and presented many of the formulae there developed, along with some others. They were used principally to aid the calculation of water-flow in the various rivers and canals of the region. He took the chance to reiterate his philosophy of science, for he claimed these formulae to be founded on a 'theory founded absolutely on a reconciliation of the facts and on the use of an analytical method by which one deduces, by observations, the laws of the phenomena' (pp. 49–50).

8.5.5 Girard and the Ourcq canal project

In various places in his 1804 study of water-flow in canals de Prony praised some recent publications by Girard. Now back in France after the Egyptian campaign, Girard had begun to publish on the Ourcq canal project, the task which was to dominate his life's work (even after 1817, when following the Restoration he lost his post as chief engineer). He wrote many reports and articles on the project, and on the related question of the Parisian water supply, between 1803 (his 39th year) and the mid 1820s, when the project was completed. He planned a comprehensive three-volume report, incorporating these works and new material; but he lived only to publish in 1831 the first volume of his *Mémoires sur le canal de l'Ourcq*. A second posthumous volume appeared in 1843, under the care of a colleague; and in 1831 and 1845 there were published two companion volumes of diagrams and plans, many drawn according to the principles of descriptive geometry. The work as a whole is cited as Girard *Ourcq*. Our principal concern is with some of his conceptual ideas, which appeared mostly in his writings of the 1800s and so are treated here. Firstly, however, a brief outline of the project is needed; his histories *1831a* of the project and *1813a* of the Parisian water supply are authoritative sources.

The Ourcq river runs from Fère-en-Tardenois, south-west of Rheims, and after 65 kilometres of a roughly western course, enters the Marne at Lizy about 50 kilometres east of Paris. In 1799 a plan was put forward to divert Pariswards some of the Ourcq water from a point above Lizy. A *classe* commission of Bossut, Cousin, Coulomb and de Prony backed the project in the following January (*PV*, *2*, 77–79), and Girard was put in charge in September 1802, soon after his return from Egypt. The scheme that was eventually adopted tapped the Ourcq at Mareuil, about 25 kilometres upsteam from Lizy, and carried the water somewhat over 100 kilometres to bring it in to Paris at the *Bassin de la Villette*, a basin specially dug during the late 1800s just outside the eastern wall of the city (today the area is in the 19th *arrondissement*). From this basin two further canals were run, each to link up with the Seine: the *Canal Saint Denis*, running eight kilometres to a village of that name; and the *Canal Saint Martin*, which passed roughly south through the city to meet the river opposite the *Jardin de Plantes* (square 15D of Figure 263.1). Thus was formed the *Canal de la Seine à la Seine* (as it was called), which cut off a large loop in the course of the river.

Hundreds of workmen, including many prisoners of war, were involved in the digging. Some of the detailed planning was given to students of the *Ecole des Ponts et Chaussées*, including Fresnel in the late 1810s; even some of the school's students were used, including Cauchy in the late 1800s.[1])

In one of his first pronouncements Girard presented to the *Assemblée des Ponts et Chaussées* in 1803 a report in which he outlined proposals concerning the size of the canal required to accommodate Ourcq river traffic, and also the sectional profile needed to cause the water to flow at the minimum velocity (0.35 metres per second) felt necessary, for reasons of public health, to prevent it from stagnating. For premises he drew on Coulomb's law (852.2) of fluid resistance (although, as de Prony pointed out in *1804f*, 53, he sometimes preferred to equate its two constants *a* and *b*), and laws (853.6) concerning mean velocity (Girard *1803a*, 15–21, 38–43). He also cited his pamphlet *Essai sur le mouvement des eaux courantes*, also written in 1803 but published the following year as Girard *1804a*.[2])

---

[1]) See *AN*, F[14] 2228[2] and 2187[2] respectively (Girard noted Cauchy's involvement in his *1831a*, 90). Among manuscripts on the Ourcq project in *AN*, F[14], see especially the file at 10116 and the collection of maps and plans at 10125.

[2]) Girard submitted the *Essai* to the *CI* in May 1803 (*PV*, *2*, 663). Lacroix and Laplace failed to report, although Lacroix's draft survives at *BI*, ms. 2400. Note Girard's manuscripts on contraction, dated 1792 and 1796, in *ENPC*, mss. 1366 and 1382.

Girard published revised versions of the *Essai* and of parts of the 1803 report in the first (1831) volume of his Ourcq *Mémoires*. A major difference between the two editions of the report is worth noting at once: in the Ourcq version of (852.2), both a and b were included (*1803a*, 330–333).

The *Essai* was Girard's main contribution to theory, and so gains most attention here. To distinguish the two editions, the Ourcq publication is cited by article numbers and the original version (when different) by page numbers. In this work he drew on analogy (a favoured method, apparently: recall from § 8.4.3 his treatment of a stone column as an 'aeriform fluid'). He analysed the motion of a flexible chain upon a surface in a state of 'stable flow', claimed it to be the same as the motion of an incompressible fluid in a flexible pipe, and then in effect interpreted the chain as a stream of flowing water.

Girard's reasoning began as follows. n masses $\{m_j\}$, linked by light inextensible strings in a sequence, moved under gravity on an apparently smooth and fixed surface. With m_j located at (x_j, y_j, z_j), distant r_j from the origin of coordinates, the mutual acceleration φ of the masses was given by

$$\varphi = g \sum_{j=1}^{n} (m_j dz_j/ds_j) \Big/ \sum_{j=1}^{n} m_j. \tag{855.1}$$

Thus, interpreting a chain as a sequence of infinitesimally neighbouring masses dM, (855.1) became a ratio of integrals:

$$\varphi = g\left(\int dM\, dz/ds + B\right)\Big/\left(\int dM + A\right), \text{ with } A \text{ and } B \text{ constants.} \tag{855.2}$$

For a chain of uniform density, $dM = kds$ and (855.2) turned into

$$\varphi = g(z - z_0)/(s - s_0), \tag{855.3}$$

where z_0 and s_0 were values given by one end of the chain (at some $t = 0$?: *1804a*, arts. 27–32). Friction, assumed proportional to the square of the velocity and inversely proportional to 'the' (*sic*) radius r of curvature, needed a fresh term adding to (855.3), which thus became

$$\varphi = g((z - z_0) - f(\sigma - \sigma_0) \mp 2fh(\theta - \theta_0))/(s - s_0), \tag{855.4}$$

$$\text{where} \quad h := u^2/2g, \quad \theta := \int ds/r, \quad \sigma := \int \sqrt{dx^2 + dy^2}\,, \tag{855.5}$$

f was the coefficient of friction, and the sign in (855.4) was chosen according as the curvature of the surface was concave or convex (arts. 40–42).

'[E]xperience has made known that moving objects sometimes suffer another resistance which is proportional to a certain function of velocity' u, continued Girard (art. 42). He claimed that 'asperities' that inevitably lay on surfaces caused resistance proportional to u^2 (a situation surely covered already in obtaining (855.4)). Thus (855.4) had to be modified further to

$$\varphi = [g((z - z_0) - f(\sigma - \sigma_0) \mp 2fh(\theta - \theta_0))/(s - s_0)] - bu^2 \tag{855.6}$$

(pp. 15-17, omitting some inessential details). In the Ourcq version he also included the resistance due to the 'degree of adherence' of even a perfectly polished chain on the surface, which he claimed to be proportional to u; so $(855.6)_2$ gained the term $-au$, with a constant (art. 45).

Girard now considered various curves upon the surface which a chain C, with fixed ends A and B, could adopt. If C exercised pressure p on the surface, then its 'state of permanence' was given by optimal total pressure (pp. 20-29, assuming a differential model); in the Ourcq version variational methods were used to produce

$$\delta \int_{\mathrm{C}} p ds = 0, \quad \text{where also} \quad \delta \int_{\mathrm{C}} ds = 0 \tag{855.7}$$

(art. 56). The curve satisfying (855.6) would be that of 'stable flow' for the chain (art. 62).

Equations (855.6-7) together gave Girard the route that he sought to the theory of water-flow. Citing de Prony's *Mécanique philosophique* as the general source for hydrodynamical theory, and two of Euler's papers (*1761a* and *1771b* on fluid flow, but not *1767c* on the motion of rivers), and (in effect) invoking the hypothesis of parallel slices, he claimed that the continuity condition reduced to

$$u_s = 0, \quad \text{so that, therefore} \quad \delta u = 0 \quad \text{and} \quad du/dt = \varphi \tag{855.8}$$

(where u and φ were functions of t only), while the pressure equilibrium equation became

$$0 = \delta p = -g\delta z - \varphi \delta s \tag{855.9}$$

(arts. 66-70: in the original version this was given in an appendix on pp. 50-56). Integrating (855.9) brought

$$\varphi = g(z - z_0)/(s - s_0) \tag{855.10}$$

(art. 72), of the same form as (855.4), while considerations of equilibrium yielded (855.7) and thus the condition of 'stable flow' for a thin stream of water (art. 75). Further, Coulomb's law (852.2) of fluid resistance, obtained for stagnant water, could be applied (for Girard) to uniform motion in open canals (art. 86); it led to de Prony's equation (854.3) for water-flow down a sloping canal (art. 98). Girard then related these and other results to experimental data obtained by Bossut, Dubuat, de Prony and others (arts. 91-101), and to the Ourcq version he added a section on the corrosive effect of water on the banks of a canal (arts. 114-152).

Girard did not set theoretical considerations aside after this essay; for example, around 1807 he drew on de Prony's formula (854.1) relating mean to surface velocity instead of Dubuat's (853.6) (*1807a*, 541–542). But his attention must have became dominated by the practical and professional concerns attendant upon the construction of the canal, until he was able to compose the *Mémoires*. As a mathematical analysis of water-flow his procedures are at times optimistic, even hair-raisingly so: I pointed out a few curiosities *en route* through my summary, and it is not hard to find several more. However, in 1800 the mathematical theory of water-flow *guided by practical needs* (§ 132.4) was at its early stages, and Girard—like Coulomb and de Prony—has to be recognised as a pioneer as well as a *savant*. His calculations played their role in the construction of the canal, and appraisal of its performance; and while they may seem naive, the canal is still there, and still flowing.

8.5.6 Girard on the flow of fluids in capillary tubes, 1814–1817

Despite the onerous duties connected with the Ourcq canal project, Girard found some time to carry out the first extensive series of experiments on fluid flow in straight capillary tubes. He submitted three papers to the *classe* in December 1814 and the following January and February (*PV*, *5*, 436, 457, 462); they were approved in May by de Prony's report *1815a*, which was published in Delambre's notice *1816a* of the year's work of the *classe*. A month after the report Girard was elected to the *classe* (§ 8.2.2) and so was now able to present his papers. Accordingly, several more came in between April 1816 and October 1817 (*PV*, *6*, 49, 51, 138, 139, 148, 229). Three papers were assembled for publication out of all this for the *Mémoires* (Girard *1816a*, *1818a* and *1817b*): they comprised 240 pages in all. The dating code of the third paper reflects the fact that it first appeared in the new (§ 10.1.2) *Annales de chimie et de physique*, which also took a summary *1817a* of the second paper. The first news in print was a short note *1815a* put out by the *Société Philomatique* and in the *Journal des mines*.

The contrast between theoretical and practical inclinations, a recurring theme in our tale, shows itself again here; for Girard concerned himself with the flow of fluids within capillary tubes, rather than the equilibrium of capillary surfaces which had been Laplace's interest (§ 7.2.2). Girard appealed to the formulae, discussed by himself and de Prony a decade earlier, relating the (constant) velocity of flow u in a tube of length l and internal diameter D, with a drop in height h over the flow. He gave (854.3) as the most general version:

$$gDh = 4lu(a + bu), \tag{856.1}$$

where a and b were to be determined by experiment (*1816a*, 264-265): but in this first paper he concentrated on the cases of '*linear motion properly said*', as he called it, where the fluid adhered to the (smooth) wall of the tube and produced steady flow (p. 288). Since the resistance term bu^2 was held to be due to the 'asperities' of the wall, it was removed from (856.1), which thus became

$$gDh/4lu = a. \tag{856.2}$$

He used copper tubes specially manufactured for the experiments.

Although in his theory Girard allowed for flow in sloping tubes (pp. 263-264), he experimented only on horizontally mounted tubes, taking their fluid from a reservoir. He calculated u by measuring the time taken for a given quantity of fluid to flow out into a receptacle, and then calculated the values, constant over the experiments or not, of a. He summarised his findings on pp. 296-299; de Prony gave a more extended statement in the *classe* report. Among other results, for a given head of water and value of the diameter D of the tube, linear flow obtained for tubes greater than a certain determinable length, which increased with D and with the head of fluid. As regards a, its value decreased with D and increased with temperature, though its rate of increase declined as temperature went up. These last findings showed that the general theory of fluid flow could not apply to capillary flow—doubtless a disappointment to this lover of analogy (§ 8.2.2).

In order to obtain explanations of his findings, Girard examined (856.2) and related formulae, and the experimental situation itself (pp. 306-333); but he was not able to obtain close quantitative corroboration. An interesting speculation was his somewhat Laplacian analysis of the 'sphere of activity' of a molecule on a surface upon the molecules of a neighbouring parallel surface, as a means of measuring the 'adherence' of the surfaces (pp. 316-321).

In his second paper Girard extended his theory by relating the velocity V of descent of the surface of the fluid in the reservoir at level Y to its cross-sectional area X^2 there by means of the continuity condition at the opening of the orifice:

$$VX^2 = uD^2. \tag{856.3}$$

He then took various shapes of the reservoir, giving certain relationships between Y and X, and found the appropriate form for (856.2) and related equations. For example, for a vertical cylinder of constant cross-section A^2 the momentum gained by the fluid (of unit density) in falling in height from h' to h'' was given by

$$-\int_{h'}^{h''} VA^2\,dY = (gD^3/8la)\,(h'^2 - h''^2) \quad \text{(from (856.2-3))} \tag{856.4}$$

$$= A^2V'(h' - h''), \tag{856.5}$$

where V' was defined as the mean velocity of the flow. Assuming that V' was related to the mean velocity u' of flow in the tube by (856.3), the rate of flow Q of fluid in the tube was given by

$$Q := \tfrac{1}{4}\pi D^2 u' = \tfrac{1}{4}\pi A^2 V' = g\pi D^3(h' + h'')/32la, \tag{856.6}$$

thanks to (856.4-5). Since Q was measurable as volume per given time, a could now be calculated (*1818a*, 190-193).

In this paper Girard also reported his experiments on the flow in glass tubes of various fluids at differing temperatures, and found (once again) that the rate of increase of flow itself decreased as the temperature T rose. He sought for explanation by adopting from his first paper (*1816a*, 333-340) the expansions, in power series of T, of the diameter D and of the density of the fluid (introduced into (856.2) by modifying a itself): the coefficients in the series were to be determined by experimental data (*1818a*, 231-236). The short third paper, *1817b*, described a similar series of experiments using ether. Further, a long sequel essay, presented to the *Académie* throughout April 1819 (*PV*, *6*, 437-438) and published in the *Mémoires* as *1824d*, used some of the same ideas in attempting to theorise on the 'liquid atmospheres' that adhered to a body when it was immersed in that liquid; for example, he expanded the 'retarding force' as a power series of the velocity of the body in the fluid (p. 3). He again reported a long series of experiments, in which clay and water were mixed in varying proportions and the rate of precipitation measured in terms of the level of the boundary layer separating the two substances in the vessel.

The last part of Girard's second paper on capillary flow dealt with fluids which left dry the walls of the tube, and for which (856.1) was held to take the form, complementary to (856.2), given by

$$gDh/4l = bu^2. \tag{856.7}$$

He then repeated the argument for flow from a cylindrical reservoir to produce a formula analogous to (856.6), and then reported a few experiments with mercury at various temperatures—rather too few to draw any general conclusions (*1818a*, 239-256). The main thrust of his research lay in the earlier papers, and their emphasis on the parameters affecting capillary flow. While theoreticians were apparently uninterested (Laplace ignored this work in his survey *1819e* of capillary theory, a paper noted in § 12.2.4), it was to catch the eye of Navier, as we shall see in § 15.2.7.

8.5.7 Poisson and Cauchy versus Hachette on the flow of water out of orifices, 1816

In contrast to the mutual uninterest of the theoreticians and the engineers just exemplified once again, here is a case where work produced from one group was considered by members of the other. To complete this section, we consider now some studies made by Hachette, and their reception in the hands of Poisson and Cauchy.

In his mid forties around 1815, Hachette studied especially the rate of flow out of an orifice as a function of its shape and other factors. He submitted his first paper to the *classe* in December 1815 (*PV*, *5*, 594) and published a brief summary *1816a* early the next year with the *Société Philomatique*. Poisson sent in his report in February, and it soon appeared in the *Correspondance sur l'Ecole Polytechnique* as *1816b*, with editor Hachette's own added notes. His second paper arrived at the recently reconstituted *Académie* the next June (*PV*, *6*, 73): the newly appointed Cauchy reported in October (as his first such *Académie* duty). The report quickly appeared in part with the *Société Philomatique* as Cauchy *1816b*, and in full as *1816a* in Arago's *Annales*, where Hachette himself published a part *1817a* of the paper in the following year. In addition, Delambre gave a quite detailed account of the work in his notice of the *Académie*'s activities for 1816 (*1818a*, xxxvi-xliii).

Although both reporters recommended the papers for the *Savants étrangers*, for some reason Hachette hid them until 1828, when he described parts of their contents, and reprinted the reports, in the fourth edition of his textbook on machines.[1]) Perhaps he had been affected by the strange personal circumstances; recently stripped of his *Ecole Polytechnique* position by the Bourbons (§ 8.2.2), he had had his work judged by two former students, one now an *examinateur permanent* there and the other a *professeur*, the latter also recently appointed to the *Académie* without election (§ 10.1.3). Further, neither man was skilled in experimental techniques, with which his papers were principally concerned.

Among his results, Hachette found that, for a given head of water, the rate of outflow was the same if the orifice was horizontal and of a given area while taking various concave shapes; however, it varied if convex shapes were used or the orifice was not horizontal. He studied the improvement in flow achieved by using nozzles, confirming that their surfaces help to reduce contraction for water, mercury, oil and alcohol; but the improvement was largely lost if the nozzles

[1]) Hachette *Machines*$_4$ (1828), 73-102; this section of the book also appeared then, slightly modified, as the pamphlet *1828b*. On these and related researches, see Barlow *1829a*, 211-220; on the textbook and its significance, see § 16.3.5 below.

were dry or if the flow took place into a vacuum or at a low atmosphere. He also made use of descriptive geometry to determine the 'surfaces of fluid veins' in various cases. When the head of the fluid was very small, the stream lines took different shapes from the usual and independent of the shape of the orifice: he called them 'secondary veins'.

Cauchy ended his report *1816a* by inviting Hachette to look at still further factors: for example, the relationship between the size of the orifice and the head of fluid for the occurrence of secondary veins. However, neither reporter discussed the hydrodynamical theory which might describe the phenomena—a surprising feature, since both men had been deeply involved with such studies in 1815 (§ 10.3). Apparently Hachette did not discuss theory much in his papers, either; but some related material came from his and Poisson's pens, and appeared in his *Correspondance sur l'Ecole Polytechnique*.

Hachette wrote a summary *1815c* of de Prony's monograph *1804f* in measuring water-flow, which was described in § 8.5.4; and at the end he cited a paper that Poisson had published in the *Correspondance* as *1807a*, on the hypothesis of the parallelism of slices. As we saw in § 8.5.3, in this approach the fluid body was dissected into rectilinear differential slices and the flow was interpreted as the parallel motion of these slices. Figure 857.1 shows the diagram upon which Pois-

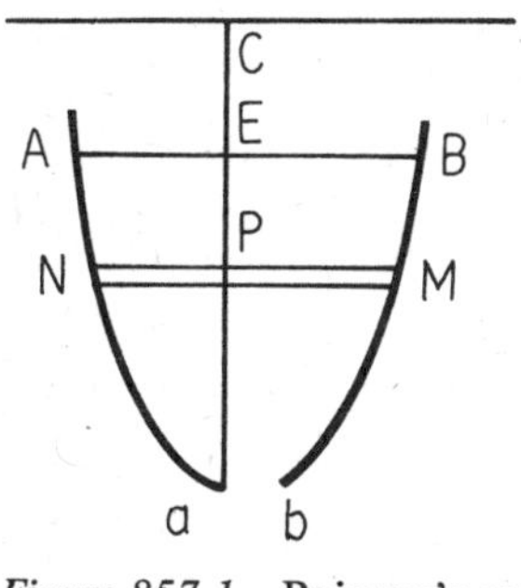

$CP = x, \quad CE = x_0$

y = area of section at NPM

Ea = h (depth of fluid)

Figure 857.1 Poisson's model of water-flow (after *1807a*)

son analysed the flow of fluid out of a horizontal orifice at the base of a container. The slice NPM, of constant thickness dx, moved towards the orifice ab with (variable) velocity v. By taking the balance of 'motive forces' caused by gravity and the acceleration $\dot{v}$ he found for the pressure p at NPM given by the weight of the slices above that level the expression

$$p = Ay + gx_0y - y\int_{x_0}^{x} \dot{v}\,dx, \tag{857.1}$$

where A was atmospheric pressure and the density of fluid was taken to be unity. Further, assuming the fluid to be incompressible furnished the continuity condition

$$vy = ku, \tag{857.2}$$

where k was the area of the orifice ab and u the value of v there (*1807a*, 289-291).

From these two equations Poisson manipulated to obtain, as an energy equation,

$$0 = gh - kN\dot{u} - \frac{1}{2}u^2(1 - k^2/c^2), \quad \text{where} \quad N := \int_{x_0}^{h} dx/y, \tag{857.3}$$

which reduced for small orifices to Torricelli's theorem

$$u^2 = 2gh(+O(k^2)). \tag{857.4}$$

'But it is good to observe that the fluid does not take this final velocity instantaneously'; for, by integrating (857.3) he obtained the equation

$$\left(\sqrt{2gh} - u\sqrt{1 - \frac{k^2}{c^2}}\right) = \left(\sqrt{2gh} + u\sqrt{1 - \frac{k^2}{c^2}}\right)\exp\left(-\frac{t\sqrt{2gh}\ \sqrt{1 - k^2/c^2}}{kN}\right), \tag{857.5}$$

in which the exponential term took time (albeit little of it, given the smallness of k) to reduce to negligible measure, thus yielding (857.4) (including the approximation term: pp. 293-294). He repeated the analysis in his texts on mechanics for the *Ecole Polytechnique* (*1810a*, *2*, 335-342; *Mechanics*$_1$ (1811), *2*, 444-456).

In a short note added to Poisson's paper, editor Hachette *1807a* noted the novelty of Poisson's calculation of time-lapse; but he seems not to have drawn on it in his experimental work of 1815. However, Poisson himself remarked, in his report on Hachette's first paper, that when the orifice was not tiny the 'time of flow cannot no longer be determined by the theorem of Toricelli [*sic*]' (*1816b*, 397) but required gamma functions; and in an adjoint paper *1815f* published in the *Correspondance* he furnished the details. He took (857.3) in the form

$$(k^2N/y)\,d\zeta/dh + h - (1 - h^2/y^2)\,\zeta = 0, \quad \text{where} \quad \zeta := u^2/2g. \tag{857.6}$$

and considered the special case of a vertical cylinder of constant cross-sectional area b, when $(857.3)_2$ became

$$N = h/b; \quad \text{setting also} \quad m := b^2/k^2 \tag{857.7}$$

converted (857.6) into

$$d\zeta/dh + m - (m-1)\zeta/h = 0, \tag{857.8}$$

with solution

$$\zeta = m(h - ch^{m-1})/(m-2), \quad c \text{ constant.} \tag{857.9}$$

The initial conditions

$$u = 0 \quad \text{and} \quad h = H \quad \text{when} \quad t = 0 \tag{857.10}$$

gave for u, via $(857.6)_2$,

$$u^2 = 2gmh(1 - (h/H)^{m-2})/(m-2), \tag{857.11}$$

and it was the second term in $(857.11)_2$ which led to gamma functions when u and related variables were substituted back into (857.1, 6) (pp. 284-285). The rest of the paper was taken up with producing these solutions for different values of m in $(857.7)_2$; he took his values of gamma functions from the recently published first volume of Legendre's *Exercices du calcul intégral* (§ 10.2.2). In several cases he found that the time taken to evaluate the cylinder was proportional to $\sqrt{H/g}$, equal to the period of oscillation of a simple pendulum of length given by the appropriate proportion of H. This connection between hydrodynamics and the pendulum was to recur later in Poisson's thought, as will be noted briefly in § 18.2.4.

So the contrast between the theoreticians' approach and that of the engineers is exemplified after all, especially between the analysis of Poisson just rehearsed and Girard's reasoning at (856.6) on the descent of water in a vessel. Both men took the fluid to be incompressible, and used the hypothesis of the parallelism of slices. However, Girard made no attempt to seek a differential equation, but instead gave formulae suitable to relate to his masses of experimental data, of which Poisson had none.

8.6 Concluding comment: the spectrum of engineering mathematics

The points just made concerning Poisson and Girard could be extended in various ways with regard to other studies recorded in this chapter. As mathematicians, the engineer *savants* did not – maybe could not—extend their analyses as far as, for example, Poisson's (857.5) on the fluid not achieving its final velocity immediately. On the other hand, they were adept at constructing feasible models of the phenomena under examination, models sufficiently close to the known effects to be worth developing, and even responding to suitable desimplifications. Coulomb, who has been the father figure for this chapter in the way that Lagrange and Laplace have served for the previous five, showed this skill well in his models of friction, rupture and fluid resistance (§ 8.4.1-2, § 8.5.2); and in addition he brought to bear some careful experimental evidence (for example, on the effects of torsion in § 8.4.1). His achievements were recognised not only in France but also in other countries: on engineering there, see Interlude 860.1.

Interlude 860.1 On engineering outside France

Like the chapter as a whole, this interlude is confined to the mathematical aspects of engineering. It seems that France was especially dominant in these areas, since theorising was much more strongly cultivated than elsewhere. (Of course the resultant engineering product was not necessarily rendered thereby more worthwhile.) Some nervous comments are offered concerning Britain, Germany and Italy; a few aspects of Russian engineering appear in § 17.1.2, when the connections developed with the French are described. For a general survey of technology in the 18th century, see Cardwell *1980a*.

1. *Britain* The country which led the Industrial Revolution tended to concentrate in the early years of the process more on the calculation of equity and wage than of equilibrium and water; empirical laws were preserved as "strict" rules of thumb rather than as starting-points for theoretical calculation. Even Young's *Course of lectures* (*1807a*) contained little mathematical dress in its extensive accounts of friction and hydraulics (compare the second footnote in § 7.4.5). Among other English figures, the most relevant around 1800 were Atwood, Olinthus Gregory and Hutton, who all published textbooks and/or treatises on aspects of engineering as well as research papers. Among the Scots, Robison contributed many relevant articles to the *Encyclopaedia Britannica*.

2. *Germany* As with the calculus (see Interlude 480.1), Germans were fairly well aware of developments in France: German translations of de Prony's *Hydraulique*, Girard's book on the strength of materials, and Français's essay on embankments were mentioned in the chapter, and other

works were also translated. The various encyclopaedias mentioned in Interlude 480.1 contain a fairly good representation of engineering mechanics. Among individual figures, Eytelwein is worth recording, as the German equivalent to de Prony. A French translation of one of his works is noted in § 17.7.3. From the 1790s through to the 1830s he wrote extensively on water-flow, materials and constructions, and tried to tailor his mathematics to the practical needs of calculation.

3. *Bohemia* The efforts of Gerstner are worth recording, for he founded a 'polytechnical school' in Prague in 1806 and modelled some of his research on French work. See an excellent obituary written by a former student, himself now famous: Bolzano *1837a*.

4. *Italy* A distinguished tradition in the engineering of architecture and water flow attends the history of this land of ancient culture, and landscape ribbed by mountains running down to the plains the lakes, and the coasts. The *Società Italiana*, whose existence was noted in Interlude 480.1, published regularly on engineering mechanics. Hydraulics was a major concern of some of their leading figures: Fossombroni, Lorgna and Venturi. Of these Venturi was the best-known in France. In particular, an essay on experiments on orifices, contraction, and eddies in rivers was submitted to the *CI* in August 1797 (*PV*, *1*, 261), quickly praised by de Prony in his report *1797a*, and published in Paris as Venturi *1798a*: an English translation appeared a year later. This work was to catch the attention of Hachette, in connection with his studies of water-flow described in § 8.5.7.

Following Coulomb's tradition, de Prony and Girard, the two most important figures in this chapter after him, made important contributions to the study of fluid-flow (§ 8.5.3–6) and also to structures (§ 8.4.3–5). Some differences of philosophy are evident between these two friends, however: de Prony was explicitely a truthbound inductivist, as he stressed after (845.2), whereas Girard liked the bold (but, by implication, more speculative) analogy, such as the stone column analysed as an 'aeriform fluid' (§ 8.4.3), or on the flow of fluid in a pipe treated (at (855.1)) like a flexible line lying in equilibrium on a surface. By and large Girard's philosophy seems to be the more realistic!

Overall, the range and quality of the work seems impressive, and with its tireless advocacy by de Prony, already in his treatise of the 1790s (§ 8.1.1) and then in his teaching at the *Ecole Polytechnique*, it gained much prominence in French science. By contrast, the other major component of engineering mathematics taught at that school (and elsewhere), Monge's descriptive geometry, gives a poor impression. We saw in § 8.3.2 that the subject drifted towards the notional in the gap between its intentions and the material presented in Monge's textbook. Disciples such as Brisson and Dupin could deploy techniques with good effect (§ 8.3.3-4), but by and large the subject fossilised and rightly lost its dominance in the instruction at the *Ecole Polytechnique*. It will not be visited again in this book.

However, the engineer *savants* still appear frequently in other contexts. They have Chapter 16 almost entirely to themselves in connection with their general theory of *forces vives* and work, and already before then Navier makes a major mark in the account of elasticity theory in Chapter 15 – where, once again, the engineers and theoreticians were to clash.